NMR as a Structural Tool for Macromolecules

Current Status and Future Directions

NMR as a Structural Tool for Macromolecules

Current Status and Future Directions

Edited by

B. D. Nageswara Rao and
Marvin D. Kemple

Indiana University-Purdue University, Indianapolis (IUPUI)
Indianapolis, Indiana

Plenum Press • New York and London

Library of Congress Cataloging-in-Publication Data

NMR as a structural tool for macromolecules : current status and
  future directions / edited by B.D. Nageswara Rao and Marvin D.
  Kemple.
       p.   cm.
    "Proceedings of an International Symposium on NMR as a Structural
  Tool for Macromolecules: Current Status and Future Directions, held
  October 30-November 1, 1994, in Indianapolis, Indiana"--T.p. verso.
    Includes bibliographical references and index.

    1. Nuclear magnetic resonance spectroscopy--Congresses.
  2. Macromolecules--Structure--Congresses.   I. Nageswara Rao, B. D.
  II. Kemple, Marvin D. III. International Symposium on NMR as a
  Structural Tool for Macromolecules: Current Status and Future
  Directions (1994 : Indianapolis, Ind.)
  QP519.9.N83N383 1996
  574.19'285--dc20   96-21442
                                                               CIP

Proceedings of an International Symposium on NMR as a Structural Tool for Macromolecules: Current
Status and Future Directions, held October 30 – November 1, 1994, in Indianapolis, Indiana

ISBN-13: 978-1-4613-8029-0   e-ISBN-13: 978-1-4613-0387-9
DOI: 10.1007/978-1-4613-0387-9

© 1996 Plenum Press, New York
Softcover reprint of the hardcover 1st edition 1996
A Division of Plenum Publishing Corporation
233 Spring Street, New York, N. Y. 10013

PREFACE

The contemplation of truth and beauty is the proper object for which we were created, which calls forth the most intense desires of the soul, and of which it never tires
-Hazlitt

In his Nobel lecture Purcell commented that when he saw snow in New England after the discovery of NMR, it appeared like "heaps of protons quietly precessing in earth's magnetic field." If he were to make the comment in the context of how NMR is being used today, he could have conjured up an image of hydrogen, carbon, and nitrogen nuclei in proteins of an earthbound organism subtly orchestrating a quiet symphony of frequencies, from 150 Hz to 2 kHz, carrying clues to the three-dimensional structure of the macromolecules. The manner in which the basic discoveries of Bloch and Purcell have led to the emergence of NMR, several decades later, as a major technique of biological and medical physics (and chemistry) is a striking example of the power of basic research. It is also a fascinating saga whereby whenever it was felt that the field had reached a plateau, new directions, new technologies, and sometimes serendipity produced new developments that revolutionized the technique and enhanced its capability. As Richard Ernst points out "NMR is intellectually attractive, ...the practical importance of NMR is enormous, and can justify much of the playful activities of an addicted spectroscopist" (Nobel lecture). Specifically, in the case of high resolution (HR) NMR, the 2D-revolution of the mid-70s accompanied by the development of NOESY, launched the technique as a macromolecular structural tool by the early 80s. This led to a feverish activity in the next decade with a bewildering array of methodological refinements which established HRNMR as a powerful method for structural investigations. Once again questions are being raised concerning whether the technique has reached another methodological plateau, and about what new directions it might take.

Thus, when one of us (BDNR) received the Glenn Irwin, Jr. Award from our university, the decision was made to organize a symposium with the title "NMR as a Structural Tool for Macromolecules: Current Status and Future Directions." The organization was off to a flying start when Richard Ernst accepted the invitation to be the keynote speaker. Practically every one of the scientists we approached agreed to speak. We were thus able to assemble an impressive array of accomplished practitioners of HRNMR as plenary speakers. The meeting was 3 days long, with a panel discussion at the end of the lectures on each day, and poster sessions on the first two evenings after dinner. The goal was that the symposium not be a massive presentation of the latest results in biomolecular structures, but rather be a comprehensive evaluation of the technique along the lines of "where do we stand, and where do we go from here?" The three panel discussions addressing topics of current interest were designed to make an important contribution in this regard. The symposium ended with a panel discussion on "NMR *vis-à-vis* Other Structural Methods" moderated by Mildred Cohn.

Since this was a one-time only symposium on a specific theme, we felt that it would be useful to publish a compendium of the lectures and discussions from the symposium. Thus, it was decided to make a verbatim record of all the discussions following each talk and of the three panel discussions. All of the invited speakers agreed to provide manuscripts that summarized or included the context of their lectures. It took about nine months to collect all of the manuscripts and another three months to put everything together. Abstracts of all of the posters presented at the symposium are also included in the book. We sincerely hope that this book will be useful to people in the field, especially to the young scientists who aspire to become established investigators in the area of biological NMR.

There were over 200 participants at the meeting other than the invited speakers and panel members. The conferees were enthusiastic, the discussions were lively, and the atmosphere at the University Place Conference Center at IUPUI during those three days was exhilarating. We received a number of letters from attendees generously commending us for organizing the meeting. We are deeply grateful to all of the speakers and participants for making the symposium such a success. We are especially indebted to Richard Ernst for opening the symposium and to Mildred Cohn for closing it. The presence of these two distinguished scientists, who touched the careers of all the participants in one form or another, made the symposium most memorable.

We wish to thank David Gorenstein, Allen Kline, and Stephen Wassall for working with us on the Organizing Committee of the symposium, the IUPUI NMR group for diligently recording the discussions, and Padmini Nallana who worked tirelessly as Coordinator of the symposium and in compiling the lectures and discussions that are included in this book. Financial support for the entire endeavor, received from the Glenn Irwin, Jr. Award to BDNR, the School of Science and the Office of Faculty Development of IUPUI, the National Institutes of Health (GM R13 51260), Eli Lilly, Varian Associates, Isotec, Biosym Technologies, Cambridge Isotope Laboratories, and Nalorac, is gratefully acknowledged.

B.D. Nageswara Rao
Marvin D. Kemple

CONTENTS[*]

The Theme of the Symposium 1
    B.D. Nageswara Rao

Keynote Lecture: Intramolecular Dynamics of Biomolecules. Possibilities and
    Limitations of NMR 15
    R.R. Ernst, M.J. Blackledge, T. Bremi, R. Brüschweiler, M. Ernst,
        C. Griesinger, Z.L. Mádi, J.W. Peng, J.M. Scmidt, and Ping Xu

Discussion 31

Structural, Dynamic, and Folding Studies of SH2 and SH3 Domains 35
    Julie D. Forman-Kay, Steven M. Pascal, Alex U. Singer, Toshio Yamazaki,
        Ouwen Zhang, Neil A. Farrow, and Lewis E. Kay

Discussion 49

NMR Studies of Proteins Involved in Cell Adhesion Processes 51
    Gerhard Wagner, Daniel F. Wyss, Johnathan S. Choi, Antonio
        R.N. Arulanandam, Ellis L. Reinherz, Andrzej Krezel, and Robert
        A. Lazarus

Discussion 63

Combining ^{2}H and ^{13}C Selective Enrichment to Probe Protein Dynamics 65
    David M. LeMaster

Incorporating Motional Properties into the Interpretation of Three-Dimensional
    Solution Structures 77
    Walter J. Chazin

Discussion 87

Phosphotyrosyl Peptide-Enzyme Complexes: How Much Structure Can We Get from
    Transferred NOE's? 91
    Carol Beth Post and Michael L. Schneider

Discussion 101

---

[*]Names of moderators of panel discussions and authors who delivered the lectures are shown in bold face.

Panel Discussion on Structural Refinement and Dynamics 103
David Case, Marvin D. Kemple, N. Rama Krishna, Carol B. Post, and
Gerhard Wagner

Recent Developments in Protein NMR Spectroscopy 117
Stephan Grzesiek, Geerten W. Vuister, Andy C. Wang, Frank Delaglio, and
Ad Bax

Discussion 121

Field-Cycling NMR Applied to Macromolecular Structure and Dynamics 123
Alfred G. Redfield

Discussion 133

Cross-Correlations: Obstacles or Tools for Structure Determination of Biomolecules . 135
Anil Kumar

Discussion 143

Towards the Accurate Measurement of Internuclear Distances in Biological
Macromolecules by Suppression of Spin Diffusion 145
Sébastien J. F. Vincent, Catherine Zwahlen, and Geoffrey Bodenhausen

Discussion 165

NMR of Symmetrical Assemblies of Self-Recognizing Oligonucleotides 167
Maurice Guéron, Kalle Gehring, and Jean-Louis Leroy

Discussion 173

Protein-DNA Interaction from NMR and Monte Carlo Docking 175
R. Kaptein, M. Slijper, V.P. Chuprina, J.A.C. Rullmann, R.M.A. Knegtel, and
R. Boelens

Discussion 189

Dynamic Structure of Nucleic Acid Duplexes 191
Thomas L. James, Carlos González, He Liu, Uli Schmitz, and
Nikolai B. Ulyanov

Discussion 205

Panel Discussion on Extension of Techniques to Larger Molecules 207
G. Marius Clore, Stephen W. Fesik, David G. Gorenstein, David M. LeMaster,
and **John L. Markley**

NMR Structures of Proteins Involved in Signal Transduction 221
S.W. Fesik, R.P. Meadows, E.T. Olejniczak, A.P. Petros, P.J. Hajduk,
H.S. Yoon, J.E. Harlan, T.M. Logan, M.-M. Zhou, D.G. Nettesheim, H.
Liang, and L. Yu

Discussion 235

Structures of Multimeric Proteins by NMR 237
   G. Marius Clore and Angela M. Gronenborn

Discussion 243

NMR Structural Studies of Flexible Molecules 245
   Peter E. Wright and H. Jane Dyson

Iron-Sulfur Proteins: Investigations of Hyperfine-Shifted Hydrogen, Carbon, and
   Nitrogen Resonances 251
   Bin Xia, Hong Cheng, Young Kee Chae, Lars Skjedal, William M. Westler, and
   John L. Markley

On the Use of NMR in Complex Biological Systems: NMR Studies of Calcium
   Sensitive Interactions amongst Muscle Proteins 275
   Brian D. Sykes, Carolyn M. Slupsky, David S. Wishart, Frank D. Sönnichsen,
   and Stéphane M. Gagné

Discussion 285

The Structure of Lentiviral Tat Proteins in Solution 287
   Paul Rösch, Peter Bayer, Andrzej Ejchart, Rainer Frank, Arnona Gazit,
   Franz Herrmann, Margot Kraft, Rina Rosin-Arbesfeld, Heinrich Sticht,
   Dieter Willbold, and Abraham Yaniv

Discussion 305

A Structural Biologist's View of Precision and Accuracy of Structural Models of
   Proteins Based on NMR Data 307
   A. Joshua Wand

Discussion 325

Panel Discussion on NMR vis-à-vis Other Structural Methods 327
   Bernard Brooks, Mildred Cohn, Thomas L. James, Franklyn G. Prendergast,
   and Janet L. Smith

Poster Abstracts 339

Index 379

B. D. Nageswara Rao and Richard R. Ernst

THE THEME OF THE SYMPOSIUM

B. D. Nageswara Rao

Department of Physics, IUPUI
402 North Blackford Street
Indianapolis, IN 46202

INTRODUCTION

I wish to describe how I arrived at the theme of this Symposium, viz., Current Status and Future Directions of NMR as a Structural Tool. This was an outgrowth of my own interpretation of various events that impacted on the evolution of high resolution NMR spectroscopy during the last 35 years or so. Many a time I was quite intrigued, if not fascinated, by how major advances in a scientific field occur when seemingly unrelated ideas and experimental technologies converge at a propitious time. The timing seems to be crucial. There are times when good ideas do not come to fruition. Sometimes a productive confluence of ideas and experimental methods occurs in a somewhat predictable fashion. Some other times serendipity brings about a totally unforeseen change, which then takes us in a inevitable and seemingly irreversible new direction. NMR has gone through all these types of phases during the last 35 years. In order to illustrate this point, I need to take you through a little bit of history.

HIGH RESOLUTION NMR IN EARLY 60'S

State of the Art

Spectrometers. The commercial high-resolution (HR) NMR spectrometers of the day operated in the CW (continuous wave) mode at proton resonance frequencies of 60

NMR As a Structural Tool for Macromolecules: Current Status and Future Directions
Edited by B.D. Nageswara Rao and Marvin D. Kemple, Plenum Press, New York, 1996

1

MHz or 100 MHz (referred to as megacycles per second). The static magnetic fields of the requisite stability and homogeneity were produced with electromagnets. The spectra were recorded using carefully controlled slow sweeps of the static field of about one milligauss per minute (or a few nanoTesla per second).

Spectra and Information. A typical spectrum of a strongly coupled two-spin system (AB) is shown in Figure 1 (Nageswara Rao et al, 1965). The "wiggles" following each line are a result of the transient response of the spin system to the slow sweep. The high-resolution NMR spectroscopist of the sixties almost had a romantic relationship with these wiggles. The more they were seen the more exciting it was because they showed that the magnet was shimmed well.

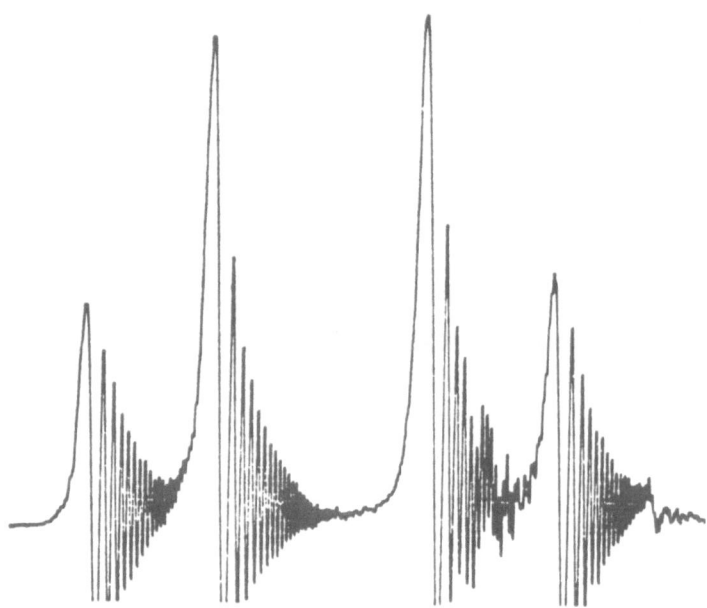

Figure 1. Proton High-Resolution NMR spectrum of a strongly-coupled two-spin system (AB) recorded on a 60 Mc/sec(!) spectrometer operating in a continuous wave (CW) mode. (Nageswara Rao et. al., 1965)

A lot of effort was expended in analyzing strong coupling effects in the spectra. When the analysis was complete the interpretable experimental data emerged as chemical shifts (δ's) and spin-spin coupling constants (J's). Sometimes chemical shifts were given as τ's where $\tau = 10 + \delta$. This was apparently done because organic chemists did not like to deal with negative numbers! Since the interpretation of δ's and J's in terms of "structure" was difficult (HR) NMR was more an "analytical" rather than a "structural" tool. The CW spectrometers were not particularly suitable for relaxation studies.

Physicists more or less ignored (HR) NMR as evidenced by the following quote from p. 405 of Abragam's book (1961):

> *The technique of high resolution necessary for the detailed observation of these spectra has, during the last few years, developed as a tool for the investigation of chemical structures, and a whole book could (and very likely will) be † devoted to all the problems connected with the analysis and interpretation of these spectra.*

> *† And indeed has been: High Resolution Nuclear Magnetic Resonance, J. A. Pople, W. G. Schneider and H. J. Bornstein (1959).*

Abragam came to know about the book by Pople et. al.(1959) only about two years after it was printed. Physicists were obviously more attracted to the possibility of studying spin systems as a means to further the fundamental understanding of statistical physics, as suggested on p.152 of a book on Paramagnetic Resonance by G. E. Pake (1962):

> *-----, the approach of nuclear or electronic spin assemblies to equilibrium provides perhaps the best vehicle yet for the study of irreversible statistical mechanics. Even for 10^{22} spins, there are still only a finite - if large - number of spin states accessible to the system, and it is possible experimentally to prepare the assembly in a variety of reasonably well-defined initial nonequilibrium states. It should not be forgotten that the derivation of equations describing irreversible dissipative behavior by beginning with the Schrödinger equation is a fundamental problem of physics not yet fully understood. For a recent attempt and discussion, inspired in large measure by magnetic relaxation effects, see the work of Sher and Primakoff.[15]*

Relaxation Studies

Relaxation time (T_1 and T_2) measurements on liquids were mostly performed at *low resolution* using (home-built) pulsed spectrometers. However, those of us familiar with the rich multiplicity of high-resolution NMR spectra fancied the possibility of using double-resonance methods to gain insight into relaxation mechanisms operating in coupled spin systems. In these experiments we used to, for example, selectively irradiate one of the transitions in a spectrum with a rather strong radiofrequency (rf) field of controllable amplitude, and then sweep through and observe the spectrum with the usual weak rf field.

[15]*A. Sher and H. Primakoff, Phys. Rev. , **119**, 178 (1960)*

The idea was that by perturbing the population dynamics of the spin energy levels, the spectra will depict the competition between the known perturbation and unknown relaxation mechanism, and will, therefore, provide information on the relaxation processes (Nageswara Rao, 1965, 1970). Figure 2 shows an example of what used to be called "weak-irradiation" spectra (Anil Kumar et al, 1970). The irradiation frequency in the spectra is at the center of the beat pattern and the amplitude of irradiation is smaller than the linewidths of the resonances. The intensity changes in the other transitions are obvious.

At larger irradiation strengths the coherence induced in the spin system by the irradiating rf field manifests itself by increasing the multiplicity of the spectrum, and by generating "inverted" transitions. An example of this is shown in Figure 3 (Anil Kumar et al, 1970). Under the experimental conditions of this spectrum, the density matrix of the two-spin system is full, in the sense that the irradiation generated all the single-, double- and zero-quantum coherences. These coherences are not separable, and their effects are lumped together in the observed spectrum. Elaborate theoretical procedures were devised based on density-matrix equations along with the use of Redfield's relaxation matrices (super-operators) to analyze the spectra and extract information on the spin relaxation process (Nageswara Rao, 1965, 1970).

The results of the double resonance effort, in so far as the relaxation processes were concerned, may be summarized as follows:

1. Almost all of the time $\omega \tau_c \ll 1$, where ω is the nuclear Larmor frequency and τ_c is the correlation time for the relaxation mechanism i.e., "extreme-narrowing" prevailed.

2. Chemical shift anisotropy (CSA) was "primarily" of academic interest.

3. Spin-rotation interaction as a relaxation mechanism captured the attention for a while. However, this mechanism is only significant for (very) small molecules at high temperatures.

4. Spin-lattice relaxation was often due to dissolved oxygen. If oxygen was carefully removed from the sample, T_1's were long (larger than 1 min. sometimes). After afew days the T_1's shortened again because (paramagnetic) metal ions leached out of glass tubes! Linewidths were mostly due to magnetic field inhomogeneity (even if the field was shimmed to about 0.2 - 0.5 Hz).

5. Dipolar interaction, which carried structural information, was difficult to *isolate* in a reliable fashion. The 1D-CW NMR methodology was too inefficient for this purpose.

The double resonance effort for elucidating relaxation processes in coupled spin systems is an example of good ideas not coming to fruition because the experimental technology has not yet acquired the necessary sophistication.

Overhauser Effect

Al Overhauser's prediction (1953) of polarization transfer from conduction electrons in metals to nuclei was initially received with great skepticism. However, the

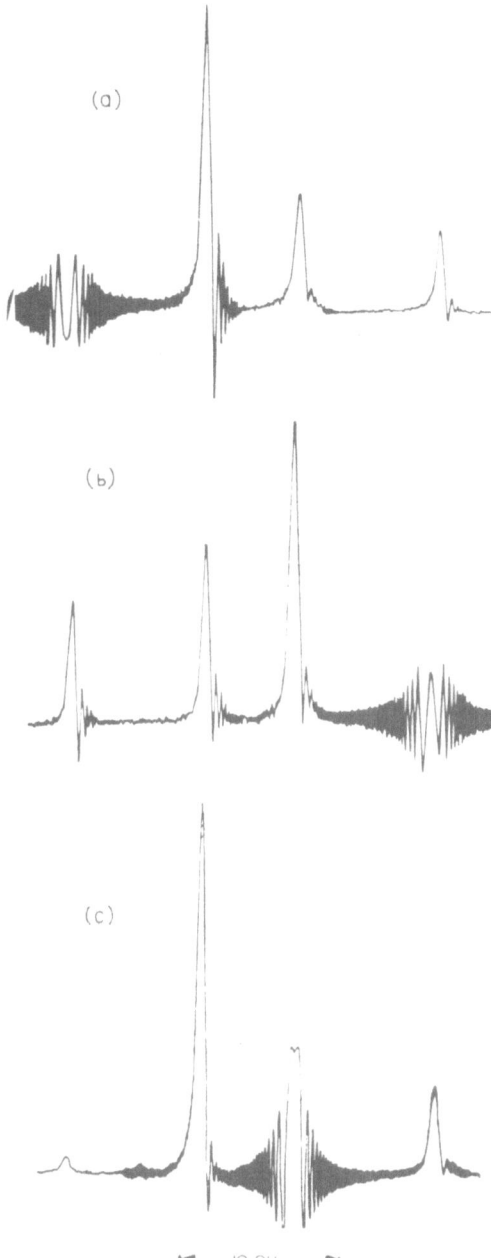

Figure 2. Homonuclear proton double resonance spectra of an AB spin system under "weak-irradiation" conditions. (Anil Kumar et al, 1970)

effect was experimentally verified (Carver and Slichter, 1953) in the first serious attempt made to do so. The nuclear version of Overhauser effect (NOE) was established as a straightforward consequence of relaxation mechanisms bilinear in spin operators leading to coupled equations for the magnetizations of the participating spins (Abragam, 1961). Steady-state measurements of NOE led to measurements of internuclear distances giving us a glimpse of the potential of NMR as a structural tool rather than a mere analytical tool (Noggle and Schirmer, 1973).

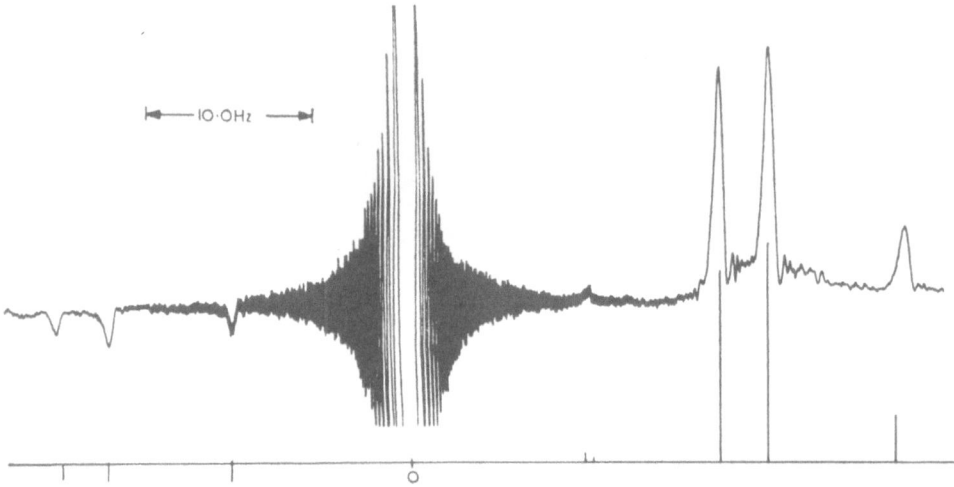

Figure 3. Homonuclear proton double resonance spectra of an AB system under "strong-irradiation" conditions. (Anil Kumar et al, 1970)

Theory

Most of the theoretical apparatus being used to explain (HR) NMR in the 90's were formulated by this time. These include density matrices, relaxation matrices (Redfield, 1957), and even the so-called "cross-correlations." For example, the cross-correlation between dipole-dipole interaction and chemical shift anisotropy in NMR was formulated by Hiroshi Shimizu (1964) in analogy with that between g-tensor anisotropy and hyperfine interaction, which was theoretically predicted by McConnell (1956), and experimentally observed by Rogers and Pake (1960), in ESR of free radical solutions. Some of the cross-correlation effects were "rediscovered' in the 80's and the 90's.

NMR and Biochemistry

(HR) NMR percolated into biochemistry. For instance Mildred Cohn (Cohn and Hughes, 1960, 1962) recorded a ^{31}P NMR spectrum of adenosine triphosphate (ATP) shown in Figure 4 (top), and investigated the effects of cations and pH on the spectral parameters. When she communicated this paper to *J. Am. Chem. Soc.*, it was rejected! The paper was eventually published in the *Journal of Biological Chemistry*. It is now a citation classic! The paper signalled the potential of NMR in biochemistry. Mildred Cohn used a 500 mM concentration of ATP which was too large for studying the biochemically significant complexes of this important molecule with many enzymes for which it is a substrate. The need for improving the methodology that will allow a significant decrease in the concentrations required for NMR was obvious.

CW-FT Equivalence

Linear response theory, which established the equivalence of CW and Fourier Transform (FT) spectroscopies, was well known. Lowe and Norberg (1957) verified this experimentally in solid-state NMR.

The time was ripe for a methodological advance in NMR. Perhaps the delay occurred because Richard Ernst was still a graduate student at the time!

FOURIER TRANSFORM NMR

This was a predictable methodological advance. Ernst and Anderson (1965), working at Varian Associates, took this important step of developing the FT technology for (HR) NMR. I understand that their paper was rejected for publication in the *Journal of Chemical Physics*. It was eventually published in *Review of Scientific Instruments*. This paper is also a citation classic!

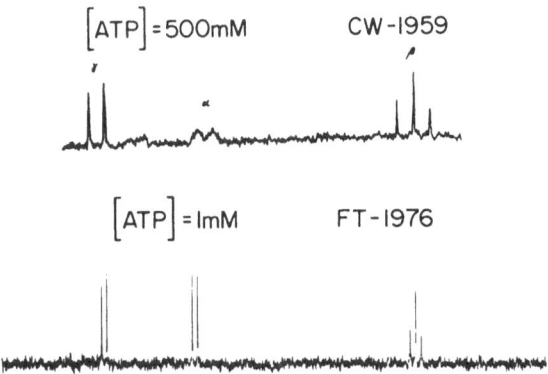

Figure 4. ^{31}P high-resolution NMR spectra of ATP in 1959 and 1976. The 1959 spectrum was recorded on a CW spectrometer with a 500mM sample (Cohn and Hughes, 1960), and the 1976 spectrum on an FT-spectrometer with a 1 mM sample.

This step came along with advances in superconducting and computer technologies, and the fast Fourier Transform algorithm. Spectrometer frequencies went up to 300 - 360 MHz. Sample concentrations came down to a few mM (See Figure 4 (bottom)). NMR was firmly entrenched as a tool for studying biochemically interesting systems.

An equally important consequence of the FT-advance was *versatility*. Spectral information *and* relaxation information could be obtained on the *same* spectrometer under high resolution conditions.

Spectral crowding was still a problem for large molecules. Pushing up the operating frequency was not enough to overcome this problem.

THE 2D REVOLUTION - BOLT FROM THE BLUE

This was not a predictable advance. Picking up on a proposal by Jeener (1971), Richard Ernst (Aue et. al., 1976) master-minded this revolution in NMR methodology. Assignment of resonances in crowded spectra was overcome — at least in principle.

However, the decisive step in launching NMR as a "structural tool" for macromolecules came through the Nuclear Overhauser Effect Spectroscopy (NOESY) experiment (see Figure 5) by Anil Kumar, Ernst and Wüthrich (1980) which evolved out of the 2D-exchange experiment by Jeener, Meier, Bachmann and Ernst (1979). In the 2D-game NOE resembles "chemical exchange!" The distance information from dipolar interactions could be isolated and picked up for all proton pairs in one experiment.

There was an inevitability to what followed, such as isotope editing by the use of ^{13}C and ^{15}N labelled materials and the consequent nD (n > 2) methods. High Resoultion NMR arrived as a *comprehensive* structural tool.

The manner in which the 2D-revolution has altered the way we do NMR, and has taken the field to its current level of sophistication and versatility is all too familiar to all of us, including those who recently entered the field (Ernst et al., 1987). It has all the elements of serendipity leading to irreversible and dramatic changes in a field of scientific endeavour. Reflecting upon these developments, as I thought about a theme for the symposium raised in my mind the possibility of assessing the HR NMR as it stands today and visualizing what the future might hold.

CURRENT STATUS AND FUTURE DIRECTIONS

I wish to, therefore, request this gathering to address the general question for high-resolution NMR spectroscopy as a technique for the elucidation of macromolecular structures: Where are we and where do we go from here? Specifically, the following kinds of questions come to mind.

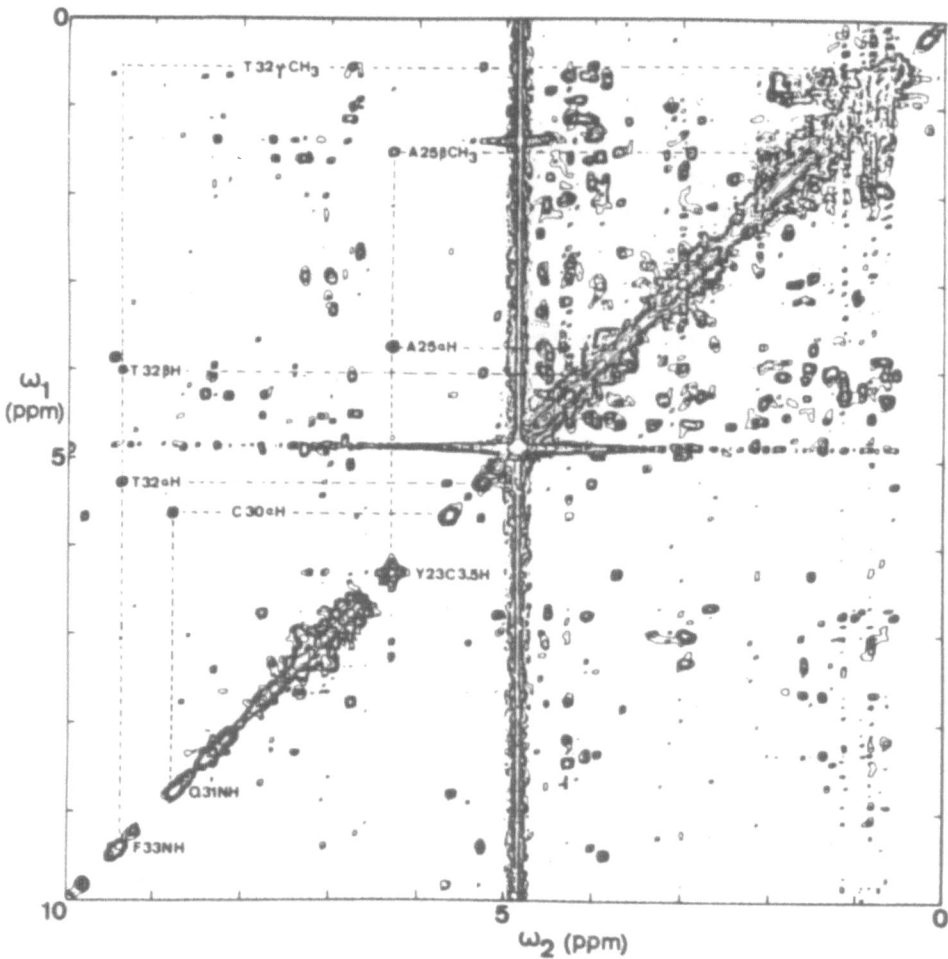

Figure 5. Proton 2D-NOESY spectrum of bovine pancreatic trypsin inhibitor (BPTI). (Anil Kumar et al, 1980)

Methodology

1. It is clear that NMR methodology as a structural tool has matured. Procedures for automated structure determinations are being talked about. Have we reached a methodological plateau?

2. Are there any predictable advances in the offing?

3. The frequency has gone up to 750 MHz. Perhaps 1 GHz is round the corner. We probably begin to see isotope effects on chemical shifts in a routine fashion. CSA will perhaps be more important than ever before. We may also see alignment effects of the large fields of 18 - 20 T, leading to incomplete averaging of some anisotropic tensors. Are these features going to offset the increased sensitivity and dispersion? Or will they become potential sources of new information?

Structures

1. How accurate and unique are the NMR-determined structures?

2. Do we need some standardization in the data presentation, analysis and refinement procedures?

3. How much confidence do the NMR-determined structures inspire in colleagues from other spectroscopies?

4. What are the effects of molecular motion on the structures determined?

5. How high can we go in molecular weight?

6. Can we answer specific biochemical questions regarding macromolecular functions, such as enzyme catalysis in structural terms? Here NMR has to be optimized along with the biochemistry to make sure that we are indeed measuring what we think we are measuring.

I wish to show an example from our recent work to illustrate the last point above. The transferred nuclear Overhauser effect spectroscopy (TRNOESY) method has been known for over a decade as a useful means of determining the conformations of small molecules bound to macromolecules. We wanted to use this method to determine the

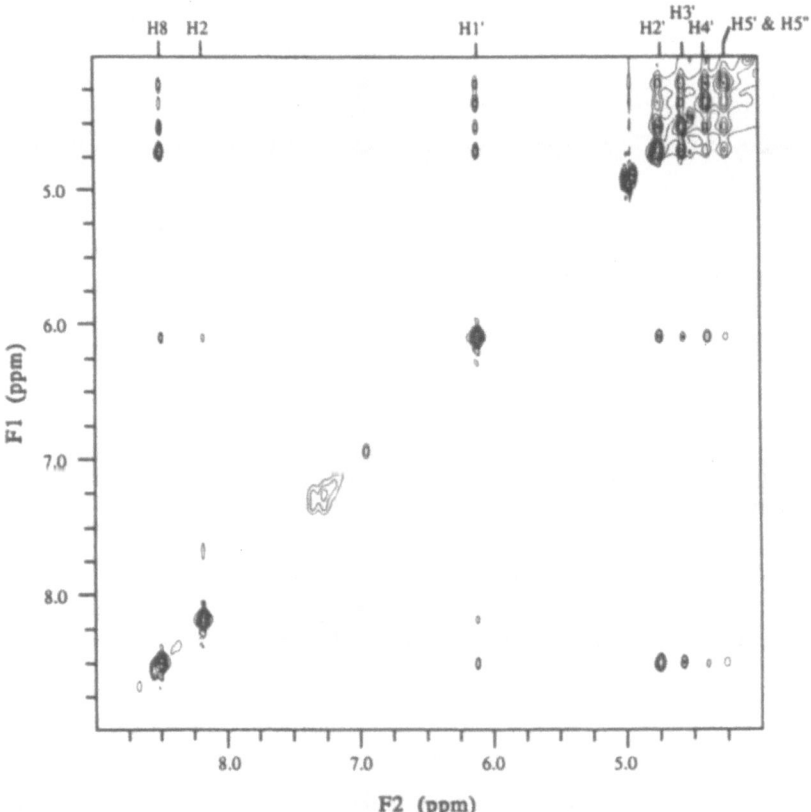

Figure 6. 500 MHZ proton TRNOESY spectrum of a sample containing 1.03 mM creatine kinase, 10.2 mM ADP, and 20 mM $MgCl_2$ in 50 mM Tris-d_{11}-Cl buffer at pH 8.0 and 10°C.

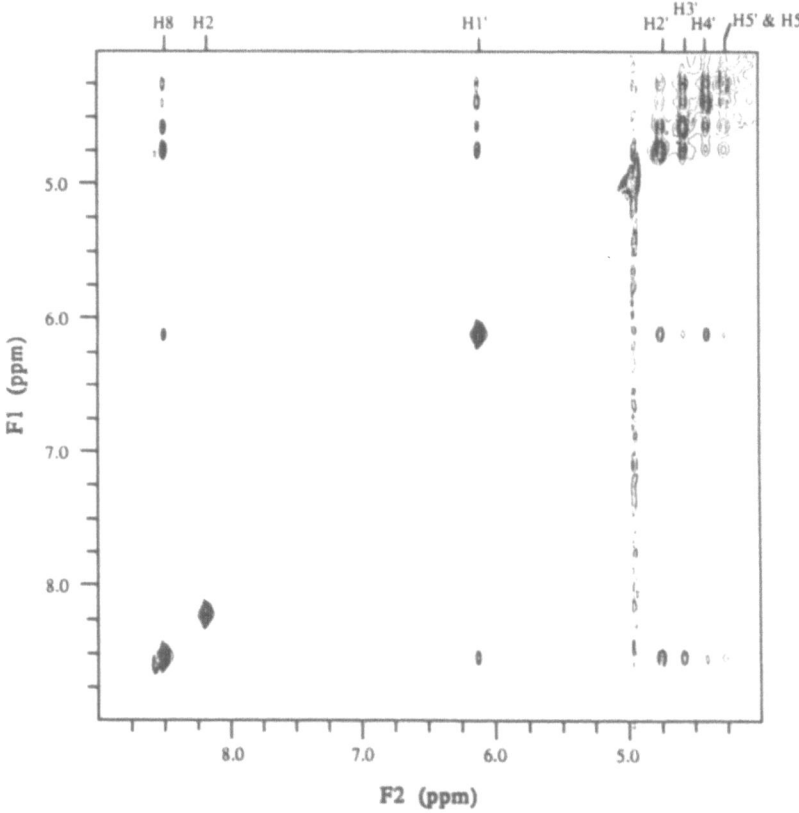

Figure 7. 500 MHZ proton TRNOESY spectrum of a sample containing 0.45 mM γ-globulin, 10.18 mM ADP, and 20 mM MgCl₂ in 50 mM Tris-d₁₁-Cl buffer at pH 8.0 and 10°C. (Murali et al, 1993)

conformations of the adenosine moeity of ATP and ADP bound to various enzymes for which they are substrates. Figure 6 shows the ^{1}H TRNOESY spectrum of 10 mM MgATP in the presence of 1mM creatine kinase. This concentration ratio was typical for TRNOESY measurements performed till then. The TRNOESY data looked very good and standard analysis generated a 'conformation'. However, we asked the question whether the data represents MgADP bound *only* at the active site of the enzyme. Therefore, we ran independent experiments in which creatine kinase was replaced by γ-globulin or bovine serum albumin (BSA) which do not possess specific binding sites for MgADP. Figure 7 shows the TRNOESY data obtained with γ-globulin, which is virtually indistinguishable from the data of Figure 6! It soon became clear that the data in Figure 6 is primarily coming from weak non-specific binding of MgADP perhaps at positively charged regions on the surface of creatine kinase, as it is in Figure 7. A careful investigation of the ligand concentration dependence of the TRNOESY data led us to choose a sample protocol of about 1.5 mM MgATP and 0.4 mM creatine kinase which minimizes the adventitious binding effects, and ensures that the NOE's arise from the active site (Murali et al, 1993).

Most of the TRNOE-determined "conformations" of ATP *bound* to various enzymes, previously published, are likely to be beset by the contributions from adventitious binding, and may, therefore, be incorrect. This is a simple example to illustrate that reliability of the structures deduced requires optimization both from NMR and biochemical points of view.

SUMMATION

A number of questions like those I raise above can be formulated to assess the current status and future directions of (HR) NMR as a structural tool. Or perhaps it may become unnecessary to ask these questions in this manner, because, there could be, among this audience or elsewhere, a youngster, as yet unheralded, isomorphic to an "Ernst" or a "Jeener" who might change the way we should look for the answers and render the current thinking obsolete. Whether something like that could come about is beyond my abilities to foretell. However, if we recognize such things might happen and leave that side, it seemed like a great idea to address these questions I raised above and thoroughly discuss them, if not find answers for them.

REFERENCES

Abragam, A., 1961, "Principles of Nuclear Magnetism," Oxford University Press, New York.

Anil Kumar, Ernst, R.R., and Wüthrich, K., 1980, *Biochem. Biophys. Res. Comm.* 95:1.

Anil Kumar, Rama Krishna, N., and Nageswara Rao, B.D., 1970, *Mol. Phys.* 18:11.

Aue, W.P., Bartholdi, E., and Ernst, R.R., 1976, *J. Chem. Phys.* 71:4546.

Carver T.R., and Slichter, C.P., 1953, *Phys. Rev.* 92:212.

Cohn, M., and Hughes, T.R., 1969, *J. Biol. Chem.* 235:3250.

Cohn, M., and Hughes, T.R., 1969, *J. Biol. Chem.* 237:176.

Ernst, R.R., and Anderson, W.A., 1965, *Rev. Sci. Instr.* 36:1696

Ernst, R.R., Bodenhausen, G., and Wokaun, A., 1987, "Principles of Nuclear Magnetic Resonance in One and Two Dimensions," Clarendon Press, Oxford.

Jeener, J., 1971, "Ampère International Summer School," Basko Polje, Yugoslavia.

Jeener, J., Meier, B.H., Bachman, P., and Ernst, R.R., 1979, *J. Chem. Phys.* 71:4546.

Lowe, I.J., and Norberg, R.E., 1957, *Phys. Rev.* 107:46.

McConnell, H.M., 1956, *J. Chem. Phys.* 25:709.

Murali, N., Jarori, G.K., and Nageswara Rao, B.D., 1993, *Biochemistry* 32:12941.

Nageswara Rao, B.D., 1965, *Phys. Rev.* 137:A467.

Nageswara Rao, B.D., 1970, *Adv. Magn. Reson.* 4:271.

Nageswara Rao, B.D., Anderson, J.M., and Baldeschwieler, J.D., 1965, *Phys. Rev.* 137:A1477.

Noggle, J.H., and Schirmer, R.E., 1971, "The Nuclear Overhauser Effect," Academic Press, New York.

Overhauser, A.W., 1953, *Phys. Rev.* 92:212.

Pake, G.E., 1962, "Paramagnetic Resonance," Benjamin Inc., New York.

Pople, J.A., Schneider, W.G., and Bernstein, H.J., 1959, "High Resolution Nuclear Magnetic Resonance," Mc Graw Hill, New York.

Rogers, R.N., and pake, G.E., 1960, *J. Chem. Phys.* 33:1107.

Sher, A., and Primakoff, H., 1960, *Phys. Rev.* 119:178

Shimizu, H., 1964, *J. Chem. Phys.* 40:3357.

....

INTRAMOLECULAR DYNAMICS OF BIOMOLECULES, POSSIBILITIES AND LIMITATIONS OF NMR

R.R. Ernst, M.J. Blackledge, T. Bremi, R. Brüschweiler,
M. Ernst, C. Griesinger, Z.L. Mádi, J.W. Peng,
J.M. Schmidt, and Ping Xu

Laboratorium für Physikalische Chemie
Eidgenössische Technische Hochschule
ETH-Zentrum
8092 Zürich, Switzerland

INTRODUCTION

Currently, the description of chemical and biochemical phenomena is heavily based on the conception of a molecular "structure", i.e. of a rigid arrangement of atoms in space, defining the conformation of a molecule. This view matches our limited means of representation and visualization with pen and pencil. Chemical reactions are conceived as instantaneous conversions of an initial into a final structure, occurring with a certain probability per unit time. Molecules are imagined to hop from one minimum to another on the potential-energy hypersurface.

Some highly successful experimental techniques have been invented for determining molecular structures. Especially X-ray diffraction (1) and, to a lesser extent, neutron diffraction (2) provide very accurate "images" of molecules and revealing insights into the secrets of nature. In the course of the past two decades, nuclear magnetic resonance (NMR) has developed into a powerful complement to the diffraction techniques. In particular in those cases where no single crystals of sufficient quality can be grown, two-dimensional NMR techniques proved to be of great value for the structural investigation of small and medium-size biomolecules, as shown first by the research group of Kurt Wüthrich (3,4). For obtaining solution structures, NMR is today unmatched in its

NMR As a Structural Tool for Macromolecules: Current Status and Future Directions
Edited by B.D. Nageswara Rao and Marvin D. Kemple, Plenum Press, New York, 1996

15

information content. It may be complemented by scattered information from other sources, such as optical spectroscopy and electron paramagnetic resonance. Of great help are the computer procedures of molecular modeling that allow one to substitute missing experimental information (5).

Clearly, a rigid molecular structure represents at best a snap-shot of a molecule on the move. In many cases, the experiments deliver merely an average structure, often under not too well defined constraints, sometimes even trying to unify conflicting pieces of information in an inappropriate picture. Indeed, molecules with n atoms are flexible objects, and their intramolecular dynamics should be described by an ensemble of dynamic trajectories in a (3n-6)-dimensional conformation space. Often their flexibility is of great importance for their functionality. Rigid molecules are less capable of undergoing intimate interactions with partner molecules, and flexibility is required for achieving an optimum match. Obviously, a detailed understanding of molecular reactivity requires the characterization of the dynamical features of molecules as well.

A simple estimate of the dynamical degrees of freedom convinces oneself that in the vast majority of cases a full characterization of the intramolecular dynamics is out of question. While a molecular conformation is defined by a point in the (3n-6)-dimensional conformation space, dynamics must be characterized by the motional invariants of a complete ensemble of trajectories in this space. This amounts up to (3n-6) autocorrelation functions, $(3n-6)(3n-7)/2$ pair correlation functions, and a large number of higher-order correlation functions. Obviously, in practice, one has to be satisfied with a small subset of dynamic characteristics.

Several simplified treatments have been described. Perhaps the most frequently publicized and used one is the model-free approach by Lipari and Szabo (6) where the intramolecular motion is characterized by a set of internuclear vectors or axes of chemical shielding or quadrupolar-interaction tensors. Most conveniently, one uses ^{15}N-^{1}H and ^{13}C-^{1}H vectors for this purpose. In this way, it is possible to obtain an impression of the local flexibility, say along a polypeptide chain.

Another possibility is to consider particular motional models, for example of the segmental mobility of side chains. For each considered degree of freedom, a model, such as rotational diffusion or random discrete rotational jump motion, is selected and characterized by a minimal set of parameters, often just by a single correlation time, sometimes combined with an angular restriction angle or a Gaussian angular distribution function.

SOURCES OF INFORMATION ON DYNAMICS

NMR is, in principle, an exceptionally powerful tool that allows the investigation of processes between tens of picoseconds up to seconds and covers about 11 orders of magnitude. This section presents a brief summary of available techniques covering the various ranges of rate constants.

16

Flow and Stopped Flow Techniques

For the investigation of slow chemical reactions, NMR can be used as a convenient monitor in flow and stopped flow experiments (7,8). Here, non-equilibrium chemical concentrations are set up to put the reaction rates into evidence. No limits exist on the slow reaction rate end, while the minimum time resolution is in the order of 1 ms. Flow experiments are useful for the study of one-sided reactions that proceed virtually to completion and cannot be studied in the dynamic equilibrium state.

Saturation Transfer Experiments

Equilibrium state reactions with sizable steady-state concentrations of all components can be investigated by saturation-transfer experiments (9) where the exchanging components are labeled by selectively saturating their magnetization. It is then possible to monitor the transfer of the saturation to other species and explore the chemical reaction pathways. Saturation labeling functions only when the reaction proceeds within the nuclear spin-lattice relaxation time T_1. On the other hand, the reaction should not be too fast to cause severe line broadening, reducing the spectral resolution that is necessary to distinguish the various sites. This sets the following limits for the reaction rate constant k:

$$1/T_1 < k \leq \Delta\omega \tag{1}$$

where $\Delta\omega$ is the required spectral resolution.

Transient Nuclear Overhauser Effect

Variants of the saturation transfer technique are the transient NOE method (10) and the truncated driven NOE scheme (11,12) where the time evolution after a transient selective perturbation is observed. They have the same limitations for the observable rate constants as the saturation transfer technique.

Two-Dimensional Exchange Experiments

Two-dimensional exchange spectroscopy (EXSY) (4,13-15) can be considered as an efficient variant of saturation transfer spectroscopy in which the labeling of the various magnetization components is achieved simultaneously. Instead of, so to say, binary labeling by saturation, a much more versatile frequency-labeling principle is used. But with regard to the range of measurable exchange rate constants, EXSY is again equivalent to saturation transfer.

Line-Shape Studies

The quantitative analysis of exchange-broadened and exchange-narrowed line shapes, usually by computer simulation (16-18), allows one to measure rate constants in the limits

$$1/T_2^* \le k \le T_2^*(\Delta\omega)^2/8 \tag{2}$$

where $\Delta\omega$ is the chemical shift difference between the sites in a two-site exchange situation. The lower limit is given by the requirement that the exchange line broadening should at least be comparable to the line width in the absence of exchange. The upper limit originates from the requirement that the exchange line narrowing should not reduce the line broadening effect below the line width in the absence of exchange. The total range of measurable rate constants thus depends on the expression $\left(T_2^*\Delta\omega\right)^2\!\big/8$.

Rotating-Frame Relaxation Measurements

Slightly faster processes can be investigated by rotating-frame relaxation measurements (19,20). One finds that in the presence of a strong radio frequency field with the strength ω_1 the relaxation along the direction of the radio frequency field is given by the expression

$$T_{1\rho}^{-1} = T_{1\rho d}^{-1} + \frac{p_1 p_2 \tau_{ex}}{1+\omega_1^2 \tau_{ex}^2}(\Delta\omega)^2, \tag{3}$$

provided that the dynamics can be described by a two-site jump process with the populations p_1 and p_2, the exchange-time constant τ_{ex}, and the chemical shift difference $\Delta\omega$. $T_{1\rho d}^{-1}$ is the dipolar relaxation-rate constant due to the rotational molecular tumbling. As long as the second term in Eq.[3] is larger than or comparable to the first one, the exchange dynamics can be monitored. Obviously, the larger the chemical shift difference between the two sites is the larger is the accessible range of the exchange dynamics. An analogous expression applies when a change of a J-coupling constant is involved in the exchange process.

Laboratory-Frame Relaxation Measurements

The fastest intramolecular dynamics processes can be monitored by various types of relaxation measurements in the laboratory frame (4,21-23). Here, invariably, a competition between the relaxation-active molecular tumbling and the intramolecular motion takes place which sets limits to the observability of the intramolecular modes. The motional processes are characterized by spectral density functions $J(\omega)$ of which specific function values enter the expressions for the relaxation rate constants. For longitudinal cross relaxation induced by a dipolar interaction between spins I_k and I_ℓ, for example, one finds the rate constants of the two-by-two cross relaxation matrix

$$\Gamma_{kk} = \frac{1}{2}q_{k\ell}\left[J(\omega_{0k}-\omega_{0\ell})+3J(\omega_{0k})+6J(\omega_{0k}+\omega_{0\ell})\right]$$

$$\Gamma_{\ell\ell} = \frac{1}{2}q_{k\ell}\left[J(\omega_{0k}-\omega_{0\ell})+3J(\omega_{0\ell})+6J(\omega_{0k}+\omega_{0\ell})\right]$$

$$\Gamma_{k\ell} = \Gamma_{\ell k} = \frac{1}{2}q_{k\ell}\left[-J(\omega_{0k}-\omega_{0\ell})+6J(\omega_{0k}+\omega_{0\ell})\right] \tag{4}$$

while for T_2 relaxation of spin I_k, in the absence of resolved J couplings, the expression

$$1/T_{2k} = \frac{1}{4}q_{k\ell}\left[4J_{k\ell}(0)+J_{k\ell}(\omega_{0k}-\omega_{0\ell})+3J_{k\ell}(\omega_{0k})+6J_{k\ell}(\omega_{0\ell})+6J_{k\ell}(\omega_{0k}+\omega_{0\ell})\right] \tag{5}$$

holds with

$$q_{k\ell} = \frac{1}{10}\gamma_k^2\gamma_\ell^2\hbar^2 r_{k\ell}^{-6}\left(\frac{\mu_0}{4\pi}\right)^2. \tag{6}$$

The spectral density function is influenced by both the molecular tumbling with the correlation time τ_c and by the intramolecular processes, here assumed to have a correlation time τ_i. In addition, also the extent of motional modulation by the intramolecular process is relevant. It is in the simplest approach, proposed by Lipari and Szabo (6,24), represented by the order parameter S. This leads then to the Lipari-Szabo equation (6,24)

$$J(\omega) = S^2\frac{2\tau_c}{1+\omega^2\tau_c^2}+(1-S^2)\frac{2\tau_{tot}}{1+\omega^2\tau_{tot}^2} \tag{7}$$

where

$$1/\tau_{tot} = 1/\tau_c + 1/\tau_i \tag{8}$$

Only when the intramolecular correlation τ_i is in the interval

$$\tau_c/10 \le \tau_i \le 10\tau_c, \tag{9}$$

it will affect significantly the relaxation behavior. On the upper side where $10\tau_c < \tau_i$, the intramolecular process will become relaxation-inefficient and is not observable via relaxation measurements in the laboratory frame. On the other hand, at the lower end when $\tau_i < \tau_c/10$, only the order parameter S remains active which will reduce the efficiency of relaxation in a τ_i-independent manner. In this limit, intramolecular motion may be inferred, however without a possibility to determine its rate constants.

This leads to an observability window of intramolecular motional rate constants as shown in Fig.1 when laboratory-frame relaxation is considered as a means of observation and the relaxation mechanism in isotropic solution is due to an orientational modulation of an anisotropic interaction, such as dipolar interaction, chemical shielding anisotropy (CSA) interaction, or quadrupolar interaction. Whenever, in addition, an isotropic parameter is modulated by the intramolecular process, additional means of observation of slower processes arise, as mentioned under (v) and (vi).

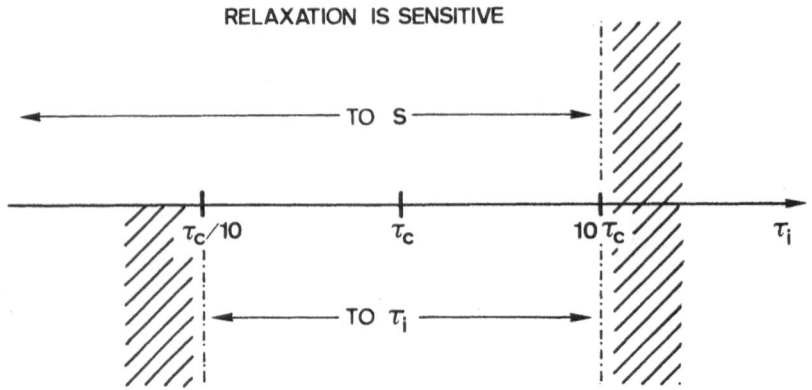

Figure 1. Observability window of intramolecular motional rate constants based on laboratory-frame relaxation measurements as described in the text.

In order to enhance the power of relaxation measurements towards a wider range of measurable rate constants, one might consider to shift the observability window on the frequency axis. It is well-known that the viscosity η of the solvent affects the rotational molecular correlation time via the Debye formula for a spherical molecule of radius r:

$$\tau_c = 4\pi\eta r^3 / kT \tag{10}$$

The viscosity could be increased by the addition of thickening agents (25) or it could be decreased by using low-viscosity supercritical solvents. Of course shifting the observability window makes sense only when the intramolecular mobility is not or is less affected by the viscosity than the overall tumbling. This has to be checked on a case-to-case basis.

INTRAMOLECULAR DYNAMICS OF ANTAMANIDE

Antamanide, a cyclic decapeptide [-Val[1]-Pro[2]-Pro[3]-Ala[4]-Phe[5]-Phe[6]-Pro[7]-Pro[8]-Phe[9]-Phe[10]-] (26) shows a variety of dynamic features that can be used to demonstrate the mentioned principles. A slow conformational backbone-dynamics mode can be investigated by rotating-frame relaxation measurements (27). Phenylalanine side chains have two degrees of freedom that lead to longitudinal relaxation effects (28). The proline

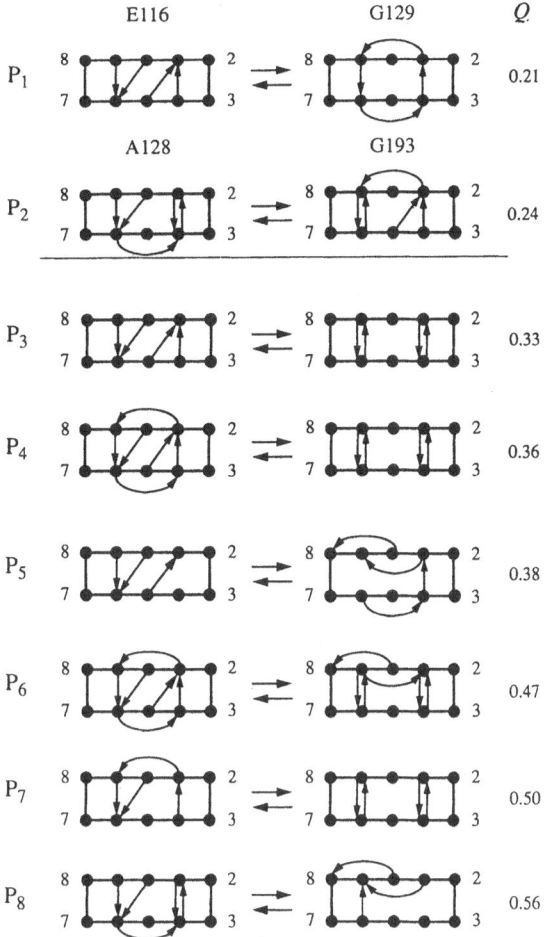

Figure 2. Types of conformational pairs of antamanide which appear among the structures that fit best the experimental data. The arrows indicate the hydrogen bonds (H→O) between the numbered amino acid residues. Q is the least squares fitting error of the structural data (from Ref. 27).

ring puckering is apparent from averaged J-coupling parameters and from a reduced dipolar relaxation efficiency (29). Finally water-binding exchange dynamics can be studied by NOESY experiments and by rotating frame relaxation measurements (30).

Peptide Backbone Dynamics

Cyclic peptides can show a variety of backbone motions. Some peptides have a single well-defined conformation. Others exhibit slowly interconverting conformers, while antamanide belongs to a class that shows conformational dynamics on the microseconds time scale. The existence of dynamics in this range has been inferred already from ultrasonic absorption measurements (31) and from the incompatibility of the distance constraints with a single conformation, noted by Kessler et al. (32).

21

The time scale of the dynamics can be determined by $T_{1\rho}$ measurements. It has been found that in particular the $T_{1\rho}$ values of Val[1]NH and Phe[6]NH are dependent on the radio-frequency field strength ω_1. This can be interpreted in terms of a two-site exchange model, leading to Eq.(3). An evaluation of this expression, assuming $p_1 = p_2 = 1/2$, led to an exchange time constant of ~27 µs at 320 K and an activation energy of $E_a \cong 21$ kJ/mol.

This result seems to indicate that the molecule undergoes dynamically a conformational change where hydrogen bonds involving the protons Val[1]NH and Phe[6]NH are broken and formed. The other NH protons, which do not show such an effect, are either in stationary hydrogen bonds or their chemical shift changes $\Delta\omega$ are sufficiently small not to appreciably affect $T_{1\rho}^{-1}$. Although it is not too likely, it cannot be excluded that more than two conformations contribute to a significant extent. It can also not be excluded that traces of water might influence the hydrogen-bond exchange dynamics.

More detailed structural information can be obtained from an analysis of the conflicting dipolar distance information among the backbone protons. It shall also be interpreted in terms of two exchanging conformations, leading to average cross-relaxation rate constants $R_{k\ell}$ between spins k and ℓ:

$$R_{k\ell} = C\left\{ p_1 r_{k\ell 1}^{-6} + p_2 r_{k\ell 2}^{-6} \right\}. \tag{11}$$

Here the distances between the two spins in the conformations 1 and 2 are denoted by $r_{k\ell 1}$ and $r_{k\ell 2}$, respectively. Obviously the rate constant $R_{k\ell}$ is dominated by the conformation with the smaller distance $r_{k\ell}$. In many cases it is thus possible to divide the distance constraints into two sets, one dominated by the conformation 1 and one by conformation 2. Of course, there is no way around a trial and error search for the best choice of the two subsets.

A special procedure, called MEDUSA, has been developed (27,33) to determine a large set of trial structures that fulfill a subset of the distance constraints while violating the rest. For this purpose, the constraints are applied in a random order, rejecting those which are incompatible with the earlier ones. Each arrangement of constraints leads to a specific conformation, most of them being different from each other. In this way, more than 900 conformations have been obtained for antamanide. Finally, they are combined pairwise with weights that are optimized to obtain a best fit of all experimental data, including dihedral-angle constraints originating from J-coupling measurements. The various pairs are then ranked according to their fitting error Q. Those with a small error have a higher probability to correspond to the actual backbone dynamics of antamanide.

The best two and some further characteristic conformational pairs are shown in Fig.2 in terms of their hydrogen-bonding patterns. All of the shown pairs involve, as required, hydrogen-bond dynamics at the protons Val[1]NH and Phe[6]NH. P_1 and P_2 have the smallest fitting error and may be termed "syn" and "anti pair" as the two essential dynamic hydrogen bonds Val[1]NH-Phe[9]O and Phe[6]NH-Ala[4]O are formed either on the same or on

Figure 3

E116

G129

A128

G193

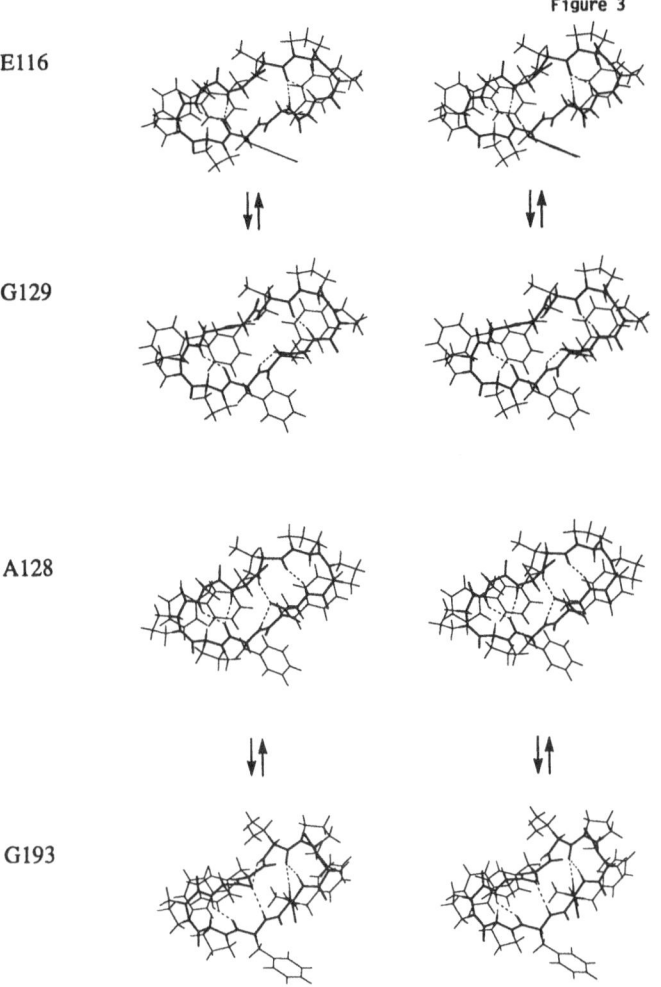

Figure 3. Two conformational backbone-structure pairs of antamanide that lead to the best fit of the experimental NMR data, shown as stereo plots. The numbering of the conformations corresponds to Fig.2 (from Ref.27).

opposite conformers. Their 3D structures are shown in Fig.3. It is apparent that the peptide-ring dynamics are limited to the residues ^{4}Val-^{5}Phe and ^{9}Phe-^{10}Phe. In particular, the four proline residues provide conformational stability to the remainder.

It is tempting to assume that either P_1 or P_2 is the proper pair. But it is also feasible that a four-state equilibrium occurs, combining the states of P_1 and P_2 with a somewhat lower probability for the two P_2 states. It cannot be excluded that also further conformations contribute, again most likely with smaller populations. The available data are not sufficient to gain more detailed insight as only 23 backbone distance constraints, 82 "absent distance constraints", and 28 angular constraints have been measured.

Proline Ring Puckering

It is well known that proline can show dynamics in peptides (34,35). First of all, it can undergo a *cis-trans* conformational change. It is found that in antamanide no such transition occurs. The peptide bonds Pro²-Pro³ and Pro⁷-Pro⁸ are cis, while the bonds Val¹-Pro² and Phe⁶-Pro⁷ are trans (32). However, a puckering dynamics of the proline rings cannot be excluded.

The proline ring conformation is easier to access through the vicinal J-coupling constants, using Karplus-type relations for the deduction of dihedral angle constraints (36), than through distance constraints. For this reason, 21 vicinal proton-proton J-coupling constants have been determined for each of the four proline rings (29) by least-squares fitting from an E.COSY spectrum (37-39).

At first, it was attempted to satisfy the 21 angular constraints for each proline residue by a single ring conformation. The fitting errors for Pro³ and Pro⁸ were by a factor ~4 smaller than for Pro² and Pro⁷. To reduce the errors, a fit by two rapidly exchanging conformations with free weight factors was attempted. While for Pro³ and Pro⁸ the errors insignificantly decreased, the errors for Pro² and Pro⁷ were reduced by a factor ~4, bringing all errors in the same range. The resulting population ratios were for Pro² 0.35/0.65 and for Pro⁷ 0.45/0.55. The two conformations for Pro² are shown in Fig.4. It is seen that they correspond to envelope-type conformations where C_γ on the flap of the envelope is mobile. The same type of motion is also found for Pro⁷.

One might argue that also multi-state pseudo-rotation could take place instead of just a two-state process. First of all, the success of the fitting speaks against such a process and, secondly, more than two states are unlikely in a peptide environment where the basis of the five-ring is held fixed by the peptide backbone.

By means of carbon-13 T_1-relaxation measurements, it is also possible to determine the puckering rate constant. First of all it is found that, in agreement with expectations, C_γ of Pro² and Pro⁷ have the longest T_1's due to partial averaging of the one-bond C-H dipolar interaction by the ring puckering motion. This already indicates that the ring puckering is quite rapid in comparison with the ^{13}C Larmor frequency of 75 MHz. A very careful analysis of the relaxation times in terms of a two-state puckering model with known puckering angle allowed a rough estimation of the puckering time constants, leading to $\tau_{pucker} \cong 30\,ps$ for Pro² and $\tau_{pucker} \cong 36\,ps$ for Pro⁷. The values carry large uncertainties because T_1 is, following the discussion in the previous section, rather insensitive to the rate constant of rapid intramolecular processes.

Obviously, there is little dynamic coupling between the proline and the backbone motion. However, it is the backbone conformation that puts constraints upon the proline rings through the - angle and causes the different behavior of Pro², Pro⁷ and Pro³, Pro⁸.

Two computational studies have been appended to the experimental investigation in order to investigate the capability of molecular dynamics simulation programs in reproducing the experimental findings (40,41). In a first attempt, the program CHARMM

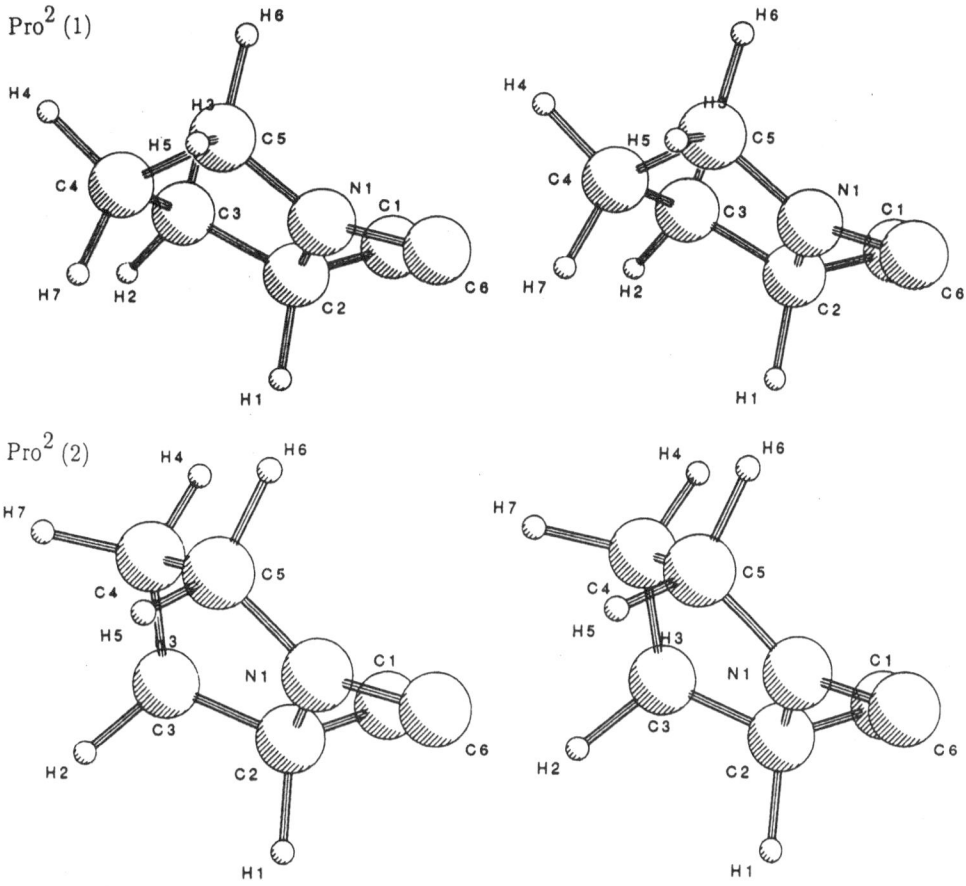

Figure 4. The two rapidly interconverting ring conformations of proline 2 in antamanide, shown as stereo plots (from Ref.29).

19 was used with an explicit representation of the solvent molecules. Over the simulation period of 800 ps, no large-amplitude ring puckering motion was observed (unpublished work). This was attributed to a rather stiff force field that led to too high energy barriers between the different states.

In a second attempt, the improved atomic force field parameter set CHARMM 22.0ß was applied (41). Because no major solvent effects on the proline dynamics was expected, the solvent chloroform was represented in all subsequent simulations by stochastic random forces and a friction coefficient that mimics the viscosity of the solvent (Langevin dynamics (44)). The results were of surprising good coincidence with the experimental findings. Figure 5 shows that indeed Pro^2 and Pro^7 undergo a two-state puckering process while Pro^3 and Pro^8 remain preferably in one state with occasional attempts to jump into the second energetically less favored state. The life times deduced from the simulation are with 12.5 ps for Pro^2 and 15.7 ps for Pro^7 slightly shorter than the experimental values. Considering the subtlety of the effects and the global representation of the solvent no better agreement was expected. It was also possible to show that the backbone conformation determines the proline puckering motion. By fixing the dihedral bond angles ϕ_3 and ϕ_8 to -60^0, all four proline rings pucker with the same rate and similar occupancy numbers.

Similar results were obtained by molecular dynamics simulations using the GROMOS program (40). The agreement with the experimental data was somewhat less perfect, both regarding the differentiation between Pro^2, Pro^7 and Pro^3, Pro^8 as well as the life times of the pucker states.

Phenylalanine Side-Chain Rotation

The four phenylalanine side chains with their two rotational degrees of motion about χ_1 and χ_2 are good systems for testing the sensitivity of the NMR data for distinguishing different motional models. In a detailed study, four motional models for the χ_1 motion and

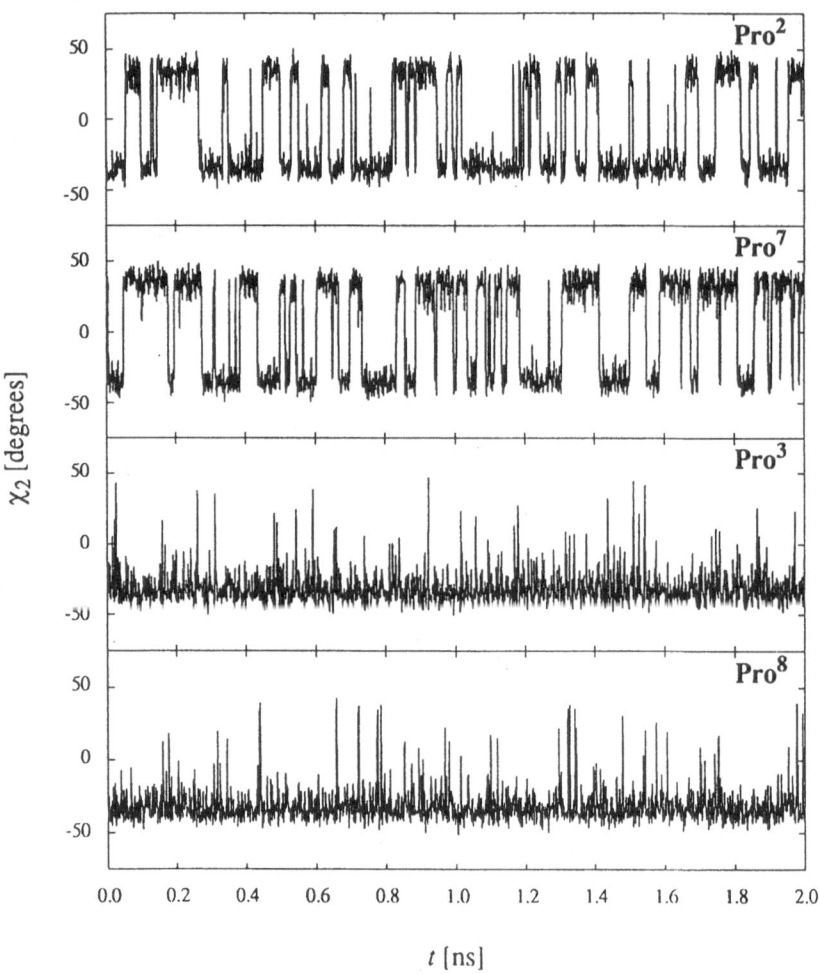

Figure 5. CHARMM 22.0 ß computer simulation of the proline ring dynamics represented by the time dependence of the χ_2 angles of the four proline residues (from Ref.41).

two models for the χ_2 motion, leading to a total of eight combinations, were compared (28). The models were for χ_1: A_1 = three-site jump motion with equal probabilities and rates, C_1 = three-site jump motion with equal probabilities and one forbidden transition, D_1 = unrestricted rotational diffusion, and E_1 = rotational diffusion with an angular restriction. For χ_2, the models B_2 = two-site jump motion and D_2 = unrestricted rotational diffusion were explored.

For the characterization of the process with correlation times in the picoseconds range, carbon-13 T_1, NOE, and T_2 values were measured. In addition, to take account of the chemical-shielding-anisotropy (CSA) relaxation of the aromatic carbons, it was necessary to measure also the cross-relaxation time to two-spin order and its decay-time constant. The measurements were performed at 200, 400, and 600 MHz proton resonance frequency, respectively, leading to 31 relaxation parameters for each residue.

The investigation revealed that the possibilities to differentiate between the eight motional models are marginal and a clear-cut distinction of them is not possible due to the weak influence of the motional character on the measurements. This is due to the dominant influence of the overall tumbling on the relaxation. In addition, it was found that the fitting error remained larger than expected from the estimated measurement errors and might indicate that the relaxation model is not fully appropriate. For example, it is conceivable that the assumption of an isotropically tumbling molecule is not justified. It could also be that the measurements are subjected to systematic error sources.

Nevertheless, it is found that the restricted diffusion model in χ_1 leads to the best fits with motion in a sector of the widths 64^0, 80^0, 64^0, and 44^0 for Phe[5], Phe[6], Phe[9], and Phe[10], respectively. For the χ_2 motion, it is only possible for Phe[9] to decide with some certainty for an unrestricted rotational diffusion model. For the other three Phe's, both models provide comparable fitting quality. The results are shown in graphical form in Fig.6.

Obviously, there is a demand for higher measurement accuracy in order to better exploit the inherent sensitivity of NMR experiments for discriminating motional models.

The rotation of the methyl groups of Val[1] and Ala[4] has also been investigated (42). The rotational time constants are in the range of 15-25 ps at 310 K.

Water-Binding Properties of Antamanide

It is known from X-ray diffraction studies that antamanide is binding in the crystalline form eight water molecules, four of them within the peptide ring. It could be imagined that also in chloroform solution, antamanide maintains some of its water-binding properties. This was investigated by preparing samples with a well defined water content. Water solubility in chloroform is very low (0.006% weight at 248 K and 0.014% weight at 272 K (43)). At higher temperatures ($\geq$ 300 K), we encountered difficulties in maintaining an equilibrium in the sealed sample tube. Slight temperature gradients led to a nearly quantitative transport of water to the top of the sample. For this reason, the experiments

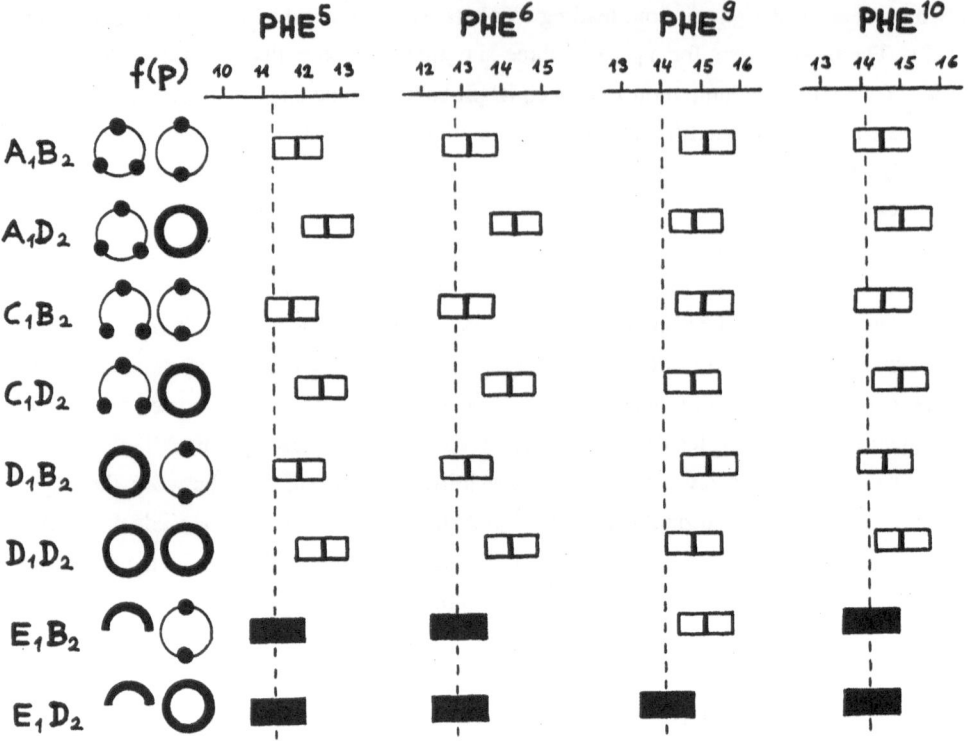

Figure 6. Graphical representation of the normalized fitting errors f(p) for the eight models described in the text fitted to the experimental data of Ref. (28). The uncertainty of the expected fitting errors is indicated by rectangles with a side length of 2 σ(f(p)); p is the parameter vector as defined in Ref.(28).

were restricted to a temperature of 250 K. Under saturation conditions, 1.5 water molecules are complexed to one antamanide molecule, obviously much less than in the crystalline environment.

The bound water is in a rapid dynamical equilibrium with the free water in chloroform solution (0.34 molecules per antamanide molecule at the used concentration). A characteristic $T_{1\rho}$ effect points to an exchange time constant in the microseconds region (30).

It was found that the intramolecular dynamics of the NH protons is influenced by the presence of water. In fact, water appears to have a stabilizing effect on the backbone and leads to a reduction of the rf-field dependence of $T_{1\rho}$. It can also be concluded that the sample used for the studies described in the previous sections was not completely water-free.

Computer molecular dynamics simulations with different amounts of water allowed the study of possible binding configurations. They also revealed that the rigidity of the peptide backbone increases with the water content (30).

CONCLUSIONS

NMR is indeed a very powerful technique for the study of dynamical processes in molecules. Nevertheless, it is not feasible to explore the dynamical properties to such an extent as it is possible to characterize a rigid geometry. The complexity of the motional quest is by order of magnitudes greater than a structure determination.

In addition, there are "blind spots" on the accessible time scale that render certain processes unobservable. This is mainly due to the dominance of the overall molecular tumbling in determining the relaxation-rate constants. It requires very high measurement accuracy for quantifying the effects caused by additional intramolecular motion. Improved instrumentation and improved measurement techniques could significantly enhance the information content of relaxation measurements. The present trend of instrumentation is certainly going in this direction, giving hope for future relaxation studies.

ACKNOWLEDGMENTS

The first author is grateful to Prof. Dr. B.D. Nageswara Rao for the organization of an extremely inspiring symposium in Indianapolis. The research has been supported by the Swiss National Science Foundation, the Cancer Research Fund of the Damon Runyon-Walter Winchell Foundation, by the Deutsche Forschungsgemeinschaft, and by Spectrospin AG, Fällanden. The manuscript has been processed by Mrs. Irène Müller.

REFERENCES

1. J.P. Glusker, M. Lewis, and M. Rossi, "Crystal Structure Analysis for Chemists and Biologists", VCH Publ., New York, 1994.

2. G.E. Bacon, "Neutron Diffraction", 3rd Edn., Oxford University Press, Oxford, 1975.

3. K. Wüthrich, "NMR of Proteins and Nucleic Acids", Wiley, New York, 1986.

4. R.R. Ernst, G. Bodenhausen, and A. Wokaun, "Principles of NMR in One and Two Dimensions", Clarendon Press, Oxford, 1987.

5. W.F. van Gunsteren, P.K. Weiner, and A.J. Wilkinson (ed.), "Computer Simulation of Biomolecular Systems, Theoretical and Experimental Applications", Vol.2, ESCOM, Leiden, 1993.

6. G. Lipari and A. Szabo, J. Am. Chem. Soc. 104, 4546 (1982).

7. D.W. Jones and T.F. Child, Adv. Magn. Reson. 8, 123 (1976).

8. R.O. Kühne, T. Schaffhauser, A. Wokaun, and R.R. Ernst, J. Magn. Reson. 35, 39 (1979).

9. S.H. Forsén and R.A. Hoffmann, J. Chem. Phys. 39, 2892 (1963); 40, 1189 (1964); 45, 2049 (1966).

10. A. Kalk and H.J.C. Berendsen, J. Magn. Reson. 24, 343 (1976).

11. A. Dubs, G. Wagner, and K. Wüthrich, Biochem. Biophys. Acta 577, 177 (1979).

12. G. Wagner and K. Wüthrich, J. Magn. Reson. 33, 675 (1979).

13. J. Jeener, B.H. Meier, P. Bachmann, and R.R. Ernst, J. Chem. Phys. 71, 4546 (1979).

14. R. Willem, Progr. NMR Spectrosc. 20, 1 (1987).

15. C.L. Perrin and T.J. Dwyer, Chem. Rev. 90, 935 (1990).

16. L.M. Jackman and F.A. Cotton, "Dynamic NMR Spectroscopy", Academic Press, New York, 1975.

17. J.I. Kaplan and G. Fraenkel, "NMR of Chemically Exchanging Systems", Academic Press, New York, 1980.

18. J. Sandström, "Dynamic NMR Spectroscopy", Academic Press, New York, 1982.

19. C. Deverell, R.E. Morgan, and J.H. Strange, Mol. Phys. 18, 553 (1970).

20. K.D. Kopple, K.K. Bhandary, G. Khartha, Y.-S. Wang, and K.N. Parameswaran, J. Am. Chem. Soc. 108, 4637 (1986).

21. A. Abragam, "Principles of Nuclear Magnetism", Clarendon Press, Oxford, 1961.

22. A.G. Redfield, Adv. Magn. Reson. 1, 1 (1964).

23. R. Tycko (ed.), "NMR Probes of Molecular Dynamics", Kluwer Academic Publ., Dordrecht, 1994.

24. G. Lipari and A. Szabo, J. Am. Chem. Soc. 104, 4559 (1982).

25. A.L. Luck and R.C. Landis, Organometallics 11, 1003 (1992).

26. T. Wieland and H. Faulstich, Crit. Rev. Biochem. 1978, 185.

27. M.J. Blackledge, R. Brüschweiler, C. Griesinger, J.M. Schmidt, P. Xu, and R.R. Ernst, Biochem. 32, 10960 (1993).

28. T. Bremi, M. Ernst, and R.R. Ernst, J. Phys. Chem. 98, 9322 (1994).

29. Z.L. Mádi, C. Griesinger, and R.R. Ernst, J. Am. Chem. Soc. 112, 2908 (1990).

30. J.W. Peng, C.A. Schiffer, P. Xu, W.F. van Gunsteren, and R.R. Ernst (in preparation).

31. W. Burgermeister, T. Wieland, and R. Winkler, Eur. J. Biochem. 44, 311 (1974).

32. H. Kessler, J.W. Bats, J. Lautz, and A. Müller, Liebigs Ann. Chem. 1989, 913.

33. R. Brüschweiler, M. Blackledge, and R.R. Ernst, J. Biomol. NMR, 1, 3 (1991).

34. R.E. London, J. Am. Chem. Soc. 100, 2678 (1978).

35. S.C. Shekar and K.R.K. Easwaran, Biopolymers 21, 1479 (1982).

36. C.A.G. Haasnoot, F.A.A.M. De Leeuw, H.P.M. De Leeuw, and C. Altona, Org. Magn. Reson. 15, 43 (1981); Biopolymers 20, 1211 (1981).

37. C. Griesinger, O.W. Sorensen, and R.R. Ernst, J. Am. Chem. Soc. 107, 6394 (1985).

38. C. Griesinger, O.W. Sorensen, and R.R. Ernst, J. Chem. Phys. 85, 6837 (1986).

39. C. Griesinger, O.W. Sorensen, and R.R. Ernst, J. Magn. Reson. 75, 474 (1987).

40. R.M. Brunne, W.F. van Gunsteren, R. Brüschweiler, and R.R. Ernst, J. Am. Chem. Soc. 115, 4764 (1993).

41. J.M. Schmidt, R. Brüschweiler, R.R. Ernst, R.L. Dunbrack, D. Joseph, and M. Karplus, J. Am. Chem. Soc. 115, 8747 (1993).

42. M. Ernst and R.R. Ernst, J. Magn. Reson. A 110, 202 (1994).

43. C.W. Gibby and J. Hall, J. Chem. Soc. 1931, 691.

44. M.P. Allen and D.J. Tildesley, "Computer Simulations of Liquids", Clarendon Press, Oxford, 1987.

DISCUSSION

Alfred Redfield - I have two questions. Your work cries out for comparison with perhaps among other things solid state NMR studies of frozen glasses or something like that which I guess is easier said than done. Do you think solid-state studies in frozen solutions could confirm your conclusions in any way? Would you like to express an opinion about such studies?

Richard Ernst - Yes. Being interested in peptide dynamics, I would be interested in investigating antamanide also in the solid state. Of course one has to expect an effect from the solid environment but still it would be very interesting to see what actually happens in the solid state. We started to do some experiments but they are by no means conclusive yet.

Redfield - What about going to low temperatures and just seeing if there are two species or would you not be able to see it?

Ernst - No, you can't freeze these motions. You can't get down enough in temperature. The processes are too fast. You can see some line broadening but you cannot really go below the coalescence point.

Redfield - The other question was that if I understand it correctly, which I may not, the relaxation studies you described have to do with the diagonal elements of the density matrix and I wondered if that understanding is correct.

Ernst - Yes.

Redfield - Have you considered off-diagonal elements i.e. T_2 processes.

Ernst - For cross relaxation, I obviously concentrated on the diagonal elements - on the I_z and $I_z S_z$ term. But of course there is also some relaxation of the off-diagonal elements which we considered in some T_2 and $T_{1\rho}$ relaxation measurements. The rf dependence of $T_{1\rho}$ proved to be particularly revealing. One could still do a more complete investigation. There is still room for that.

Redfield - Thank you.

B. D. Nageswara Rao - Several people are using the so-called model free approach of Attila Szabo in deriving this internal motion information using order parameters and the like. On the basis of your detailed analysis using molecular dynamics and so on would you care to comment upon how your results would compare with a model free approach type of fitting the data.

Ernst - I think it's certainly a valid approach to determine Lipari and Szabo parameters and then in a second step to interpret these parameters in terms of a model. This is essentially equally good. If you know which models you would like to compare, then you can do it directly. If you like to remain model free from the beginning then you better do it via Lipari

Szabo parameters. That's a little bit a matter of taste. If one interprets the data in terms of a particular model, then it doesn't make much difference whether one does it in one step or two steps. The significance of comparison of two models is just given by the measurement errors which propagate through the two-step procedure as well. So both approaches are feasible.

Rao - Do you think you will reach the same conclusions?

Ernst - If you go down to specific models, yes. Of course you can't just remain with the Lipari and Szabo parameters and say that's what you want to publish and not say something about specific models, that's up to you. But if you want to speak about specific models, then it's equivalent.

Marc Adler - In terms of broad design, multiple conformations are often frowned upon because the theory is that there is only one conformation that binds to the enzyme. Therefore, the loss of entropy upon binding is not a disfavorable thing. Now, I was wondering with your molecule whether you might design methyl groups into it to freeze out the proline ring flipping and see if that was a more potent molecule.

Ernst - Yes, I think that's exactly the type of information one would like to get if one does this type of dynamics studies in order to understand what is going on in a binding process, and how important dynamics of a flexible molecule is. If one understands this, one can start to tailor a molecule to a particular application. I think that's the final philosophy behind it. Of course, we don't go as far as that because we are biologically innocent.

Carol Post - About the prolines, there are two states for two prolines and one state for two of the other four prolines. What makes the two that are dynamic, different? Did you have any ideas about any coupling of that motion to conformational transitions?

Ernst - Yes. What we have done for example is to very slightly restrict the backbone angles in our peptide and then suddenly all the four prolines started to pucker. So it has to do with the conformation of the backbone and there is a slight asymmetry in the molecule. The molecule is almost symmetric but the asymmetry induces sufficient strain to hinder the puckering of two of the prolines. Of course, to understand this in simple terms is difficult. It's just a complicated system. There is a lot of interactions but after all it has to do with the conformation of the backbone which restricts the possibility of motion of the sidechain.

Post - And then a point of clarification about the same system. From NMR you measured the correlation time for the jump between the two states of 40 ps and you showed CHARMm gave something a little bit shorter, more dynamic and you also had a correlation time for the other two that were not dynamic.

Ernst - Of course, there is also fast motion going on, local motion within one energy minimum, this means fluctuations about the equilibrium position with correspondingly short correlation times. But this is not really conformational motion.

Post - Were those numbers obtained from some proton-proton vector in the computation?

Ernst - There is a small-angle motion which is very rapid. The parameters are lumped together in the same table. They don't refer to the same process. You are correct. One shouldn't do that.

Maurice Guéron - I might have misunderstood about the computations. You showed that you were getting effects which were fitted with a given correlation time for the whole molecule and then a longer correlation time for rotation of the phenylalanine, I think it was τ_2. In both cases these motions are affecting the dipolar interaction if I am not mistaken. Now if you have a fast τ_c with isotropic or nearly isotropic motion affecting the dipolar relaxation, how can you detect anything which corresponds to slower modulation of the dipolar interaction?

Ernst - It depends on how much slower the intramolecular motion is. If it is within the same order of magnitude, within a factor of 10 smaller or larger, then it will still affect the relaxation. Even for anisotropic motion of the whole molecule the sensitivity to these processes might not be great. One has to do a lot of measurements. But there are weak influences.

Bernard Brooks - In the analysis of models of χ_1 and χ_2 distributions for the phenyl alanines, looking at the various models, did you try looking at the distributions of the states obtained from molecular dynamics and see if that has any improvement or does it give you any differences in the simpler model?

Ernst - Molecular dynamics simulations are very difficult to apply in this case since the motions are relatively slow, especially the χ_2 motion where the correlation times are about 1000 ps or so. One has to perform enormously long computations to get any reliable results, and we didn't do that. We were investigating the potential energy surface to see if it is compatible. But to obtain real sidechain dynamics from molecular dynamics is very difficult as one has to calculate way up into the nanoseconds, 100 ns or so to get reliable results. So that is the limitation at the moment.

Yuan Xu - I have two questions. The first one is about the multiple conformations. For instance, in this molecule you have multiple conformations, so you see the several conformations and put those that contribute together. In doing so, the experimental data can be better explained than a single conformation. So in this case how did you determine the contribution for each individual conformation. For the proline, one contributes 40% and the other one contributes 50%, so could you explain how this was done?

Ernst - It's very simple in principle. You just have a fit parameter which is the relative population of the two conformations and you vary this fit parameter till you find the best fit. The only question is that how sensitive are the results to this fit parameter. We found that for proline 3 and proline 8, when we added the second conformation in order to try to make the fit better, the second conformation contributed only about 5-10% which is hardly a significant contribution. So we just concluded that the residue is rigid and the second conformation can be ignored. But of course, there is a certain error in this population and we don't want to say much more than either the equilibrium is much on one side or it is approximately in the middle.

Xu - What happens if you have more conformations to put together and more parameters to vary?

Ernst - From the fitting, one cannot decide whether there are even more conformations present. So we also performed some MD computer simulations. All the computations which we have done using different MD programs invariably gave us two rigid conformations, two well defined conformations which did not differ much between the different programs. Only the timescale of interconversion was a little different due to the different force fields.

Xu - My second question is when you investigate the multiple conformations using the molecular dynamics approach, is that as good as molecular dynamics if you use Monte Carlo simulations to investigate the multiple conformations?

Ernst - We did some Monte Carlo simulations as well. One obtains similar results.

Yuan Xu - So the two approaches are pretty much the same.

Ernst - Yes.

Xu - Thank you.

STRUCTURAL, DYNAMIC, AND FOLDING STUDIES
OF SH2 AND SH3 DOMAINS

Julie D. Forman-Kay,[1] Steven M. Pascal,[1,2] Alex U. Singer,[1]
Toshio Yamazaki,[2] Ouwen Zhang,[1,2] Neil A. Farrow,[2] and Lewis E. Kay[2]

[1]Biochemistry Research Division
Hospital for Sick Children
Toronto, Ontario M5G 1X 8 CANADA
[2]Departments of Medical Genetics, Biochemistry and Chemistry
University of Toronto
Toronto, Ontario M5S 1A8 CANADA

INTRODUCTION

Recent advances in NMR methodology have made it a powerful approach for the study of biomolecular structure and dynamics (Bax and Grzesiek, 1993; Bax, 1994; Farrow et al., 1994a). In addition to NMR being a structural tool, as a solution spectroscopy it is exquisitely sensitive to dynamic processes - not only fast, low amplitude motions which can often be described by analysis of X-ray crystallographic B factors (Ringe and Petsko, 1985), but also slower, larger amplitude motions. These may include conformational exchange on millisecond time-scales or longer between states as dissimilar as folded and unfolded states of proteins and reflecting motions of tens of angstroms. We have exploited this distinguishing capability of NMR spectroscopy to describe the dynamic processes observed during structural studies of two isolated domains of signal transduction proteins, a Src Homology 2 (SH2) domain of phospholipase Cγ in complex with a phosphopeptide from the platelet-derived growth factor receptor (PDGFR) and an isolated Src Homology 3 (SH3) domain from the drosophila protein Drk[1].

SH2 and SH3 domains were first identified as regions of significant sequence similarity in the non-catalytic portion of the src family of protein tyrosine kinases (Pawson et al., 1993). These small, modular, independently folding domains have since been found

---

[1]The results described in this lecture are presented in greater detail in three manuscripts: S.M. Pascal, T. Yamazaki, A.U. Singer, L.E. Kay, and J.D. Forman-Kay, 1995, Structural and dynamic characterization of the phosphotyrosine binding region of an SH2 domain-phosphopeptide complex by NMR relaxation, proton exchange and chemical shift approaches, in preparation; O. Zhang and J.D. Forman-Kay, 1995, Structural characterization of folded and unfolded states of an SH3 domain in equilibrium in aqueous buffer, *Biochemistry*, submitted; N.A. Farrow, O. Zhang, J.D. Forman-Kay and L.E. Kay, 1995, Comparison of the backbone dynamics of a folded and an unfolded SH3 domain existing in equilibrium in aqueous buffer, *Biochemistry* 34:868.

in a number of other proteins involved in signaling, including enzymes and the so-called adapter molecules which function primarily by binding to multiple proteins. SH2 domains, of approximately 100 residues, and SH3 domains, of about 60 residues, function not through catalytic activity but by mediating protein-protein interactions in signal transduction. SH2 domains bind to specific sites of phosphorylated tyrosine on their biological targets, while SH3 domains bind to poly-proline type II helices (Yu et al., 1994).

C-TERMINAL SH2 DOMAIN OF PHOSPHOLIPASE Cγl

When a growth factor receptor binds to the extracellular region of a receptor protein tyrosine kinase, it induces dimerization of these receptors and subsequent cross-phosphorylation by the kinase domains on exposed tyrosine residues (Ullrich and Schlessinger, 1990). This creates specific phosphotyrosine binding sites for SH2 domains in a number of signaling proteins. In the case of phospholipase Cγ (PLCγ), the Tyr-1021 site in the C-terminal region of the β-platelet-derived growth factor receptor is an important biological target (Valius and Kazlauskas, 1993; Valius et al., 1993). Upon binding of the C-terminal SH2 domain of PLCγ to the phosphorylated Tyr-1021 site, the kinase domain of the receptor can phosphorylate PLCγ, activating its catalytic metabolism of phosphoinositol-bis-phosphate to the second messengers diacylglycerol and inositol-triphosphate (Rhee, 1991). In order to understand at the atomic level the mechanism of the sequence specificity in the biological recognition taking place between the PLCγ SH2 domain and this phosphorylated tyrosine site, we have studied the isolated SH2 domain, labeled with ^{15}N and ^{13}C, in complex with a synthetic 12-residue phosphopeptide derived from the sequence around the Tyr-1021 site of the receptor (Asp-Asn-Asp-pTyr-Ile-Ile-Pro-Leu-Pro-Asp-Pro-Lys; gift of Dr. Steve Shoelson). Binding studies with this peptide have shown that the interaction of PLCγ with peptide has similar affinity to the *in vivo* receptor interaction, with a K_D of approximately 50-100 nM (Piccione et al., 1993).

NMR is particularly useful for the investigation of protein-protein interactions because we can employ isotopic labeling strategies to isolate intramolecular and intermolecular interactions, enabling us to probe the structure of the individual components (the SH2 domain and the peptide) and the SH2-peptide contacts separately (W. Lee et al., 1994). The SH2 domain is isotopically labeled with ^{15}N and ^{13}C and the synthetic peptide contains the natural abundance isotopes ^{14}N and ^{12}C. We utilize three types of NMR experiments to assign the resonances and structurally characterize the system: (1) isotope-edited experiments which probe interactions between protons bound to ^{15}N and ^{13}C for assignment and structure determination of the labeled SH2 domain, (2) isotope-filtered experiments which probe interactions between protons bound to ^{14}N and ^{12}C for the assignment and structure of the peptide and (3) half-filtered experiments which probe interactions between protons bound to either ^{15}N or ^{13}C and protons bound to either ^{12}C or ^{14}N to detect specific interactions between the SH2 domain and peptide. An example of this third class of experiments is the 3D ^{13}C-half-filtered NOESY experiment with proton and ^{13}C resonances of the SH2 domain along the F_1 and F_2 axes, respectively, correlated to proton resonances of the phosphorylated peptide along the F_3 axis. This experiment allowed us to identify over one hundred NOEs between the peptide and the protein, including numerous strong NOEs observed between the H^δ and H^ϵ resonances of the phosphotyrosine (pTyr) and residues of the SH2 domain.

In addition, NMR allows us to describe the dynamics of the interaction. Tight binding does not always lead to rigid complexes. It is of interest to observe any motions within this tight binding complex in order to better understand the potential entropic role in maintaining high affinity binding. The dynamic studies described below focus on the motion of residues of the SH2 domain and are of the isotope-edited class of experiments.

Structural Basis of Specificity in Protein Recognition

Utilizing NOE data along with coupling constants and stereospecific assignments we have refined our previously published structures of the PLC-phosphopeptide complex (Pascal et al., 1994) and are now able to define the backbone coordinates to a precision of 0.6 Å overall (residues 11-99) and to 0.3 Å within the binding interface. The electrostatic potential surface of the SH2 domain (Nicholls et al., 1991) displays a very deep strongly positive binding pocket for the pTyr residue created by four arginine residues. The invariant arginine found in all SH2 domains (Koch et al., 1991), defined using the Eck nomenclature for SH2 residues (Eck et al., 1993) as Arg-βB5 (or Arg-37 in our numbering), is located at the bottom of this pocket with three additional arginines, Arg-αA2 (Arg-18), Arg-βB7 (Arg-39), and Arg-βD6 (Arg-59) surrounding it and contributing to the intense positive electrostatic potential. In addition to this deep pocket, a long hydrophobic groove is observed which has significant interactions with the residues C-terminal to the pTyr extending from isoleucine at the +1 position, isoleucine at the +2 position, proline at the +3 position, leucine at the +4 position and proline at position +5. There are also NOEs observed to the backbone of the aspartic acid at the +6 position. The extent of the hydrophobic binding groove for these hydrophobic residues C-terminal to the phosphotyrosine is quite significant and helps to explain the sequence specificity of the binding of this SH2 domain of PLCγ to the Tyr-1021 site of the PDGFR. The root-mean-squared deviation of all non-hydrogen atom coordinates within this hydrophobic binding interface to the mean coordinates is 0.6 Å. Thus, we are able to clearly describe the structural mechanisms of sequence specific recognition for this PLCγ-Tyr-1021 interaction. The structure also demonstrates our current general understanding of SH2 domain-target binding specificity. Not all phosphorylated tyrosine residues bind to all SH2 domains. Binding is strongly dependent on the particular amino acid sequence context of the pTyr, notably the residues C-terminal to it (Songyang et al., 1993, 1994).

SH2 domains, since they are very important molecules in signal transduction, have been the subject of a number of other structural studies (reviewed in Kuriyan and Cowburn, 1993; Yu and Schreiber, 1994). Complexes of the src and lck tyrosine kinase SH2 domains with the high-affinity binding peptide containing the sequence pTyr-Glu-Glu-Ile have been determined in the laboratories of John Kuriyan (Waksman et al., 1993) and Steve Harrison (Eck et al., 1993), respectively. These structures demonstrate a two-pronged plug model of SH2 binding with two deep pockets, one for the pTyr and one for the hydrophobic sidechain of the isoleucine residue at the +3 position C-terminal to the pTyr. Our structure of the PLCγ-PDGFR peptide complex shows a very different mode of binding where all of the residues from +1 to +5 are important for sequence-specific recognition. A complex with a high-affinity binding peptide to the N-terminal SH2 domain of the syp protein tyrosine phosphatase was recently solved in the laboratory of John Kuriyan (C.-H. Lee et al., 1994). This structure also demonstrated interactions of residues at the +1 through +5 positions of the peptide to the SH2 domain.

Dynamic Studies of Arginines

During the course of our structural studies, we found numerous resonances which were broadened or doubled. These observations stimulated our studies of the dynamic processes within this complex. We have specifically focused on the study of arginine dynamics (Pascal et al., 1995) because of the important electrostatic role of the four arginines lining the pTyr binding site in the SH2-peptide interaction. We have interpreted the results using the following principles to guide us: (1) conformational averaging in the intermediate exchange regime on the NMR time-scale leads to broadening of resonances, (2) fast rotation about the Nε-Cζ bond of an arginine leads to degeneracy of the Nη resonances and fast rotation about the Cζ-Nη bond leads to degeneracy of the Hη proton resonances and (3) hydrogen bonding of the Hη or Hε protons causes distinctive chemical shifts. In a 3D H(CCO)NH-TOCSY spectrum (Grzesiek et al., 1992; Logan et al., 1992; Montelione et al., 1992) of the SH2-peptide complex we observe that a number of sidechain resonances of the arginines in the pTyr binding site, especially the δ and γ proton resonances, have extremely broad linewidths. In the ^{15}N-^{1}H HSQC spectrum in the upfield region where Nη and Nε resonate (Figure 1) most of the arginines display degenerate Nη and Hη resonances at the expected chemical shifts for arginines that are interacting only with solvent and many are significantly broadened, indicative of intermediate time-scale rotation about their Nε-Cζ bonds. However, Arg-37 which is located at the base of the pTyr binding pocket has two distinct Nη chemical shifts and four distinct Hη chemical shifts, indicative of significantly slowed rotation about the Nε-Cζ and Cζ-Nη bonds. Two of the Hη resonances are extremely down-field shifted, one on each of the two Nη, which suggests involvement in two phosphate hydrogen bonds.

The Nε and Hε resonances observed in the upfield region of the HSQC spectrum are much better resolved than the Nη and Hη resonances. In order to better understand the dynamics and the pTyr-arginine interactions in the binding site, we exploited the chemical shift dispersion of these resonances and performed ^{15}N T_1, T_2 and heteronuclear ^{15}N-^{1}H NOE relaxation studies of the arginine Nε sites in the SH2 domain. We analyzed this data using the model-free approach of Lipari and Szabo (1982a,b) and found that three of the four arginines in the pTyr binding site, Arg-αA2 (Arg-18), Arg-βB5 (Arg-37) and Arg-βB7 (Arg-39) have order parameters in the range of 0.7 to 0.8, while Arg-βD6 (Arg-59) has an order parameter of 0.3. Utilizing these relaxation results in conjunction with our previous structural data and interpretation of chemical shifts of the guanidino group resonances, we can construct a model for the interactions of the arginines in the pTyr binding site. The highly conserved Arg-37 at the base of the binding site appears to be very rigid and donates two phosphate hydrogen bonds, one from each of the two Nη NH$_2$ groups. Arg-18 and Arg-39 also are stabilized by interactions within the binding site, probably with a phosphate and potentially with the electrons of the tyrosine ring itself in an aromatic-amino interaction. These interactions may be more transient, with an off-rate on a millisecond time-scale, leading to broadening of the aliphatic resonances, such that a number of different potential hydrogen bonding arrangements could be sampled. Arg-59 appears not to be involved in significant stabilizing hydrogen bonding interactions with the pTyr, but the role of the guanidino group may be for electrostatic attraction of the pTyr to the binding site. This suggestion correlates well with surface plasmon resonance experiments that have been used to investigate SH2-pTyr peptide binding (Felder et al., 1993). These studies show high affinities but also high off-rates which can be explained by compensating high on-rates which approach or exceed the diffusion limit. On-rates of this magnitude may be due to an electrostatic drawing mechanism which would require the presence of a large positive field in the pTyr binding site, with contributions from a number of positively charged residues.

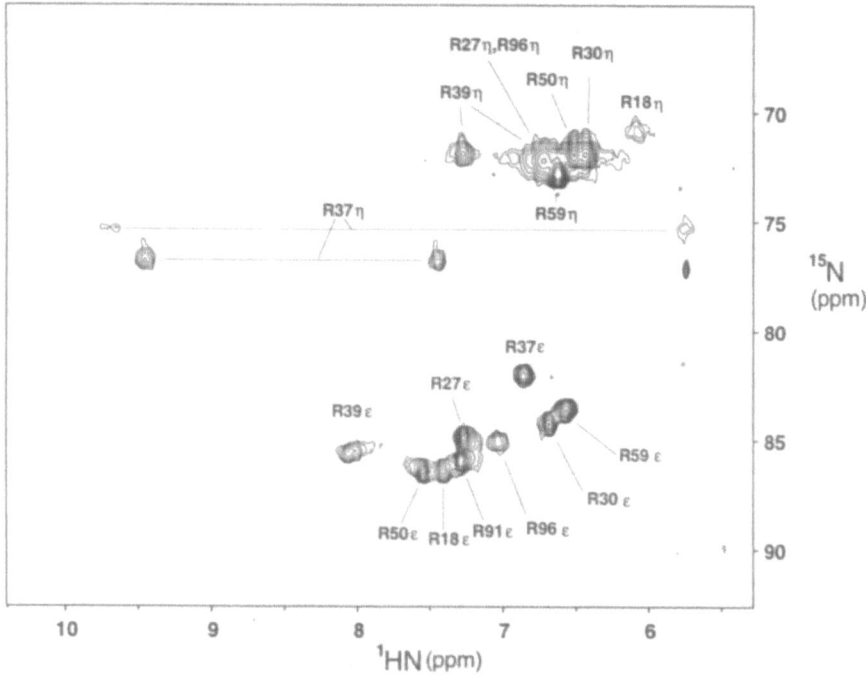

Figure 1. NεHε and NηHη region of the ^{1}H-^{15}N HSQC spectrum of the complex between the C-terminal SH2 domain of PLCγ and a 12-residue phosphopeptide from the Tyr-1021 high-affinity binding site of the PDGFR. The figure is modified from Yamazaki et al. (1995) where assignment methods are described.

The conclusions that we have drawn about the different kinds of motion within the binding site are currently qualitative. We hope to interpret them more quantitatively in order to understand the role of entropy in high-affinity binding. This may be important in the rational design of specific inhibitors of SH2 binding, since drug design requires an understanding of not just the enthalpic interactions derived from structural studies of sequence specific recognition but also entropic components of the free energy of binding.

THE N-TERMINAL SH3 DOMAIN OF DRK

Drk, the Drosophila homologue of *C. elegans* SEM-5 and mammalian GRB2, is composed of three domains, an N-terminal SH3 domain, a central SH2 domain and a C-terminal SH3 domain (Olivier et al., 1993; Clark et al., 1992; Simon et al., 1993). We have expressed the isolated N-terminal SH3 domain of the protein, including the first 59 residues of this molecule up to the conserved tryptophan which is the start of the SH2 sequence. We expected to perform structural studies similar to those described for the PLCγ SH2 domain. However, when we analyzed our first NMR spectrum of the sample in 50 mM phosphate buffer, pH 6, we observed an equilibrium between folded and unfolded states. Since we had discovered a system which enables the study of an unfolded state under the identical conditions as the folded state, at near physiological conditions, we have exploited it instead for protein folding. NMR is a very useful technique for probing folding since unfolded and partially folded states, which cannot be crystallized, can be characterized in solution (Shortle, 1993; Neri et al., 1992). Analysis of these states to elucidate the interactions which are formed very early in folding can aid our understanding of the protein folding pathway. Characterization of the unfolded state is also important since knowledge of the beginning state, as well as any intermediates, is necessary for truly understanding any thermodynamic transition. The end states, folded proteins, have been very well characterized in many instances, but much less is known about unfolded states. A significant problem for interpretation of any results in this area is the fact that unfolded states are highly sensitive to the conditions of the denaturation (Shortle, 1993). Our equilibrium between folded and unfolded states of the SH3 domain, enabling us to study both states under the identical conditions, is a very unique one.

Structural Characterization of Folded and Unfolded States Existing in Equilibrium

The ^{15}N-^{1}H HSQC spectrum of this 59 residue SH3 domain in 50 mM sodium phosphate, pH 6, displays twice as many peaks as one would expect for a single folded state (Figure 2a). In addition, there are a number of amide proton resonances clustered between 8.0 and 8.5 ppm. In a homonuclear TOCSY we observed many amide-amide correlations, with resonances in one of the dimensions also clustered around 8-8.5 ppm. On the basis of previous structural studies the SH3 domain was known to be composed predominantly of β-strands (reviewed in Kuriyan and Cowburn, 1993; Yu and Schreiber, 1994). Thus, we would not expect artifactual amide-amide NOE effects in our TOCSY, especially since we utilized a clean TOCSY sequence (Griesinger et al., 1988). Therefore, these peaks must be indicative of conformational exchange.

In order to confirm that we were observing slow exchange on the NMR time-scale between folded and unfolded states, we performed titration experiments with agents known to destabilize or stabilize folded structure including pH, temperature, guanidine hydrochloride (Gdn) and Hofmeister series salts (Arakawa and Timasheff, 1985). The sidechain indole of the single tryptophan in this molecule displays two isolated downfield resonances, one for each of the folded and unfolded states. Thus, we could utilize 1D NMR spectra in the titrations and we observed the intensity of the more downfield of the two increase at high temperature, low pH or high Gdn concentration but completely disappear in the presence of 400 mM sodium sulfate. In an ^{15}N-^{1}H HSQC spectrum in 2 M Gdn (Figure 2b), approximately one peak per residue is observed, clustered between 8 and 8.5 in the proton dimension with very good dispersion in the ^{15}N dimension. In 400 mM sodium sulphate (Figure 2c), we observe a typical, well-dispersed spectrum of a β-sheet protein also containing about one peak per residue. If these two spectra are summed, the resulting spectrum is remarkably similar to the original HSQC in 50 mM sodium phosphate, pH 6.

We assigned the equilibrium folded and unfolded states simultaneously in these original conditions using 3D ^{15}N-edited gradient/sensitivity enhanced TOCSY- and NOESY-HSQC experiments (Kay et al., 1992; Zhang et al., 1994). The assignment could be performed relying on the cross-peaks between only folded or only unfolded resonances, as well as on TOCSY or NOESY exchange peaks between the two states. The secondary structure of the folded state of the SH3 domain (Figure 3a) based on $^3J_{HN\alpha}$ coupling constants, which are related to phi torsion angles, and patterns of sequential NH_i-NH_{i+1} and $C\alpha H_i$-NH_{i+1} NOES is very similar to that of other SH3 domains that have been solved. The domain consists largely of β-strands and can topologically be described as a β-barrel.

We also performed a titration experiment using a fragment of the biological target of this SH3 domain, the C-terminal tail of SOS, which contains a number of proline rich regions that form poly-proline helices (Olivier et al., 1993). There were significant chemical shifts of resonances of the folded state of the SH3 domain, demonstrating specific binding of the folded state in this equilibrium to its biological target.

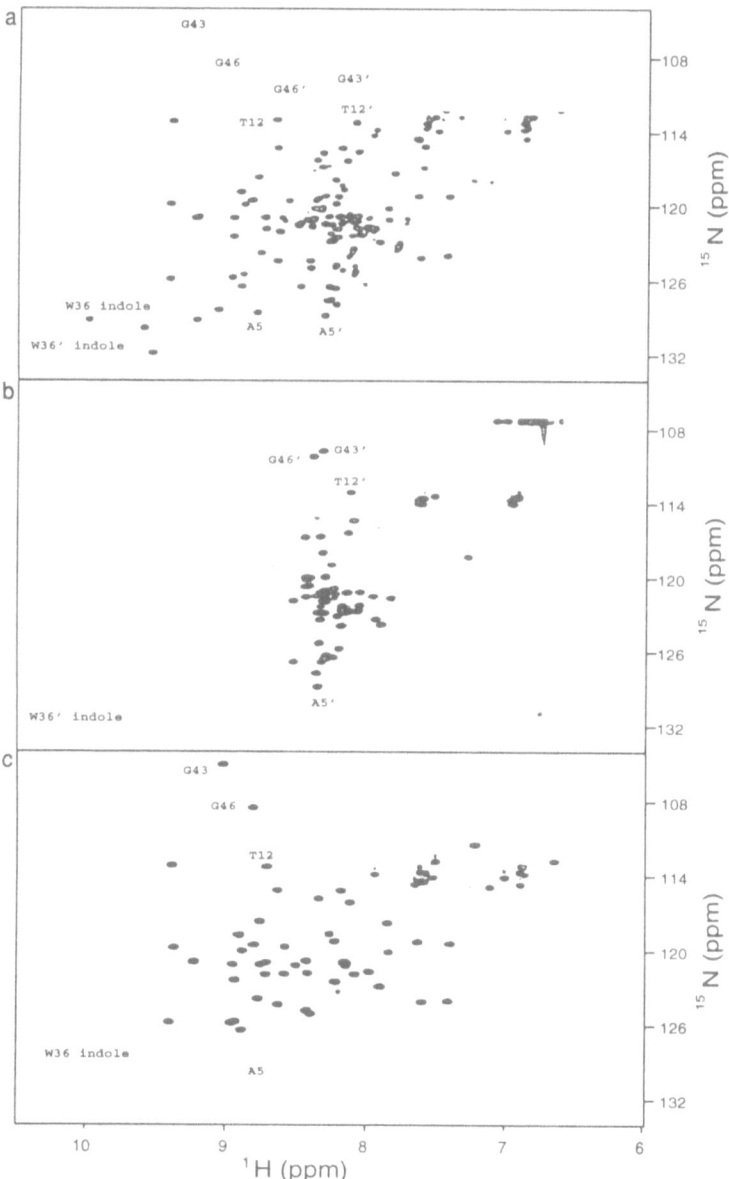

Figure 2. ^{1}H-^{15}N HSQC spectra of the N-terminal SH3 domain of drk in (a) 50 mM sodium phosphate, pH 6, (b) 2M Gdn, 50 mM sodium phosphate, pH 6 and (c) 400 mM sodium sulfate, 50 mM sodium phosphate, pH 6. Figure was modified from Zhang and Forman-Kay (1995).

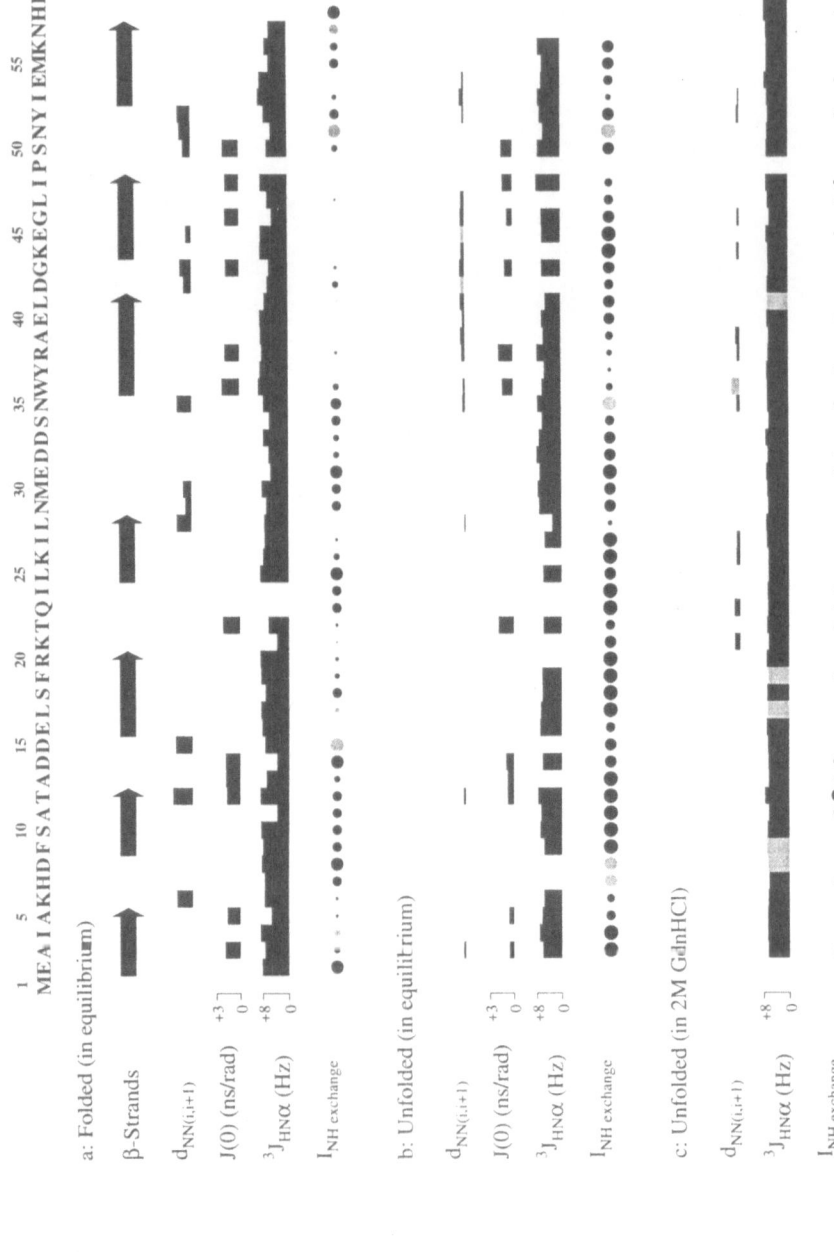

Figure 3. Sequence, secondary structure, sequential NH_i-NH_{i+1} NOEs, $J(0)$ values, $^3J_{HN\alpha}$ couple constants and intensities of amide-water exchange peaks for the N-terminal SH3 domain of drk (a) in the equilibrium folded state, (b) in the equilibrium unfolded state and (c) in the unfolded state in 2M Gdn.

The unfolded state, however, is very different from the folded states of other SH3 domains (Figure 3b). The changes in CαH proton chemical shift from random coil values (Wishart and Sykes, 1994) are quite small and the $^3J_{HN\alpha}$ coupling constants cluster around 6 and 7 Hz, indicative of conformational averaging, except for Leu-28 which has a very small coupling of less than 4 Hz. Amide proton exchange with solvent water in this state is much greater than in the folded state which is protected by hydrogen bonds within stable β-structure. Of special interest is a stretch of NH_i-NH_{i+1} sequential NOES in the region from Asn-35 to Ile-48, spanning a highly stable β-strand, β-turn, β-strand region in the folded structure and for which an extended stretch of NH_i-NH_{i+1} NOES is not observed in the folded state. While deviation from purely random structure in unfolded proteins is typically described as "residual structure," since these NOEs are not present in the folded state it is not accurately described as such.

Relaxation Studies

It is important to note that many of the NMR observables that were used for interpretation of structure in the unfolded state average in a complex manner over the rapidly exchanging ensemble of multiple unfolded protein conformations. Chemical shift is the easiest to interpret in this regard because it is a simple linear, population weighted average, while $^3J_{HN\alpha}$ coupling constants are averaged as a cosine modulated function of the phi torsion angle and NOEs are either an $<r^{-6}>$ or $<r^{-3}>$ average (Kessler et al., 1988). In addition, the value of the NOE is not merely a reflection of interproton distances, since it also contains contributions from dynamic behavior. In the macromolecular limit the NOE is directly proportional to the J(0) spectral density term. Thus, the extended stretch of sequential NOEs can not be interpreted as a clear preference of these residues for sampling conformations in the α-region of (φ,ψ) conformational space. If the J(0) spectral density values in this region are much higher than those in other regions, these NOEs must be interpreted simply as a consequence of motion. In order to establish that the NOEs are reflective of structural preferences, we performed ^{15}N relaxation experiments to measure the values of J(0) as a function of residue.

Relaxation experiments are typically analyzed by quantitating and fitting the intensity of peaks in T_1, T_2 and ^{15}N-^1H NOE spectra recorded with various delay times. In this particular system with exchange between folded and unfolded states, we have an additional complication. New pulse sequences were developed (Farrow et al., 1994b) to enable the separation of the exchange rate and the relaxation rates. However, in order to fit the data, the intensity of the two diagonal peaks (i.e. the folded and the unfolded peak) and at least one of the two exchange cross-peaks must be unambiguously quantified. This limits the number of resonances that can be characterized because of the poor spectral dispersion and overlap of many peaks in the unfolded state. We were able to analyze 12 backbone amides and the tryptophan sidechain indole (Farrow et al., 1995). The folded SH3 domain behaves as expected for a molecule of its size, with an overall correlation time of 5.5 ns. Interestingly, the unfolded domain also has the same average correlation time, which suggests that this state is rather compact as opposed to an extended random coil. Detailed analysis of the values of the relaxation rates for the unfolded state demonstrate that the residues towards the center of the molecule display very similar behavior to that observed for the folded state, while residues towards the ends of the molecule deviate much more significantly from the behavior of the folded state. The forward and backward exchange rates can also be determined from the fitting procedure. Since the transition is from folded to unfolded states of the protein, the exchange rates are identical to the rates of unfolding and folding. The rate of folding of this isolated SH3 domain is approximately 0.9 sec^{-1}, which is very similar to a value measured for the rate of folding of interleukin-1β, another all β-sheet protein (Varley et al., 1993). The rate of folding of the SH3 domain can be measured at each of the 12 residues. We observe that the rates at each site are not identical and the differences in the measurements are greater than our experimental error.

Rather than analyzing this relaxation data in terms of the Lipari and Szabo model-free formalism and assuming isotropic tumbling which may not hold in the case of an unfolded protein, we have mapped the spectral density functions in a similar way to that described by Peng and Wagner (1992a,b). We have grouped the three high frequency terms into one, $J(\omega^h)$, and determined $J(0)$, $J(\omega^N)$ and $J(\omega^h)$. The spectral density mapping approach that we employed shows the following relationships for the folded state, $J(0) \gg J(\omega^N) \gg J(\omega^h)$. The values of $J(0)$ are relatively constant over the protein, providing confidence that NOE intensities in the folded state may be interpreted directly in terms of structural properties. It is noteworthy that similar trends were observed for the unfolded state, with $J(0) > J(\omega^N) > J(\omega^h)$ and $J(0)$ values for some residues approaching their values in the folded state. This suggests that the unfolded state is not well described by a random coil model having non-interacting residues and that the motion in this state may be closer to that of a folded protein.

Addressing our original question regarding the distribution of the $J(0)$ values in the unfolded state, we measured $J(0)$ values for four residues in the region from Asn-35 to Ile-48 (Trp-36, Arg-48, Glu-43 and Gly-46) where NH_i-NH_{i+1} sequential NOEs are observed. These four values vary greatly from about 0.7 to 1.8 ns/rad, from close to the lowest and highest of all measured values. There is also a very high $J(0)$ value of 1.9 ns/rad for residue Thr-22 for which no NH_i-NH_{i+1} sequential NOE is observed. This is evidence supporting a significant role for structural preferences (i.e. interproton distances) giving rise to these NOEs since there appears to be no direct correlation between the presence of the NOEs and the value of $J(0)$. Thus, these NOEs can be interpreted as reflecting preferential sampling of the α-region of (ϕ,ψ) conformational space.

Implications for Protein Folding

We have also analyzed the equilibrium constants for the unfolding/folding transition as a function of residue by performing fully relaxed HSQC experiments and quantitating the size of the folded and unfolded peaks. Again, we noted differences across the protein. Although it is difficult to interpret, the fact that we see differences in exchange rates as a function of residue and differences in equilibrium constants as a function of residue implies that there is not perfect cooperativity in this folding reaction. We have explored this further by measuring the heat capacity change during thermal unfolding of the SH3 domain in high salt and observed what is, at a global level, a two-state cooperative transition (P. Morin and E. Friere, personal communication). Thus, the NMR results may be providing higher resolution data than is often seen using differential scanning calorimetry or from thermal melting curves using CD spectroscopy.

In order to compare the unfolded state that is present in denaturant (2 M Gdn) with the unfolded state that exists in equilibrium with the folded state, we assigned and analyzed the chemical shifts, sequential NOEs, $^3J_{HN\alpha}$ coupling constants and amide exchange with solvent for a 2 M Gdn sample of the N-terminal SH3 domain (Figure 3c). Some sequential NH_i-NH_{i+1} NOEs are present in the region from Asn-35 to Ile-48, but the long contiguous stretch in that region is not preserved. Again very little difference between the CαH proton chemical shift and random coil values (Wishart and Sykes, 1994) is observed as expected. All of the $^3J_{HN\alpha}$ coupling constants reflect averaging of phi torsion angles and the very low coupling at Leu-28 has disappeared. Very interestingly, the observed amide exchange rates with solvent water, with the exception of a region near Thr-12, is much less than that in the aqueous equilibrium unfolded state. We also assigned and performed similar comparisons of the fully stabilized folded state of the SH3 domain in 400 mM sodium sulfate with the folded state in equilibrium with the unfolded in low ionic strength. Comparison of the backbone chemical shifts shows that these two folded states are virtually identical. However, the unfolded state in equilibrium with the folded is not identical to the unfolded state in Gdn based on chemical shift comparisons. There are significant deviations near Leu-28, the residue which has a coupling constant of less than 4 Hz in low ionic strength and a much larger one in Gdn, reflecting conformational averaging. While both unfolded states show differences in backbone chemical shifts from the random coil values (Wishart and Sykes, 1994; Braun et al., 1994), currently these differences are not easily interpretable. With the recent rapid advances in our understanding of chemical shifts (de Dios et al., 1993a,b), however, it is likely that we will be able to learn many things in the future by such comparisons.

In summary, the unfolded state of the N-terminal SH3 domain in low ionic strength aqueous buffer is not a true random coil state. It appears to be more compact, with non-random structure. It is also significantly different in describable ways from the unfolded state in 2 M Gdn. Of particular interest, this state contains a stretch of residues demonstrating rapidly averaging conformations with a structural preference for the helical region of (ϕ,ψ) space. These residues span a β-strand, β-turn, β-strand region in the folded state, so their structural preference in the unfolded state is decidedly non-native. In addition, we have measured the exchange rates as a function of residue, which are identical to the folding and unfolding rates. The average rate agrees well with previous measurements of the rate of folding of β-sheet proteins but differences in the rates as a function of residue imply deviations from exact cooperativity. We are currently pursuing this issue of microscopic non-cooperativity.

ACKNOWLEDGMENTS

We thank our collaborator Dr. Tony Pawson at the Samuel Lunenfeld Research Institute, Mount Sinai Hospital, Toronto for the original clones of the PLCγ SH2 and Drk SH3 domains and for his support of these studies and Dr. Steve Shoelson at the Joslin Diabetes Center and Department of Medicine, Brigham and Womens Hospital and Harvard Medical School who provided the phosphorylated peptide from the Tyr-1021 site of the PDGFR.

REFERENCES

Arakawa, T., and Timasheff, S.N., 1985, Theory of protein solubility, *Methods Enzymol.* 114:49.

Bax, A., 1994, Multidimensional nuclear magnetic resonance methods for protein studies, *Curr. Opin. Str. Biol.* 4:738.

Bax, A., and Grzesiek, S., 1993, Methodological Advances in Protein NMR, *Acc. Chem. Res.* 26:131.

Braun, D., Wider, G., and Wüthrich, K., 1994, Sequence-corrected ^{15}N "random coil" chemical shifts, *J. Am. Chem. Soc.* 116:8466.

Clark, S.G., Stern, M.J., and Horvitz, H.R., 1992, C. elegans cell-signalling gene sem-5 encodes a protein with SH2 and SH3 domains, *Nature* 356:340.

de Dios, A.C., Pearson, J.B., and Oldfield, E., 1993a, Secondary and tertiary structural effects on protein NMR chemical shifts: an ab initio approach, *Science* 260:1491.

de Dios, A.C., Pearson, J.B., and Oldfield, E., 1993b, Chemical shifts in proteins - an ab initio study of carbon-13 nuclear magnetic resonance chemical shielding in glycine, alanine and valine residues, *J. Am. Chem. Soc.* 115:9768.

Eck, M.J., Shoelson, S.E., and Harrison, S.C., 1993, Recognition of a high-affinity phosphotyrosyl peptide by the src homology-2 domain of p56[lck], *Nature* 362:87.

Farrow, N.A., Muhandiram, R., Singer, A.U., Pascal, S.M., Kay, C.M., Shoelson, S.E., Pawson, T., Forman-Kay, J.D., and Kay, L.E., 1994a, Backbone dynamics of a free and phosphopeptide-complexed src homology 2 domain studied by ^{15}N NMR relaxation, *Biochem.* 33:5984.

Farrow, N.A., Zhang, O., Forman-Kay, J.D., and Kay, L.E., 1994b, A heteronuclear correlation experiment for simultaneous determination of ^{15}N longitudinal deacy and chemical exchange rates of systems in slow equilibrium, *J. Biomol. NMR* 4:727.

Farrow, N.A., Zhang, O., Forman-Kay, J.D., and Kay, L.E., 1995, Comparison of the backbone dynamics of a folded and an unfolded SH3 domain existing in equilibrium in aqueous buffer, *Biochem.* 34:868.

Felder, S., Zhou, M., Hu, P., Ureña, J., Ullrich, A., Chaudhuri, M., White, M., Shoelson, S.E., and Schlessinger, J., 1993, SH2 domains exhibit high-affinity binding to tyrosine-phosphorylated peptides yet also exhibit rapid dissociation and exchange, *Mol. Cell. Biol.* 13:1449.

Griesinger, C., Otting, G., Wüthrich, K., and Ernst, R.R., 1988, Clean TOCSY for proton spin system identification in macromolecules, *J. Am. Chem. Soc.* 110:7870.

Grzesiek, S., Anglister, J., and Bax, A., 1992, Correlation of backbone amide and aliphatic side-chain resonances in ^{13}C/^{15}N enriched proteins by isotropic mixing of ^{13}C magnetization, *J. Magn. Reson. Series B* 101:114.

Kay, L.E., Keifer, P., and Saarinen, T., 1992, Pure absorption gradient enhanced heteronuclear single quantum correlation spectroscopy with improved sensitivity, *J. Am. Chem. Soc.* 114:10663.

Kessler, H., Griesinger, C., Lautz, J., Muller, A., van Gunsteren, W.F., and Berendsen, H.J.C., 1988, Conformational dynamics detected by nuclear magnetic resonance NOE values and J coupling constants, *J. Am. Chem. Soc.* 110:3393.

Koch, C.A., Anderson, D., Moran, M.F., Ellis, C., and Pawson, T., 1991, SH2 and SH3 domains: elements that control interactions of cytoplasmic signaling proteins, *Science* 252:668.

Kuriyan, J., and Cowburn, D., 1993, Structures of SH2 and SH3 domains, *Curr. Opin. Str. Biol.* 3:828.

Lee, C-H., Kominos, D., Jacques, S., Margolis, B., Schlessinger, J., Shoelson, S.E., and Kuriyan, J., 1994, Crystal structures of peptide complexes of the amino-terminal SH2 domain of the Syp tyrosine phosphatase, *Structure* 2:423.

Lee, W., Revington, M., Arrowsmith, C.H., and Kay, L.E., 1994, A pulsed-field gradient isotope-filtered 3D ^{13}C HMQC-NOESY experiment for extracting intermolecular NOE contacts in molecular complexes, *FEBS Lett.* 350:87.

Lipari, G., and Szabo, A., 1982a, Model-free approach to the interpretation of nuclear magnetic relaxation in macromolecules: 1. Theory and range of validity, *J. Am. Chem. Soc.* 104:4546.

Lipari, G., and Szabo, A., 1982b, Model-free approach to the interpretation of nuclear magnetic relaxation in macromolecules: 2. Analysis of experimental results, *J. Am. Chem. Soc.* 104:4559.

Logan, T.M., Olejniczak, E.T., Xu, R., and Fesik, S.W., 1992, Side chain and backbone assignments in isotopically labeled proteins from two heteronuclear triple resonance experiments, *FEBS Letts.*, 314:413.

Montelione, G.T., Lyons, B.A., Emerson, S.D., and Jashiro, M., 1992, An efficient triple resonance experiment using carbon-13 isotopic mixing for determining sequence-specific resonance assignments of isotopically-enriched proteins, *J. Am. Chem. Soc.* 114:10974.

Neri, D., Billeter, M., Wider, G., and Wüthrich, K., 1992, NMR determination of residual structure in a urea-denatured protein, the 434-repressor, *Science* 257:1559.

Nicholls, A., Sharp, K.A., and Honig, B., 1991, Protein folding and association: insights from the interfacial and thermodynamic properties of hydrocarbons, *Protein Struct. Funct. Genet.* 11:281.

Olivier, J.P., Raabe, T., Henkemeyer, M., Dickerson, B., Mbamalu, G., Margolis, B., Schlessinger, J., Hafen, E., and Pawson, T., 1993, A Drosophila SH2-SH3 adaptor protein implicated in coupling the Sevenless tyrosine kinase to an activator of Ras guanine nucleotide exchange - Sos, *Cell* 73:179.

Pascal, S.M., Singer, A.U., Gish, G., Yamazaki, T., Shoelson, S.E., Pawson, T., Kay, L.E., and Forman-Kay, J.D., 1994, NMR Structure of an SH2 Domain of Phospholipase C-γ1 Complexed with a High-Affinity Binding Peptide, *Cell* 77:461.

Pascal, S.M., Yamazaki, T., Singer, A.U., Kay, L.E., and Forman-Kay, J.D., 1995, Structural and dynamic characterization of the phosphotyrosine binding region of an SH2 domain-phosphopeptide complex by NMR relaxation, proton exchange and chemical shift approaches, in preparation.

Pawson, T., Olivier, P., Rozakis-Adcock, M., McGlade, J., and Henkemeyer, M., 1993, Proteins with SH2 and SH3 domains couple receptor tyrosine kinases to intracellular signaling pathways, *Phil. Trans. R. Soc. Lond. B* 340:279.

Peng, J.W., and Wagner, G., 1992a, Mapping of spectral density functions using heteronuclear NMR relaxation measurements, *J. Magn. Reson.* 98:308.

Peng, J.W., and Wagner, G., 1992b, Mapping of the spectral densities of N-H bond motions in Eglin c using heteronuclear relaxation experiments, *Biochem.* 31:8571.

Piccione, E., Case, R.D., Domchek, S.M., Hu, P., Chaudhuri, M., Backer, J.M., Schlessinger, J., and Shoelson, S.E., 1993, Phosphatidylinositol 3-kinase p85 SH2 domain specificity defined by direct phosphopeptide/SH2 domain binding, *Biochem.* 32:3197.

Rhee, S.G.,1991, Inositol phospholipid-specific phospholipase C: interaction of the γ_1 isoform with tyrosine kinase, *TIBS* 16:297.

Ringe, D., and Petsko, G.A., 1985, Mapping protein dynamics by X-ray diffraction, *Prog. Biophys. Molec. Biol.* 45:197235.

Shortle, D., 1993, Denatured states of proteins and their roles in folding and stability, *Curr. Opin. Str. Biol.* 3:66.

Simon, M.A., Dodson, G.S., and Rubin, G.M., 1993, An SH3-SH2-SH3 protein is required for p21[Ras1] activation and binds to Sevenless and Sos proteins in vitro, *Cell* 73:169.

Songyang, Z., Shoelson, S.E., Chaudhuri, M., Gish, G., Pawson, T., Haser, W.G., King, F., Roberts, T., Ratnofsky, S., Lechleider, R.J., Neel, B.G., Birge, R.B., Fajardo, J.E., Chou, M.M., Hanafusa, H., Schaffhausen, B., and Cantley, L.C., 1993, SH2 domains recognize specific phosphopeptide sequences, *Cell* 72:767.

Songyang, Z., Shoelson, S.E., McGlade, J., Olivier, P., Pawson, T., Bustelo, R.X., Barbicid, M., Sabe, H., Hanafusa, H., Yi, T., Ren, R., Baltimore, D., Ratnofsky, S., Feldman, R.A., and Cantley, L.C., 1994, Specific motifs recognized by the SH2 domains of csk, 3BP2, fes/fps, Grb-2, SHPTP1, SHC, Syk and vav, *Mol. Cell. Biol.* 14:2777.

Ullrich, A., and Schlessinger, J., 1990, Signal transduction by receptors with tyrosine kinase activity, *Cell* 61:203.

Valius, M., Bazenet, C., and Kazlauskas, A., 1993, Tyrosines 1021 and 1009 are phosphorylation sites in the carboxy terminus of the platelet-derived growth factor receptor β subunit and are required for binding of phospholipase Cγ and a 64-kilodalton protein, respectively, *Mol. Cell. Biol.* 13:133.

Valius, M., and Kazlauskas, A., 1993, Phospholipase C-γ-1 and phosphatidylinositol 3 kinase are the downstream mediators of the PDGF receptor's mitogenic signal, *Cell* 73:321.

Varley, P., Gronenborn, A.M., Christense, H., Wingfield, P.T., Pain, R.H., and Clore, G.M., 1993, Kinetics of folding of the all-β sheet protein interleukin-1 β, *Science* 260:1110.

Waksman, G., Shoelson, S.E., Pant, N., Cowburn, D., and Kuriyan, J., 1993, Binding of a high affinity phosphotyrosyl peptide to the src SH2 domain: crystal structures of the complexed and peptide-free forms, *Cell* 72:779.

Wishart, D.S., and Sykes, B.D., 1994, Chemical shifts as a tool for structure determination, *Methods Enzymol.* 239:363.

Yamazaki, T., Pascal, S.M., Singer, A.U., Forman-Kay, J.D., and Kay, L.E., 1995, NMR pulse schemes for the sequence-specific assignment of arginine guanidino ^{15}N and ^{1}H chemical shifts in proteins, *J. Am. Chem. Soc.*, in press.

Yu, H., and Schreiber, S.L., 1994, Signalling an interest, *Nature Str. Biol.* 1:417.

Yu, H., Chen, J.K., Feng, S., Dalgarno, D.C., Brauer, A.W., and Schreiber, S.L., 1994, Structural basis for the binding of proline-rich peptides to SH3 domains, *Cell* 76:933.

Zhang, O., and Forman-Kay, J.D., 1995, Structural characterization of folded and unfolded states of an SH3 domain in equilibrium in aqueous buffer, *Biochem.*, submitted.

Zhang, O., Kay, L.E., Olivier, J.P., and Forman-Kay, J.D., 1994, Backbone ^{1}H and ^{15}N resonance assignments of the N-terminal SH3 domain of drk in folded and unfolded states using enhanced-sensitivity pulsed field gradients NMR techniques, *J. Biomol. NMR* 4:845.

DISCUSSION

Weixing Zhang - If you mutate the position +4 to +6 of your phosphopeptide, what is the difference in SH2 binding compared to non-mutated phosphopeptide?

Foman-Kay - It is important to realize that you cannot make generalizations from one SH2 complex to another and I think that this is one of the significant points that has been emerging. It is difficult to classify SH2 domains and to say that for this class only certain residues are important for binding while for a second class other residues are important. It's necessary to do binding studies and specific truncation studies or alanine scans in order to get at the data for each individual system. Our collaborator, Steve Shoelson, has done some very interesting work where he has shown that you can actually remove all of the residues other than the -1, pTyr and +1 positions (i.e. removing +2 and beyond) and still maintain a significant binding affinity in the case of C-terminal SH2 of PLCγ, whereas you absolutely need positions +2, +3 and +4 for binding in the structurally similar Syp phosphatase N-terminal SH2 domain. However, our collaborator Tony Pawson has done studies showing that if you change the +4 position in the pγ1021 peptide, which is a leucine, to a serine, then PLCγSH2 will bind but in addition the SH2 domain of the p85 subunit of P13K will bind. Then, if you make an additional substitution of the +3 position to a methionine, you can convert the peptide so that p85 is the only SH2 domain that will bind and the PLCγSH2 does not bind. What I am saying is that there are no generalizations and mutagenesis studies and specific binding studies have to be done in each case; it's complicated. We're trying to understand this complexity.

Zhang - Thank you.

Marc Adler - The folding data suggested that there are two models which could potentially explain the data. Could the non-cooperativity be explained by there being a third state? You analyzed it according to a two state model. Could there be a third state which is in rapid equilibrium with an "unfolded" one and could that explain the various rates? Would that be consistent with the data?

Forman-Kay - I think it would. It would be impossible for us to confirm, however, if that third state is in rapid exchange with the folded state or the unfolded state. It's impossible to actually get an experimental handle on it, but it's a model which we are pursuing to try to understand the current data.

Adler - Another one, I'll try to keep it quick. You said the N-N connectivities suggest the α region of the $\varphi - \psi$ space. Could you have a local formation of an oil drop which included a lot of tight turns and therefore gave rise to the N-N connectivities? This could be exchanging rapidly.

Forman-Kay - When I say that the region of conformational space is preferentially sampled, I am not talking about a cooperative formation of helices at all. I am talking about local helical turns. You can call them ß turns or helical turns but that region of conformational space which includes turn conformations on a local level is what's being sampled. I don't think there's any cooperativity between the individual residues. I don't think there is a formation of an actual helix.

Adler - Thank you.

NMR STUDIES OF PROTEINS INVOLVED IN CELL ADHESION PROCESSES

Gerhard Wagner[1], Daniel F. Wyss[2], Johnathan S. Choi[1],
Antonio R, N. Arulanandam[2], Ellis L. Reinherz[3],
and Andrzej Krezel[1], Robert A. Lazarus[4].

[1]Department of Biological Chemistry and Molecular Pharmacology, Harvard Medical School, 240 Longwood Avenue, Boston, MA 02115
[2] Procept, Inc., 840 Memorial Drive, Cambridge, MA 02139
[3] Dana Farber Cancer Institute
[4] Department of Protein Engineering, 460 Point San Bruno Boulevard, Genentech, South San Francisco, CA 94080

INTRODUCTION

Many biological processes on the molecular level are associated with membrane-bound receptors or other integral membrane proteins. We have been working with proteins that are domains of receptors, or inhibit their function. The first topic are the T-cell surface glycoprotein receptor CD2 and its counter receptor CD58. The other topic is about antagonists of the integrin adhesion receptor glycoprotein IIbIIIa (GPIIbIIIa) which is found on platelet surfaces. Human CD2 is a glycoprotein, and the carbohydrate of its adhesion domain is crucial for adhesion function. The platelet receptor GPIIbIIIa, a Ca^{2+} dependent heterodimeric glycoprotein from the integrin family, binds fibrinogen and mediates the aggregation of platelets to form a blood clot. Natural protein antagonists of this receptor have primarily been found in the venum of various snakes, which have been termed disintegrins, and in the saliva of blood-sucking leeches. They contain an Arg-Gly-Asp (RGD) sequence in their active site. Due to the potent antiplatelet effect of these RGD proteins, the structures of their active sites have been of considerable interest for the design of antithromotic drugs.

HUMAN CD2

An important step of immune reactions is the interaction of T-cells with antigen-presenting cells. One aspect of this process is the adhesion of the T-cell glycoprotein receptor CD2 with the counter receptor CD58 of antigen presenting cells. CD2 is a 50-55 kDa surface glycoprotein which is found on virtually all T lymphocytes as well as on natural killer cells. It initiates the adhesion of T lymphocytes to infected target cells and antigen presenting cells (APCs). At the N-terminus, it consists of two extracellular domains followed by a single transmembrane domain and a proline-rich cytoplasmatic tail of 113 residues. The N-terminal extracellular domain has been shown early on to have sequence homology to variable domains of the immunoglobulin superfamily (IgSF V-set). All adhesion function of CD2 is mediated via this N-terminal adhesion domain. The second extracellular domain has sequence homology to constant domains of the immunoglobulin superfamily (IgSF C-set). Indeed, the expectations from the sequence homologies have been confirmed as NMR solution studies of

NMR As a Structural Tool for Macromolecules: Current Status and Future Directions
Edited by B.D. Nageswara Rao and Marvin D. Kemple, Plenum Press, New York, 1996

51

the adhesion domains of rat CD2 (Driscoll et al., 1991) and human CD2 (Withka et al., 1993) revealed conformations typical of immunoglobulin variable domains. Furthermore, X-ray structures of the whole extracellular portions of rat CD2 (Jones et al., 1993) and human CD2 (Bodian et al., 1994) confirmed this observation and also showed that the structures of the second extracellular domains are similar to constant domains of immunoglobulins. The two extracellular domains form a head-to-tail assembly in which the long axes of the two domains form an angle of ca 40°. However, the relative orientation of the two domains differs by ca. 20° between the rat and the human protein (Bodian et al., 1994) indicating that the two domains may have a certain mobility relative to each other in solution. The extracellular portion of CD2 has similarities to domains 1 and 2 of either chain of the growth hormone receptor (deVos et al., 1992), domains 1 and 2 of CD4 (Wang et al., 1990; Ryu et al., 1990), and domains 3 and 4 of CD4 (Brady et al., 1993; Lange et al., 1994).

The extracellular portion of hCD2 contains three N-linked carbohydrates, one in domain 1 (at Asn65) and two in domain 2 (at Asn117 and Asn126). The carbohydrates of the second domain can be removed without affecting the adhesion function of hCD2. Removal of the single glycan of the adhesion domain, however, either by treatment with Peptide:N-glycosidase F (PNGase F), or by mutation of the N-glycosylation sequence Asn65-Gly66-Thr67 results in loss of adhesion function (Recny et al., 1992), or at least a severe destabilization of the protein (Davis et al., 1995). Here, we have focused on solving the spatial structure of the carbohydrate and investigating its functional significance.

Polypeptide Structure of the hCD2 Adhesion Domain

As a start to study the role the N-linked glycan plays we have initially solved the polypeptide structure of the fully glycosylated adhesion domain of human CD2 by NMR spectroscopy (Wyss et al., 1993; Withka et al., 1993). This was achieved with standard homonuclear NMR experiments. Because the protein requires the glycan linked to the side chain of Asn65, the protein could not be expressed in a bacterial expression system. Thus, efficient labeling with ^{15}N and ^{13}C was not possible, and the NMR work relied mostly on homonuclear experiments. Fig. 1 shows a stereo diagram of the polypeptide backbone of an ensemble of 18 NMR structures. The polypeptide has the typical fold of a the immunoglobulin superfamily variable domain, constituted by a sandwich of two β-sheets. The spread of the structures indicate that there is higher mobility in the N-terminal hexapeptide, the three C-terminal residues, and the BC, C'C" and C"D loops (Fig. 1). The core of the protein is well defined.

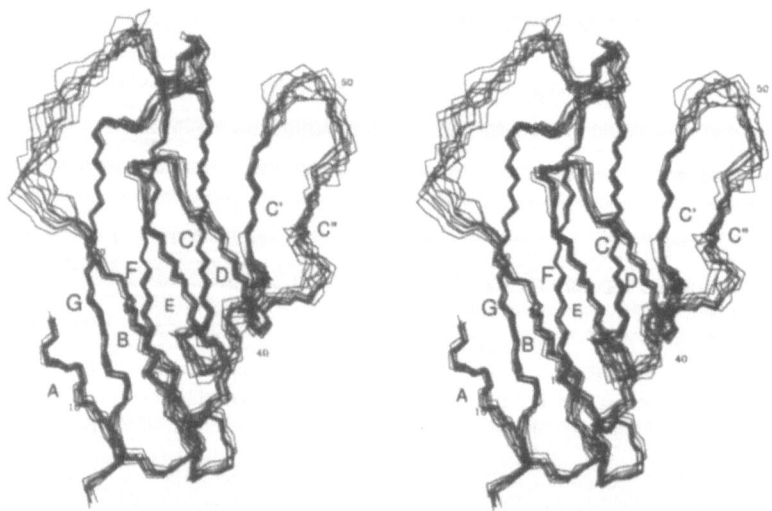

Figure 1. Stereo view of an ensemble of NMR structure of the adhesion domain of hCD2.

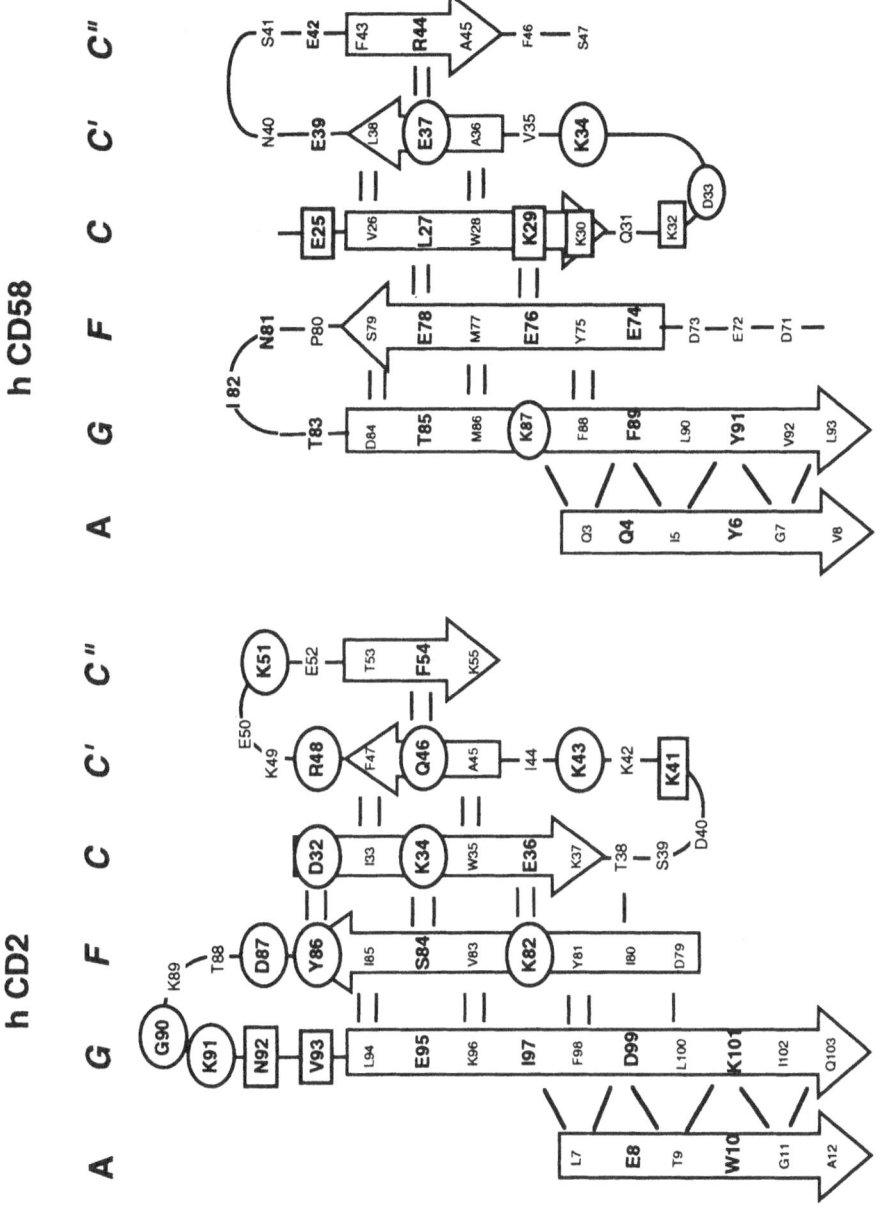

Figure 2. Schematic diagram of the AGFCC'C" faces of CD2 and CD58. Residues with exposed side chains are indicated by ellipses, knockout Ala mutation sites are drawn in bold, knockout Ala mutation sites are indicated by ellipses, Ala-mutations that reduce adhesion are indicated by rectangles.

Identification of the CD58 binding site on CD2

Based on the hCD2 structure, residues with surface exposed side chains were mutated to alanines, and the mutants were tested for adhesion function. The results showed that the binding site is located on a highly charged surface area consisting of the GFCC'C" β-sheet (Arulanandam et al., 1993). Fig. 2a shows a schematic representation of this face of hCD2. Knock-out mutations are indicated with ellipses, whereas mutations that reduce adhesion are marked with rectangles. Bodian et al. (1994) have recently solved the crystal structure the two-domain extracellular region of human CD2. They have expressed residues 1-182 in Chinese hamster ovary cells in the presence of a glycosidase I inhibitor. In this case, oligosaccharide processing is blocked and glycoproteins are secreted with predominantly $Glc_2-Man_9GlcNAc_2$ oligosaccharides (Karlson et al., 1993).

These preparations are highly sensitive to endoglycosidase H, and treatment with this enzyme yields protein with a single GlcNAc at each glycosidation site. The protein produced this way maintains adhesion function.

Identification of the CD2 binding site on CD58

CD58 mediates the interactions of antigen presenting cells and target cells, with activated T-cells and thymocytes. Binding of CD58 to CD2 enhances T-cell stimulation. CD58 is homologous to CD2 and CD48. It consists of two extra cellular domains. The N-terminal adhesion domain has homology to variable domains of the immunoglobulin superfamily; the second domain has characteristics of constant domains. CD58 is more extensively glycosylated than CD2 although the extent of the glycosylation is not yet fully elucidated.

Recently, a structure of CD58 has been modeled, based on the sequence homology and the known 3D structures of human CD2, rat CD2, CD4 and CD8 (Arulanandam et al., 1994). Based on this modeling, residues with surface exposed side chains were mutated to alanines (E25, K29, K30, K32, D33, K34, E37, E39, R44, K50, R52, D56, E72, D73, E76, E78, D84, K87) and the effect on the adhesion function was tested. The results showed that the CD2 binding site includes the C-strand (E25, K29, K30, the CC' loop (K32, D33, K34, the C' strand (E37), and the G-strand (K87).

These results indicate that the receptor interface between CD2 and CD58 is constituted of two highly charged surfaces. This is surprising considering related interfaces with known structures. These include the interfaces between the variable domains of heavy and light chains of immunoglobulins (Chothia et al., 1985) or the dimer interface of CD8 (Leahy et al., 1992). In these cases, the interface is composed of predominantly apolar side chains. However, one has to consider that the structural model derived by homology modeling from CD2 and CD4 may miss some crucial features of the CD58 structure. For example, there might be a β-bulge in the G-strand, starting with K87 or F88, and a second β-bulge in the C' strand involving residues D33 to V35. These β-bulges could expose several hydrophobic side chains that could be involved in the receptor co.-receptor contacts. Bulges in these positions have generally been observed in immunoglobulins and are conserved structural motifs. This face of hCD2 contains two β -bulges as well but they are at different positions than in immunoglobulins (Wyss et al., 1993; Withka et al., 1993). If these bulges were present in CD58 they could contribute a number of hydrophobic contacts.

Resonance assignments of the CD2 glycan and preliminary structure

The high mannose glycan in our samples of hCD2 as produced in Chinese hamster ovary (CHO) cells is composed of heterogeneous isomeric structures (glycomers) containing predominantly Man5 (~20%), Man6 (~34%), Man7 (~40%) and Man8 (~6%) glycoforms as determined by electrospray ionization mass spectrometry (ESI-MS) (Withka et al., 1993). The glycomers present in our sample arise from trimming the Man_8 oligosaccharide (Fig. 3) at different branch points. The assignment of carbohydrate resonances in glycoproteins is complicated by the fact that the vast majority appear in a very small spectral width (~3.5-4.0 ppm) with consequent overlap problems. Exceptions are the anomeric ^1H signals which resonate at lower field (~4.4-5.2 ppm) due to the electron-withdrawing properties of the ring

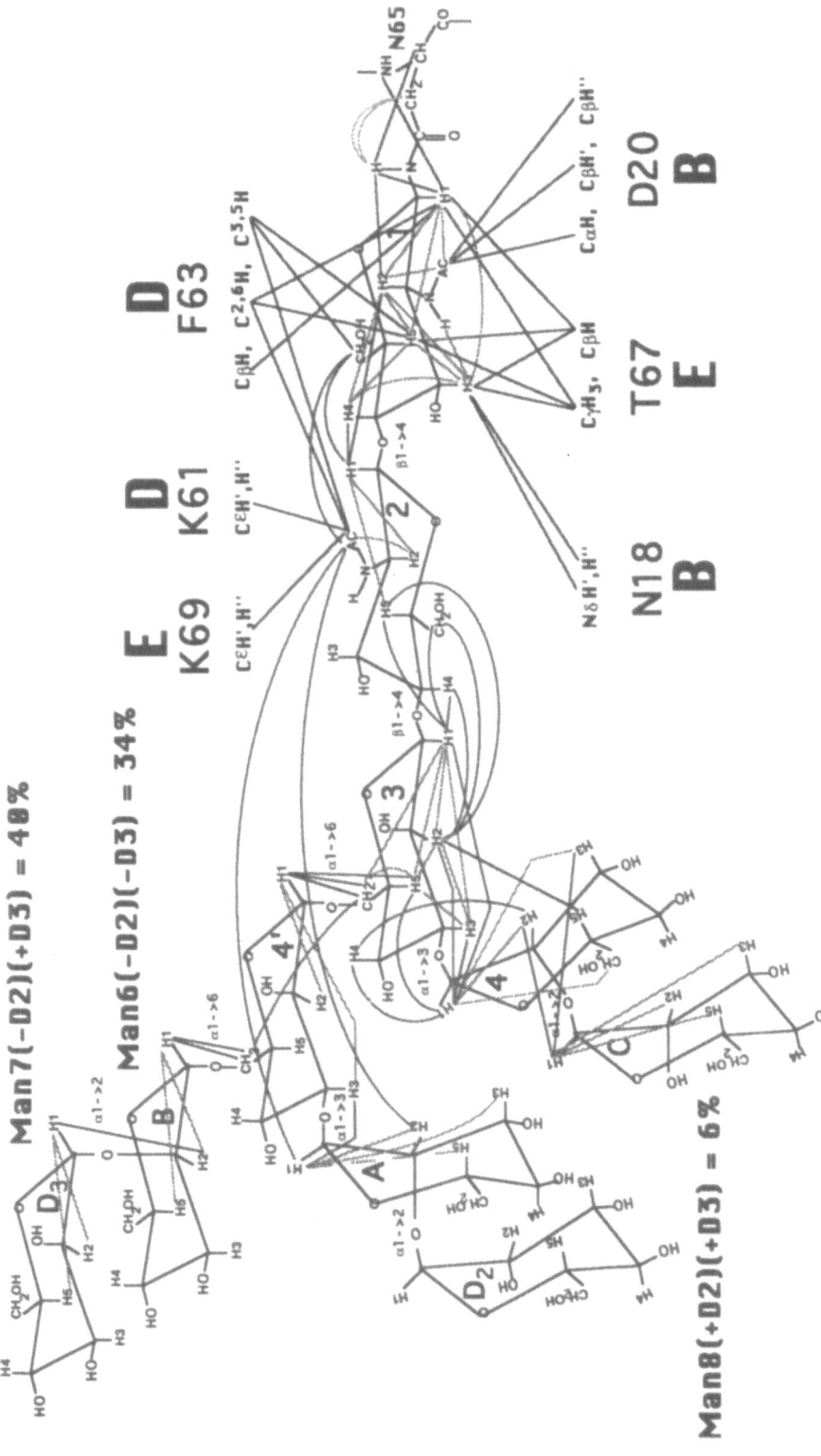

Figure 3. Covalent structure of the N-linked glycan of hCD2. The residues that make contact with the glycan are listed as well as the corresponding strands of the β-sheet. NOEs observed are also indicated.

oxygen. As a consequence, the assignment strategy for oligosaccharides is usually based on these resolved anomeric ^{1}H resonances (see for example Homans, 1990). However, in glycoproteins these anomeric ^{1}H resonances are usually overlapped with H$^\alpha$ resonances from the protein and hence, the anomeric region in 2D ^{1}H-^{1}H spectra is complicated by the appearance of large numbers of cross peaks arising from the spin systems of amino acid residues, such as serines, threonines, and glycines. However, the anomeric carbon resonances of carbohydrates are well separated from all polypeptide carbon signals. Thus, our assignment strategy made extensive use of proton-carbon heteronuclear experiments at natural abundance of ^{13}C (Wyss et al., 1995a).

First, we recorded ^{1}H-^{13}C HSQC experiments to identify glycan cross peaks, in particular the resonance positions of the anomeric protons. However, most of the other cross peaks of the glycan are also well resolved from polypeptide cross peaks, and reports in the literature about resonance assignments of free high-mannose glycans (Wormald et al., 1991) gave some clues in which spectral regions typical cross peaks should be expected. Next, we recorded homonuclear TOCSY experiments with different mixing times between 50 and 250 ms and analyzed the cross peaks lined up at the positions of the anomeric protons identified in the HSQC spectra. With these experiments we could establish connectivities between the anomeric protons and most other protons of a particular monosaccharide. The H^6 protons were identified by selecting for CH$_2$ groups in a DEPT-HMQC spectrum. Generally, the GlcNAc cross peaks were observed already in TOCSY spectra with short mixing times because of the relatively large H^1-H^2 coupling constants. In contrast, mannose residues have small H^1-H^2 and H^2-H^3 coupling constants, and thus the cross peaks in these residues were primarily observed in spectra with longer mixing times in which many cross peaks of the polypeptide were no longer observed due to faster transverse relaxation.

Sequential assignments of the carbohydrate units were obtained by observing cross peaks between the H^1 and H^4 protons of the β1->4 glycosidic bonds (GlcNAc2- GlcNAc1; Man3- GlcNAc2), between the H^1 and H^6 protons of the two α1->6 glycosidic bonds (Man4'-Man3; ManB-Man4'),between the H^1 and H^3 protons of the two α1->3 glycosidic bonds (ManA-Man4'; Man4->Man3) and between the H^1 and H^2 protons of the α1->2 glycosidic bonds (ManD3-ManB; ManD2-ManA; ManC-Man4).

The heterogeneity of the glycan is clearly manifested in the NMR spectra. In particular, the ^{1}H-^{13}C HSQC spectra show different sets of cross peaks for the different glycoprotein forms. Fortunately, the heterogeneity of the glycan effects primarily the ends of the carbohydrate branches. In particular, the resonances of the penultimate mannose residues ManA and ManB are doubled representing the forms with and without the terminal ManD2 and ManD3 residues, respectively.

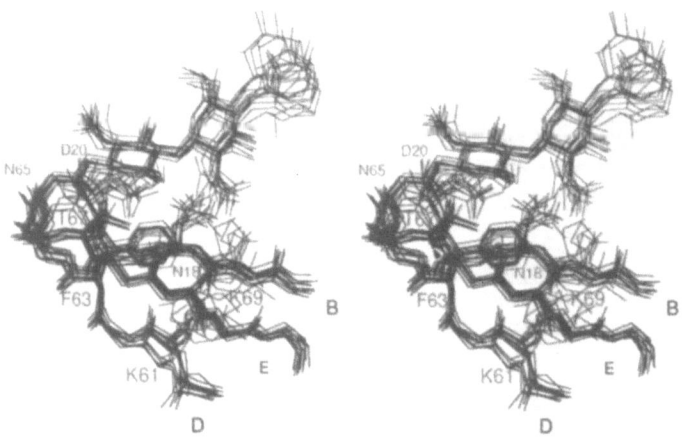

Figure 4. Stereo diagram of the carbohydrate core covering the BDE face of hCD2. Backbone atoms of strands D, E, and B of the polypeptide are shown. Side chains of residues that have NOE contacts with the carbohydrate are also displayed.

The tertiary structure of the glycan was characterized from analysis of highly resolved NOESY spectra. We identified more than 30 NOEs between the first two carbohydrate residues (GlcNAc1 and GlcNAc2) and side chains of residues on the B, D and E β-strands of the protein. Numerous intraglycan NOEs define the conformation of the branch ending in ManC. Interestingly, we observe several NOEs between ManA and GlcNAc2 indicating that the ManA branch of the glycan is folded back to the stem of the carbohydrate. However, this is only observed for those glycan forms that do not have the ManD2 residue attached. Thus, cleaving off the terminal ManD2 causes the ManA branch to adopt a relatively well defined structure. Fig. 4 shows some representative distance geometry structures of the first three carbohydrate residues. Details of the structure analysis of the carbohydrate and a discussion of its functional significance have been described elsewhere (Wyss et al., 1995b).

DECORSIN

The blood clotting process is extremely complex and involves both platelet adhesion and aggregation as well as the coagulation pathway. In response to injury, platelets become activated, first adhering to the site of injury, then aggregating to form a hemostatic plug. The aggregation process is mediated by the integrin GP IIb-IIIa, the glycoprotein receptor for fibrinogen on the platelet surface. Fibrinogen contains the RGD tripeptide sequence, which is thought to be important for binding to GP IIb-IIIa. The coagulation pathway involves a complex cascade of zymogen activation in both the extrinsic and intrinsic pathways, resulting in the formation of Factor Xa, which activates prothrombin to thrombin, which then cleaves fibrinogen to fibrin and leads to the formation of a clot.

Many proteins affecting the blood clotting process have been purified and characterized from natural sources including leeches, snake venoms, vampire bats, ticks and others (Bang & Clayman, 1992; Lazarus & McDowell, 1993; Markwardt, 1994; Niewiarowski et al., 1994; Ribeiro, 1987; Sawyer, 1986; Sawyer, 1991; Seymour et al. 1990; Stocker, 1990; Waxman, 1990). In view of the complexity of blood coagulation, it is not unexpected that the agents isolated from these natural sources mediate their hemostatic effects through a variety of mechanisms, which include antagonism of GP IIb-IIIa receptor binding to fibrinogen as well as the inhibition of thrombin and Factor Xa.

Solution structure of the GPIIbIIIa antagonist decorsin

We have studied the small protein decorsin from the North American leech *Macrobdella decora* (Seymour et al., 1990). It contains the RGD adhesion site recognition sequence, which is found in fibrinogen, fibronectin, vitronectin, and other GP IIb-IIIa ligands, and is thought to be essential for binding to many integrins (Ruoslahti & Pierschbacher, 1992; Hynes, 1992). Decorsin is related to the disintegrins, a family of RGD-containing GP IIb-IIIa protein antagonists isolated from snake venoms (Lazarus & McDowell, 1993; Niewiarowski et al., 1994). Previous NMR structural studies on the disintegrins kistrin, echistatin, and flavoridin have shown that the RGD sequence, which is the only critical epitope required for binding, lies at the apex of a conformationally ill-defined extended loop (Adler et al., 1991; Saudek et al., 1991; Senn and Klaus, 1993). The solution structure of this 39-residue protein was solved with standard homonuclear NMR methods. Due to the good quality of the NMR spectra, the structure could be highly refined (Krezel et al., 1994). The average pairwise root mean square difference (rmsd) for the backbone atoms (N, C^α, C') of residues 4 through 39 was 0.35 Å, the average rmsd to the mean coordinates was 0.24 Å. The dominant features of the decorsin structure are two β-sheets which are linked by three disulfide bonds. The three disulfides form the core of the protein, which contains no bulky hydrophobic side chains. The backbone conformation of the RGD-containing recognition loop is well-defined (Fig. 5). There are two proline residues (Pro-30 and Pro-36) that flank the RGD sequence. The limited flexibility of the φ torsion angle of proline residues may contribute to the observed rigidity of the recognition site. Whereas the side chain of Asp-33 is conformationally restricted to some extent, the side chain of Arg-31 is ill-defined. The latter shows NMR spectral characteristics similar to those of Arg-3, but very different from the well-defined side chain of Arg-28. The conformation of the recognition sequence places the side chains of Arg-31 and Asp-33 in almost opposite directions (Fig. 5).

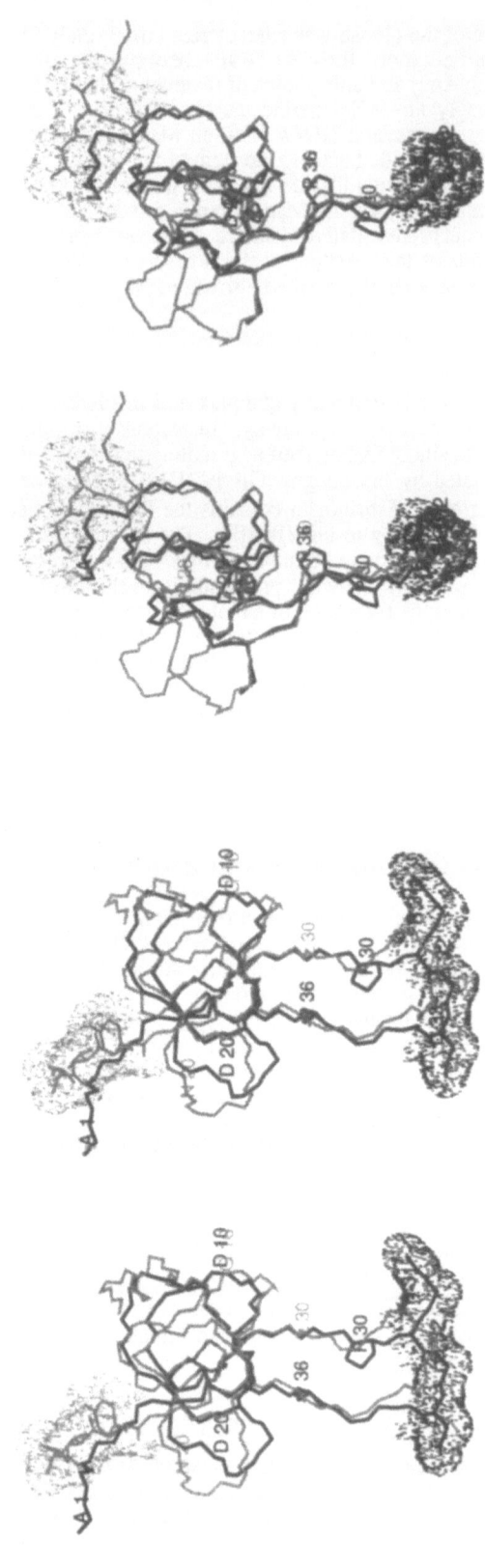

Figure 5. Comparison of the NMR structures of decorsin (black) and hirudin (gray) (Krezel et al., 1994). Two orientations are shown in stereo representations. The active sites-residues of both proteins are indicated with dotted surfaces.

The LAP motif

While we were working on the structure of decorsin, we were not aware of any obvious sequence homology to other proteins. However, after the structure was solved we discovered that the fold of decorsin closely resembles the fold of the first 40 residues of the thrombin inhibitor hirudin, a 65-residue protein from the leech *Hirudo medicinalis*, (Krezel et al., 1994; Markwardt, 1994; Rydel et al., 1990; Clore et al., 1990; Haruyama and Wüthrich, 1989; Szyperski et al., 1992). The active site of hirudin is at the N-terminus, opposite to the site of the RGD loop in decorsin (Fig. 5). Thus, leeches have used the same structural motif to inhibit two different factors of the blood clotting pathway. The sequence identity between hirudin and decorsin is 24%, and the cysteines alone account for 15%. Thus, the cysteine spacing appears to be the dominant sequence motif. With this in mind, we searched the Dayhoff protein database for the following motif:

$$Cys\text{-}X_{6\text{-}12}\text{-}Cys\text{-}X\text{-}Cys\text{-}X_{3\text{-}6}\text{-}Cys\text{-}X_{3\text{-}6}\text{-}Cys\text{-}X_{8\text{-}14}\text{-}Cys$$

A total of 181 out of the 140,192 sequences searched fit this criterion. Almost all of these werer variants of hirudin, however, we found 9 sequences that were ornatins, and 9 that were antistasins. Ornatin is a GP IIb-IIIa antagonist isolated from leech *Placobdella ornata* (Mazur et al., 1991). Several isoforms exist which share about 40 percent sequence identity with decorsin. The position of the RGD sequence in ornatin is the same as in decorsin. The sequence and functional similarity of ornatin and decorsin suggest that their structures are also similar. Antistasin is a 119-residue cysteine-rich protein from the leech *Haementaria officinalis*. Antistasin is a potent anticoagulant that inhibits Factor Xa, another serine protease in the coagulation cascade (Tuszynski et al., 1987). It contains a two-fold internal repeat which represents two closely related disulfide-bonded structures (Nutt et al., 1988). To our knowledge, its 3D structure has not been solved yet. Based upon the conservation of cysteines and some of the key turn residues, we speculate that each domain of antistasin shares a similar fold to that of decorsin and hirudin. We have propose the term 'LAP motif' (for Leech Antihemostatic Proteins) to refer to the general tertiary fold of this protein family[1].

The LAP motif apparently provides a framework for entirely different, yet highly specific activities of leech salivary proteins that potently affect the blood clotting process. Proteins that share the LAP motif likely arose via divergent evolution. Although decorsin, hirudin, and antistasin are all from leeches and all affect the hemostatic process, their sequences, activities, and binding epitopes differ. Decorsin binds to GP IIb-IIIa, presumably via its RGD sequence thus preventing fibrinogen binding. Decorsin does not contain the binding epitopes for Factor Xa or thrombin and thus does not inhibit either enzyme. Hirudin forms a tight complex with thrombin, having numerous significant binding contacts; residues at the amino terminus bind at the active site and residues at the carboxy terminus have interactions with the exosite (Rydel et al., 1990). Hirudin does not contain an RGD sequence and has no binding affinity for GP IIb-IIIa; it does not inhibit Factor Xa. Antistasin is a slow tight-binding inhibitor of Factor Xa and inhibits by the standard mechanism common to many serine protease inhibitors. The P1 active site residue in antistasin is Arg34 (Nutt et al.,1988) which corresponds to Pro23 in decorsin and Gly23 in hirudin (Krezel et al., 1994). Antistasin does not inhibit thrombin; no reports of any GP IIb-IIIa antagonist activity exist.

In summary, the fact that various species of leeches can use the same protein scaffold - the LAP motif, but totally different binding epitopes to interact with their respective targets and affect hemostasis by totally different mechanisms is both surprising and enlightening. This diversity of function residing in similar structures may lend insight into the evolutionary relationships between various leech species.

---

[1] During the lecture presented in Indianapolis, on October 31, 1994, I reported that the LAP sequence motif had been found recently in proteins of a giant bat living at the Indonesian island of Sumatra, and that the LAP motif was found in a seven-fold repeat. This, as well as the claim that the scientist Dr. Faust, who reported his findings in the Journal of Blood Research, deceased after being bitten by one of these blood sucking animals could not be confirmed since the article in that issue is covered with a big red stain and is unreadable.

Acknowledgments

This work was supported by NIH (grants GM 38608 and GM 47467 to GW and AI321226 to ELR) and by a grant from Sandoz.

REFERENCES

Adler, M, Lazarus, R, Dennis, M and Wagner G: Solution structure of kistrin a potent platelet aggregation inhibitor and integrin GPIIbIIIa antagonist, Science, 253, 445-448 (1991).

Arulanandam, ARN, Kister, A, McGregor, MJ, Wyss, DF, Wagner, G and Reinherz, EL.: Interaction between CD2 and CD58 involves the major β-sheet surface of each of their respective adhesion domains. J. Exp. Med., 180, 1861-1871 (1994).

Arulanandam, ARN, Withka, JM, Wyss, DF, Wagner, G, Kister, A, Pallai, P, Recny, MA and Reinherz, EL.: The CD58 (LFA3) binding site is a localized and highly charged surface area on the AGFCC'C" face of the human CD2. Proc. Natl. Acad. Sci. USA. 90, 11613-11617 (1993).

Bang, NU and Clayman, MD: Antithrombotic agents from salivary glands of hematophagous animals. Trends Cardiovasc. Med. 2, 183-188 (1992).

Bodian, DL, Jones, EY, Harlos K, Stuart, DI and Davis, SJ. Crystal structure of the extracellular region of the human cell adhesion molecule CD2 at 2.5 Å resolution. Structure 2, 755-766 (1994).

Brady RL, Dodson EJ, Dodson GG, Lange G, Davis SJ, Williams AF and Barclay AN: Crystal Structure of Domains 3 and 4 of Rat CD4: Relation to the NH2-Terminal Domains. Science 260, 979-983(1993).

Chothia, C, Novotny, J, Bruccoleri, R and Karplus, M: Domain association in immunoglobulin molecules. The packing of variable domains. J. Mol. Biol. 186, 651-663 (1985).

Clore, GM, Sukumaran, DK, Nilges, M, Zarbock, J and Gronenborn, A: The conformation of hirudin in solution: a study using nuclear magnetic resonance, distance geometry and restrained molecular dynamics. EMBO J. 6, 529-539 (1987).

Davis SJ, Davis EA, Barclay AN, Daenke S, Bodian DL, Jones EY, Stuart DI, Butters TD, Dwek RA and van der Merwe PA. Ligand bindingby the immunoglobion superfamily recognition molecule CD2 is glycosylation-independent. J. Biol. Chem. 270, 369-375 (1995).

deVos, AM, Ultsch, M and Kossiakoff, AA: Human growth hormone and extracellular domain of its receptor: Crystal structure of the complex. Science 255, 306-312 (1992).

Driscoll, PC, Cyster, JG, Campbell, ID and Williams, AF: Structure of domain 1 of rat T lymphocyte CD2 antigen. Nature, 353, 762-765 (1991).

Haruyama, H and Wüthrich K: Conformation of recombinant desulfatohirudin in aqueous solution determined by nuclear magnetic resonance. Biochemistry 28, 4312-4317 (1989).

Homans SW: Oligosaccharide conformations: Application of NMR and energy calculations. Progr. NMR Spectrosc. 22, 55-81 (1990).

Hynes, R O: Integrins: versatility, modulation, and signaling in cell adhesion. Cell 69, 11-25 (1992).

Jones, EY, Davis, SJ, Williams, AF, Harlos, K and Stuart, DI: Crystal structure at 2.8Å resolution of a soluble form of the cell adhesion molecule CD2. Nature, 360, 232-239, (1992).

Karlsson GB, Butters TD, Dwek RA and Platt FM. Effects of the imino sugar N-butyldeoxynojirimycin on N-glycosylation of recombinant gp120. J. Biol. Chem. 268, 570-576 (1993).

Krezel, AM, Gerhard Wagner, G, Ulmer, JS, and Lazarus, RA: Structure of the RGD protein decorsin: conserved motif and distinct function in leech proteins that affect blood clotting. Science 264, 1944-1947 (1994).

Lange G, Lewis SJ, Murshudov GN, Dodson GG, Moody PCE, Turkenburg JP, Barclay AN and Brady RL: Crystal Structure of an Extracellular Fragment of the Rat CD4 Receptor Containing Domains 3 and 4. Structure 2, 469-481 (1994).

Lazarus, R A and McDowell, R S: Structural and functional aspects of RGD-containing protein antagonists of glycoprotein IIb-IIIa. Curr. Opin. Biotechnol. 4, 438-445 (1993).

Leahy, DJ, Axel, R and Hendrickson, WA: Crystal structure of a soluble form of the human T cell coreceptor CD8 at 2.6Å resolution. Cell 68, 1145-1162 (1992).

Markwardt, F: The development of hirudin as an antithrombotic drug. Thromb. Res. 74, 1-23 (1994).

Mazur, P, Henzel, WJ, Seymour, J L and Lazarus, R A: Ornatins: potent glycoprotein IIbIIIa antagonists and platelet aggregation inhibitors from the leech Placobdella ornata.Eur. J. Biochem. 202, 1073-1082 (1991).

Niewiarowski, S, McLane MA, Kloczewiak, M, and Stewart, GJ: Disintegrins and other naturally occurring antagonists of platelet fibrinogen receptors. Semin. Hematol. 31, 289-300 (1994).

Nutt, E, Gasic, T, Rodkey, J, Gasic, GJ, Jacobs, JW and Friedman, PA: The amino acid sequence of antistasin. A potent inhibitor of Factor Xa reveals a repeated internal structure. J. Biol. Chem. 263, 10162-10167 (1988).

Recny, MA, et al., & Reinherz, EL: N-Glycosylation is required for human CD2 immunoadhesion functions. J. Biol. Chem. 267, 22428-22434 (1992).

Ribeiro, JMC: Role of saliva in blood-feeding arthropods. Ann. Rev. Entomol. 32, 463-478 (1987).

Ruoslahti, E and Pierschbacher, M D: New perspectives in cell adhesion: RGD and integrins. Science 238, 491-497 (1987).

Rydel, TJ, Ravichandran, KG, Tulinsky, A, Bode, W, Huber, R, Roitsch, Carolyn, Fenton II, JW: The structure of a complex of recombinant hirudin and human a-thrombin. Science 249, 277-280 (1990).

Ryu, SE, Kwong, PD, Truneh, A, Porter, TG, Arthos, J, Rosenberg M, Dai, X, Xuong, N, Axel, R, Sweet, RW and Hendrickson, WA: Crystal structure of an HIV- binding recombinant fragment of human CD4. Nature, 348, 419-426 (1990).

Saudek, V, Atkinson, R A, Pelton J T: Three-dimensional structure of echistatin, the smallest active RGD protein. Biochemistry 30, 7369-7372 (1991)

Sawyer, RT: Leech Biology and Behaviour. Clarendon Press, Oxford (1986).

Sawyer, RT: Thrombolytics and anticoagulants from leeches. Biotechnol. 9, 513-518 (1991).

Senn, H. and Klaus, W.: The nuclear magnetic resonance solution structure of flavoridin an antagonist of the platelet GPIIbIIIa receptor. J. Mol. Biol. 232, 907-925 (1993).

Seymour, JL, Henzel, WJ, Nevins, B, Stults, JT, and Lazarus, RA: Decorsin: A potent glycoprotein IIb-IIIa antagonist and platelet aggregation inhibitor from the leech Macrobdella decora. J. Biol. Chem. 265, 10143-10147 (1990).

Stocker, KF : Snake venom proteins affecting hemostasis and fibrinolysis. Medical Uses of Snake Venom Proteins (Stocker, K. F. ed) pp. 97-160, CRC Press, Boca Raton (1990).

Szyperski, T, Güntert, P, Stone, SR and Wüthrich, K: Nuclear magnetic resonance solution structure of hirudin (1-51) and comparison with corresponding three-dimensional structures determined using the complete 65-residue hirudin polypeptide chain. J. Mol. Biol. 228, 1193- (1992).

Tuszynski, GP, Gasic, T B, Gasic, GJ: Isolation and characterization of antistasin. J. Biol. Chem. 262, 9718-9723 (1987)

Wang, J, Yan, Garrett, TPJ, Liu, J, Rodgers, DW, Garlick, RL, Tarr, GE, Husain, Y, Reinherz, EL and Harrison, SC: Atomic structure of a fragment of human CD4 containing two immunoglobulin-like domains. Nature 348, 411-418 (1990).

Waxman, L, Smith, DE, Arcuri, KE, and Vlasuk, GP: Tick anticoagulant peptide (TAP) is a novel inhibitor of blood coagulation Factor Xa. Science 248, 593-596 (1990).

Withka, JM, Wyss, DF, Wagner, G, Arulanandam, ARN, Reinherz, EL and Recny, MA: Structure of the glycosylated adhesion domain of human T lymphocyte glycoprotein CD2. Structure, 1, 69-81 (1993).

Wormald, MR, Wooten, EW, Bazzo, R, Edge, CJ, Feinstein, A, Rademacher, TW, and Dwek, RA: The conformational effects of N-glycosylation on the tailpiece from serum IgM, Eur. J. Biochem. 198, 131-139 (1991).

Wyss, DF, Choi JS, andWagner G: Composition and sequence specific resonance assignments of te heterogeneous N-linked glycan in the 13.6 kDa adhesion domain of human CD2 as determined by NMR on the intact glycoprotein. Biochemistry, 34, 1622-1634 (1995a).

Wyss, DF, Choi JS, Li, JL, Knoppers, MH, Willis, KJ, Arulanandam, ARN., Smolyar, A, Reinherz, EL andWagner G: Conformation and function of the N-linked Glycan in the Adhesion Domain of Human CD2. Science, in press (1995b).

Wyss, DF, Withka, JM, Knoppers, MH, Sterne, KA, Recny, MA & Wagner, G: ^{1}H resonance assignments and secondary structure of the 13.6 kDa glycosylated adhesion domain of human CD2. Biochemistry 32, 10995-11006 (1993).

DISCUSSION

Ad Bax - First, I would like to complement you on this very fantastic work that you presented. I also have a question about somewhat of a paradox about the glucose of the sugar unit on your protein where you observe obviously very strong increased mobility because of the difference in relaxation rates. Nevertheless, you see this loop folding back. Could this be in a small fraction of the time, the thing folds back and causes the NOE or what is going on?

Gerhard Wagner - That's a good question, eternal question about the NMR structures. The structure we recorded was of proteins that have an rmsd of 0.3 $\overset{0}{\text{A}}$. On the other hand, we know on the interior of the protein, the aromatic rings rotate. Every rotation of an aromatic ring requires an opening of the structure by at least 1.5 $\overset{0}{\text{A}}$. So it's hard to question how much of the protein is in the well-defined state. In the interior of the protein, it might just be 99% and 1% may be a more open structure. I think it is just a gradual decrease of the folding structure and it may well be possible that 80% of the carbohydrate is folded and that the 20% is in a more open state but it is very difficult to identify by experiment but you just have to be aware of it. You should also know that at room temperature proteins have a small fraction of completely unfolded state but that is known from energy exchange experiments. You just have to be aware of this but it is difficult to find out the fraction of the unfolded part of the carbohydrate.

Bax - On a very qualitative level couldn't you argue that if the T_2 is 5 or 10 times longer then it could be only a very small fraction of the time. Let's say 20 % of the thing would be folded up.

Wagner - The problem is to find out if it is 25% or 60%. It is a good point but it's difficult to answer.

Bax - In your case it is almost an order of magnitude, isn't it?

Wagner - That's possible since these carbohydrates have much fewer proteins to relax with.

COMBINING ^{2}H AND ^{13}C SELECTIVE ENRICHMENT TO PROBE PROTEIN DYNAMICS

David M. LeMaster

Department of Biochemistry, Molecular Biology and Cell Biology
Northwestern University
Evanston, IL 60208

INTRODUCTION

The dipolar interactions between ^{13}C nuclei and the directly attached ^{1}H nuclei offer a potentially rich set of spectroscopic monitors of the dynamical behavior of biological macromolecules. However, in contrast to the more widely exploited ^{1}H-^{15}N dipoles of the mainchain amide resonances of proteins, additional technical complexities have impeded the more extensive use of ^{13}C relaxation analysis. As for the case of ^{15}N studies, the low sensitivity of natural abundance samples have limited ^{13}C relaxation studies to quite small highly soluble systems. On the other hand uniform enrichment introduces scalar and dipolar interactions between directly bonded ^{13}C nuclei which confound the interpretation of the conventional pulse sequences used in relaxation analysis. Progress has been reported in the development of pulse sequences to circumvent some of the complications arising from geminal ^{13}C nuclei,[1,2] but a general solution is not yet available. An alternate approach which allows for extensive high levels of site specific enrichment with minimal occurrence of geminal ^{13}C-^{13}C couplings can be achieved via protein expression in bacterial strains carrying suitable metabolic lesions so as to direct the flow of isotopic label from selectively enriched carbon sources.

The interacting ^{1}H-^{13}C dipoles present at methlyene and methyl positions introduce additional complications for relaxation analysis due to the cross-correlation (interference) effects. For the case of methyl resonances the cross-correlation artifacts can be largely suppressed by exploiting the symmetry of the equivalent spins.[3,4] For the methylene positions no such general solution is available. In principle these cross correlation effects can provide significant dynamical information, particularly with regards to motional asymmetries. However, the few such analyses reported for peptide and protein systems (eg.[5,6]) have emphasized the desirability of simultaneous autocorrelation experiments. Deuterium substitution provides an alternate means of circumventing the problems associated with cross correlations as in practice each proton bearing

NMR As a Structural Tool for Macromolecules: Current Status and Future Directions
Edited by B.D. Nageswara Rao and Marvin D. Kemple, Plenum Press, New York, 1996

65

carbon can be reduced to an IS spin system. However, deuteration is not truly a "silent" substitution. Various of the specific spectroscopic effects of deuterium substution are considered below.

DIFFERENTIAL DEUTERIUM ISOTOPE SHIFTS AND SCALAR COUPLINGS IN THE CONFORMATIONAL ANALYSIS OF PROTEIN GLYCINE POSITIONS

Von Philipsborn and coworkers[7] originally pointed out the conformational dependence of $^1J_{H\alpha C\alpha}$ in a set of small peptides. More recently, the anticipated trigonometric dependence of this scalar coupling on the mainchain dihedral angles has been extensively verified experimentally using isotopically enriched samples.[8,9] These studies did not include the glycine residues in part due to the ambiguities of the stereochemical assignments. Such assignments are readily ascertained by the use of chirally deuterated ^{13}C enriched glycine samples.[10,11] When these selectively labeled amino acids are incorporated into protein samples, the corresponding ^2H-decoupled ^{13}C-coupled HSQC spectra exhibit crosspeaks as illustrated in Figure 1 of staphlococcal nuclease.[12] As expected two ^{13}C-coupled doublets exhibit different $^1J_{H\alpha C\alpha}$ values. Indeed when such scalar couplings are analyzed against the corresponding high resolution x-ray structures for 16 glycine residues of staphlococcal nuclease and E. coli thioredoxin,[12] a trigonometric parameterization is obtained which corresponds quite closely to one earlier reported for the methine C_α positions.[9]

Unanticipated was the substantial difference in the ^{13}C chemical shifts for the ^2H_R and ^2H_S crosspeaks. Deuterium substution is well known to cause an upfield shift for the directly attached ^{13}C nucleus, 0.25 ppm in this case. However, for the crosspeaks of Figure 1 each ^{13}C nucleus is substituted by a single deuterium differing only in the chirality of the substitution. A conformational analysis of these differential isotope shifts was conducted in terms of a trigonometric dependence on the mainchain dihedral angles for the same 16 glycine residues. The resultant parameterization exhibited striking

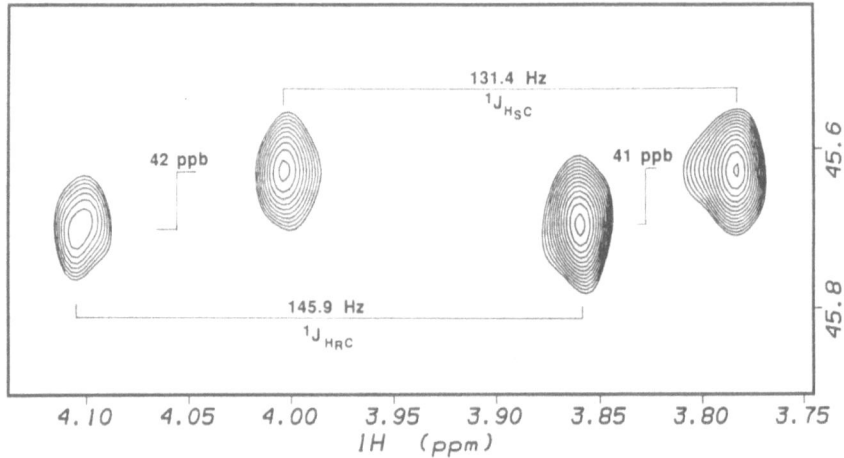

Figure 1. ^2H-decoupled, ^{13}C-coupled HSQC spectrum of [^2H_R, 2-^{13}C]glycine + [^2H_S, 2-^{13}C]glycine enriched staphylococcal nuclease H124L in the $Ca^{2+}\cdot pTp$ ternary complex at 14.1 T. The crosspeaks corresponding to Gly 20 are illustrated. A 2 kHz ^2H decoupling field was used. Reprinted with permission from ref. 12. Copyright ESCOM Science Publishers B.V. 1994.

similarities to those obtained for the glycine and methine C_α $^1J_{HC}$ analyses. In particular the sign of the dependence is opposite for the $\cos^2\phi$ and $\cos^2\psi$ dependencies and the ψ dependency is approximately three-fold larger. In Figure 2 is presented the linear regression analysis of the experimental differential isotope shifts and the values predicted from the parameterized trigonometric fit function. A linear regression coefficient $r^2 = 0.85$ is obtained. It should be noted that in one respect there is a substantial difference between the variations of the $^1J_{H\alpha C\alpha}$ values and the differential isotope shifts. As a function of (ϕ,ψ) the $^1J_{H\alpha C\alpha}$ values vary over a range of the average $^1J_{H\alpha C\alpha}$ +/- 10%. In contrast the range of differential isotope shifts cover +/- 30% of the average one-bond deuterium isotope shift value.

The clear similarity in the conformational dependence of the $^1J_{H\alpha C\alpha}$ values and the differential isotope shifts strongly suggest a similar physical basis. Following the standard interpretation of geometric dependencies of scalar couplings, Von Philipsborn and coworkers[7] interpreted their data in terms of hyperconjugation giving rise to an increased coupling resulting from the alignment of the H-C bond with the p_z orbital of the attached nitrogen. Conversely, alignment of the H-C bond with the π orbital of the carbonyl carbon results in a smaller coupling constant. On the other hand the role of hyperconjugation in isotope shift effects remains controversial. Recent reviews [13,14] discuss the argument that since the isotope shift fundamentally arises from

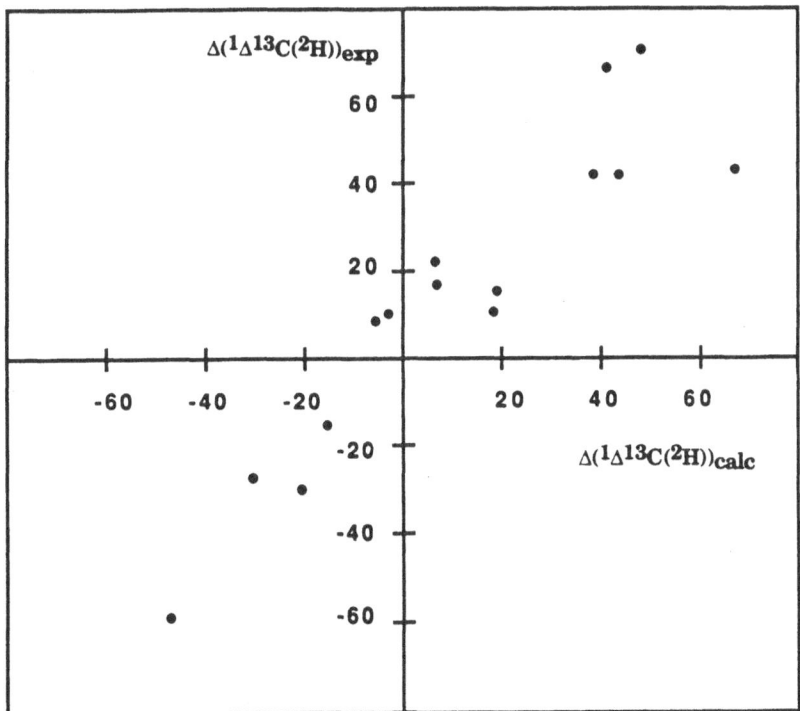

Figure 2. Linear regression analysis of 16 glycine methylene experimental differential isotope shifts vs. those calculated using the trigonometric fit function relating these data (in ppb) to the corresponding (ϕ, ψ) dihedral angles of the high resolution x-ray structures. Reprinted with permission from ref. 12. Copyright ESCOM Sciences Publishers B.V. 1994.

a differential vibrational effect explaining isotope shift effects via hyperconjugation effects implies a breakdown of the Born-Oppenheimer approximation. This concern does not appear valid for the one bond isotope shift data presented herein. Both theoretical and experimental studies (eg. [15-18]) have demonstrated bond length variations as a function of the orientation of adjacent lone pair and π orbitals. These geometrically dependent bond length variations are correlated with corresponding variations in the vibrational frequencies. As the isotope shift arises from the difference in the zero point vibrational frequencies for the proton and deuteron, conformationally dependent changes in the vibrational manifold will give rise to differential isotope shifts. As the two geminal glycine H-C bonds in general occupy two different orientations with respect to the adjacent π orbitals, the differential ^{13}C shifts result.

The similarity of the conformational dependencies for the $^1J_{H\alpha C\alpha}$ values and the differential isotope shifts argue for the benefit of their combined use in analysis of mainchain conformation. Indeed predictions based on their use in combination appear to be substantially more robust than those based on either set alone. However, it is anticipated that the more significant application of the combined use of these spectroscopic values may be in the prediction of stereochemical assignments. Given only the $^1J_{H\alpha C\alpha}$ values and the differential isotope shifts for the 16 glycine residues analyzed above, a determination of chirality can be readily obtained in all cases using only the additional constraint that the sign of ϕ is known. A simple protocol can be given as follows:

1. If $|\Delta^1J_{HC}|$ is greater than 6 Hz, the pro-R hydrogen has the larger coupling for $\phi < 0$ and the smaller coupling for $\phi > 0$.
2. Otherwise, pro-R hydrogen is the most upfield shifted resonance for $\phi < 0$ and most downfield shifted for $\phi > 0$.

Although the glycine residues of the two proteins considered represent only a modest subset of all observed glycine conformations, these results suggest that reliable stereochemical assignments can be obtained from the isotope shift and scalar coupling data with only fairly limited additional structural information without recourse to more involved chiral labeling techniques.

In this context it is worthwhile to note that such one bond differential isotope shifts are not limited to hyperconjugation with π systems. Saturated hydrocarbons have also been argued to exhibit comparable effects resulting from the periplanar alignment of vicinal H-C bonds.[19,20] We note that differential isotope shifts can be observed for various methylene types in protein sidechains.

DYNAMIC FREQUENCY SHIFTS IN ^2H-COUPLED ^{13}C SPECTRA

The heteronuclear correlation spectra illustrated in Figure 1 was collected with a 2 KHz ^2H decoupling field present during the t_1 evolution in order to suppress the effects of the residual ^2H-^{13}C scalar coupling. When the corresponding ^2H-coupled spectra is collected, the contour plots exhibited an unexpected teardrop shape indicating an asymmetry in the ^2H interactions. In Figure 3 is shown an f_1 slice of a ^2H-coupled ^1H-^{13}C HSQC spectrum drawn through the resonance of Gly 74 of E. coli thioredoxin.[21] Included as well is the corresponding f_1 slice with ^{2}H decoupling applied. Rather than the symmetric triplet pattern expected of coupling to a spin-1 nucleus in the extreme narrowing limit, the frequency of the m=0 spin state is shifted relative to those of the m= +/-1 spin states. This dynamic frequency shift has been shown to arise from cross correlation between the dipolar and quadrupolar interactions for slow tumbling ^{2}H-^{13}C vectors.[21,22] This interference gives rise to a contribution from the normally negligible imaginary component of the complex spectral density function.

The resultant dynamic frequency shift is defined as

$$\delta_{\omega m} = 4(2-3m^2)\, L^{D\text{-}Q}(\omega) \tag{1}$$

with $L^{D\text{-}Q}(\omega)$ representing the corresponding spectral density for the dynamic frequency shift. While the more familiar spectral densities for the dipolar and quadrupolar interactions have the form

$$J(\omega) = G^2\, \tau/(1+\omega^2\tau^2) \tag{2}$$

where G is the dipolar or quadrupolar coupling constant, respectively, and τ is the molecular correlation time. In contrast

$$L^{D\text{-}Q}(\omega) = {}^DG \cdot {}^QG\, \omega\tau/(1+\omega^2\tau^2) \tag{3}$$

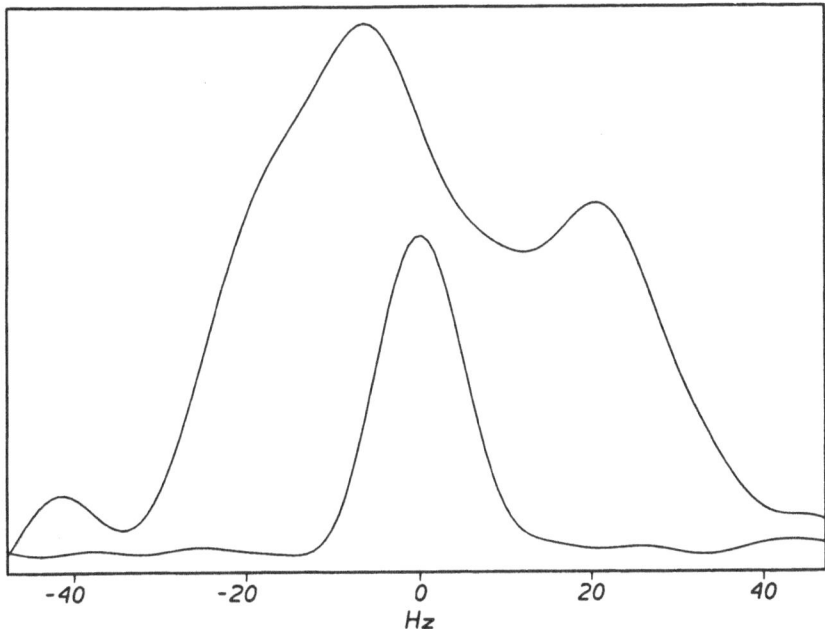

Figure 3. Upper trace; f_1 slice of ^{2}H-coupled ^{1}H-^{13}C HSQC spectrum of [2-^{2}H$_R$,2-^{13}C]glycine enriched E. coli thioredoxin drawn through the Gly 74 resonance. Lower trace: The corresponding f_1 slice obtained using broadband ^{2}H decoupling (attenuated 6-fold relative to the ^{2}H-coupled spectrum). These spectra were obtained at 14.1 T under conditions exhibiting an 8.1 ns correlation time. Reprinted with the permission from ref. 22. Copyright American Chemical Society, 1994.

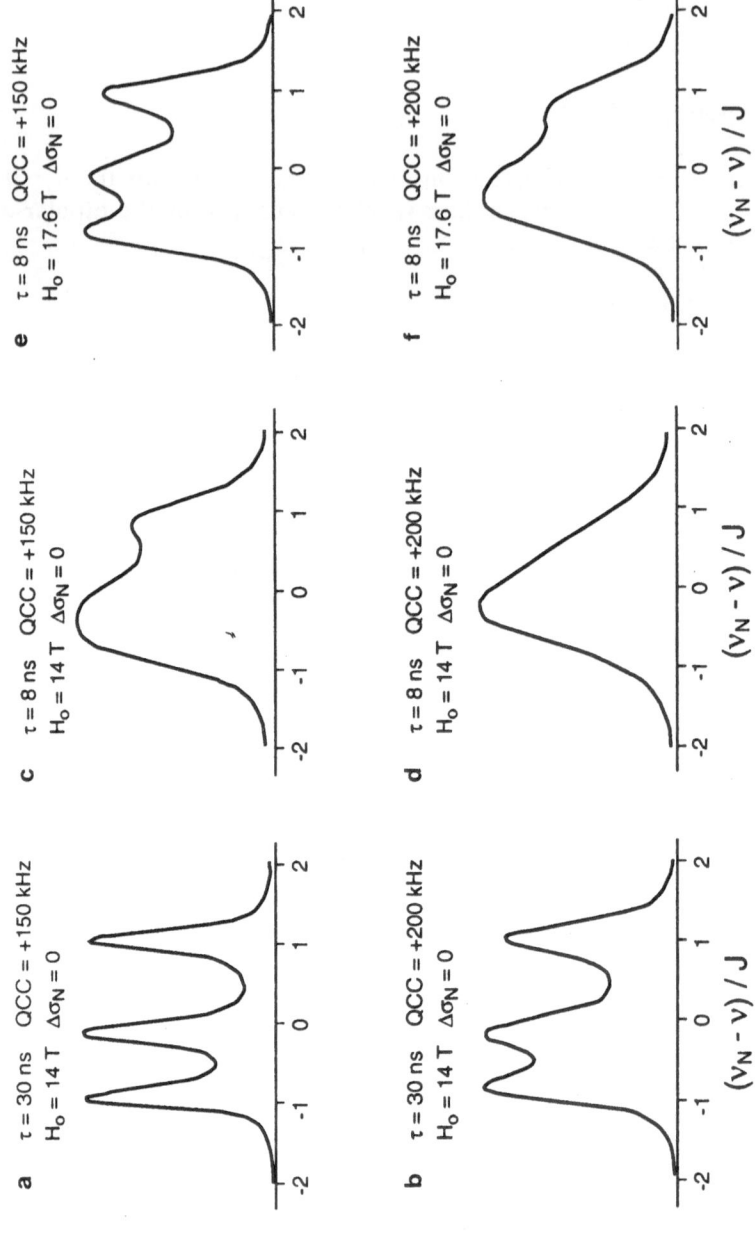

Figure 4. Plot of spin ½ line shape function vs. reduced frequency assuming $J = 136\,s^{-1}$ (21.6 Hz); $(\gamma_I \gamma_S h / r_{IS}^3) = 2.2 \times 10^4\,s^{-1}$; $(e^2 qQ_S/h)$ = $1.1 \times 10^6\,s^{-1}$; $\omega_S = 5.8 \times 10^8\,s^{-1}$. The simulations a, b, c and d correspond to $\tau = 0.1$, 1.0 8.1 and 30 ns, respectively. Reprinted with permission from ref. 22. Copyright American Chemical Society, 1994.

As a result of the additional ω dependence the frequency response of the dynamic frequency shift is substantially different than that of the more familiar spectral densities. In particular the values of both J^D and J^Q become frequency independent at the maximum of the Lorentzian when the molecular correlation time matches the Larmor frequency of the spin-1 nucleus. In contrast the sensitivity to variations in the molecular correlation time is maximal for $L^{D-Q}(\omega)$ at this frequency. Using molecular parameters corresponding to the spectra of Figure 3, the effect of the interference between the dipolar and quadrupolar interactions as a function of molecular correlation time is modeled in Figure 4.[21]

Note that molecular correlation time used for Figure 4c corresponds to that of the experimental protein sample of Figure 3. At longer correlation times not only does the dynamic frequency shift become more apparent, the differential relaxation behavior of the central vs. outer components is increasingly evident as well. Although these simulations do not include the linewidth contributions due to the ^1H-^{13}C dipole used to "spy" on the ^2H-^{13}C interaction for the spectra of Figure 3, the general appearance of the spectra is not substantially affected by their inclusion.

Two practical questions which arise in considering the utility of the dynamic frequency shift for studying protein dynamics are the field strength dependence and the applicability to other relevant pairs of nuclei. In Figure 5 is given simulations for the ^2H-^{15}N pair as a function of field strength and of the quadrupole coupling constant. Qualitatively the results are quite similar to

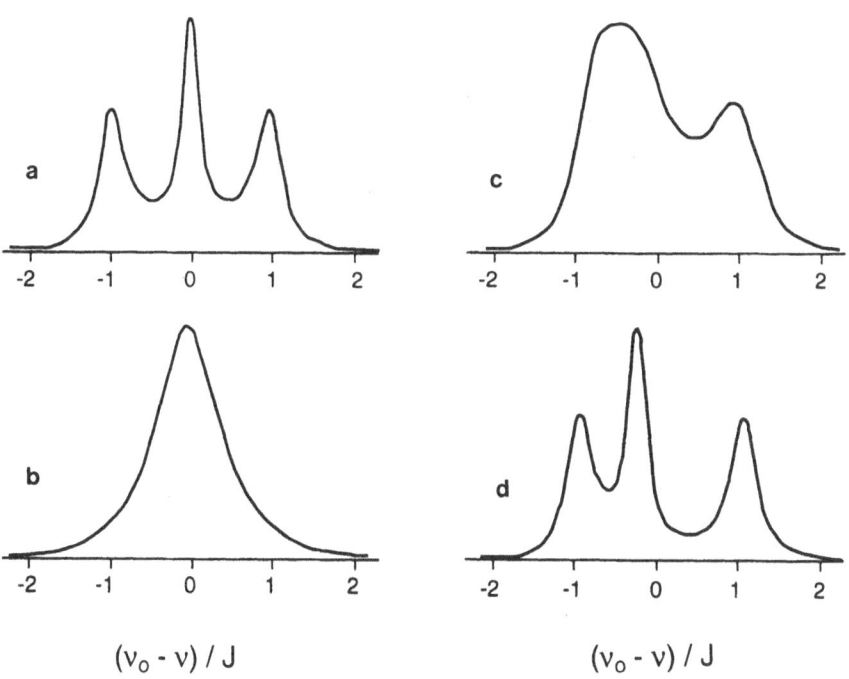

$(\nu_o - \nu) / J$ $(\nu_o - \nu) / J$

Figure 5. Plot of spin 1/2 line shape function vs. reduced frequency analogous to that of Figure 4 with the parameters adjusted to those appropriate for the ^2H-^{15}N spin system. Varied parameters are as given in the figure. Reprinted with permission from ref. 22. Copyright American Chemical Society, 1994.

those obtained for the ^{2}H-^{13}C pair reflecting the approximately proportional decrease in both the ^{1}J coupling and the slow tumbling limit dynamic frequency shifts. The potential benefit of component narrowing at higher field strengths is partly mitigated by the inverse dependence of the dynamic frequency shift on the field strength.[22]

COMBINING RANDOM FRACTIONAL DEUTERATION WITH ALTERNATING CARBON ENRICHMENT FOR PROTEIN RELAXATION STUDIES

The primary goal for the selective isotope labeling approaches described herein is to obtain the relaxation contributions due to isolated ^{1}H-^{13}C dipoles throughout a protein structure. Despite the substantially smaller magnetogyric ratio the relaxation contribution of ^{2}H nucleus must be explicitly considered. In the extreme narrowing limit the contribution of the ^{2}H-^{13}C dipole is 6.3% that of the ^{1}H-^{13}C dipole for both T_1 and T_2 as expected from the dependence of the spectral density functions on $I(I+1)\gamma^2$. Such a contribution is well outside the level of precision commonly expected for relaxation measurements on biological macromolecules. Furthermore, at longer correlation times the relative contribution of the ^{2}H-^{13}C dipole is increased. For T_1 this increased contribution reflects the presence of the $J(\omega_{13C}-\omega_{2H})$ spectral density term which represents a significantly lower frequency of sampling than any of the ^{1}H-^{13}C dipole spectral density terms. As a result at typical field strengths the fractional contribution to T_1 relaxation due to the ^{2}H-^{13}C dipole rises to ~13% for $1/\tau$ near ω_{2H} and then rises to ~22% in the slow tumbling limit. In an analogous fashion the fractional contribution to T_2 relaxation due to the ^{2}H-^{13}C dipole rises to ~13% near the Larmor frequency of ^{2}H before dropping back down to 6.3% in the slow tumbling limit as the $J(0)$ term becomes dominant.

The relaxation contribution due to the ^{2}H-^{13}C dipole can be straightforwardly subtracted out on the assumption of similar correlation behavior for the geminal ^{1}H-^{13}C and ^{2}H-^{13}C dipole vectors. However, for the most interesting case of the methylene positions it should be noted that the validity of this assumption can be directly examined by consideration of the relaxation behavior of the crosspeak corresponding to the conjugate ^{1}H-^{13}C - ^{2}H-^{13}C dipole pair. Indeed it is anticipated that the differential relaxation of the ^{2}H$_R$ and ^{2}H$_S$ crosspeak pairs should provide a useful monitor for motional anisotropy.

As is apparent in the f_1 slices presented in Figure 2, ^{2}H decoupling serves to dramatically collapse the distorted linebroadening induced by the directly bonded ^{2}H nucleus. However, it has been pointed out that narrowing to the dipolar broadened limit is not achieved by ^{2}H decoupling.[22] At a correlation time of 8 ns an additional linebroadening of ~0.5 Hz due to the interference between the dipolar and quadrupolar interactions is anticipated. This corresponds to 4% of the total linewidth predicted for a monodeuterated methylene position. For correlation times above the inverse Larmor frequencies the additional linewidth contribution due to the interference between the dipolar and quadrupolar interactions scales with τ as does $J(0)$. As a result neglect of this linebroadening contribution in the standard dynamical analysis could give rise to a slightly increased apparent correlation time or else a quite small apparent chemical exchange contribution.

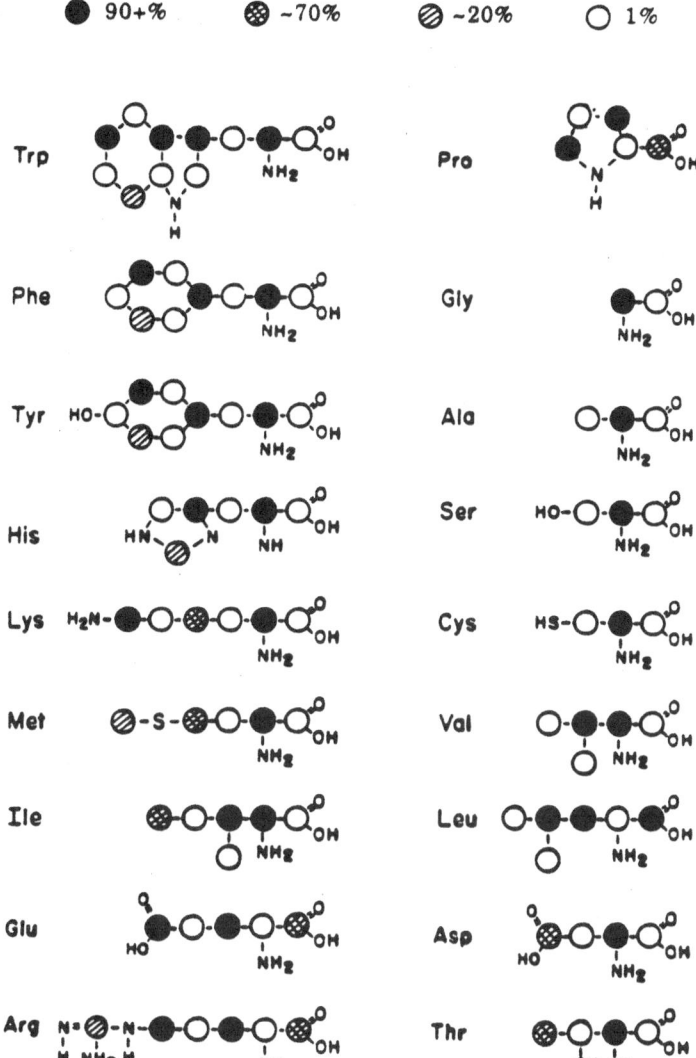

Figure 6. ^{13}C enrichment pattern obtained for the constituent amino acids of E. coli thioredoxin when expressed in an alternate carbon enrichment bacterial expression strain grown on [2-^{13}C]glycerol.

The potentially conflicting demands of high ^{13}C enrichment at sites throughout the protein structure and the desire to avoid the complications arising from enrichment of adjacent carbon sites can be largely resolved by the use of bacterial expression systems bearing a suitable combination of metabolic lesions to suppress scrambling of the selective labeling pattern utilized in the primary carbon source. As earlier reported[23] such an E. coli strain can faithfully preserve that labeling selectivity during the carbon incorporation into the constituent amino acids.

The resultant ^{13}C labeling pattern obtained for E. coli thioredoxin expressed in this bacterial system grown on [2-^{13}C]glycerol + H^{13}CO$_3^-$ is illustrated schematically in Figure 6. For most positions the enrichment is essentially

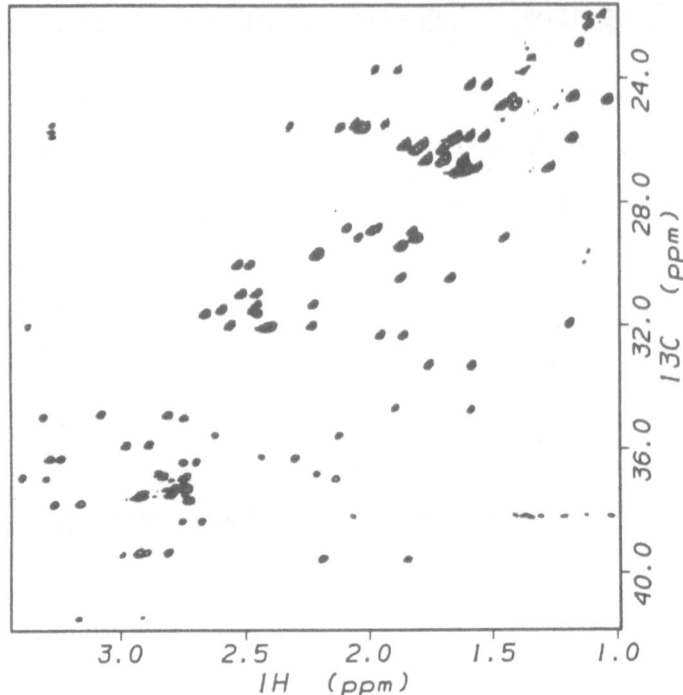

Figure 7. The methylene region of E. coli thioredoxin labeled from [1,3-$^{13}C_2$]glycerol. This ^1H-^{13}C HSQC experiment incorporated refocused INEPT components to select signals from IS spin systems.

complete or else negligible. Modest isotopic dilution was observed in the bicarbonate incorporation into the aspartate family of amino acids and a limited amount of label scrambling was observed for the aromatic rings arising from the pentose phosphate pathway intermediates. Most importantly for all of the proton-bearing carbon sites, only at the branch sites of valine, leucine and isoleucine are directly bonded ^{13}C nuclei observed.

When combined with 50% random fractional deuteration this alternating carbon enrichment scheme provides a means of generating the isolated IS spin systems desired for the dynamical analysis of nearly all proton bearing carbon sites in a protein molecule. The signals from these monoprotiated carbon sites are readily selected using the INEPT transfer components already incorporated into the standard ^1H detected relaxation experiments. Furthermore, particularly for the case of the nonequivalent methylene positions, the resultant elimination of the geminal ^1H-^1H couplings yield a substantial enhancement in crosspeak resolution. These combined spectral benefits are illustrated by the ^1H-^{13}C correlation spectrum from [1,3-$^{13}C_2$]glycerol labeled E. coli thioredoxin for which the methylene spectral region is given in Figure 7. The good spectral resolution and sensitivity combined with the comparatively straightforwardly interpretable relaxation characteristics provide a means for obtaining the extensive dynamical mapping of a protein molecule.

ACKNOWLEDGEMENTS

This work was supported by the National Institutes of General Medical Sciences (GM-38779), the National Science Foundation (DMB-8957336). The calculations for figures 4 and 5 carried out by Robert London are gratefully acknowledged.

REFERENCES

1. T. Yamazaki, R. Muhandiram and L. E. Kay, NMR Experiments for the Measurement of Carbon Relaxation Properties in Highly Enriched, Uniformly $^{13}C,^{15}N$-Labeled Proteins: Application to $^{13}C_\alpha$ Carbons,*J. Am. Chem. Soc.* 116:8266 (1994).

2. J. Engelke and H. Ruterjans, Determination of $^{13}C_\alpha$ Relaxation Times in Uniformly $^{13}C/^{15}N$-Enriched Proteins,*J. Biomolec. NMR.* 5:173 (1995).

3. A. G. Palmer, P. E. Wright and M. Rance, Measurement of Relaxation Time Constants for Methyl Groups by Proton-Detected Heteronuclear NMR Spectroscopy,*Chem. Phys. Lett.* 185:41 (1991).

4. L. E. Kay, T. E. Bull, L. K. Nicholson, C. Greisinger, H. Schwalbe, A. Bax and D. A. Torchia, The Measurement of Heteronuclear Transverse Relaxation Times in AX_3 Spin Systems via Polarization-Transfer Techniques, *J. Magn. Reson.* 100:538 (1992).

5. V. A. Daragan, M. A. Kloczewiak and K. H. Mayo, ^{13}C Nuclear Magnetic Relaxation-Derived Φ,Ψ Bond Rotational Energy Barriers and Rotational Restrictions for Glycine $^{13}C_\alpha$-Methylenes in a GXX-Repeat Hexapeptide,*Biochemistry.* 32:10580 (1993).

6. M. Ernst and R. R. Ernst, Heteronuclear Dipolar Cross-Correlated Cross Relaxation for the Investigation of Side-Chain Motion, *J. Magn. Reson. Ser. A* 110:202 (1994).

7. H. Egli and W. von Philipsborn, Conformational Dependence of One-Bond $C_\alpha H_\alpha$ Spin Coupling in Cyclic Peptides,*Helv. Chim. Acta.* 64:976 (1981).

8. D. F. Mierke, S. G. Grdadolnik and H. Kessler, Use of One-Bond C_α-H_α Coupling Constants as Restraints in MD Simulations,*J. Am. Chem. Soc.* 114:8283 (1992).

9. G. W. Vuister, F. Delaglio and A. Bax, The Use of $^1J_{C_\alpha H_\alpha}$ Coupling Constants as a Probe for Protein Backbone Conformation,*J. Biomolec. NMR.* 3:67 (1993).

10. R. J. Homer, M. S. Kim and D. M. LeMaster, The Use of Cystathionine γ-Synthase in the Production of α and Chiral β Deuterated Amino Acids,*Anal. Bioch.* 215:211 (1993).

11. D. M. Kushlan and D. M. LeMaster, Resolution and Sensitivity Enhancement of Heteronuclear Correlation for Methylene Resonances via ^2H Enrichment and Decoupling,*J. Biomolec. NMR.* 3:701 (1993).

12. D. M. LeMaster, J. C. LaIuppa and D. M. Kushlan, Differential Deuterium Isotope Shifts and One-Bond ^1H-^{13}C Scalar Couplings in the Conformational Analysis of Protein Glycine Residues,*J. Biomolec. NMR.* 4:863 (1994).

13. P. E. Hansen, Isotope Effects in Nuclear Shielding,*Prog. NMR Spectrosc.* 20:207 (1988).

14. S. Berger, Chemical Models for Deuterium Isotope Effects in ^{13}C and ^{19}F NMR,*NMR Basic Princ. Prog.* 22:1 (1990).

15. D. J. DeFrees, M. Taagepera, B. A. Levi, S. K. Pollack, K. D. Summerhays, R. W. Taft, M. Wolfsberg and W. J. Hehre, Role of Hyperconjugation in

Secondary β-Deuterium Isotope Effects, *J. Am. Chem. Soc.* 101:5532 (1979).

16. K. B. Wiberg, V. Walters and S. D. Colson, A Vibrational Force Field for Acetaldehyde, *J. Phys. Chem.* 88:4723 (1984).

17. W. B. Farnham, B. E. Smart, W. J. Middleton, J. C. Calabrese and D. A. Dixon, Crystal and Molecular Structure of $[(CH_3)_2]_3S^+CF_3O^-$. Evidence for Negative Fluorine Hyperconjugation, *J. Am. Chem. Soc.* 107:4565 (1985).

18. A. Rathna and J. Chandrasekar, The Influence of Lone-pair Repulsions on C-C Bond Lengths, *J. Chem. Soc., Perkins Trans II.* 1661 (1991).

19. I. H. Williams, Calculated Conformatônal Equilibrium Iostope Effect for $[^2H_1]$Cyclohexane, *J. Chem. Soc. Chem. Commun.* 627 (1986).

20. R. Aydin and H. Gunther, Chemical Applications of NMR Isotope Effects: Ring Inversion and Conformational Equilibrium in $[^2H_1]$Cyclohexane, *Angew. Chem. Int. Ed.* 985 (1981).

21. R. E. London, D. M. LeMaster and L. G. Werbelow, Unusual NMR Multiplet Structures of Spin 1/2 Nuclei Coupled to Spin 1 Nuclei, *J. Am. Chem. Soc.* 116:8400 (1994).

22. S. Grzesiek and A. Bax, Interference between Dipolar and Quadrupolar Interactions in the Slow Tumbling Limit: A Source of Line Shift and Relaxation in ^2H-Labeled Compounds, *J. Am. Chem. Soc.* 116:10196 (1994).

23. D. M. LeMaster and J. E. Cronan, Jr., Biosynthetic Production of ^{13}C-Labeled Amino Acids with Site-Specific Enrichment, *J. Biol. Chem.* 257:1224 (1982).

Incorporating Motional Properties into the Interpretation of
Three-dimensional Solution Structures

Walter J. Chazin

Department of Molecular Biology
The Scripps Research Institute (MB-2)
10666 North Torrey Pines Road
La Jolla, California 92037

INTRODUCTION

A detailed understanding of biological systems at the molecular level requires characterization of structure and dynamics in atomic detail.   NMR spectroscopy is uniquely suited to the determination of high resolution solution structures, and has vast potential as a means for examining molecular dynamics because a wide array of nuclear sites within a molecule can be assayed at a variety of different time scales. However, at present we are in the midst of a search to identify how the structural and dynamic information can be fused to generate a unified and comprehensible view for analyzing biomolecular function.

Our current perception of biomacromolecular structure is dominated by an extensive database of three-dimensional structures determined by X-ray crystallography. Although a certain degree of low amplitude motion is present in the crystalline state, the general view of a macromolecule from crystallography is as a single, static structure. While this usually can be safely regarded as *an* accurate representation of the molecule, little insight into the intrinsic flexibility of the biopolymer is provided. The most important point in this context is that the motional dynamics are an integral property of biomacromolecules that must be factored into any attempt at understanding their biological activity. In this respect, NMR spectroscopy stands out as a technique that is sensitive to both structure and dynamics.

From the viewpoint of the "future directions" theme of the symposium, it is clear that one critical area of future development is the ability to integrate motion into the classical static manner in which we view molecular structures. Of course, the great difficulty in dealing with dynamics is the problem of motions occurring on multiple timescales that span a range greater than 10 orders of magnitude! Using a typical small protein, calbindin D_{9k}, as a case in point, and considering only those dynamical properties that can be measured in a direct and relatively easy manner, there are nonetheless numerous events occurring at a multitude of timescales. These include: picosecond to nanosecond motions of the backbone detected by ^{15}N relaxation measurements, calcium-binding events on the microsecond to millisecond timescale, reduced rates of aromatic ring flips on the millisecond timescale detected by exchange broadening of resonance lines, and even slower events such as backbone amide proton exchange and proline *cis-trans* isomerism. Each of the above-named processes has been examined in separate experiments. The challenge now remains to integrate these data and further, to incorporate these properties of the molecule's dynamics into the time-averaged structure.

Calbindin D_{9k} is a member of the EF-hand superfamily of Ca^{2+}-binding proteins (CaBPs) that are well known for their role in calcium signal transduction pathways, but which also function in Ca^{2+} buffering, transport and other aspects of calcium homeostasis. These proteins are characterized by the high homology of the helix-loop-helix calcium-binding motifs (EF-hands), which are typically paired in an integrally packed globular domain. The solution structure and internal dynamics of the small 75 residue calbindin D_{9k} protein have been extensively characterized by a variety of biophysical techniques (for a review see Forsén et al., 1993). High resolution three-dimensional structures have been determined for various states of the protein and corresponding molecular motions have been characterized by several different NMR techniques. Insights into the molecular basis for the cooperativity in Ca^{2+}-binding have been obtained from a combined view of these conformational and dynamical responses to ion binding (Akke et al., 1991; Carlström & Chazin, 1993; Linse & Chazin, 1995; Wimberly et al., 1995). In addition, these results for calbindin D_{9k} have provided a foundation for a comparative analysis of the EF-hand CaBPs, that is designed to elucidate how the fine-tuning of the response to calcium-binding is used to generate the extremely wide functional diversity within this family of proteins (Skelton et al., 1994).

To obtain a complete understanding of the ion binding process, it is necessary to characterize not only the ion-free and fully ion-bound states of the protein, but also the intermediate half-saturated states (Fig. 1). Thus, in addition to the apo and $(Ca^{2+})_2$ states, we have studied the $(Cd^{2+})_1$ state, in which the ion is bound exclusively in the C-terminal EF-hand. The high resolution solution structures determined for the Ca^{2+}-free, the $(Cd^{2+})_1$ half-saturated, and $(Ca^{2+})_2$ states of calbindin D_{9k} have been described in detail along with the changes associated with the progressive stepwise binding of ions (Kördel et al., 1993; Skelton et al., 1995; Akke et al., 1995). NMR measurements of internal dynamics, including ^{15}N relaxation and amide proton exchange measurements (Kördel et al., 1992; Skelton et al., 1992; Akke et al., 1993), have been used to characterize internal motions, as well as assess the sources of uncertainties in the solution structures. Here, we discuss these results while focussing on current methods for integrating certain aspects of dynamics and structure, to improve the accuracy of the description of the structures.

THREE-DIMENSIONAL STRUCTURES OF CALBINDIN D_{9k}

The family of 32 final $(Cd^{2+})_1$ calbindin D_{9k} structures refined by restrained molecular dynamics (rMD; Clore et al., 1986) is shown in Fig. 2. It is clear that the helical regions of

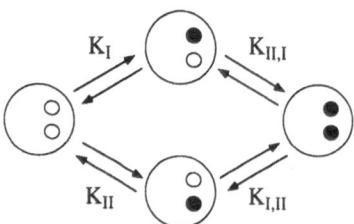

Figure 1. Schematic diagram of the calcium-binding events in calbindin D_{9k}. The microscopic binding constants are identified for each binding step.

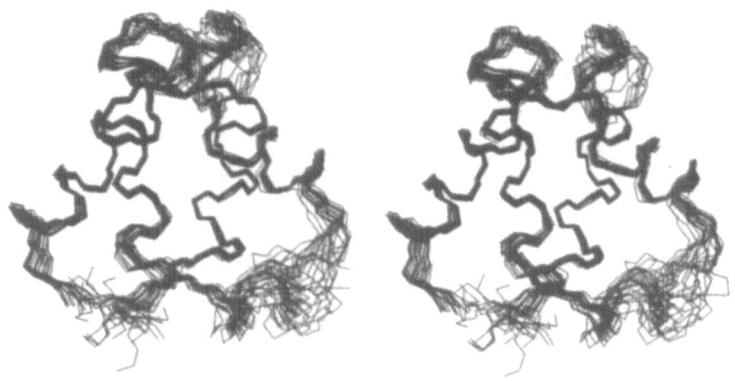

Figure 2. High resolution structure of $(Cd^{2+})_1$ calbindin D_{9k}. Family of 32 rMD refined structures reported in Akke et al. (1995).

the protein are quite well-defined, whereas the two binding-loops and the linker between the two EF-hands sample a greater range of conformations, as is also observed in the NMR structures calculated for the apo and $(Ca^{2+})_2$ states of the protein. An alternate means to visualize the spread in the NMR structural ensemble is shown in Fig. 3, where the smoothed C^α trace of the mean $(Ca^{2+})_2$ structure is shown. The diameter of the tube has been made proportional to the root mean square deviation (RMSD) among the members of the ensemble. The differences in the uncertainties in different regions of the molecule are associated with the lack of constraints in the poorly defined regions. This

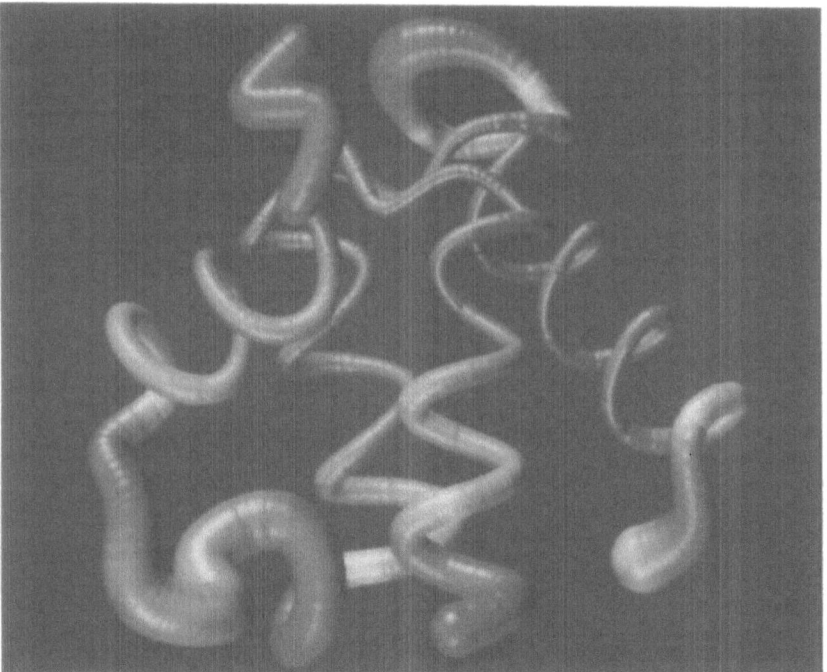

Figure 3. Tube representation of the structural ensemble of $(Ca^{2+})_2$ calbindin D_{9k}. The smoothed C^α trace of the mean structure is displayed with the diameter adjusted to be proportional to the square root of that residue's RMSD from the mean. The same parameter is also utilized for shading.

can be caused by geometric factors (e.g. less interresidue contacts at the surface versus in the interior), the lower proton density associated with the smaller amino acids, inability to identify or assign NOEs due to resonance degeneracies, or intrisically greater flexibility.

Figure 4 utilizes the variable tube display to show a comparison of the apo and $(Ca^{2+})_2$ states of calbindin D_{9k}. The path of the tube follows the C^α trace of the apo state coordinates, with the diameter proportional to the structural changes that occur upon ion binding. The shading of the tube has been made proportional to the sum of relative uncertainties in the two structural ensembles. The regions with very high uncertainties (N- and C-termini and in the linker between the two EF-hands) are not displayed, to facilitate observation of the changes that occur in the two EF-hands.

Comparisons between similar structures are in many cases difficult to make because the results will be dependent on the manner in which the structures are fit to eachother. Thus, comparing structures by calculating RMSD after best-fit superposition can be misleading. In contrast, utilization of a distance difference matrix approach allows for an unbiased comparison of structures (Akke et al., 1995). Figure 5 shows the application of the difference distance matrix methodology to the intercomparison of the apo, $(Cd^{2+})_1$ and $(Ca^{2+})_2$ states of calbindin D_{9k}, revealing the far greater change in structure associated with the first ion binding step. This finding has important implications regarding the cooperativity in ion binding for the pathway where the ion binds first in the C-terminal EF-hand.

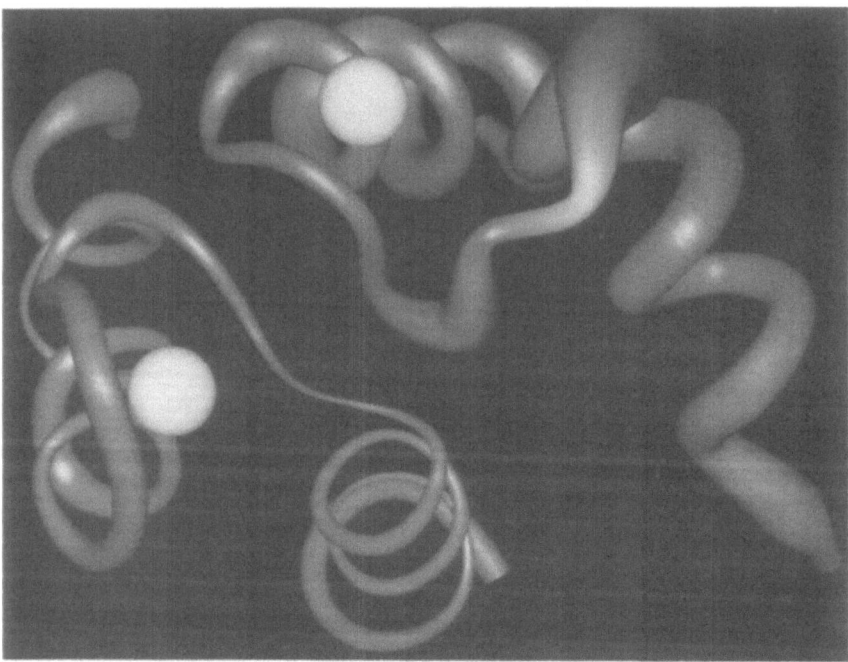

Figure 4. Tube representation of the structural changes that occur upon ion binding in calbindin D_{9k}. The smoothed C^α trace of the apo state is displayed with the diameter adjusted to be proportional to the RMSD between the mean apo structure and the mean $(Ca^{2+})_2$ structure. The N-terminal EF-hand is below and to the left and the C-terminal EF-hand is above and to the right. Reproduced with permission from Journal of Molecular Biology [Vol. 249(2), cover, 1995].

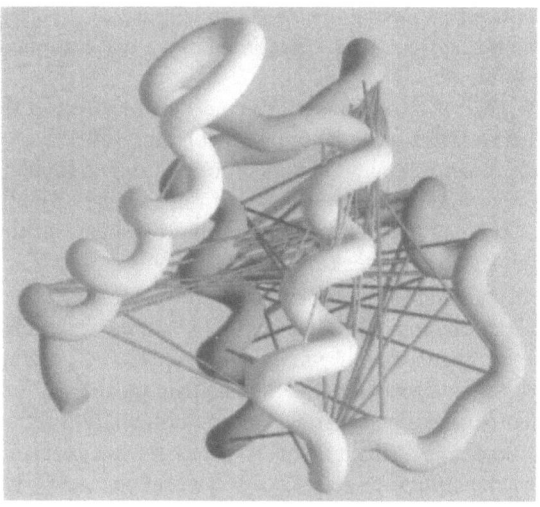

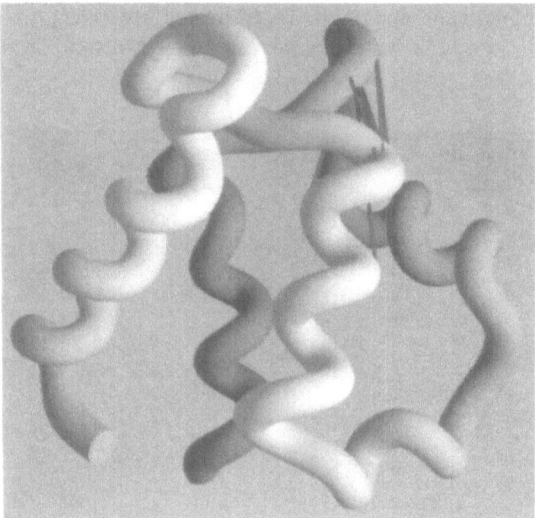

Figure 5. Distance difference matrix analysis of the conformational response to the two ion binding steps in calbindin D_{9k}. Lines are drawn between pairs of atoms having significant elements in the matrix. Comparison of apo and $(Cd^{2+})_1$ states at top, and $(Cd^{2+})_1$ and $(Ca^{2+})_2$ states at bottom. [Adapted from Akke et al., 1995].

TOWARDS HIGHER RESOLUTION STRUCTURES

The ultimate goal in the analysis of calbindin D_{9k} is to compare structures that are globally very similar to eachother. Consequently, it is absolutely essential to utilize all possible means to obtain the highest resolution and precision in these structures. The results from the comparative analysis of structure and dynamics of the protein indicates that certain regions, in particular the ion binding loops, have high uncertainties that do not arise from flexibility (*vide infra*). Certainly, one contributing factor to this uncertainty is the absence of ions in the rMD simulations; consequently, we have recently undertaken a series of refinements with the ions included. In order to incorporate the ions properly, it is necessary to also incorporate explicit solvent and a more complete treatment of electrostatics than is used in simulations *in-vacuo* (Figure 6). This greatly increases the number of atoms in the system, and correspondingly, the amount of time required to complete a cycle of refinement.

In these simulations, it is also important to consider the effect of the higher viscosity of water (relative to the vacuum environment) on the effective sampling properties achievable with a given set of simulation conditions. Longer simulation times or heating to higher temperatures is required for the solvated refinements. The primary benefit of the more sophisticated simulations is the desired improvement in the resolution of the structures in the binding loop regions (Figure 7).

One very significant additional pay-off from the inclusion of solvent addresses the issue of the approximations made in refinements *in-vacuo*, which can cause distortion of the conformations and reduced conformational sampling of certain surface sidechains. This has been observed as a tendency for the aliphatic portion of long sidechains to fold back onto the protein surface, maximizing the non-bonding attractive interaction energy. This effect is also reflected in a smaller than expected solvent accessible surface area. The surface properties of the structures refined in solvent appear to be significantly improved (Kördel, D.A. Pearlman & W.J. Chazin, unpublished results).

The addition of solvent and more sophisticated treatment of electrostatics in the refinement, allows a more reliable evaluation of the electrostatic properties of the protein.

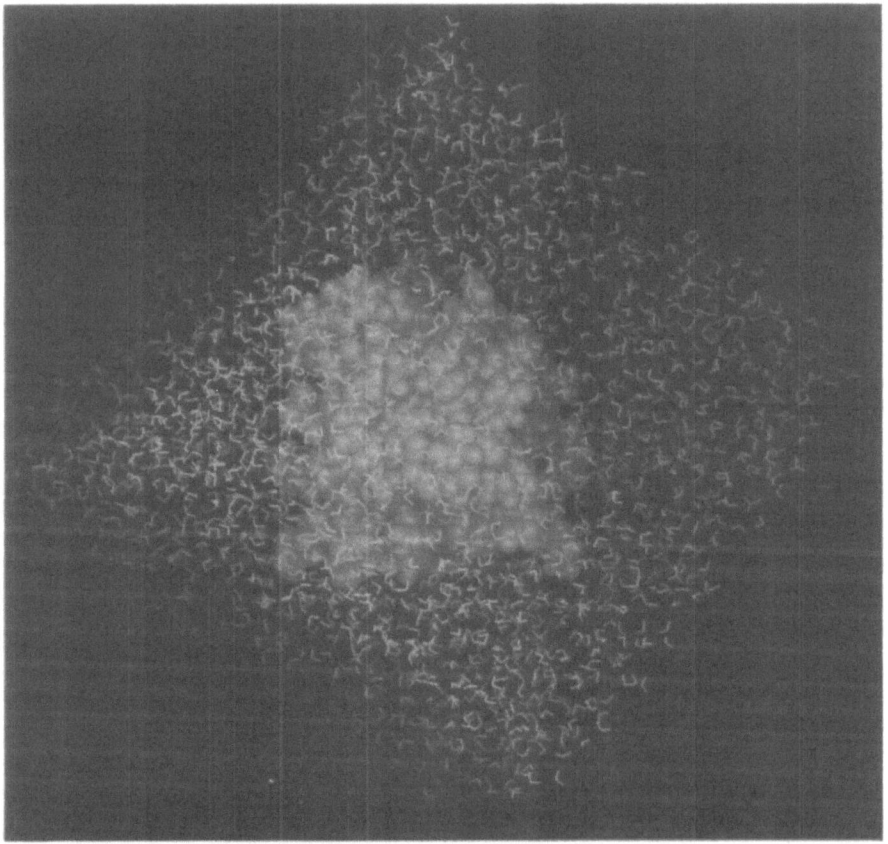

Figure 6. The calbindin D_{9k} molecule embedded in a bath of ~3600 water molecules after minimization of the solvent and a 50 ps restrained molecular dynamics simulation using the program AMBER (Pearlman et al., 1991a, 1991b).

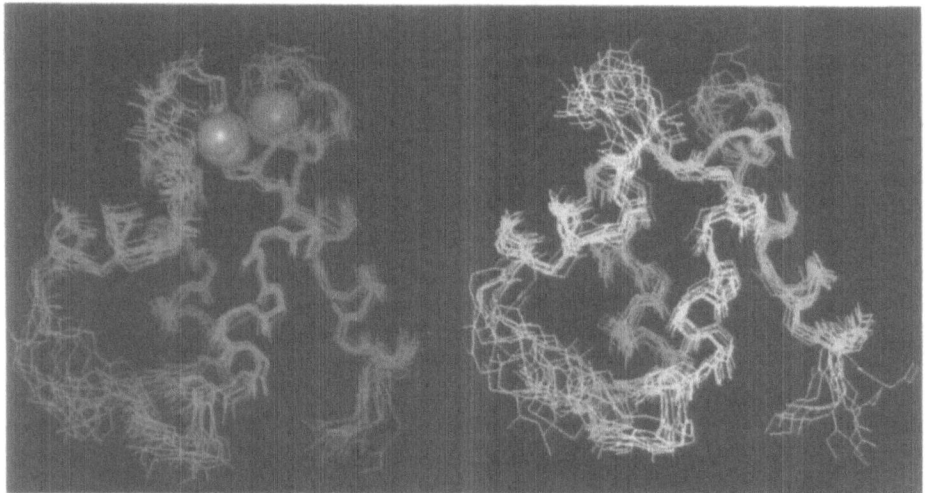

Figure 7. Comparison of rMD-refined structural ensembles calculated *in-vacuo* (Kördel et al., 1993; at left) versus when Ca^{2+} ions, a solvent bath, and an electrostatic treatment with full charges are incorporated in the simulation (J. Kördel, D.A. Pearlman & W.J. Chazin, unpublished; at right).

This is particularly important for the calbindin D_{9k} system, because the protein is highly charged, the charges are distributed in a very asymmetric manner, and the charge neutralization effects associated with the binding of one or two cations are expected to be considerable. Figure 8 shows preliminary evaluations of the structures refined *in-vacuo* using the program GRASP (Nicholls et al., 1991). While these results have raised some interesting questions, there remains a considerable uncertainty regarding the results of these calculations, given the approximate nature of the treatment of electrostatics in the refinement of the structures. Clearly, these observations must be verified by re-evaluating the electrostatic field distribution in the structures refined with ions and explicit solvent.

INTEGRATING DYNAMICS INTO BIOMOLECULAR STRUCTURE

The availibility of NMR parameters that report on the motional properties of the protein allows the assessment of whether increased flexibility is likely to be the cause of larger structural uncertainties. For example, ^{15}N relaxation has been used in a number of

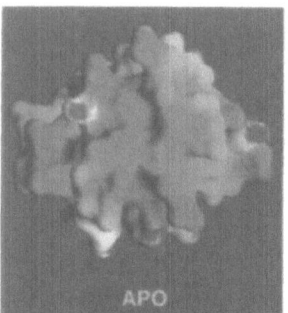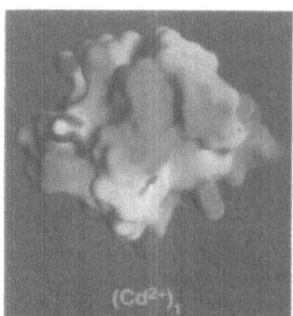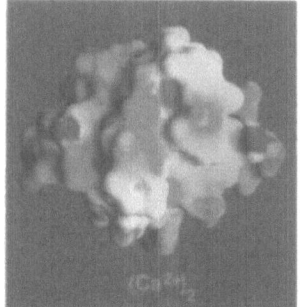

Figure 8. Electrostatic surface of calbindin D_{9k} in the apo, $(Cd^{2+})_1$ and $(Ca^{2+})_2$ states. This figure was generated using the program GRASP (Nicholls et al., 1991).

studies to assess relative flexibility of the protein backbone on the picosecond to nanosecond timescale (reviewed in Palmer, 1993), including the apo, half-saturated, and Ca^{2+}-loaded states of calbindin D_{9k} (Kördel et al., 1992; Akke et al., 1993). The values of the ^{15}N order parameter for the $(Ca^{2+})_2$ state are plotted in Figure 9, along with the RMSD of the structural ensemble. The variation in the RMSD is seen to have a general, but not strict, inverse correlation with the motional order parameter. Most of the ill-defined regions of the protein are seen to exhibit above-average high frequency internal motions. Despite this correlation, it is important to also consider the possibility that the structural uncertainty arises from sampling of multiple conformations due to motions that occur on a slower timescale, or a combination of both effects. Information about slower timescale motions can be obtained from a variety of different NMR measurements, including certain relaxation parameters (e.g. T_2 and $T_{1\rho}$) and amide proton exchange rate measurements.

The integration of structure and dynamics requires a survey of motional properties on multiple timescales because the various motional modes are not necessarily correlated. The values of the backbone amide proton exchange rates and ^{15}N order parameters in calbindin D_{9k} have been compared (Akke et al., 1993). There is a clear correlation between these two parameters in the helical regions, where slow amide proton exchange is correlated with high (>0.8) values of the order parameter. There is also a general correlation in the loop regions where amide exchange is rapid and order parameters are low (<0.8). However, there are also residues for which this correlation is clearly not evident. Furthermore, the comparison of these data for the three states of the protein

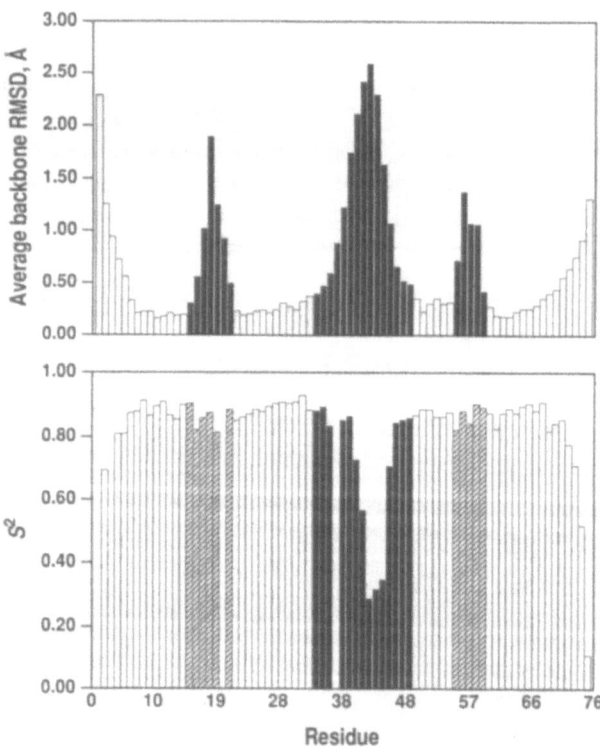

Figure 9. Comparison of the structural and dynamical parameters at the backbone amide nitrogen atoms of $(Ca^{2+})_2$ calbindin D_{9k}. The lower panel shows the ^{15}N order parameters determined from T_1, T_2 and NOE measurements (Kördel et al., 1992). The upper panel shows the average RMSD from the mean structure.

reveals that the differential effects associated with ion binding are spread throughout the protein, and that the changes that do occur are different at different motional timescales. These results stress that to obtain accurate insights into biological phenomena, it is critical to assay motional properties over a range of timescales.

CONCLUDING REMARKS

Using calbindin D_{9k} as an example, some insights have been provided into current efforts to increase the resolution and accuracy of high resolution NMR structures of metalloproteins. Assuming the description of the coordinate positions is sufficient, the next phase in attempting to correlate this information with biological function requires the incorporation of the molecule's motional properties. While only rudimentary examples can be provided here, this will certainly become an area of intense future development.

ACKNOWLEDGMENTS

This research was supported by an operating grant from the National Institutes of Health (RO1 GM40120). I thank the various graduate and postdoctoral students who have contributed to the studies of calbindin D_{9k} described in this chapter, in particular Drs. Nicholas J. Skelton, Mikael Akke and Johan Kördel. Beth Larson is acknowledged for assistance in the preparation of the manuscript.

REFERENCES

Akke, M., Forsén, S. & Chazin, W. J. (1991). Molecular basis for cooperativity in Ca^{2+} binding in calbindin D_{9k}. J. Mol. Biol. 220, 173-189.

Akke M, Forsén, S. & Chazin W. J. 1995. Three-dimensional solution structure of $(Cd^{2+})_1$ calbindin D_{9k} reveals details of the stepwise structural changes along the apo→ $(Ca^{2+})_1$ II → $(Ca^{2+})_2$ I,II binding pathway. J. Mol. Biol. In press.

Akke, M., Skelton, N. J., Kördel, J., Palmer, A. G. & Chazin, W. J. (1993). Effects of ion binding on the backbone dynamics in calbindin D_{9k} determined by ^{15}N NMR relaxation. Biochemistry 32, 9832-9844.

Carlström, G. & Chazin, W. J. (1993). Two-dimensional ^1H nuclear magnetic resonance studies of the half-saturated $(Ca^{2+})_1$ state of calbindinD_{9k}. J. Mol. Biol. 231, 415-430.

Clore, G. M., Brünger, A. T., Karplus, M. & Gronenborn, A. M. (1986). Application of molecular dynamics with interproton distance restraints to three-dimensional protein structure determination. A model study of crambin. J. Mol. Biol. 191, 523-551.

Forsén, S., Kördel, J., Grundström, T. & Chazin, W. J. (1993). The molecular anatomy of a calcium-binding protein. Acc. Chem. Res. 26, 7-14.

Kördel, J., Skelton, N. J., Akke, M., Palmer, A. G. & Chazin, W. J. (1992). Backbone dynamics of calcium-loaded calbindin D_{9k} studied by two-dimensional proton-detected NMR spectroscopy. Biochemistry 31, 4856-4866.

Kördel, J., Skelton, N. J., Akke, M. & Chazin, W. J. (1993). High-resolution solution structure of calcium-loaded calbindin D_{9k}. J. Mol. Biol. 231, 711-734.

Linse, S. & Chazin, W.J. (1995). Quantitative measurements of the cooperativity in an EF-hand protein with sequential calcium binding. Prot. Sci. In press.

Nicholls, A., Sharp, K. A. & Honig, B. (1991). Protein folding and association: Insights from the interfacial and thermodynamic properties of hydrocarbons. Proteins: Structure, Function, Genetics 11, 281-296.

Palmer, A.G. III (1993). Dynamic properties of proteins from NMR spectroscopy. Curr. Opin. Biotech. 4, 385-391.

Pearlman, D. A., Case, D. A., Caldwell, J. C., Seibel, G. L., Singh, U. C., Weiner, P. & Kollman, P. A. (1991a). AMBER 4.0. San Francisco, University of California.

Pearlman, D. A., Case, D. A. & Yip, P. (1991b). SANDER/AMBER 4.0. San Francisco, University of California.

Skelton, N. J., Kördel, J., Akke, M. & Chazin, W. J. (1992). Nuclear magnetic resonance studies of the internal dynamics in apo, $(Cd^{2+})_1$, and $(Ca^{2+})_2$ calbindin D_{9k}. The rates of amide proton exchange with solvent. J. Mol. Biol. 227, 1100-1117.

Skelton, N. J., Kördel, J., Akke, M., Forsén, S. & Chazin, W. J. (1994). Signal transduction versus buffering activity in Ca^{2+}-binding proteins. Nature Structural Biology 1, 239-245.

Skelton, N. J., Kördel, J. & Chazin, W. J. (1995). Determination of the solution structure of apo calbindin D_{9k} by NMR spectroscopy. J. Mol. Biol. 249, 441-462.

Wimberly, B., Thulin, E. & Chazin, W. J. (1995). Characterization of the N-terminal half-saturated state of calbindin D_{9k}: NMR studies of the N56A mutant. Prot. Sci. In press.

DISCUSSION

Yuan Xu - I have one question about your time average constraints. What is the number of conformations that you collected in your time average constraints?

Walter Chazin - What we did in that case was ten 100-picosecond simulations, each was averaged over the last 80 picoseconds, then the ten averaged structures were superimposed.

Xu - Do you assign equal probability to those ten conformers?

Chazin - They all have equal probability. The memory function is set so the restraints are satisfied on average over the memory period as opposed to at any particular instance; it's in essence a mathematical trick.

Xu - When you do the time averaged constraints, is it also possible to do the same thing using Monte Carlo simulations?

Chazin - Yes, I don't see why not.

Brian Sykes - I am interested in the liganding sidechains in the apo structure. You gave the impression or showed that they really did not move very much. But with all those negative charged sidechains in the loops, wouldn't they face away from each other in the apo state? Correspondingly, you would expect a real increase in flexibility in the apo structure.

Chazin - That's a multi-tiered question. The first point is that in calbindin D_{9K}, when you look at one site, you have a different answer than the other site. In our preformed site, if it's really going to be preformed, things need to be oriented in the proper way. The key issue in the N-terminal site is that it doesn't ligate in the same way as a typical calcium binding protein in that it utilizes a number of neutral backbone carbonyl atoms as opposed to charged sidechain carboxyl atoms. The other EF hand is a more traditional site, and there we see the larger changes. We also see a change in flexibility, at least on the picosecond to nanosecond timescale, that is much more significant. So, there are two stories, I think that this is very integrally related with the nature of calbindin and the fact that it's an S-100 protein; it has an unusual and atypical binding loop in the N-terminal EF-hand.

Sykes - I have one supplementary question. I really have no information but I am just interested in the fact that there can be protons on the carboxylic acids; maybe they have a much higher pKa and there are protons in there stabilizing them much like that of water molecules were stabilizing this morning's story. Have you done anything with water binding or anything like that?

Chazin - We have looked at the waters of hydration in calbindin in the calcium loaded state only and there we do see the two waters of hydration that are involved in ligation. In the C-terminal archetypal site it's at residue number 9, which is a water mediated ligand, and that water is there. There's another water that is seen in the crystal structure of calcium-loaded calbindin D_{9K} in the N-terminal ψ-EF-hand and that water is there also. With regard to evidence for unusual pKa's, we haven't done any measurements. Well, actually we've done measurements but we have not had a chance to analyze those measurements since that was the work of a high school student intern. The data were collected but not really analyzed.

Mengli Cai - I would like to ask an experimental detail. When you study the structure of the apoprotein, half-saturated and fully saturated, how do you prepare the half saturated protein?

Chazin - The trick there is that we're using cadmium instead of calcium. Cadmium it turns out, from an inorganic chemistry point of view, has a propensity to bind more covalently than calcium. So it requires a much more specific geometry for ligating than calcium and it doesn't fit so well in the rigid N-terminal EF hand. It is not optimized and so the magnitude of affinity, the binding constant, is three orders of magnitude lower for cadmium in the N-terminal site than the C-terminal site. So we have sequential binding. The first equivalent of cadmium goes into the C-terminal EF hand and the other hand remains empty. We have recently determined that cadmium still binds cooperatively though and that's a very important aspect in studying these half saturated states. I should mention someone's name, Sara Linse, who has worked a lot in this area and who really has done a wonderful job in looking at these cooperative phenomena by doing binding constant measurements.

Cai - All right, Thank you.

Gerhard Wagner - Walter, you had quite some discussion about your rmsd's regarding the comparison with the relaxation parameters. Now, rmsd depends very much on how you align the structures. How do you align the structures?

Chazin - When we do a fitting like for that diagram, where we compare against the order parameters, we try to be as general as possible. However, we try to use well-defined regions of the protein. In particular in the calbindin case, we have used the four helices and the well-defined sidechains for the fitting. When we try and look at the conformational consequences of binding in calbindin's case, we have an N-terminal EF hand that does not move much and a C-terminal EF hand which does. So we fit the N-terminal EF hand. We do the best fit there and look at what the change is in the C-terminal EF hand.

Wagner - But once you start an alignment, you have to identify a single structure used to align them. You said your structure has the lowest rmsd, or least violation?

Chazin - What we would do is take the ensemble, determine the geometric average, best fit the geometric averages and then go back, and best fit the ensembles to the superimposed means. If we take two structures and just compare two structures, we take the structure which is closest to the geometric average but which is still a solution, not a refined average, or the average itself, but the single representative structure.

Wagner - Walter, it is going into much detail, but since your geometric average is again dependent on how you initially align the structures, I just want to mention an alternate approach. We have once used a method which is independent of any particular structure. So we have used a bunch of structures (Tim Havel's idea), taken the distances for each of the structures, and then calculated the averages. It is similar to what Gordon Crippen did when he compared the structures.

Chazin - Right.

Wagner - Take the average distances, use them for distance geometry calculations and you get a cannonical structure which is very much distorted or can be distorted. It is not a realistic structure, but is an independent way of getting alignment completely independent of structure. So this can be done.

Chazin - Yes, I agree most whole heartedly. It is like the distance difference matrix method. I think this is a very important thing. Right now we do ourselves a bit of disservice by relying so much on the rmsd, until some of the other less fitting specific methods do become more common. They also have to be accessible and useable so to speak. I think the rmsd is only valuable for internal comparisons, as things change during the course of refinement or something like that, but otherwise they actually can be dangerous.

PHOSPHOTYROSYL PEPTIDE-ENZYME COMPLEXES: HOW MUCH STRUCTURE CAN WE GET FROM TRANSFERRED NOE's?

Carol Beth Post and Michael L. Schneider

Department of Medicinal Chemistry and Pharmacognosy
Purdue University
Heine Pharmacy Building
West Lafayette, IN 47907-1333

INTRODUCTION:

How vital cellular processes are exquisitely coupled at the molecular level to properly regulate cell division, growth and proliferation is one of the major issues in biological research. This coupling involves the transduction of extracellular signals via membrane proteins to control an entire cascade of biochemical events in the cell. A key element of signaling cascades is the control of protein-protein association through phosphorylation of tyrosine residues.

Membrane-associated protein complexes in their entirety are not yet amenable to structural study either by crystallographic or solution nmr methods. It is necessary to simplify the system in order to obtain 3-dimensional structural information of such complexes. One approach for structural study of a complex involving a membrane protein and a cytoplasmic protein is to define the interaction using a peptide model for the membrane-associated protein combined with exchange-transferred nuclear Overhauser effect spectroscopy (ET-NOESY). This approach can provide valuable structural information on complicated systems, although it is limited to defining the 3-dimensional structure of the small molecule peptide only. If the structure of the cytoplasmic protein has been determined previously, then the complex can be modeled assuming the interaction site is known. A peptide may involve less specific interactions with the protein, leading to conformational heterogeneity. Under such circumstances, the methods for determining structure from nmr data and for modeling the complex between the protein and peptide must recognize and account for the variant conformations, unlike the methods developed for a well ordered protein structure that strive to attain extremely high precision.

In this paper we describe a transferred NOE structure determination, coupled with a clustering analysis, to define the interaction of one such peptide-protein model of a membrane-associated complex: the binding of aldolase to erythrocyte band 3. Based on the results, we propose a mechanism for mediating protein-protein interactions via tyrosine phosphorylation that is driven by intramolecular electrostatic repulsion, which, because of the nonspecific nature of an intramolecular repulsion, could occur in other systems as a common feature of phosphorylation control.

Glycolysis levels in erythrocytes can be altered 30-fold by regulating tyrosyl-phosphorylation of band 3 (Low, *et al.*, 1993). Band 3 binds three enzymes essential for carrying out glycolysis (Figure 1): aldolase, glyceraldehyde-3-phosphate dehydrogenase (G3PDH), and phosphofructokinase (PFK) (Prasanna Murthy, *et al.*, 1981; Low, 1986; Salhany, 1990). These enzymes are tetrameric and approximately 140kDa. The formation of a complex leads to full inactivation of the enzymes, and is mediated through phosphorylation of band 3 at Tyr 8 (Low, *et al.*, 1987; Harrison, *et al.*, 1991) with p72^{syk}

tyrosine kinase (Harrison, *et al.*, 1994) and a red-blood-cell acid phosphatase (Boivin, *et al.*, 1986). Phosphorylation of Tyr 8 blocks binding of the glycolytic enzymes. Thus, a new type of metabolic control, based on neither metabolite feedback nor covalent modification of the regulatory enzyme has been proposed for erythrocyte glycolysis (Harrison, 1991).

The interaction of band 3 with aldolase was defined with a synthetic pentadecapeptide of the N-terminal residues 1-15 of band 3 bound to rabbit muscle aldolase using ET-NOESY (Balaram, *et al.*, 1973; Clore and Gronenborn, 1982). This band 3 peptide (MEELQDDYEDMMEEN) is a good model for the interaction since it binds and fully inactivates aldolase (Low, 1986).

EXCHANGE-TRANSFERRED NOESY STRUCTURE DETERMINATION:

In an exchange system involving dissociated and enzyme-bound states of a small-molecule ligand (*i.e.* peptide), the nmr observables chemical shift, linewidth, and NOE cross-relaxation, are averaged between the two states according to the molar fraction of each state, f_{free} and f_{bound}, and the exchange rate relative to the appropriate timescale.

$$\text{Enz} + \text{Pep} \underset{k_{off}}{\overset{k_{on}}{\rightleftharpoons}} \text{Enz - Pep}$$

In the case of an ET-NOESY experiment (Figure 2), the cross-peaks with measurable intensity are those from ligand-ligand proton interactions that are produced in the enzyme-bound state. Because of the short correlation time of a small-molecule ligand, the NOE of the dissociated state is either negligible ($\omega \times \tau \sim 1.12$) or builds-up slowly during the short mixing time of an ET-NOESY experiment. Furthermore, NOE intensities involving enzyme protons are generally too broad to be interpreted (aldolase mol. wt. $\sim$ 160 kDa).

ET-NOESY experiments are done with a molar excess of ligand to favor the free-state values for the chemical shift and linewidth. To facilitate the interpretation of NOE intensities, fast exchange with respect to the cross-relaxation rate, σ_{ij}, is desirable (Clore and Gronenborn, 1982, 1983; Landy and Rao, 1989; Campbell and Sykes, 1991; Ni, 1992; London, *et al.*, 1992):

If,

$$k_{on}, k_{off} \gg \sigma_{ij}$$

then

$$\sigma_{ij} = f_{free}\,\sigma_{free,ij} + f_{bound}\,\sigma_{bound,ij}$$

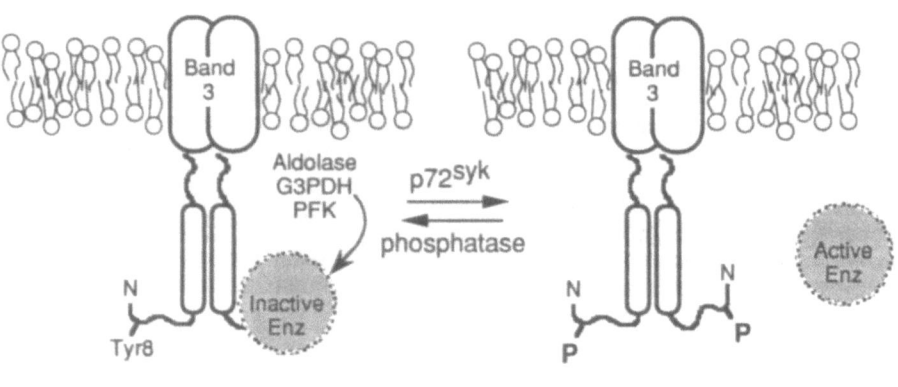

Figure 1. Schematic of band 3 association with the glycolytic enzymes, aldolase, G3PDH and PFK. Association inhibits enzymic activity and is controlled by tyrosyl phosphorylation.

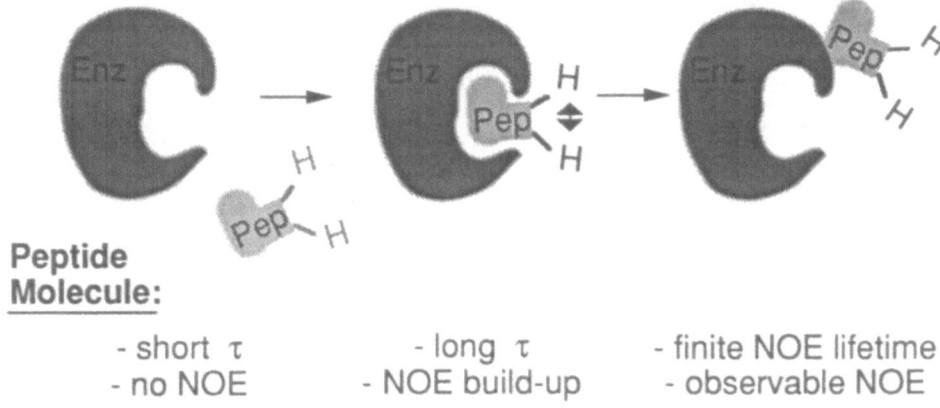

Peptide Molecule:

- short τ	- long τ	- finite NOE lifetime
- no NOE	- NOE build-up	- observable NOE

Figure 2. Cartoon of an exchange system.

The rate of dissociation for the band 3 peptide from aldolase (pH 5.5, 10 mM phosphate, 25 C) is estimated from the equilibrium inhibition constant to be approximately 520 s^{-1} (Schneider and Post, manuscript in preparation), which is in the fast exchange limit.

Three-dimensional structure determination by nmr of peptides bound to large proteins follows the same general methodolgy as that of proteins. In essence, distances and dihedral angles estimated from NOE interactions and coupling constants, respectively, are used as restraints in a search of conformational space, often implemented by molecular dynamics simulation, and the conformational solution is judged by its agreement with the nmr data. However, the nature of the problem differs. The peptide binds on the surface of the protein, and therefore experiences fewer NOE interactions than a protein molecule with a large portion of residues buried and fully surrounded by nearby spins. In addition, the conformational search by simulated-annealing molecular dynamics on a small-molecule ligand is not limited by packing restraints to the extent it is with a protein. Finally, as well as intramolecular spin diffusion leading to errors in distance estimates, *inter*molecular spin diffusion can contribute to observed ET-NOE intensities (Zheng and Post, 1993; Ni 1993). However, for most spin geometries intermolecular effects are negligible compared with other experimental sources of error (Zheng and Post, 1993). Furthermore, in this initial structural analysis of band 3 peptide, distances are estimated qualitatively such that errors due to spin diffusion are mostly irrelevant (Post, *et al.*, 1993).

The structure of band 3 peptide when bound to aldolase was determined (Schneider and Post, 1995) from ET-NOESY data and simulated annealing molecular dynamics. Several key interactions occur that determine important structural features of the peptide inhibition; examination of Figure 3 finds that there are many cross-peaks involving Tyr 8 Hδ's and Hε's (7.1 and 6.8 ppm), and Leu 4 methyl groups (0.8 and 0.9 ppm). NOE interactions between peptide protons were measured with 150 or 200 ms mixing times. Distance restraints were estimated from 68 NOE interactions between peptide protons, classified as strong (1.8 to 2.7 Å), medium (1.8 to 3.3 Å) or weak (1.8 to 5.0 Å), and enforced using a force constant of 50 kcal/mol/Å^2. Additional nmr restraints included fourteen φ dihedral angle restraints (-180 deg $\leq \phi \leq$ -30 deg) to maintain φ values in the range normally observed in proteins, and to aid in the convergence of the nmr structures. Molecular dynamics, including the nmr restraint terms in the potential energy function, was carried out at high temperature with X-PLOR 3.1 (Brünger, 1993). 200 structures were generated starting with initial coordinates for the peptide either in a right-handed α helix or an extended structure. These two very different sets of initial coordinates produced indistinguishable sets of structures, indicating that the results do not suffer from initial-coordinate bias, and conformational sampling is adequate. Further tests of sampling radius using molecular dynamics in four spatial dimensions (van Schaik, *et al.*, 1993) are planned.

The set of 200 structures was analyzed in terms of the number of NOE violations, the pairwise rms deviation between structures, and the X-PLOR energy value after energy-minimization. The 200 structures are conformationally heterogeneous, but an acceptable

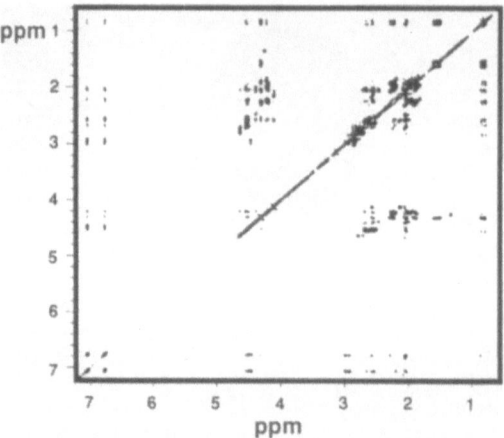

Figure 3. Exchange-transferred NOESY spectrum in D$_2$O of band 3 peptide and aldolase at a molar ratio of approximately 10:1, respectively.

subset of structures could not be defined based on the number of NOE violations alone. A large number of structures had no nmr violations greater than 0.2 Å (Figure 4), yet the average of the rms pairwise deviation among the 200 structures is ~ 4.0 Å. The conformational heterogeneity occurs in the ends of the peptide. That there are a large number of variant peptide conformations in agreement with the nmr data requires further consideration to define the structural characteristics of the bound band 3 peptide.

CLUSTERING ANALYSIS OF BAND 3 PEPTIDE NMR STRUCTURES:

To handle a large number of equally valid structural solutions, we have used a clustering analysis (Karpen, et al, 1993) on the bundle of 200 coordinate sets. In such an analysis, structures are clustered, or grouped, based on a set of criteria for the purpose of categorizing a large number of structures into a few clusters with members which are conformationally similar. Further evaluation of the nmr structures then may proceed within a manageable framework of a few distinct conformations.

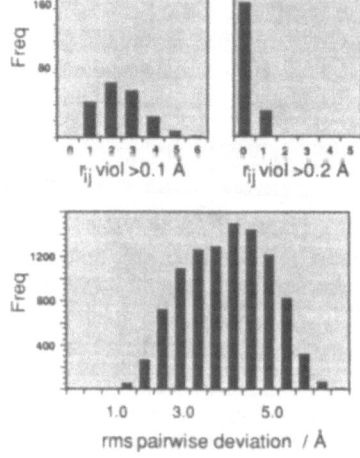

Figure 4. The distribution in the number of NOE violations greater than 0.1 Å or 0.2 Å, and the pairwise rms deviation in mainchain atom positions from the set of 200 initial, band 3 peptide structures.

Table 1. Clustering of 200 band 3 peptide structures generated with X-PLOR using 68 r_{ij} and fourteen ϕ restraints. Only the clusters with the largest number of members are listed. The top six clusters have low energy values and few NOE violations.

Cluster No.	No. of Members	Average[a] XPLOR Energy	Average No.[b] of Violations
1	20	2.61 kcal/mol	0.85
14	16	3.09	1.06
8	12	3.35	1.08
11	15	3.77	1.40
3	12	4.07	1.33
24	15	4.19	1.33
15	20	4.64	1.90
4	8	5.25	2.25
7	8	9.34	1.62

[a] The energy value averaged over the cluster after energy minimization of the X-PLOR potential.

[b] The average value over the cluster for the number of NOE violations where the actual distance is greater than 0.1 Å outside the restraint limits.

The choice of criteria on which to cluster is not unique, but determined from the specific application of clustering. After some trial-and-error of various internal coordinates, we found that for the band 3 peptide a suitable clustering was based on seven well-chosen Cα-Cα distances. With this clustering an apparently random set of configurations for the ends sorted into six major clusters (see below). The criteria were suitable as judged by cluster sizes of 12 to 20 members each (criteria that are too restrictive give a large number of clusters all with fewer than five members), small rms coordinate deviations within a cluster, and a small value for the cluster average of nmr violations and energy (Table 1). Future plans include more detailed evaluation of the structural results, within the framework of the clusters, in terms of agreement with the ET-NOESY data (predicted cross-peaks which do not exist), quality of the structure, and goodness of fit to aldolase in modeling the complex.

BAND 3 PEPTIDE STRUCTURE DETERMINED BY ET-NOESY:

The six major clusters (see Table 1) are shown in Figure 5. Representative structures of the full set of 200 structures are shown in Figure 5A and 5C with the mainchain atoms of residues 4 to 9 superpositioned. The peptide ends appear to be randomly oriented. However, the results of the cluster analysis (Figure 5B and 5D) show that there are six primary types of band 3 peptide conformations. A loop formed from residues 4 to 9 is a common feature of all the clusters (1.2 Å rms deviation between mainchain atoms). This loop is stabilized by a small hydrophobic patch comprising the Leu 4 methyl groups, the ring of Tyr 8 and the Met 12 methyl group (Figure 6B), and is well defined by the large number of NOE interactions involving Tyr 8 (see Figure 3). However, it should be noted that a general correspondence between the number of NOEs and the precision with which the local structure is determined need not hold in the case of peptide structure determination in exchange systems. Unlike many examples cited from protein nmr structural studies in which a small atomic-positional deviation for a region corresponds to a large number of NOE interactions, in band 3 peptide the entire loop structure 4-9 is well determined by Tyr 8 NOE interactions with Leu 4 or Met 12, while Asp 6 and Asp 7 have very few NOE cross-peaks (Schneider and Post, 1995).

The primary difference in the structure of the clusters for band 3 peptide bound to aldolase is the conformation of the N and C termini. Importantly, differential broadening of the band 3 peptide resonances in the presence of aldolase (Schneider and Post, 1995) shows that some degree of structural heterogeneity actually exists in solution. That is, the effective correlation times as a result of averaging the dissociated and enzyme-bound peptide states are

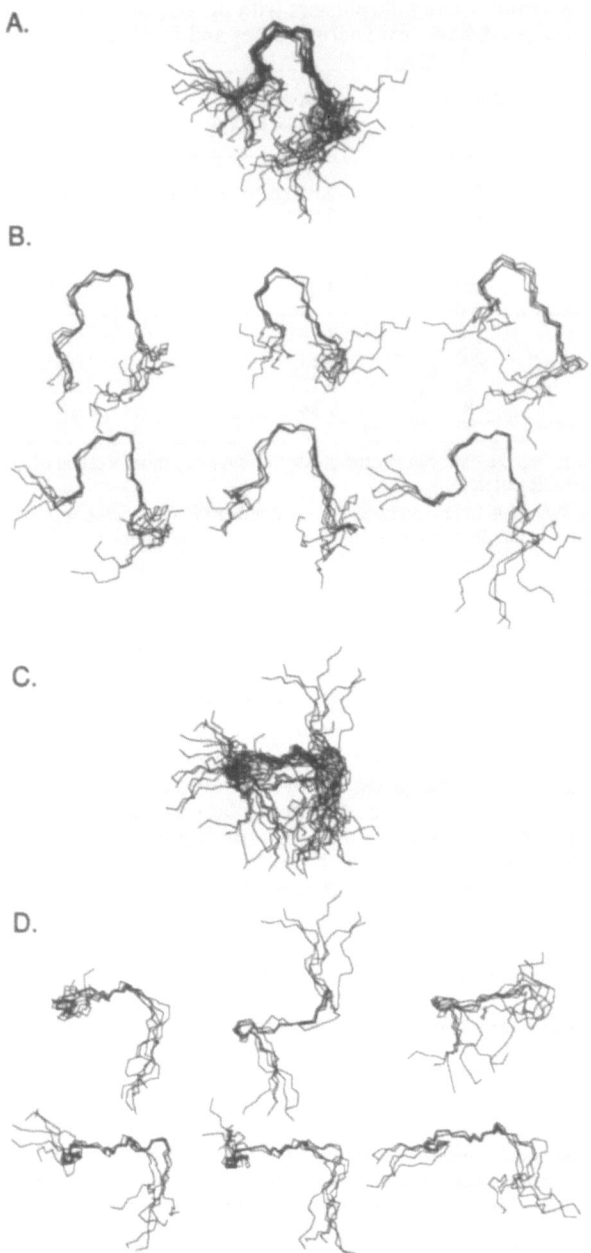

Figure 5. Nmr structures of band 3 peptide determined from ET-NOESY data and simulated annealing molecular dynamics. A and C: Superposition based on mainchain atoms for residues 4 to 9 of 30 structures out of the full set of 200 structures. B and D: A cluster analysis of the 200 structures gives the six major clusters shown. Structures in C and D are the same as those in A and B, respectively, viewed from the top, *i.e.*, after 90° rotation about a horizontal axis in the plane of the paper. The large heterogeneity apparent in A and C is simplified to a manageable set of six primary conformations shown in B and D.

unequal for different parts of the band 3 peptide. These variations in averaging indicated by the differential line broadening are due to either weaker binding interactions or multiple binding modes for the terminal residues.

MODEL OF THE BAND 3 PEPTIDE-ALDOLASE COMPLEX:

Although ET-NOESY provides structural information on the bound state of only the band 3 peptide, a picture of the whole protein-peptide complex can be obtained by modeling the peptide nmr structure with the crystallographic structure of unligated aldolase. Modeling of this type requires a proposal for the binding site, and assumes no significant changes in aldolase conformation from the unligated structure. We propose that band 3 binds the active site of aldolase since inhibition by band 3 peptide is competitive with the substrate, and the stoichiometry of binding is 4:1 for peptide molecules to aldolase tetramer (Schneider and Post, manuscript in preparation).

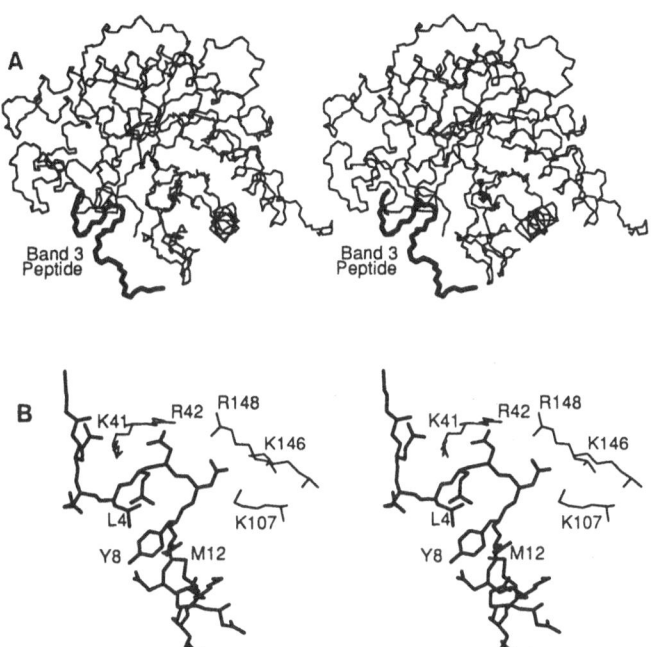

Figure 6. The band 3 peptide-aldolase complex modeled from manual docking of the peptide in the active site, followed by molecular dynamics and energy minimization. Initial coordinates were the peptide coordinates averaged over the members of cluster no. 1 (Table 1) and crystallographic coordinates for human muscle aldolase (REF). A. Mainchain tracing of one monomeric subunit of the aldolase tetramer (thin lines) and the band 3 peptide (thick lines) modeled in the active site groove. B. Hydrogen bond interactions between the aldolase residues K41, R42, K107, K146 and R148 (thin lines) and the band 3 peptide (thick lines). The hydrophobic patch stabilizing the peptide structure comprises L4, Y8, M12.

In Figure 6 we show a preliminary model of the complex between band 3 peptide and aldolase. The complex was modeled from the average structure of the cluster with the largest number of members (cluster no. 1, see Table 1) and the crystallographic coordinates of human muscle aldolase (REF). A more complete study to model the complex will include evaluation of band 3 peptide structures from all clusters consistent with the NOE distances and with good geometry. For this model, the common loop residues 4 to 9 were manually docked in the active site of aldolase using the graphics program QUANTA. Energy minimization and a short period of molecular dynamics were performed on the full-length pentapeptide, plus aldolase residues having an atom within 10 Å of any band 3 peptide atom.

Two points should be noted regarding the complex. Firstly, the loop of band 3 packs into a groove at the aldolase active site, while the ends of the peptide lie along the surface of aldolase (Figure 6A). Secondly, the complex includes a number of favorable intermolecular interactions (Figure 6B). In particular, Asp 6 and Asp 7 in the peptide hydrogen bond with Lys 41, Arg 42, Lys 107, Lys 146 and Arg 148 of aldolase. Lys 107, Lys 146 and Arg 148 are highly conserved in all type I aldolases and are needed to orient the substrate during the cleavage reaction (Sygusch, et al., 1987). Thus, there is good structural and energetic complementarity in this complex. Moreover, the modeled complex is consistent with the structural hetergeneity evident in the nmr data; fewer packing restraints on the peptide ends lying on the surface of aldolase is consistent with the disorder apparent in the nmr structures and the differential peptide linewidths, as discussed above.

The interaction of band 3 peptide with aldolase is unlike that in SH2 complexes where tyrosyl phosphorylation is required for the formation of the complex. The tyrosyl peptides bind SH2 domains with an extended structure and involve direct interaction between the phosphotyrosyl group and the SH2 domain (Waksman, et al., 1992; Xu, et al., 1995; Stanfield and Wilson, 1995). In contrast, the complex described here has a folded loop structure for the tyrosine containing peptide, and little direct interaction occurs between the tyrosyl group and aldolase.

DISRUPTION OF PROTEIN-PROTEIN COMPLEXES BY TYROSYL-PHOSPHORYLATION:

Phosphorylation of Tyr 8 of band 3 blocks the inhibition of aldolase activity (Low et al., 1987). The structure shown in Figure 5 would not explain exclusion of a phosphoryl moiety on Tyr 8 on the basis of steric conflict; indeed there is space to fit this bulky group. However, addition of a phosphoryl moiety is energetically unfavorable due to electrostatic repulsion between the phosphate oxygens and the acidic residues of band 3. That is, there are strong *intra*molecular repulsive interactions that destabilize the band 3 structure in the complex. Formation of the complex is thereby controlled by acidic residues remote to the tyrosine (peptide residues Glu 3, Asp 10 and Glu 13), while recognition for binding aldolase occurs through the adjacent residues Asp 6 and Asp 7. Such an intramolecular mechanism for control of protein interactions by tyrosyl phosphorylation is attractive by being a generalizable mechanism that could occur as a common feature in other systems.

REFERENCES

Balaram, P., Bothney-By, A.A. and Breslow, E., 1973, "nuclear Magnetic Resonance Studies of the Interaction of Peptides and Hormones with Bovine Neurophysin," *Biochemistry* 12:4695.

Boivin, P. and Galand, C., 1986, "The Human Red Cell Acid Phosphatase is a Phosphotyrosine Protein Phosphatase which Dephosphorylates the Membrane Protein Band 3," *Biochem. Biophys. Res. Com.* 134:557.

Brünger, A.T., 1993, *X-PLOR 3.1 Manual*, Yale University, New Haven, CT.

Campbell, A.P. and Sykes, B.D., 1991, "Theoretical Evaluation of the Two-Dimensional Transferred Nuclear Overhauser Effect," *J. Magn. Reson.* 93:77.

Clore, G.M. and Gronenborn, A.M., 1982, "Theory and Applications of the Transferred Nuclear Overhauser Effect to Determine the Structure of the Inhibitory Troponin I Peptide when Bound to Skeletal Troponin C," *J. Mol. Biol.* 48:402.

Clore, G.M. and Gronenborn, A.M., 1983, "Theory of the Time Dependent Transferred Nuclear Overhauser Effect: Applications to Structural Analysis of Ligand-Protein Complexes in Solution," *J. Mag. Res.* 53:423.

Harrison, M.L., Isaacson, C.C., Burg, D.L., Geahlen, R.L. and Low, P.S., 1994, "Phosphorylation of Human Erythrocyte Band 3 by Endogenous p72[syk]," *J. Biol. Chem.* 269:955.

Harrison, M.L., Rathinavelu, P., Arese, P., Geahlen, R.L. and Low, P.S., 1991, "Role of Band 3 Tyrosine Phosphorylation in the Regulation of Erythrocyte Glycolysis," *J. Biol. Chem.* 266:4106.

Karpen, M.E., Tobias, D.J. and Brooks, C.L., III, 1993, "Statistical Clustering Techniques for the Analysis of Long Molecular Dynamics Trajectories: Analysis of 2.2-ns Trajectories of YPGDV," *Biochemistry* 32:412.

Landy, S.B. and Rao, B.D.N., 1989, "Dynamical NOE in Multiple Spin Systems Undergoing Chemical Exchange," *J. Magn. Reson.* 81:371.

London, R.E., Perlman, M.E. and Davis, D.G., 1992, "Relaxation-Matrix Analysis of the Transferred Nuclear Overhauser Effect for Finite Exchange Rates," *J. Magn. Reson.* 97:79.

Low, Philip S., 1986, "Structure and Function of the Cytoplasmic Domain of Band 3: Center of Erythrocyte Membrane-peripheral Protein Interactions," *Biochimica et Biophysica Acta* 864:145.

Low, P.S., Allen, D.P., Zioncheck, T.F., Chari, P., Willardson, B.M., Geahlen, R.L., and Harrison, M.L., 1987, "Tyrosine Phosphorylation of Band 3 Inhibits Peripheral Protein Binding," *J. Biol. Chem.* 262:4592.

Low, P.S., Rathinavelu, P. and Harrison, M.L., 1993, "Regulation of Glycolysis via Reversible Enzyme Binding to the Membrane Protein, Band 3," *J. Biol. Chem.* 268:14267.

Ni, F., 1992, "Complete Relaxation Matrix Analysis of Transferred Nuclear Overhauser Effects," *J. Magn. Reson.* 96:651.

Ni, F. and Zhu, Y., 1994, "Accounting for Ligand-Protein Interactions in the Relaxation Matrix Analysis of Transferred Nuclear Overhauser Effects," *J. Magn. Reson.* 103:180.

Post, C.B., Meadows, R.P. and Gorenstein, D.G., 1990, "On the Evaluation of Interproton Distances for Three-Dimensional Structure Determination by NMR Using a Relaxation Rate Matrix Analysis," *J. Am. Chem. Soc.* 112:6796.

Prasanna Murthy, S.M., Liu, Theresa, Kaul, R.K., Kohler, H. and Steck, T.L., 1981, "The Aldolase-binding Site of the Human Erythrocyte Membrane is at the NH_2 Terminus of Band 3," *J. Biol. Chem.* 256:11203.

Salhany, J.M., 1990, "Erythrocyte Band 3 Protein," CRC Press, Inc., Boca Raton, FL.

Schneider, M.L. and Post, C.B., 1995, "Structure of a Phospho-tyrosyl Band 3 Peptide When Bound to Aldolase from Exchange-Transferred Nuclear Overhauser NMR," submitted.

Stanfield, R.I. and Wilson, I.A., 1995, "Protein-Peptide Interactions," *Curr. Opin. Struct. Biol.* 5:103.

Sygusch, J., Beaudry, D. and Allaire, M., 1987, "Molecular Architecture of Rabbit Skeletal Muscle Aldolase at 2.7-A Resolution," *Proc. Natl. Acad. Sci. USA* 84:7846.

van Schaik, R.C., Berendsen, H.J.C., Torda, A.E. and van Gunsteren, W.F., 1993, "A Structure Refinement Method Based on Molecular Dynamics in Four Spatial Dimensions," *J. Mol. Biol.* 234:751.

Waksman, G., Kominos, D., Robertson, S.C., Pant, N., Baltimore, D., Birge, R.B., Cowburn, D., Hanafusa, H., Mayer, B.J., Overduin, M., Resh, M.D., Rios, C.B., Silverman, L. and Kuriyan, J., 1992, "Crystal Structure of the Phosphotyrosine Recognition Domain SH2 of v-*src* Complexed with Tyrosine-phosphorylated Peptides," *Science* 358:646.

Xu, R.X., Word, M., Davis, D.G., Rink, M.J., Willard, D.H. Jr. and Gampe, R.T. Jr., 1995, "Solution Structure of the Human pp60[c-src] SH2 Domain Complexed with a Phosphorylated Tyrosine Pentapeptide," *Biochemistry* 34:2107.

Zheng, J and Post, C.B., 1993, "Protein Indirect Relaxation Effects in Exchange-Transferred NOESY by a Rate-Matrix Analysis," *J. Magn. Res.* 101:262.

DISCUSSION

Thomas James - Carol, with the final six model structures that you got which have a lot of similarity, you should be able to use your protein-ligand interactions, go back through your relaxation matrix calculations, maybe even come up with a new set of distances, perhaps even do structure calculations beyond that. Of course, if you went back through that, it should be entirely consistent with the models as you presented them. Have you done this sort of thing?

Carol Post - No, we haven't done that. We have just obtained the complexes. It is certainly the direction we're heading. The difficulty is that we know there is heterogeneity because of the titration. We know that the various amide regions have different effective correlation times, so they're either heterogeneous in conformation or they're literally bound less tightly and have faster correlation times. So, to really go back and reproduce the data will be difficult. I am somewhat worried that we do not have enough data to really distinguish all the possibilities. It is the usual thing that if you have enough parameters, you'll fit everything. I think it is quite clear though that the loop region is as we've defined it.

Thomas Hurley - I was just curious; in the molecular dynamics simulation, how is the structure of the enzyme treated during the simulation?

Post - For the complex?

Hurley - Yes.

Post - Michael fixed the region which is distant to the peptide and allowed only the region within about 15 $\overset{o}{A}$ or so of the peptide to move. Again, it was just an idea of trying to "jiggle" the system up a bit to get a better interaction, and the interactions turned out to be very nice actually. The interaction energies between the peptide and the protein are very low.

PANEL DISCUSSION

Topic: Structural Refinement and Dynamics

Members: David Case, The Scripps Research Institute, La Jolla

Marvin D. Kemple, IUPUI

N. Rama Krishna, University of Alabama, Birmingham

Carol B. Post, Purdue University, West Lafayette

Gerhard Wagner (Moderator), Harvard University Medical School, Boston

Gerhard Wagner - Let us go onto the last part of this session. It is a round table discussion which is entitled "Structural Refinement and Dynamics". Dr. Rao has asked me to moderate this session. I told him I have very little to say on this topic and then he asked me at least if I could moderate it. As a consequence, we sat together yesterday night and today at noon and planned a little as to how to organize this. We agreed that we have about five minute statements from the panel members about points that might be relevant to the topics of this discussion and after that we will have a general discussion. The panel members may wish to ask some questions and everybody from the audience is invited to ask questions and contribute to this discussion.

So I have a few thoughts that I would like to present to start the whole thing. If we have a round table discussion, then we can ask some questions and these questions can either be trivial and have answers that are known, or be difficult and we won't find the answers right now. So the goal of this discussion could be to find out what are the questions, to define the most pressing questions and then if we find such questions, one can go home and solve them.

Now I thought we can separate or divide the wide theme of this discussion into three topics and later on when we go into the general discussion, it could be divided up into these three topics as well. The first topic might be structure generation. What kind of questions are open there? For example, is modeling software that is available today able to generate structures satisfactorily? Is this satisfactory, do we have to worry about this or is this essentially something software companies should worry about? Should more convenient programs be developed? Another aspect would be, is the main emphasis now in gathering more data such as NOE's, J couplings, hydrogen bond locations or other constraints for example from spin label techniques, fluorescence or solid state NMR? How important are these different areas in gathering useful data. Another aspect that might be very important is the use of validation programs. Are any of you using validations programs? Once you've generated a structure, a validation program tells you whether it's a reasonable structure or not. How do we detect errors in structures or what are the crucial areas? Where do we have to be careful and where are we quite safe? Are the modeling programs self-correcting? If you make a mistake, will it show up in inconsistencies that allow the errors to be identified or is this not the case? I think there are some cases where you have to be very careful. Another aspect is that everyone should be encouraged to

NMR As a Structural Tool for Macromolecules: Current Status and Future Directions
Edited by B.D. Nageswara Rao and Marvin D. Kemple, Plenum Press, New York, 1996

103

submit their coordinates to data bases such as the Brookhaven Protein Data Bank (PDB) or assignments to the Madison biomolecular NMR database. Should we submit all the raw data which is apparently quite unrealistic? How much of the raw data, constraint lists, or NOE intensities should be submitted? Once you have a set of raw data and repeat the calculations with a software package of another group, can you reproduce the same structure or do you get different results? Some of you may have something to say about these aspects.

The second topic is refinement. What is a refined structure? The term actually came from x-ray crystallography where refinement refers to optimizing the structure vs. raw data. In our case it should be with respect to NMR NOE intensities and I think a very few of our structures are refined in this sense. If you use strong, medium, and weak NOE's and refine versus structure constraint lists, it is not a real refined structure in the crystallographic sense. Another aspect of this is that many of us use resolution of NMR structures which is really a crystallographic term. We should never use resolution. We should go, for example, to the number of constraints per residue that would be the term corresponding to resolution in crystallography. We should forget about resolution although it shows up in many titles of NMR papers. How important is refinement vs. raw data compared to refinement vs. constraint lists? Only a few groups do this refinement vs. raw data. Is it not relevant or is it just too complicated to do? We could discuss this. There is still no consensus about getting good R-factors. It doesn't make sense to get R-factors. Not everyone does back calculations. Are they not reasonable, or should we use them? Another aspect I would like to bring up is that if you increase the number of constraints, what is the limit of the rmsd? Does the rmsd go to zero or do we end up with a final value? So that is something one could discuss.

Finally we have something to discuss about dynamics. What have you really learned from relaxation experiments other than that there are some parts of the protein that are quite mobile. Many of you use the Lipari and Szabo model, which depends on an overall correlation time, an order parameter, and an internal correlation time. You do measurements of T_1, T_2, and NOE. The overall correlation time is not an individual parameter; it is something for the whole protein. So you have only two parameters to describe your local relaxation data. You measure the first two, T_1, and T_2, and determine S^2 and τ_e and then you measure the NOE and if it does not agree you add another parameter if you think it is relevant. In contrast to this we have developed a concept of spectral density where you measure many relaxation parameters and then you extract a spectral density function. In our experience with proteins, you can essentially see each of these amide resonances. You can describe the relaxation with a single Lorentzian spectral density function and it is not obvious that you have to involve several spectral density functions. You know very little directly about amplitudes of motions from these relaxation experiments. So these are just thoughts I wanted to bring up and maybe some of you have comments to say, so bring up other thoughts. Now I would like to go on and have Marvin

Kemple first, then Rama Krishna, Carol Post and David Case talk a few minutes and address the questions I have raised.

Marvin Kemple - My major interest is in dynamics and we're one of the groups that use Lipari and Szabo model. We're trying to see if the parameters make sense under various circumstances. I am just going to give you a summary not necessarily of what we find but of what everybody has found in general. Of course, first I flash on the famous relaxation equations. These are just the usual T_1^{-1}, T_2^{-1}, NOE and are written for ^{13}C but the ^{15}N expressions are essentially the same. Those are the quantities that most people measure. Gerhard measures several more. There are two mechanisms, the dipole-dipole interaction, and the CSA. The famous Lipari and Szabo spectral density that we use has three parameters that are somewhat hidden in these expressions but are the order parameter (S^2) and this τ which is a combination of the correlation time (τ_m) for the overall motion and that for the internal motion (τ_e). If τ_m is large compared with the internal motion, τ is dominated by the internal motion. So you have what Gerhard was mentioning, τ_m for the overall motion, S^2 which is a measure of the amplitude of the internal motion, and then τ_e which is a measure of the time scale of the internal motion. There is a similar parameterization of the fluorescence anisotropy shown at the bottom of the view graph which I will not say too much about initially.

The next view graph shows the spectral density again, just reemphasizing the three parameters. What I want to mention is that generally in the literature, τ_m values are reasonable. There is of course question of how to deal with anisotropic overall motion but in practice its neglect with proteins does not seem to have a significant effect on the analysis. Now there is one example, calmodulin, where the two lobes seem to have independent correlation times and that should comprise anisotropic motion, but it did not seem to cause much problem in the analysis. We have modeled most of the systems we have studied with just a single overall correlation time, and the numbers are good, I mean amazingly good. By this I mean consider the size of the molecule, plug its molecular weight into the Stokes-Einstein relation along with a reasonable estimate for the density and τ_m is often within 10% of what you expect. So τ_m seems to be a consistent parameter.

What about the order parameter? Most of the measurements have been made on protein backbones and the S^2 values are near one. In other words, the backbones are pretty rigid. Walter Chazin showed us some examples where there are isolated parts of a molecule such as in the calcium binding loops of calbindin where the order parameters were lower. Most of the time they are pretty large and it is not clear how significant the differences are from one site to another. For methyl groups on the other hand, the order parameters are all very small. What does that tell you? We've concentrated on aromatic moieties and those order parameter values are intermediate. By large values I mean ~ 0.9, by small I mean ~ 0.2, while aromatics are ~ 0.4 - 0.6. Tryptophan is usually near the upper end of the range. Apparently, the tryptophan ring doesn't have a great deal of room to move around in a protein. In tyrosine measurements we have made, S^2 values are consistent with a ring flip or two site jump model.

Now the other aspect of this is the comparison of fluorescence anisotropy and NMR relaxation results. We've measured fluorescence anisotropy and it can be modeled in the same way. It is sort of an arm waving model but it is reasonably consistent with the data. So we measure what's called the decay of the fluorescence anisotropy which is the difference of the parallel and perpendicular components of the fluorescence intensity, where parallel is defined by the initial excitation polarization. We can put in exactly the same parameters as noted and express the anisotropy decay in an identical form to the auto-correlation function of the C-H or N-H vector used in NMR. You then fit the anisotropy decay to a double exponential form and then make the comparison between NMR and fluorescence. What we've been consistently finding is that τ_m from fluorescence is usually smaller than τ_m from NMR. This may be primarily a concentration effect, I say on the order of 20%-50%. S^2 for tryptophan is pretty much right on. In other words the order parameter is nearly the same from the two techniques. The internal correlation time for some reason, which we do not understand yet, tends to be larger in the fluorescence measurements. So there are clearly differences here in the details. The Lipari and Szabo model is not appropriate when the internal correlation time is not in the motional narrowing limit, and it is curious that the τ_e values we're getting from fluorescence are not in the NMR motional narrowing limit. Now that could be a problem with the fluorescence measurements because there are some uncertainties such as the initial value of the anisotropy and so forth or maybe we just cannot measure short enough times in fluorescence because typically in NMR, the τ_e values are less than 100 ps which is in the motional narrowing regime. This in fact is what most everybody finds in NMR except for some isolated cases.

In summary there are still questions concerning the numerical values of the parameters. Ultimately how you can relate these values to molecular structure is not clear. Certainly there are correlations that people have seen and pointed out, such as undefined regions in NMR structures, having small order parameter values. That correlation is reasonable but somehow one would think that the fluctuations in the structure should be on a longer time scale whereas this order parameter applies to fluctuations on the subnanosecond time scale. It may be that these fluctuations encompass a large range of time scales.

Wagner - The next is Dr. Rama Krishna.

Rama Krishna - Thank you, Gerhard. The theme for this panel discussion is the refinement of structures and the dynamics and it turns out that we actually do have three different posters that cover some of the subjects. Poster number 55 deals with the effect of dynamics in 2D transfer NOE experiments. Poster number 11 that Dr. Yuan Xu is going to show this evening, deals with application of a variable target function method in refining protein structures. Poster number 21 deals with X-PLOR based calculations in studying protein structures. Dr. Jablonsky and Pat Jackson will be showing posters 55 and 21. Since I have only 5 minutes, I am going to stick to the effects of protein dynamics. Later on, if there's any question about the variable target function method, I will be happy to exchange any ideas on this method.

In many traditional 2D transfer NOESY experiments, there is an implicit assumption that the ligand is sensing basically the active site. As Carol Post mentioned in her talk, 2D transfer NOESY is a very interesting and a very powerful technique and needs no introduction. The question I want to address in this panel discussion is, what happens if you are dealing with an enzyme which is highly dynamical? For example, there are some proteins that show large hinge-bending motions. These are proteins which have very large domains and these domains are undergoing hinge-bending motions. For example, there is thermolysin which was studied by Brian Matthews' laboratory at Oregon. Thermolysin has two domains with about a 16^0 separation between them. They show the hinge-bending motion. Very recently lysozyme was a topic of interest in *Science*. A group from the University of California, Santa Barbara, performed atomic force microscopy measurements on lysozyme and showed domain motions of 5-10 $\overset{0}{A}$ lasting for 50 ms. This 50 ms is very close to some of the cross relaxation rates one expects in very large proteins. Maltodextrin binding protein is another very interesting example studied by Florante Quiocho in Texas. Here the hinge bending angle is about 35^0, which is very large. In this particular class of proteins, the motion can be slow. These motional rates may become comparable to cross relaxation rates (e.g. lysozyme). There is another class of proteins, with what I would call flexible active sites. An example is carboxypeptidase where once a substrate binds, a single tyrosine residue swings around in what has to be a very fast type of motion. Also human pnp (purine nucleoside phosphorylase) which was studied by the crystallographers at UAB, showed a swinging gate motion. So it's obvious that one really has to be careful in interpreting the 2D transfer NOESY type of experiments, and actually give some thought to the dynamics of enzymes one is dealing with.

To explain this, we have put together a cartoon showing a 3-state model for enzyme-ligand interaction. The first one is the free state. You have an enzyme with two separate domains and a ligand that is free or bound to the enzyme. In the open-state, it binds; let's say not in the active site, but at a weak binding site or a non-specific binding site, similar to the ones that Dr. Rao referred to this morning. The enzyme-ligand complex can go to the closed state once the hinge bending takes place. The ligand is now buried in the active site itself. When you look at this, you can make some predictions right away. If the rate of hinge bending is extremely slow compared to the relaxation rates, you can predict that the transfer NOESY experiment is going to measure only the conformation of the ligand in the open state, which is not really what we want. What we want is the closed state (containing the active site). On the other hand, if the hinge bending rate is very fast compared to the relaxation rates then you do get information about the active site also. So to address this problem further, we have developed what we call complete relaxation and conformational relaxation matrix analysis program. We are calling it CORCEMA and if you take the center two letters C and E out, what you are left with is the CORMA program. So this is basically an extension of the CORMA program of Tom James and his colleagues to include conformational exchange in an explicit fashion. It takes care of intermolecular interactions as well.

To demonstrate to you some of the very interesting results one finds from simulations, consider a ligand with two hydrogens A and X, separated by a distance of 5.5 $\overset{o}{A}$, and with a correlation time of 10^{-11} seconds. The ligand binds with the enzyme and the correlation time becomes 10^{-8} seconds, but still the distance is the same. When the closed state hinge bending motion takes place, the enzyme inserts a hydrogen between these two at 2.75 $\overset{o}{A}$ from each. This proton, as you can guess, has the effect of causing protein mediated spin diffusion in the closed state. Then we determine the NOE intensity as a function of hinge bending rate and the enzyme off-rate. You can see some very interesting results right away. When this hinge bending rate is extremely slow, the NOE increases somewhat but stops at close to 0.15 or so. So at this stage you are basically sampling the open state of the enzyme-ligand complex. But as the hinge bending rate increases, you begin to sample some of the closed state conformations as well, and the NOE eventually reaches a plateau. At this stage you have a weighted average of the relaxation rate matrices of all the three states as the effective rate matrix.

Additional simulations based on a thermolysin-ligand complex using the pdb files for this complex deposited by Brian Matthews' laboratory were done. This is a leucine inhibitor. It has two methyl groups M_1 and M_2, and we wanted to calculate how the M_1-M_2 NOE intensity changes if you try to simulate a hinge-bending motion in thermolysin. In the open state, alanine-113 of thermolysin is quite far from this inhibitor, but in the closed state it moves very close with distances now like 3.6 $\overset{o}{A}$ between the methyl groups. For very small hinge bending rates, the NOE is extremely small because once again you're sampling the open state of the ligand. As the hinge bending rate increases, eventually you begin to sample the closed state as well. In an example like lysozyme, where you have a 50 ms motional time scale involved, you're somewhere in the middle of the curve. You can begin to appreciate that the transfer NOESY depends quite a bit on the rate of hinge bending motions and one has to take those into account. There is a dramatic increase in the NOESY as the hinge bending rate increases.

So in conclusion, I am raising some issues. If you are dealing with a dynamical enzyme, such as thermolysin or lysozyme, it's better to use at least a three-state model, and one has to be cautious in the data interpretation. The active site conformation may not be accessible if the hinge bending rate is too slow. On the other hand, if you are dealing with a three state model where the hinge bending rate is very fast but you are trying to fit the data with a two state model, you are likely to come up with a virtual conformation. So this is an important point that one needs to consider in this kind of structural analysis.

Carol Post - I am going to keep this very brief. I am just going to pose a few questions that concern me in determining structures by NMR. First of all, I think that one has to keep in mind as I am sure many people do that the results from X-PLOR or CHARMM or GROMOS for a structure depend not only on the NMR data but also on other factors. There is a force-field dependence and the results also are sensitive to what protocol you actually use, how heavily you weigh your NMR constraints, how high a temperature that you start your simulated annealing, and how quickly or slowly you cool the system down.

My laboratory is starting to look at other techniques instead of simulated annealing such as adding an extra spatial dimension to give four dimensional dynamics as done by Crippen and then van Gunsteren. The point I am trying to make is that we have NMR data but it's an underdetermined problem as we're all aware and we need to keep in mind that these other factors are very influential in the result that we get out.

So one must have a good way to assess the quality of our structures, and again this is difficult. It is nice to say that we can take our structure, use a rate matrix approach and then calculate an NOE spectrum and then talk about the agreement between that calculated NOE spectrum and the measured NOE spectrum. If you realize the sensitivity in the distances is $1/r^6$, it doesn't take much difference in let's say the heavy atom position in your structure to get a large difference in the intensity of your NOE. Secondly, the effects of motion, again as we're aware and we have heard already, can be significant since we are not looking at a rigid molecule. How do we adequately model the motional effects? If there's not much motion going on, 80% of the structure perhaps doesn't have significant internal motion and will have small thermal vibrations; an average structure is a good representation. There are regions, albeit maybe only a small percentage of the total structure, where motions are significant. The NOE intensity is substantially affected by not just 10% but by factors of two to five. If you look at simulation results, as we have done, and other groups here have done, the internal motion, even on a picosecond timescale, in isolated cases can average σ to the extent that the observed NOE could vary by a factor of five in a few isolated cases. Even though this affects only a small number of distances, we wonder what the effect of those significant errors is on the final answer that we get. I don't think, at least for me, these questions are easy to answer and maybe some of the discussion here today can bring up the ways that we might address agreement between our structures and the measured data. Clearly it's not going to be as simple as one criterion, and perhaps John Markley can address some of these points, and what kind of information we need to put into a database of NMR structures that would allow anyone to judge the quality of the structure.

Wagner - Thanks, Carol. Finally David Case.

David Case - I like Gerhard's introduction because one of the nice things of being on a panel is that you can just ask questions and you don't have to answer them. The question I want to ask is - Can we go beyond just obtaining structures and simultaneously and, maybe even if we're lucky, automatically refine both structural and dynamics or disorder parameters? We are all used to having a number of input data and traditionally we can simultaneously work at the structural level with proton-proton NOESY's and J coupling constants. There are lots of other things that one would like to put in the pot at the same time and simultaneously refine against all the data and wind up with a model which includes both structural aspects and dynamics or disorder aspects.

There are three ways I've seen generally that this can be done. First, have an ensemble of discrete conformers which Richard Ernst talked about just this morning. There, the variables that you are trying to fit would be the coordinates of conformers, pathways to go from conformer 1 to 2 or from 1 to 3, and so forth, and maybe some overall rotational constants. If you have that information in the model (you can simulate almost all of these things), you could simultaneously try to refine everything. A second method that we have worked on, effective normal modes, is somewhat similar. Here you would refine an average structure but would add motion in the form of adjustable amplitude parameters. Again overall rotational times would have to be included. Roughly, the sum of these two methods is what crystallographers do to a greater or lesser extent. If you look at a crystallographic structure, you will see B factors (which are essentially like effective amplitudes) and for certain sidechains you'll see alternate conformations. So a pdb file for a crystal will often have alternate conformations for certain parts of the molecule and the B-factor's for almost all "heavy" atoms; that's their model. They have more data than we do. We're already near the limit of barely being able to obtain structures, so it is not obvious that we have enough good data to allow more adjustable parameters than just the structural parameters. There are also two questions that comprise a standard problem in fitting data, do you have enough data and is the better fit you get by adding more conformers worth the extra parameters you've added in the list? There are many ways we know of how to go about answering these questions. They are not easy questions.

A third method uses time averaged restraints. This can be considered somewhat differently in my view. In a way it is not a refinement procedure like the methods discussed above. It's rather a prescription that says you include all your data and interpret the restraints in a certain way. What you come out with is a trajectory whose collective properties can reproduce the initial data that you've put in. You cannot count how many extra parameters you've added in the way necessary to answer questions such as, "is the better fit you get from time average constraints worth the extra degrees of freedom that you've added?" So I think there are some very interesting questions about how to interpret results from time average restraints and, just as a little plug if you want to think about some software aspects to do all of these things, you might come look at poster number 48 in tonight's poster session.

Wagner - Thank you, Dave. Now we have been through the panel and we would like to go onto a discussion in these areas and/or to comments from the panel on what somebody else has said. At the same time we would also open discussion to the audience and you can make contributions or statements or ask questions. I would like to structure this in such a way that we first talk about general aspects of structure generation, validation programs and the like.

Anil Kumar - Are structural refinement and dynamics not contradictory?

Wagner - That is a good question, who wants to answer it? David?

Case - I don't know what the question is. I think I just said my view of it. I think we do have enough data especially if you think of simultaneously refining heteronuclear relaxation data like ^{15}N or ^{13}C data and ^1H - ^1H data. We have enough data to start making sense of that. Take for example Walter Chazin's molecule, calbindin. If there is a region of the protein where the heteronuclear order parameters are very low compared to the rest of the protein, that ought to have some implications about how we interpret ^{1}H-^{1}H NOESY intensities in that region as well. I think that we need to learn how to do that. Here is another example: we found that trying to simultaneously make sure that our structures predict correct T_1 values for protons in various regions for various types of protons adds additional constraints to what kind of structures we can get. That is not a very common thing to include in an NMR structure calculation but I think that clearly these things that we think of as order parameters and only dynamical parameters actually have structural consequences and we'll learn how to find those.

Post - I would like to make a quick comment. David made the statement that if we know there are low order parameters based on the N-H vectors, that suggests that ^{1}H-^{1}H NOE values measured in that same region might also be smaller. That's true but you also have to remember that each one of these is a very specific event to those two interacting nuclei i.e. you can have a ß proton motionally averaged in one case, let's say the ß$_1$ proton is motionally averaged with another proton but the ß$_2$ does not show motional averaging; i.e., it is a very specific pairwise interaction that is not easily generalized to a spatial regime. So it's even more complicated. It shows you that there could be a problem, but it doesn't tell you what the problem is.

John Markley - Since my name was raised and also it was suggested that I have answers, I would like to throw these questions back at the panel since we have an official panel of five experts. For those interested in a database for NMR macromolecular data and given the fact that methods for determining structures keep improving, what is the minimum set of NMR data that should be put into a database? What would be the ideal set of data to put in a data base?

Wagner - Who wants to answer this? I have an opinion. I think the database should have an ensemble of structures that has been calculated and presented in the publication. It should contain a list of restraints used for these calculations. The list should include the chemical shifts so that everybody can use the information. I think it does not make sense at the moment to put NOE intensities. This would be unrealistic. I think Tom James wants to make a comment.

Tom James - Why not the intensities?

Wagner - If you want to put in data for all mixing times, intensities are very difficult to measure because you have base line offsets and all those things. Insofar, I think it is rather unrealistic at this point to put in NOE intensities but I may be wrong.

James - I think you've raised another point regarding data. Why is it that people don't take just a little bit more care with their base lines and in turn analyze their data a little more carefully? Then you can put your intensities in there and then other people could do something with it in case they wanted to use a different protocol for example. This goes back to the question that others have raised. I think it was Carol who put out the idea that the structures we get could be protocol dependent. Well, not too many people try different protocols in terms of the structures. Actually, in our particular case we have used restrained molecular dynamics and restrained Monte Carlo calculations on the same set of data. Fortunately, we got the same structures, but, nevertheless, the data could be available for other people to try their own favorite method. Presumably if the NMR data is actually giving us structures which we are all presuming it is, then various researchers should come up with pretty much the same set of structures.

Joshua Wand - Just to go back to the issue of the Lipari and Szabo approach, not that Atilla needs defending necessarily. Actually in the original treatment in 1982, anisotropy was dealt with though it seems to be raised as an issue. There are a number of examples in the literature where the data are freely fit and it can be determined whether or not tumbling is isotropic or anisotropic. The failure actually of the simple treatment of Lipari and Szabo for internal motions strictly is so far due to the fact that you get time scale separations and the Pade approximation fails. I don't think that Atilla is wrong. It is very useful, which I'll talk about a little bit in my talk, to generalize Lipari and Szabo for pairwise interactions between protons. When you do that, you can end up actually with a rather simple measure of the variance of the distance between those two protons. That is what we are going to affectionately call the missing B-factor. So I think that there is a lot of hope for using Lipari and Szabo which on one hand is actually kind of distasteful because it gives a picture of proteins which we know not to be true. For example, if you have one internal correlation time, which is really saying that you have only one motion, such is not true and there are at least three people on the panel here who have spent their life showing that this is not true. So I think that's the next issue that we need to face. This gets finally into the idea of dynamics and structure - are they incompatible, do they mean the same thing? What I think NMR spectroscopists need to do is to fit the idea of dynamics into the context of a real working model of how proteins move. For example, we need to decide in the Fraunfelder-Wolynes language, are we in tier zero or in tier one? If we are in tier zero, we are in a nice well; and we are talking about an average structure and fluctuations about it is meaningful. If we are hotter than that in the NMR experiment, then we are in trouble because we are looking at mixing of states that have relatively large barriers between them and therefore may be structurally different.

B. D. Nageswara Rao - I am quite confused about extraction of structural data when there is more than one conformation present. When there is more than one conformation present, there are twice as many distances and you need twice as much data. It looks like this has not been very clearly stated. It appears some people have given thought to this and have found some ways of dealing with it. Twice as much data is hard to obtain in every case and they're able to make some perhaps credible models. I don't fully grasp that, but if there is a rapid exchange between the available conformations, what you get as data is also not acceptable as was discussed many times. In fact just this morning, Richard Ernst showed to what extent one has to go to collect many more sources of data even for a small peptide as he has done. Many people are dealing with much larger molecules and you see papers where models are given where some exchanges are going on. How credible are these conclusions? I would like someone in the panel to address this.

Wagner - We are not quite sure who should answer this question. As far as I have understood it, your concern is the importance of exchange between different conformations going on in larger proteins. You need many more parameters to characterize two conformations and how credible is such a description? From my understanding, in many cases for larger proteins where we have indications of multiple conformations, these are important for a very small part of the molecule. You need more constraints only for this part of the protein, and whenever such exchanges are observed, people usually don't try to characterize in detail these two conformations but just to pinpoint multiple conformations in this area. Maybe I have misunderstood the question.

Rao - Maybe I can define the question. If one can take, for example, transfer NOE type of experiments where you are using a small set of protons maybe 10, 15 or 20 and where we hope to get the structures in a more reliable fashion. There you measure 20 proton pair NOE's, you cannot possibly measure more than 20 distances even if you are lucky. But if you say there are two conformations, how are you going to get enough data to get those two conformations?

Case - The question of uniqueness of models always comes up in these cases and I think, if you look at the Medusa results on some of the things that Carol talked about, in fact, you don't get unique models in the sense that one might think. But when you do the modeling multiple times, then you find that generally there are consistent features among the various models. We tend then to believe the consistent features, e.g., how the different conformers are similar or different from one another, as being real. Those things that never come up the same two times in a row are not given much credence. That's probably a general answer to the general question.

Post - I don't think you can define the entire conformation but perhaps it would be fruitful to really spend some effort to pick particular interactions if you have some preconceived notion of what the dynamics might be. Richard Ernst has beautifully shown how well you can do if you really try hard. If you characterize specific motions then I think you have some chance of not defining necessarily the whole conformation for the two states but perhaps very specific regions. This is somewhat like what we're doing when I referred rather mysteriously to this puckering conformation where you look at only two vectors, but by the nature of the chemistry of the ring those two vectors, allow you to look at the puckering motion of that ring. I think if we start to look at specific interactions, let's say in calmodulin, now that Walter has a very beautiful start on going from apo to 1 calcium to 2 calciums, that model predicts particular motions. If you go now and look very carefully at particular NOE interactions, you might be able to define something about the dynamics. That way you start to learn something from the dynamics rather than just deriving structures.

David LeMaster - I just wanted to comment in terms of bringing up the issue of modeling the dynamics. There is a question of whether or not to consider anisotropy. One point here is that there tends to be confusion of when someone brings in the idea of anisotropic motion, what model they're using. In the Lipari and Szabo paper initially they try to model anisotropic motion with a mixing term which was, as with the rest of the analysis, independent of the details of the molecule. A decade earlier however, Woessner in particular but others had addressed the same type of questions as to how to probe anisotropy when in fact you have a model in terms of where the diffusion tensor is with respect to the relaxation parameters that you are measuring. The model was applied to a number of molecules. In the discussion earlier I was obviously referring to these approaches which have the weakness that you have to have a fairly decent model to deal with. But one shouldn't try to over-read the Lipari and Szabo system in the sense that it wasn't pretending that it could match structural information in a model when in fact it was only applying to parameters that could be applied independent of the molecular structural details and so the two anisotropic analyses are really quite distinct.

Wagner - I would like to point out one problem with Lipari and Szabo model. This initial derivation implies that internal motions are uncorrelated with the overall motions of the protein. Now let's consider an example. You are sitting in a barrel that is floating on the Niagara river and have your hands moving. These are the internal motions, so you're essentially the protein, and once you go over the falls, then the internal motions will certainly be correlated to the overall motions. This applies very well to proteins. The nature of Brownian motion is that all internal motions are essentially related to the hits/collisions of the solvent with the protein. So I am wondering how good the assumption is that internal motions are uncorrelated from the overall motions. In particular, if you have a flexible tail of a protein, the motion of this tail might certainly be correlated to the overall motion of the protein. I don't know the answer to this question.

Dale Mierke - I just had one comment about fast internal motions and dynamics between say two conformations. One approach that we have utilized involves both NOE's and coupling constants. The result is that the NOE's will average much differently than the coupling constants. The NOE's will average as r^{-6} while coupling constants will average as trigonometric functions of bond angles. We have utilized this method to unambiguously identify conformational averaging fast on the NMR time scale.

Wagner - Are there any more questions? We still have 5 minutes.

Case - It was mentioned earlier by Marvin Kemple and we've noticed, not so much me personally but people at Scripps, that there do seem to be systematic differences between overall tumbling times measured by fluorescence and by NMR. This is really a question. Does anybody really understand this? Is it real? The differences are large enough to really start to make the NMR analysis more complex than it would be if everything agreed. You can talk to me later. Nobody has to answer right now. I am still confused by this difference.

Wagner - Any comments?

Ad Bax - I would like to make a brief comment on that since we have just done some measurements for calmodulin actually. There was a discrepancy. Fluorescence anisotropy gave a correlation time at 20^0 C of about 10 ns, whereas at 35^0 C by NMR, we found a correlation time of 6.5 or 7.3 ns depending on the protein domain. We could present a hand waving argument and say that it is due to the difference in temperature but it doesn't quite get you there. If you do the NMR measurements at 21^0 C, as we did very recently, it turns out we get exactly the same from the two techniques as Walter showed from fluorescence in his anisotropy measurements within 0.2 of a nanosecond. So I believe that the systematic difference may not necessarily always be there and might be due to protein concentration effects.

Walter Chazin - Except that there have been measurements in calbindin where we looked at the concentration effects in fluorescence and the response is very flat and it is still consistently different at least in that one case.

Wand - I think if you go back to the old literature you will find that it is different sensitivity to dielectric relaxation that is important.

Carol - I was just going to ask Ad what his system was . I didn't hear. Can you explain?

Bax - This system was calcium modulated calmodulin; fully ligated calmodulin. Calbindin is a very different case. They both bind calcium and that's where the similarity ends, at least from the dynamic view point.

Yuan Xu - I just want to raise three simple questions about the NMR structure and see if I can get any definite answers. The first question is, that is it possible to determine the polypeptide conformation or protein conformation by using NMR data alone without using any force fields? The second question is, what independent variable is more beneficial when you are doing the structural refinement? Is it proton-proton coordinates or just dihedral angles? The third question is, is spin diffusion really that bad or could it help your structural refinement?

Wagner - I did not get your first question.

Xu - The first question was, is it possible to determine protein or polypeptide structures by using NMR data alone without using any force fields? That might get rid of the problems of how much weight you want to put on your energy functions whatsoever.

Case - I think that's the easy question because the answer is known.

Julie Forman-Kay - I just had an answer to that question which was the opposite of yours where I recently heard a talk from Pierre Pralice on an approach he calls 'ANSWERS' and he uses only the NOE information to do some sort of a simulated annealing of free points in space and he isn't using potential functions at all. There is no covalent geometry. There is no peptide backbone and only after these points have been annealed into some kind of a structure so to speak of random points in space then there's an equivalent to a crystallographic chain trace as one might say. Then you can even begin to talk about the effect of potential energy functions. I think that this approach is one which might also begin to answer some of the other problems that I have noticed in NMR structural refinement such as the extreme sensitivity to errors in long range NOE assignments. Even if you know the assignments of the resonances, there can be extreme ambiguities in the long range NOE assignments in an approach where you don't a priori assign but you just anneal the structure and then chain trace. If it takes into account chemical shift information in the assignments, later on it might end up to be much more accurate. I think it's an interesting approach that more groups should try.

Wagner - Pierre Pralice's approach is a concept only, and has not solved any structure as far as I know.

Krishna - I think I would like to answer Dr. Yuan Xu's first question. It seems to me it should be possible to come up with at least some sort of a crude structure just using NOE intensities in the standard equations that relate NOE intensities to distances. Once you come to a certain point, then you have to answer questions like whether there are stearic overlaps or not. At some point or the other, you need to do some energy minimization. With regard to your third question about spin diffusion. Spin diffusion is very important and one should exploit it as much as possible. If you are allowing for spin diffusion using a complete relaxation matrix treatment, then you are working with very long mixing times and sensitivity is very high but at the same time one should be very careful in the data interpretation.

Post - I just had a comment about Pierre Pralice. My understanding, and I could be wrong, but I think he is actually using the angles, and bond lengths and geometry of amino acids in order to fit his mapping together. So he does use geometric information in his work and also considers non-bonding interactions. He does not let two atoms overlap in space. But I think if you refine the structure, you're going to need a force field. The second point is I think he would also have the same errors in misassigning long range NOE's because he doesn't assign the NOE but he says proton X has an interaction to proton J. I do not know what X and J are but if he has messed up that X really interacts with K and not to J, his algorithm would also have a problem. It would be a matter of self consistency if his proton has 10 NOE's and one of them is really not to that proton.

Wagner - I have one question for the modellers. It has been pointed out that potential functions are very important to the outcome of the structures. On the other hand, I don't think that the potentials used by molecular dynamics people would be capable of predicting for example, infrared frequencies or a Raman spectrum and they are much more sophisticated potentials. Why is nobody using these programs or potentials that have been developed in the spectroscopy field, Raman or infrared? Does anybody have an answer to that?

Case - I think those spectroscopic potentials are not force fields in the way that we think of force fields. I would like to calculate the infrared spectrum of a protein from general principles but I think that the spectroscopic force fields tend to be very localized and give you local information but not to do so well with what we think is driving protein folding or macromolecular structures, which tend to be electrostatic and Van der Waals and long range interactions. I think you're right in the sense that the field needs to move toward getting these two things closer together, but if I want you to do it right now, I would not know what to do. So maybe that is why I have not done it.

Wagner - So, we are close to the end of this hour. Are there more important questions? If this is not the case, I thank everybody for contributing to this discussion and hope you have learnt something. Thank you very much.

RECENT DEVELOPMENTS IN PROTEIN NMR SPECTROSCOPY

Stephan Grzesiek, Geerten W. Vuister, Andy C. Wang, Frank Delaglio, and Ad Bax

Laboratory of Chemical Physics, National Institute of Diabetes and Digestive and Kidney Diseases, National Institutes of Health, Bethesda, Maryland 20892-0520.

INTRODUCTION

For the last fifteen years, protein NMR spectroscopy has been a field which continues to develop at a rapid pace, fueled by significant advances in NMR pulse sequence methodology and improvements in spectrometer hardware. Although the concepts on which modern NMR pulse sequences rely have been well understood for over 40 years, interesting and useful applications of these old ideas are driving the field of biomolecular NMR to a level where it is possible to study both the average three-dimensional structure and the internal motions of biological macromolecules at a very detailed level. Below, three of such methodological advances are addressed. However, as detailed accounts of the principles and applications underlying these experiments have been or are being published in the original literature, only a brief overview will be presented.

AUDIO-MODULATED NUTATION FOR ENHANCED SPIN INTERACTION

Phenylalanine residues in proteins are usually packed in the hydrophobic interior. The aromatic ring protons, and in particular the H^ζ proton, typically exhibit disproportionately large numbers of long range NOE interactions. The resonance assignment of the Phe ring protons can be difficult, however, particular when resonance dispersion is poor and when more than half a dozen such residues are present in the protein. For uniformly ^{13}C-enriched proteins, ^{13}C-^{13}C COSY-type magnetization transfer steps can be used to correlate the $^{13}C^\beta$ resonance with the aromatic ring resonances (Yamazaki et al., 1993). In practice, however, this type of procedure is inefficient for transferring magnetization to the $^{13}C^\zeta/^1H^\zeta$ position. In part, the difficulty in using COSY type relay steps stems from the strong coupling effects among the aromatic ^{13}C resonances, which give rise to non-first-order spectra and adversely affect COSY type relay of magnetization. These strong coupling effects are caused by the relatively large ^{13}C-^{13}C one-bond J coupling in aromatic ring spin systems (~55 Hz), and by the small differences in chemical shifts of the ring carbons. Isotropic mixing experiments (Braunschweiler and Ernst, 1983) are not adversely affected by strong coupling effects. In fact, they are designed to cause the spin system to be "infinitely strongly coupled" during the isotropic mixing period. For systems with little resonance dispersion and large J couplings, such isotropic mixing is therefore easy to establish. However, for Phe residues, the main problem in using isotropic mixing for transferring magnetization from $^{13}C^\beta$ to the aromatic ring resonances is caused by the very large difference in chemical shift (~100 ppm) between $^{13}C^\beta$ and $^{13}C^\gamma$. Current pulse schemes for homonuclear cross polarization or isotropic mixing are insufficiently broad-banded to achieve effective magnetization transfer over such a wide band width. A new type of modulation scheme is proposed which can achieve this type of transfer with a reasonable degree of efficiency. This new scheme uses *audio modulated nutation* for *enhancing spin interaction* and is referred to as AMNESIA. Its theory and application are described in detail elsewhere (Grzesiek and Bax, 1995). Below, only a brief sketch of the underlying idea will be presented.

The basic idea behind the AMNESIA experiment is to quantize the ^{13}C spins along an effective field, which is the vector sum of the applied RF field and the resonance offset, and to apply audio frequency pulses orthogonal to the effective field. If the ^{13}C carrier is positioned midway between the $^{13}C^\beta$ and $^{13}C^\gamma$ resonance positions, at about 88 ppm, the effective field strengths experienced by $^{13}C^\beta$ and $^{13}C^\gamma$ are roughly equal, i.e., the $^{13}C^\beta$ and $^{13}C^\gamma$ spins are close to being strongly coupled. As the spins are "close to being strongly coupled", it becomes relatively easy to remove the residual difference in the Zeeman terms, experienced by the two spins, from the Hamiltonian. In the AMNESIA experiment, this is accomplished by application of audiofrequency pulses in the spin locked frame.

In most NMR pulse sequences, spins are quantized by the static magne
tic field, and RF pulses are applied by generating an oscillating magnetic field, orthogonal to the static magnetic field. The frequency at which the second field oscillates must be close to the frequency at which the spins precess about the static field. When spins are spin-locked along the y axis of the rotating frame, by an RF field of strength v_0, a third magnetic field may be applied orthogonal to the y axis, i.e., either along x or z. If this field oscillates at a rate which corresponds to the RF field strength, v_0, it is resonant in the spin lock frame, and audio pulses can be applied to the spin-locked ^{13}C nuclei, completely analogous to the application of radiofrequency pulses to spins quantized along the static magnetic field. The phase and amplitude of the audio pulse is under operator control, and quite sophisticated pulse sequences can be applied in this spin locked frame. For transferring magnetization from Phe $^{13}C^\beta$ to the aromatic ring carbons, we apply a WALTZ-16 type phase-modulation of the audio field, which results in a relatively broad-band cross-polarization profile, and all nine Phe residues in a complex between the protein calmodulin and a target peptide fragment of skeletal muscle myosin light chain kinase could be assigned in this manner (Grzesiek and Bax, 1995).

REVERSE LABELING

Before uniform isotopic labeling of proteins with ^{15}N and ^{13}C became popular, many of the heteronuclear NMR studies of proteins relied on incorporation of selected amino acid types, labeled with ^{15}N and/or ^{13}C. Isotope editing experiments were then used to select the ^1H resonances of the labeled amino acids and to study their interaction with the remainder of the protein.(Bax and Weiss, 1987; Otting and Wüthrich., 1990) In these applications, a major problem was posed by the difficulty in unambiguously identifying the protons that interact with the labeled amino acid, as the proton shift alone is usually insufficiently unique. Although the four-dimensional $^{13}C/^{13}C$- and $^{13}C/^{15}N$-edited NOESY experiments, applied to uniformly $^{13}C/^{15}N$-enriched proteins, has largely solved this problem, application of the $^{13}C/^{13}C$-edited NOESY to the study of interactions between Phe ring protons and other protons in the protein remains difficult for several technical reasons.

As has recently been demonstrated by Vuister et al. (1994a), a procedure which we refer to as "reverse labeling" can be used to study the NOE interactions between Phe ring protons and other residues in great detail. In the Phe reverse labeling procedure, natural abundance Phe (~ 20 mg/l) is added to the growth medium of the *E. coli*, which consists of standard M9 minimal medium and contains $^{13}C_6$-glucose as the primary carbon source. The overexpressed protein is uniformly enriched in ^{13}C, except for Phe, which remains close to its natural abundance ^{13}C level. Interaction between protons attached to ^{12}C and protons attached to ^{13}C then can be studied using isotope filtering experiments. The advantage of this experiment over conventional selective labeling is that the "bulk" of the protein is enriched in ^{13}C and the protons interacting with the Phe protons can be identified by both their ^1H and ^{13}C chemical shifts. The experiment proved critical in determining the structures of a heat shock factor DNA binding domain (Vuister at al., 1994b) and in determining the structure of apo-calmodulin (Kuboniwa et al., 1995). For this latter protein, an average of nearly 25 long range NOE interactions were observed in this manner for each of the aromatic rings.

THREE-BOND J COUPLINGS

Besides three-bond ^{1}H-^{1}H J couplings, heteronuclear ^{1}H-^{13}C and ^{1}H-^{15}N J couplings also contain important structural information (Bystrov, 1976). Three-bond J couplings between ^{13}C nuclei, or between ^{13}C and ^{15}N have not been studied extensively in peptides or proteins, primarily due to the need for isotopic enrichment. With the introduction of uniform labeling protocols, measurement of all of these types of J couplings has become accessible, however (Bax et al., 1994). Measurement of $^3J_{CC}$ and $^3J_{CN}$ couplings in proteins with a correlation time larger than about 5 ns is restricted primarily to methyl carbons which, due to their rapid rotation about the three-fold symmetry axis, have a ^{13}C line width that is considerably smaller than for other protonated carbons. The J couplings can either be measured from a 2D difference spectrum (Vuister at al., 1993; Grzesiek et al., 1993), or from the intensity of cross peaks in a COSY-type spectrum (Bax et al, 1992, Bax et al., 1994). These experiments are found to be extremely valuable for determining amino acid side chain χ_1 and χ_2 angles of Ile and Leu residues, and for determining the stereospecific assignments of the methyl groups in Val residues, in addition to measurement of the χ_1 angle. Three-bond trans couplings between a methyl carbon and a backbone carbonyl are *ca* 4 Hz (~3 Hz for Thr) and gauche couplings are *ca* 1 Hz. A trans coupling between a $^{13}C^\delta$ methyl carbon and $^{13}C^\alpha$ is about 3.3 Hz, and a gauche coupling is slightly below 1 Hz. For $^3J_{CN}$, a trans coupling to a backbone ^{15}N is *ca* 2.2 Hz, and a gauche coupling is *ca* 0.6 Hz.

Measurement of these heteronuclear J couplings in apo-calmodulin indicates that many of the Leu, Ile and Val residues are subject to rotamer averaging. In contrast, in the complex between calmodulin and a 26-residue target peptide, for which it has very high affinity, many of the same residues have J values that are indicative of a single staggered rotamer conformation.

REFERENCES

Bax, A. and Weiss, M., 1987, Simplification of two-dimensional NOE spectra of proteins by ^{13}C labeling. *J. Magn. Reson.* 71: 571.

Bax, A., Max, D. & Zax, D., 1992, Measurement of multiple-bond ^{13}C-^{13}C J couplings in a 20-kDa protein-peptide complex. *J. Am. Chem. Soc.* 114: 6923.

Bax, A., Vuister, G.W., Grzesiek, S., Delaglio, F., Wang, A.C., Tschudin, R., and Zhu, G., 1994, Measurement of homo- and heteronuclear J couplings from quantitative J correlation. *Meth. Enzym.* 239:79.

Bystrov, V.F., Spin-spin coulings and the conformational states of peptide systems, 1976, *Prog. Nucl. Magn. Reson. Spectrosc.* 10:41.

Braunschweiler, L. and Ernst, R.R., 1983, Coherence transfer by isotropic mixing: application to proton correlation spectroscopy, *J. Magn. Reson.* 53: 521.

Grzesiek, S. and Bax, A., 1995, Audio-frequency NMR in a nutating frame. Application to the assignment of phenylalanine residues in isotopically labelled proteins, *J. Am. Chem. Soc.* 117:6527..

Grzesiek, S., Vuister, G.W. and Bax, A., 1993, A simple and sensitive experiment for measurement of J_{CC} couplings between backbone carbonyl and methyl carbons in isotopically enriched proteins. *J. Biomol. NMR* 3:487.

Kuboniwa, H., Tjandra, N., Grzesiek, S., Ren, H., Klee, C.B. and Bax, A., 1995, Solution structure of calcium-free calmodulin. *Nature, Struct. Biol.* in press

Otting, G. and Wüthrich, K., 1990, Heteronuclear filters in two-dimensional [^{1}H, ^{1}H]-NMR spectroscopy: combined use with isotope labeling for studies of macromolecular conformation and molecular interactions, *Quart. Rev. Biophys.* 23: 39.

Vuister, G.W., Wang, A.C. and Bax, A., 1993, Measurement of three-bond nitrogen-carbon J couplings in proteins uniformly enriched in ^{15}N and ^{13}C. *J. Am. Chem. Soc.* 115:5334.

Vuister, G.W., Kim, S.-J., Orosz, A., Marquardt, J., Wu, C. and Bax, A.: Solution structure of the DNA-binding domain of *Drosophila* heat shock transcription factor. *Nature, Struct. Biol.* 1, 605-614 (1994).

Vuister, G.W., Kim, S.-J., Wu, C. and Bax, A., 1994, 2D and 3D NMR study of phenylalanine residues in proteins by reverse isotopic labeling. *J. Am. Chem. Soc.* 116: 9206.

Yamazaki, T., Forman-Kay, J. D., and Kay, L. E., 1993, Two-dimensional NMR experimentsfor correlating $^{13}C^{\beta}$ and $^{1}H^{\delta/\varepsilon}$ chemical shifts or aromatic residues in ^{13}C-labeled proteins via scalar couplings, *J. Am. Chem. Soc.* 115: 11054.

DISCUSSION

Alfred Redfield - It's a beautiful talk. As a certified old timer, who has amnesia, I can't help observing that the last six set of experiments is similar to the discovery of spin-spin interaction which was not discovered by sticking a molecule into a machine and seeing a splitting. It was discovered by doing spin echoes when you couldn't see anything except a blob and looking at what were called at Illinois, slow beats.

Ad Bax - Yes, what we are doing is really nothing else than quantitatively measuring the amplitude of the slow beats, and trying to convert this into a size for the J coupling. So, as Prof. Rao said in his opening words, there is really very little new in NMR.

Carol Post - When you do reverse labeling for the phenylalanine and measure the NOE's then for the ζ hydrogen to other hydrogens, how do you scale that given that you have a ^{12}C now?

Bax - Yes, correctly scaling of the NOE to a ^{12}C-attached proton versus that to a ^{13}C-attached one is a tricky problem. However, most people use a very rough scale anyway: strong, medium and weak. And I believe Marius Clore has demonstrated quite convincingly that use of such a qualitative scale yields adequate results for protein structure determination. Our reverse-labeled NOESY spectrum was recorded with a relatively modest NOE mixing period of 80 milliseconds and spin diffusion effects should not be excessive for this mixing time. But I completely agree with you that determining of the scale factor used for determining which NOE's are weak, medium, or strong is rather arbitrary. This problem also exists in many of the other 3D and 4D techniques and it probably is worth while to develop an optimal and unbiased procedure for dealing with it.

Brian Sykes - In reference to the hydrophobic interior of the calcium binding protein, some people use the name molten globule to describe a partially unfolded state. Is it that unfolded?

Bax - For a long time, I and others believed that we should stay away from this calcium-free calmodulin because it was a molten globule. It certainly is not. Most of the backbone is perfectly well ordered. An order parameter of 0.85, found for most of the residues apart from a few in the empty calcium-binding loops and several other loop regions, indicates that this is a normal well structured protein and the increase in dynamics pertains primarily to the hydrophobic side chains. It is interesting to consider that if apo-calmodulin were a perfectly normal protein, with a well-packed hydrophobic core, it probably would be unable to open itself up when binding Ca^{2+}, and expose its hydrophobic patch which is essential for target binding. So I think there is some significance in the fact that the sidechains are relatively disordered in this case. However, I believe that for a molten globule we typically think of something that is a lot more disordered than what we have observed for apo-calmodulin.

Marc Adler - On the same line what does the DSC (differential scanning calorimetry) tell you about the calcium-free calmodulin? Is the transition broader than you would expect for proteins that size?

Bax - I vaguely recall that the melting transition for apo-calmodulin is relatively broad. However, as calmodulin clearly is a two-domain protein, I'm not sure whether this truly reflects the melting transition of the individual domains, or whether the two domains have slightly different melting temperatures, each with a relatively narrow transition. So, I'm sorry, I can't really answer your question.

FIELD-CYCLING NMR APPLIED TO MACROMOLECULAR STRUCTURE AND DYNAMICS

Alfred G. Redfield[1]

Department of Biochemistry
Brandeis University
Waltham, MA 02254

INTRODUCTION

Field-cycling NMR (FCNMR) is defined as use of a sequence in which the main field is changed by more than a few millitesla, in order to study resonance and relaxation at a variety of fields, usually much lower than the observation field (Figure 1). Preparation often consists simply of waiting for nuclear magnetization to approach a high value. The entire field cycle takes place in a time shorter than, or sometimes comparable to, the relaxation time of the prepared spin species at the low field, and the detection part of the cycle is often a simple free induction decay after a 90° pulse, or some form of adiabatic rapid passage.

The tremendous power of high-field multidimensional NMR has diverted attention from this venerable method in recent years. Here we consider how field cycling might be applied to determination of properties of biopolymers, in the light of recent developments in NMR. There have been virtually no such studies other than the line of work developed by S. H. Koenig and others, who have studied the T_1 of water to determine, for example, the explicit character of water-protein interactions for a variety of proteins, and the character of accessible magnetic centers (Bertini and Luchinat, 1986; Sigal, 1987).

A detailed review of field cycling was written by Noack (1986). Reviews specialized for various applications are also available: quadrupole spectroscopy (Edmonds, 1977; Emsley and Pines, 1993); imaging of free radicals using dynamic nuclear polarization (Lurie and Nicholson, 1993); relaxation of water by tissues (Koenig and Brown, 1994; 1995); and water dynamics in biological systems (Koenig, 1995). We will not repeat discussions or references in these reviews but we do note the interfaces between the topic of the last review, for example, and the general topic of the detailed nature of water binding to proteins. Topics which are not referenced below can generally be found without difficulty in one of these reviews.

After a brief survey of the methods used and of applications, we will suggest various new ways in which FCNMR might be applied to biopolymers in the future.

---

[1]Also at the Physics Department, Brandeis University.

NMR As a Structural Tool for Macromolecules: Current Status and Future Directions
Edited by B.D. Nageswara Rao and Marvin D. Kemple, Plenum Press, New York, 1996

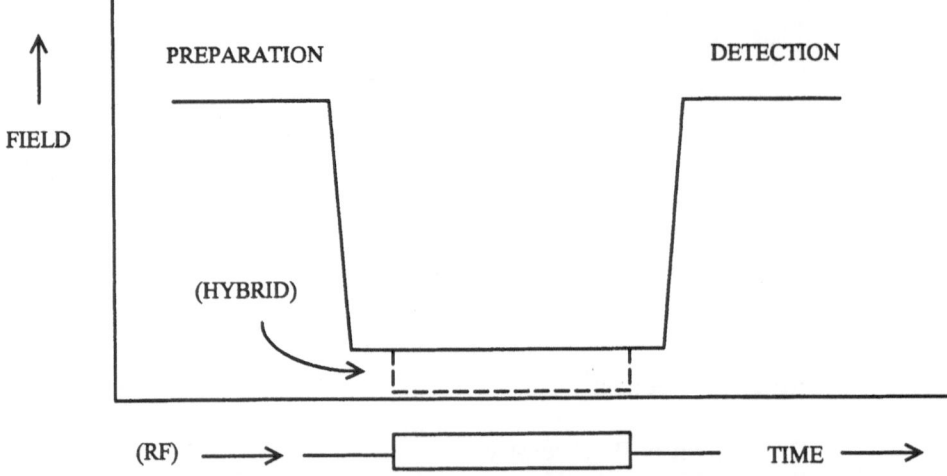

Figure 1. Typical field cycling NMR. A variety of different preparation and detection methods can be used, and they may not take place at the same field. The low field part of the cycle can include RF irradiation, or, in hybrid experiments, rapid secondary cycling to zero field.

PRIOR METHODS AND APPLICATIONS

Switched Magnets. Possibly the most interesting FCNMR experiment was the astounding discovery of increased longitudinal nuclear relaxation, just below the superconducting transition temperature of aluminum compared to just above, by Hebel and Slichter. They used a copper coil switched on and off with a large relay. Subsequently my laboratory developed the use of semiconductor switches and capacitor storage for such magnets, and we performed a variety of experiments on normal and superconducting metals. We used copper and superconducting coils (Genack, 1976), as well as a coil having a soft iron external shell, powered by a specially designed semiconductor power supply. Incidentally, my use of capacitor storage in this way was a direct consequence of my employment at the University of Illinois Betatron, as a first year graduate student in 1950-51. This accelerator was a large closed magnetic circuit in which magnetic flux was reversed about twice a second, using a noisy multicontact switch and a huge storage capacitor. When it operated, the lights all over the Champaign-Urbana area would flicker noticeably! Use of copper coils with fluorocarbon coolant was developed by F. Noack, together with a much improved high-speed switching circuit (Schweikert et al., 1988; Blanz et al., 1993). Koenig and Brown have developed a highly automated system for measuring T_1 of solvent water as a function of field, based on a liquid nitrogen cooled switched coil. Generally the field strength and homogeneity of magnets used by all these workers have not been nearly suitable for high resolution solutions studies, being in the range of less than 1.5T, with homogeneity not better than 25 mT (1000 Hz protons). It is unlikely that fields in excess of 5T are achievable in non-superconducting switched magnets, while homogeneity may be limited by thermal instability or, in the case of superconductors, nonreproducible flux trapping. On the other hand, low dead time, in the few millisecond scale, is now achievable.

Sample Shuttles. A second method of field cycling involves physical motion of the sample, as in the earliest FCNMR experiments by Ramsey, Pound, and Purcell in which the sample was moved by hand. Hahn's group pioneered moving a small sample pneumatically with gas pressure, naming this the "post office" method. A number of workers have also used various pneumatic devices to move samples indirectly using pusher rods. The obvious advantage of sample shuttling is that the best available magnets can be used, compared to field switched magnets, so that high resolution experiments are possible. The disadvantage is that the dead-time (which we define as the minimum round-trip time) is rather long, usually well over 100 msec, and that temperature control is more difficult. The shortest dead time that is likely to be achieved is well over 30 msec; this corresponds to an average speed of about one tenth of the velocity of sound, for a travel length of 40 cm.

Only two experiments have been reported, to our knowledge, that combine field cycling with genuinely high resolution methods. One was a study of the field dependence of photochemically induced DNP, detecting protons at 100 MHz (Stob et al., 1989). Another involved NOESY at 400 MHz, with a field cycle to 60 MHz in the fringe of the 400 MHz magnet during the mixing time, and a dead-time of 160 to 200 ms (Kerwood and Bolton, 1987). Positive NOE's were observed at the lower field, in contrast to the familiar negative NOE's at 300 MHz, in a relatively small molecule.

Hybrid Shuttling and Switching. A hybrid approach combining field switching and sample translation has also been used, in which the sample is polarized and observed at high field, and between is moved to a lower field coil which can be switched off and on in microseconds. In zero field, the nucleus under study, usually the deuteron, evolves usually under a pure quadrupole interaction. These experiments have been reviewed (Thayer and Pines, 1987; Emsley and Pines, 1993) and they do not have the sensitivity needed for most biologically-related problems. We will not discuss them further.

Sample Flow. Finally, field cycling can be achieved by sample flow. The first work of this type combined high pressure liquid chromatography with dynamic nuclear polarization (DNP). The sample was polarized by flowing it through a microwave spectrometer operating at lower field, where it was polarized by EPR-saturated immobilized radicals (Stevenson and Dorn, 1994). After moving the sample by flow to a 5.3 T high resolution instrument, enhanced ^{13}C NMR of small molecules could be observed.

Applications of FCNMR. Despite the number of interesting studies in the past, the recent level of activity has been low, probably because a universal commercial instrument is not available and probably could not be designed. We provide a number of references that will help the reader to investigate more recent applications to imaging (Macovski and Conolly, 1993); tissue studies (Swanson and Kennedy, 1993); imaging of radicals using DNP (Lurie et al., 1992); pure quadrupole interaction of deuterons (Seliger et al., 1994); studies of microstructures and of entangled polymers (Fatkullin and Kimmich, 1994; Kimmich et al., 1994a,b); and chemically induced DNP (Pravica and Weilekamp, 1988). For the most part we will not discuss these further, or review them.

POSSIBLE FUTURE APPLICATIONS

We will not attempt to discuss imaging applications of FCNMR, which are still tentative, or future applications of relaxometry of water in tissues or macromolecular solutions.

Other applications which we won't discuss in detail, but which deserve consideration, are brute-force increase of sensitivity by moving a sample to higher fields not suitable for spectroscopy before moving it to a typical high-resolution instrument; and use of DNP at lower field for the same reason. Moving to higher field will be limited because such fields are rarely available to the average laboratory. DNP will be limited by what power is available at high frequencies, and by relaxation rates (Lurie et al., 1992). It was suggested by M. Guéron in this Symposium that photochemically induced DNP of water might be used indirectly, via cross-relaxation (Stoesz et al., 1978), to polarize proton magnetization of proteins. Ordinary DNP of water (Overhauser effect), if it could be achieved, could be used in a similar way, but it isn't clear how much sensitivity improvement can be achieved in practice, at reasonable cost.

Solution Studies

Longitudinal Relaxation. The study of T_1 processes in the context of high resolution requires relatively simple modification of a spectrometer, as already demonstrated by Kerwood and Bolton (1987). In fact, their simple system could be inserted directly into a standard 10 mM probe. Air flow was via concentric tubes, and as a result the sample size was small, limiting sensitivity. As they suggest, an obvious improvement on their simple strategy is to construct a special purpose probe with straight-through air flow, to get greater air flow rate and filling-factor. Temperature-regulation over the +5 to 40°C range is also desirable and possible, and sample-spinning is fortunately not needed in modern magnets. The dead-time might conceivably approach, or be reduced below, the limit of 30 msec stated above, by use of high-technology position sensors and high-speed valves.

It was pointed out to the writer at this Symposium by Dr. Mark Adler that the flow technology used in stopped-flow instruments is highly developed and fast, and would require only a small increase in sample quantity, a few times as much as is used for plain pulsed NMR, to be used for field-cycling transport.

Unfortunately the information obtained from T_1 measurements is likely to be rather limited for molecules in solution. Processes occurring on a time-scale slow compared to the overall tumbling of the molecule are generally masked by the tumbling. Qualitative information about less rapid motion can often be obtained from ordinary line-width studies, at least to the extent that is justified by present-day understanding of protein motion.

The measurement of proton "T_1" processes is also exceedingly difficult and even ill-defined for larger macromolecules because longitudinal relaxation of individual spins is masked and averaged by magnetization-transfer between spins. The latter process is the familiar one, often called "spin diffusion", leading to NOE. This problem can be circumvented by various isotope-labeling strategies involving deuteron or tritium labeling, but not easily. Instead, a set of ingenious experiments has been devised, based on ^{15}N labeling, for the study of NH-group dynamics (see, for example, Peng and Wagner, 1992). The study of CH dynamics should also be possible, using ^{13}C labeling. Rates of decay of the coherences I_z, S_z, and I_zS_z were measured using different sequences, to get $J(\omega)$ for various ω. Here I and S refer to proton and ^{15}N, and J is the spectral density of the dipolar interaction between the ^{15}N and its own proton. In this way $J(\omega)$ was determined for ω equal to the nitrogen and proton frequencies and their sums and differences, and also for $\omega = 0$, by spin-echo measurements of ^{15}N line width.

There are gaps in the range of ω for which $J(\omega)$ can be measured in this way. If non-cycled measurements can be made at 300 and 600 MHz, then for ^{15}N labels the major gaps occur between zero and 30 MHz, and between 60 and 270 MHz. These gaps could be filled by incorporating field-cycled intervals into the various pulse sequences already developed for the measurements above. Study of $J(\omega)$ in these gaps could lead to useful and even unexpected information such as indication of the existence of more than one kind of internal motion.

Unfortunately some of these measurements will become difficult because the time-scale of proton relaxation, especially magnetization-transfer, may be short compared to the dead-time for field cycling. The dead time is likely to be greater than 30 msec as already mentioned. Very detailed proton dynamic measurements might be required to interpret measurements such as heteronuclear NOEs, when the proton magnetization is not constant during ^{15}N polarization build-up.

Small Molecule Probes. A high-resolution field cycler as described above could certainly observe NMR of small molecule inhibitors and substrate analogs, and measure their T_1 down to zero field in the presence of proteins. Proteins containing magnetic ions or spin labels could be probed in this way, to yield distances and correlation times, as has already been done for water (Bertini and Luchinat, 1986; Banci et al., 1991; Koenig and Brown, 1995; Koenig, 1995). Weak binding sites might likewise be probed.

Enhanced Saturation Transfer. Besides being of possible interest in themselves, some relaxation processes are changed at lower field, and may thereby enhance other experiments. The most obvious of these processes is the heteronuclear cross relaxation rate, especially between proteins and fluorine. Fluorine has long been used as a sort of proton substitute, mainly because it can be substituted regiospecifically for protons, and is then easier to identify; and also because it has a greater chemical shift range than protons (Gerig, 1994). However, NOE studies between protons and fluorine, while useful, are somewhat less sensitive than might be initially expected because the saturation-transfer rate is reduced between protons and fluorine, for the same distance between them, and despite a similar strong magnetic moment for fluorine (only 7% less than for protons). That is because at 500 MHz the energy difference required for a mutual spin flip for an H to F transfer is about 36 MHz, which is comparable to, or less than, the inverse of 2π times the rotational correlation time for larger proteins. As a result, this rate, which is the initial buildup rate for H-F NOE, is reduced because it is proportional to $J(\omega)$, for $\omega \approx 36$ MHz for an H-F transfer, in contrast to $J(0)$ for a H-H transfer. Therefore H-F NOE's are reduced in intensity and are also complicated by multistep spin-diffusion to a greater extent than are H-H NOE's, especially in larger molecules. By performing H-F NOE, with field-cycling to low field during the mixing time, this bottleneck might be alleviated. NOE involving other nuclear pairs, especially proton-phosphorous, might also be aided by field-cycling. It seems possible that NOE between carboxyl or carbonyl ^{13}C nuclei and protons could be observed by field-cycling to yield new structural and dynamic information.

Transverse Relaxation. It is unlikely that a field-cycle can be performed during any spin echo experiment with sufficient reproducibility to give a coherent signal. Therefore, it is probably necessary to study transverse relaxation by setting up a longitudinal coherence at high field, cycling to a low field, performing an echo experiment with a final pulse to produce longitudinal coherence, and reading out the latter coherence at high field. Such an experiment is more elaborate than those considered just above, because the sample must be moved to a low field of moderate quality (proton line width less than a few KHz) rather than merely to the fringe of a high field magnet.

There is also the possibility of moving to a lower field having true high-resolution quality, or to truly zero field. We can think of many experiments that might then be performed, but none which seem worth the effort. Therefore we will not discuss this possibility further, and consider only spin-echo and spin-locking experiments.

The main reason for using field-cycling in conjunction with T_2 measurements would be that transverse relaxation due to chemical shift anisotropy (CSA) would essentially disappear. Thus, by studying T_2 at various fields, the dipolar and CSA contributions could be separated more precisely.

Broadening due to conformational changes, narrowed by moderately rapid kinetics, would also be field-dependent, and disappear at low field. Thus such broadening could likewise be measured more precisely. Measurement of $T_{1\rho}$ at higher RF fields might be easier at low field because more intense resonant RF fields can be applied at lower static fields.

Transverse relaxation limits the usefulness of "spin locking" experiments, especially in larger proteins, by limiting the practical spin-lock time. These experiments include ROESY (CAMELSPIN), TOCSY, and heteronuclear TOCSY. The spin-lock section of these sequences can all be replaced by a low field spin-lock. This spin-lock would have to be preceded and followed by a 90°-(field shuttle)-90° sequence, to move the transverse magnetization to the z-axis and back, before and after the field shuttle. The extra two-fold power sensitivity gain obtained by strategies introduced by Rance (Palmer et al., 1992) presumably could not be obtained in this case. It seems likely that the "slow beat" measurement of spin-spin coupling, as outlined at this Symposium by Bax, might be performed at low field. The advantage of performing these experiments at low field is simply that the feasible spin-lock time might be significantly longer than at high field.

Solid State Studies

State of the Art Solid State NMR. One of the well-known strengths of solid-state NMR is its ability to study all time-scales of molecular dynamics. Slow processes are not masked by overall tumbling as they are in solution. Therefore, it would be of greater interest to study longitudinal relaxation versus field by solid-state methods than in solution. Unfortunately it is unlikely that a magic-angle spinning probe can be shuttled sufficiently rapidly between two magnets, although Stob et al. (1989) have shuttled a slowly-spinning sample.

Another possibility would be to avoid sample-spinning and use a pneumatic shuttle as discussed above. For sensitivity reasons, and to ease interpretation, a single crystal of the macromolecule might be desirable. A reasonable fraction of proteins can be efficiently crystallized into multimeter-sized crystals with high yield. Detailed NMR on such crystals has not yet been performed, to our knowledge.

Another possibility is to use a field-switched magnet for CPMAS solid state NMR. Currently available fields (surveyed above) are probably much too low for this purpose, and a major magnet and power supply development project would be required. Use of a (laminated) external iron core and/or a superconducting magnet should be considered, as well as high speed field regulation of types not previously used, perhaps based on solution-state NMR. Use of digital signal processing technology to analyze each free induction decay, to compensate for field instability, is a definite possibility. Homogeneity might be achieved by use of a large-diameter superconducting magnet. We have used two very different superconducting magnets for field-cycling, and they had problems, but neither was so bad that we abandoned them immediately. The technology of superconductivity has been greatly extended since those magnets were built, twenty five and nearly forty years ago.

Field cycling might also be useful for determining dynamics of small moelcules in oriented (liquid crystal) samples (Sanders et al., 1994).

Pure Quadrupole Resonance. In zero magnetic field, for nuclei with spin greater than one-half, the interaction between the nuclear quadrupolar moment and the molecular electric field gradient produces a splitting of the nuclear energy levels. Radio-frequency absorption spectroscopy among these levels, in solids is called pure quadrupole resonance (PQR). PQR is a substantial branch (or cousin) of NMR, but has never been applied to proteins because of problems of sensitivity and difficulty of labeling.

An obvious application of PQR would be the study of ligated metals like Mg, Zn, and Ca whose quadrupolar isotopes are available. If PQR spectroscopy of these metals could be developed, a lot of work would be required, based on studies of model compounds, to get experience about quadrupole interactions of ions with various ligating groups, in addition to study of biopolymers themselves. We will not attempt to discuss such a program, which might resemble similar studies for electron paramagnetic resonance.

Quadrupolar interactions can also be inferred by conventional NMR, from splittings of ordinary solid-state resonances. This possibility should be kept in mind, but such methodology would probably require the use of single crystals.

A method developed around 1962 (Edmonds, 1977; Itterman et al., 1993), which we will call field cycling pure quadrupole resonance (FCPQR), may make it possible to do pure quadrupole resonance of nuclei such as Zn, Ca, Mg in frozen metalloproteins, or ^{17}O in labelled substrates ligated to proteins. The preparation and readout parts of the experiment are similar to those for solution studies, and the low field part of the field-cycle takes place at about the earth's field. During demagnetization (and also remagnetization) entropy is preserved, and is transferred to spin-spin order in zero field. At zero field, strong RF power is applied at a pure quadrupole resonance frequency. Following ideas of Goldmann and Landesman and of Hartmann and Hahn, this strong RF field couples the quadrupolar coherence to the zero field spin order, enhancing the quadrupolar coherence. This enhanced polarization is not what is observed; instead, the strong RF field is phase-reversed with a period comparable to the time required for quadrupolar coherence to build up. This irreversible coherence transfer, occurring between 10^4 and 10^5 times per second, results in a depletion of the proton spin order in zero field, over a time comparable to the zero-field T_1 which can be seconds. As a result, the subsequent signal at high-field is decreased.

Of course the quadrupolar frequency is not known in advance, and a search has to be performed involving many field cycles while the quadrupolar frequency is varied stepwise, to find the frequency at which the final proton signal is decreased.

Of the relatively few versions of this experiment which have been performed, one by Hsieh et al. (1972) seems closest to experiments that might work for proteins. Natural abundance ^{17}O was the quadrupolar nucleus in a series of chlorquinones, and the volume density ratio of ^{17}O to protons was around 10^{-4}, similar to the ratio of metal ion to protons in a 20 Kilodalton protein. The experiment was performed at 77°C, to lengthen the proton T_1 to seconds in zero field. At high field the proton T_1 was ingeniously shortened by rotating the sample, to promote level-crossing relaxation by the ^{35}Cl nuclei.

When the applied radio frequency was at the quadrupolar frequency, a dip in the proton signal, after remagnetization, was seen whose size was in the range of 10 to 50% of the proton signal far from double resonance. The theoretical sensitivity of this experiment is exceedingly high, but in practice the problem is to see small variations in large successive proton signals. Thus, it would be desirable to design the experiment to have the highest stability while working at the lowest possible temperature, in order to avoid noise similar to "phase noise", familiar in multidimensional NMR.

It would also be of interest to apply this experiment to deuterons, perhaps on labeled substrates bound to enzymes. Unfortunately the deuteron (and also ^{14}N) is a special case, having spin S=1. For such a spin, the nuclear magnetic moment will be "quenched", as long as the electric field gradient is not axially symmetric (Leppelmeier and Hahn, 1966). This means that the zero-field spin eigenstates are linear combinations of the high field ones, such that no magnetic transitions are induced. The method described just above will not work. The hybrid zero-field methods will work, and were developed in part for the reason just given, but they are not sensitive.

A possible solution to the deuteron problem is to perform some form of "impure" quadrupole resonance with field cycling (FCIQR), by cycling to a low but non-zero magnetic field, where the nuclear magnetic moment is only partly quenched but the quadrupolar resonance width is not too large (Seliger et al., 1994).

ACKNOWLEDGEMENT

This article was prepared with the support of U.S.P.H.S. Grant GM20168. I thank P. H. Bolton, E. L. Hahn, R. Kaptein, R. Kimmich, S. H. Koenig, D. Lurie, G. Noack, J. Peng, and G. Teklemarian for useful information, and P. Murray for preparation of the manuscript.

REFERENCES

Banci, L., Bertini, I., and Luchinat, C., eds., 1991, "Nuclear and Electronic Relaxation," VCH, Weinheim.

Bertini, I., and Luchinat, C., 1986, "NMR of Paramagnetic Molecules in Biological Systems", Benjamin/Cummings, Menlo Park.

Blanz, M., Rayner, T. J., and Smith, J. A. S., 1993, A fast field-cycling NMR NQR spectrometer, *Measurement Science and Technology* 4:48.

Edmonds, D. T., 1977, Nuclear quadrupole double resonance, *Physics Reports* 29:234.

Emsley, L., and Pines, A., 1993, Lectures on pulsed NMR (2nd edition); *in*: "Nuclear Magnetic Double Resonance," B. Maraviglia, ed., North Holland, Amsterdam.

Fatkullin, N., and Kimmich, R., 1994, Nuclear spin-lattice relaxation dispersion and segment diffusion in entangled polymers. Renormalized rouse formalism, *J. Chem. Phys.* 101:822.

Genack, A. Z., 1976, Dipole energy dissipation and nuclear spin diffusion in mixed-state superconducting vanadium, *Phys. Rev.* B13:68.

Gerig, J. T., 1994, Fluorine NMR of proteins, *Prog. NMR Spectr.* 26:293.

Hsieh, J. C., Koo, J. C., and Hahn, E. L., 1972, Pure nuclear quadrupole resonance of naturally abundant ^{17}O in organic solids, *Chem. Phys. Lett.* 13:563.

Itterman, B., Bürkmann, K., Diehl, E., Dippel, R., Fischer, B., Frank, H.-P., Jäger, E., Seelinger, W., Sulzer, G., Achermann, H., and Stöckmann, H. J., 1993, Lattice locations and electric field gradients of boron implanted into vanadium, *Zeit. f. Physik.* B91:7.

Kerwood, D. J., and Bolton, P. H., 1987, A sample-shuttling device suitable for two-dimensional low-field NMR, *J. Magn. Reson.* 75:142.

Kimmich, R., Stapf, S., Callaghan, P., and Coy, A., 1994a, Microstructure of porous media probed by NMR techiques in sub-micrometer length scales, *Magnetic Resonance Imaging* 12:339.

Kimmich, R., Stapf, S., Moller, M., Out, R., and Seiter, R. O., 1994b, Field cycling NMR relaxation spectroscopy of poly(di-N-alkylsiloxanes) in solid, mesomorphic liquid, and isotropic liquid-phases, *Macromolecules* 27:1505.

Koenig, S. H., 1995, The dynamics of water in biological systems, *in*: "Encyclopedia of NMR," D. M. Grant, ed., John Wiley, New York.

Koenig, S. H., and Brown, R. D., 1994, Relaxometry and the source of contrast in MRI, *in*: "NMR in Physiology and Biomedicine," R. J. Gillis, ed., Academic Press, New York.

Koenig, S. H., and Brown, R. D., 1995, Relaxometry of tissue, *in*: "Encyclopedia of NMR," D. M. Grant, ed., John Wiley, New York.

Leppelmeier, G. W., and Hahn, E. L., 1966, Nuclear dipole field quenching of integer spins, *Phys. Rev.* 141:724.

Lurie, D. J., and Nicholson, I., 1993, Proton electron double-resonance imaging of exogenous and endogenous free radicals in vivo, *in*: "Nuclear Magnetic Double Resonance," B. Maraviglia, ed., North Holland, Amsterdam.

Lurie, D. J., Nicholson, I., McLay, J. S., and Mallard, J. R., 1992, Spin-trapped hydroxyl free radicals studied at low field by field-cycled dynamic nuclear polarization, *Appl. Magn. Reson.* 3:917.

Macovski, A., Conolly, S., 1993, Novel approaches to low-cost MRI, *Magnetic Resonance in Medicine* 30:221.

Noack, F., 1986, Field cycling spectroscopy: principles and applications, *Progr. NMR Spectrosc.* 18:171.

Palmer, A. G. III, Cavanaugh, J., Byrd, A., and Rance, M., 1992, Sensitivity improvement in three-dimensional heteronuclear correlation NMR spectroscopy, *J. Magn. Reson.* 96:416.

Peng, J., and Wagner, G., 1992, Mapping of the spectral densities of nitrogen-hydrogen bond motions in Eglin C using heteronuclear relaxation experiments, *Biochemistry* 31:8571.

Pravica, M. G., and Weilekamp, D. P., 1988, Net NMR alignment by adiabatic transport of parahydrogen addition products to high magnetic field, *Chem. Phys. Lett.*, 145:255.

Sanders, C. R., Hare, B. J., Howard, K. P., and Prestegard, J. H., 1994, Magnetically-oriented phospholipid micelles as a tool for the study of membrane-associated molecules, *Prog. NMR Spectr.* 26:421.

Schweikert, K. H., Kreig, R., and Noack, F., 1988, A high field air-cored magnet coil design for fast-field-cycling NMR, *J. Magn. Reson.* 78:77.

Seliger, J., Zagar, V., and Blinc, R., 1994, A new highly sensitive ^{1}H-^{14}N nuclear-quadrupolar double-resonance technique, *J. Magn. Reson.* A106:214.

Sigal, H., ed., 1987, "Applications of Nuclear Magnetic Resonance to Paramagnetic Species," *Metal Ions in Biological Systems* <u>4</u>.

Stevenson, S., and Dorn, H. C., 1994, ^{13}C dynamics nuclear polarization: a detector for continuous-flow, on-line chromatography, *Anal. Chem.* 66:2993.

Stob, S., Kemmink, J., and Kaptein, R., 1989, Intramolecular electron transfer in flavin adenine dinucleotide. Photochemically induced dynamic nuclear polarization study at high and low magnetic fields, *J. Am. Chem. Soc.* 111:7036.

Stoesz, J. D., Redfield, A. G., and Malinowski, D., 1978, Cross relaxation and spin diffusion effects on the proton NMR of biopolymers in H_2O, *FEBS Lett.* 91:320.

Swanson, S. D., and Kennedy, S. D., 1993, A sample-shuttle nuclear-magnetic-relaxation-dispersion spectrometer, *J. Magn. Reson.* A102:375.

Thayer, A. M., and Pines, A., 1987, Zero field NMR, *Acc. Chem. Res.* 20:47.

Addendum: Two interesting developments not covered above are presented in the following papers, and references therein.

Clough, S., 1994, The Rotational Spectra of Tunnelling Methyl Groups, <u>Zeit. fur Naturforschung</u> A49:1193.

Notter, M., Konzelmann, K., Majer, G., and Seeger, A., 1994, Investigation of Point Defects in Metals by Nuclear Quadrupole Double Resonance, <u>Zeit. fur Naturforschung</u> A49:47.

DISCUSSION

Brian Sykes - I think Bob Bryant talked in recent times of building a machine like this.

Alfred Redfield - I don't know what he is doing, but he does excellent work in the general area that I covered. As I mentioned, Swanson and Kennedy also developed a nice way of mapping out T_1's by imaging techniques in a fringe field. There are also a lot of experiments on diffusion, diffusion of things which you can think of involved in various steps relating to macromolecules.

B. D. Nageswara Rao - I wonder if you could comment on the angular momentum quenching that you talked about.

Redfield - You have three levels and you have a strong quadrupole interaction. You have a non-symmetric quadrupole interaction with three axes, I guess, which are perpendicular and have different quadrupole interactions. Semiclassically the three levels correspond to the quadrupole moment being pointed in those three directions and not having any spin attached to it or any angular momentum. So, if you try to do any kind of experiment where you go down to zero field, you irradiate and you do not get any interaction with an rf field (See Leppelmeier and Hahn, 1966). So the reason for example that Pines does his experiments as he does is that he starts with his spins in a magnetic-type Hamiltonian and they are quantized along the external field in the way that we all think. Now you turn off the field instantly and you suddenly discover that they are in a pure quadrupole interaction, so they then evolve, and then you go back up in field so you can measure the evolution frequencies. The problem is doing this in a repetitive way a la Hartmann and Hahn, as I described, where you have only one spin and you have to affect a whole bunch of spins which act as antennae is difficult. That's the kind of problem that Blinc is working on. He had a paper in the Journal of Magnetic Resonance (See Seliger et al, 1994) but what he has done really is not high sensitivity in the sense of a large change in the signal of a lot of protons interacting with one deuteron or nitrogen-14. That's a problem with deuterons. Does that answer your question? It was a very long answer.

Rao - I understand a little bit better than I did. Thank you. The other thing is the 180^0 phase shift that is applied for repetitive transfer (See Hartmann and Hahn, referenced in Hsieh et al, 1972). You have to adjust that with the pulse spacing I believe. Is that right?

Redfield - No, it's not very critical. Actually, it was a version of, what is it, amnesia? Actually the way I did it was not amnesia. I actually did it with a sine wave. I just applied two frequencies with a 180^0 phase shift, but it is really the same experiment. The idea is that you get the spins cross polarized but since you have only a dinky spin system of one spin, it gets polarized and then you make a negative temperature, if I may use that language which people may or may not be familiar with. Now you repolarize back to a positive temperature and you sap the entropy or order from the protons. That really was the ingenious thing about the Hartmann and Hahn paper which I bet most of you have never read.

Rao - Thank you very much.

Ad Bax - With the development of pulsed field gradients supposedly as you get stronger and stronger you can make basically a very small magnet inside your superconductor.

Have you ever thought about the possibility that it may be possible to make something that is more than 10 kGauss for one of those things. Since it's not a gradient anymore, it can be very inhomogeneous. Would it be feasible you think?

Redfield - Well, I am worried about quenching the magnet. I am not sure what you're talking about.

Bax - That's basically like taking a very small magnet, with very high current just like what you do for pulsed field gradient.

Redfield - Yeah, you just get down to zero field.

Bax - Well, not quite zero field but a low field.

Redfield - The problem with that kind of scheme is that you worry in a superconducting magnet about changing the distribution of trapped flux and permanently distorting the field. You may ask why is it so hard to make a field switched magnet out of a superconductor. It's the same reason, the field is not necessarily reproducible. I don't know if that answers your question. There have been schemes like this where you turn off the field with two magnets, one of which opposes the other. I don't think those are any better than just turning off the field.

Bax - I know you've thought about technical problems.

Redfield - I didn't mention experiments in the rotating frame. I like them in the sense that you have a zero field; Richard Ernst mentioned that. The problem with those experiments is that even by operating off resonance as Tom James did, you can't get to very large effective fields. So you can't map out the spectral density very far. Obviously, it is a useful experiment. It would be interesting to figure out how to extend it to higher effective fields, and you probably can.

Sykes - Al, I think, just as you said, that you have to go home and think about some of the things that Ad talked about. Al, I think you have challenged us to really think seriously about some of the things you've presented.

CROSS-CORRELATIONS: OBSTACLES OR TOOLS FOR STRUCTURE DETERMINATION OF BIOMOLECULES

Anil Kumar

Department of Physics and
Sophisticated Instruments Facility
Indian Institute of Science
Bangalore-560 012, INDIA

INTRODUCTION

Cross-correlations, which are cross-terms between different mechanisms of relaxation of a spin, have been known in NMR for a long time. Their significance in relaxation of methyl groups as well as in other multispin relaxation has been well established (Hilt and Hubbard, 1964; Band and Hubbard, 1968; Runnels, 1964; Werbelow and Grant, 1977; Vold and Vold, 1978). In relaxation studies using double resonance experiments it has been known that the cross-correlations play significant role and cannot be ignored (Anil Kumar and Nageshwara Rao, 1968).

Recent developments of the algorithm for structure determination of biomolecules using two and multidimensional NMR spectroscopy utilizing the distance information provided by the two dimensional Nuclear Overhauser Effect (NOE) spectroscopy has been extremely successful and have yielded structures of biomolecules with high degree of reproducibility and confidence. Most analyses of biomolecular structure calculations using NOE data ignore cross-correlations. Several calculations have indicated that this neglect of cross-correlations in NOE experiments is justified (Bull, 1987; James, 1991). Since this assessment is in contrast with earlier experience, the effects of cross-correlations in NOE studies are re-examined here.

THEORY

In the absence of an r.f field the longitudinal and transverse relaxation are not coupled and the longitudinal relaxation of N relaxation coupled spins, whether J coupled or not, is described by the rate equation (Ernst et al, 1983)

$$\frac{d\vec{P}}{dt} = W(\vec{P} - \vec{P_o}) \qquad [1]$$

where P_i are the populations of various levels and W_{ij} are transition probabilities connecting states i and j. For spin 1/2 nuclei the dimension of P is 2^N and that of W is $2^N * 2^N$.

For uncoupled or weakly J coupled spins, in the absence of rf fields, it is easy to show that only those transition probabilities between two eigenstates in which one of the spins changes its state depend on cross-correlations. In other words, only W_1 have contributions from cross-correlations and W_o and W_2 do not. Indeed the various W_1 of a spin differ only due to cross-correlations. An alternate and for cross-correlations a more convenient description of longitudinal relaxation is through magnetization modes. One defines, single spin modes ($A_z, M_z, X_z \ldots$) and multi-spin modes ($2A_zM_z, 2A_zX_z, 2M_zX_z \ldots, 4A_zM_zX_z \ldots$ upto N spin modes). These modes form a complete basis set in which the populations of various levels can be expressed as a linear combination of these modes and vice-versa, given by (Werbelow and Grant, 1977, Fagerness et al, 1975) .

$$\vec{M} = V\vec{P} \qquad [2]$$

where V contains elements of magnitude $\pm 1/2^N$.

The intensity of a transition (i,j) is proportional to $(P_i - P_j)$, the difference in the populations of the levels i and j. Representing these population differences in terms of modes yields, for example for a three spin system (AMX, each of spin 1/2), the intensities of the four A transitions as

$$A_1 = P_{\alpha\alpha\alpha} - P_{\beta\alpha\alpha} = \tfrac{1}{4}(A_z + 2A_zM_z + 2A_zX_z + 4A_zM_zX_z)$$

$$A_2 = P_{\alpha\alpha\beta} - P_{\beta\alpha\beta} = \tfrac{1}{4}(A_z + 2A_zM_z - 2A_zX_z - 4A_zM_zX_z)$$

$$[3]$$

$$A_3 = P_{\alpha\beta\alpha} - P_{\beta\beta\alpha} = \tfrac{1}{4}(A_z - 2A_zM_z + 2A_zX_z - 4A_zM_zX_z)$$

$$A_4 = P_{\alpha\beta\beta} - P_{\beta\beta\beta} = \tfrac{1}{4}(A_z - 2A_zM_z - 2A_zX_z + 4A_zM_zX_z)$$

The sum of all these four intensities ($=A_z$) represents the total magnetization of spin A, while the presence of higher spin modes indicates that the various lines in the multiplet of spin A will have different intensities. We call these different intensities among the various lines of a multiplet as a "multiplet effect". Thus presence of higher spin modes indicates presence of a multiplet effect and vice-versa. A symmetry in the intensities of the transitions indicates absence of certain modes. For example if $A_1 = A_4$ and $A_2 = A_3$ but $A_1 \neq A_2$ indicates presence of the three spin mode and absence of both the two spin modes. If $A_1 = A_2 \neq A_3 = A_4$ indicates presence of the two spin mode $2A_zM_z$ and absence of $2A_zX_z$ and $4A_zM_zX_z$.

Equation of motion of the magnetization modes is obtained from Eqs (1) and (2) as

$$\frac{d\vec{M}}{dt} = \Gamma(\vec{M} - \vec{M}_o) \tag{4}$$

where

$$\Gamma = VWV^{-1} \tag{5}$$

The structure of Γ matrix turns out to be extremely interesting as seen in the expansion of Eq(4) into Eq(6)

$$-\frac{d}{dt}
\begin{pmatrix}
E \\
A_z \\
M_z \\
X_z \\
2A_zM_z \\
2A_zX_z \\
2M_zX_z \\
4A_zM_zX_z
\end{pmatrix}
=$$

$$
\begin{pmatrix}
1 & 0 & 0 & 0 & 0 & 0 & 0 & 0 \\
0 & \rho_A & \sigma_{AM} & \sigma_{AX} & \Delta_{AM}^A & \Delta_{AX}^A & 0 & \delta_A \\
0 & \sigma_{AM} & \rho_M & \sigma_{MX} & \Delta_{AM}^M & 0 & \Delta_{MX}^M & \delta_M \\
0 & \sigma_{AX} & \sigma_{MX} & \rho_X & 0 & \Delta_{AX}^X & \Delta_{MX}^X & \delta_X \\
0 & \Delta_{AM}^A & \Delta_{AM}^M & 0 & \rho_{AM} & \delta_A + \sigma_{MX} & \delta_M + \sigma_{AX} & \Delta_{AX}^A + \Delta_{MX}^M \\
0 & \Delta_{AX}^A & 0 & \Delta_{AX}^X & \delta_A + \sigma_{MX} & \rho_{AX} & \delta_X + \sigma_{AM} & \Delta_{AM}^A + \Delta_{MX}^X \\
0 & 0 & \Delta_{MX}^M & \Delta_{MX}^X & \delta_M + \sigma_{AX} & \delta_X + \sigma_{MX} & \rho_{MX} & \Delta_{AM}^M + \Delta_{AX}^X \\
0 & \delta_A & \delta_M & \delta_X & \Delta_{AX}^A + \Delta_{MX}^M & \Delta_{AM}^A + \Delta_{MX}^X & \Delta_{AM}^M + \Delta_{AX}^X & \rho_{AMX}
\end{pmatrix}
\begin{pmatrix}
E \\
A_z - A_z^0 \\
M_z - M_z^0 \\
X_z - X_z^0 \\
2A_zM_z \\
2A_zX_z \\
2M_zX_z \\
4A_zM_zX_z
\end{pmatrix} \tag{6}
$$

Here the diagonal elements ρ, represent the self relaxation of each mode and σ represent cross-relaxation of one mode into another of the same order. Both ρ and σ contain exclusively auto-correlation spectral densities. Cross-correlations between CSA of spin i and dipolar interaction of ij (Δ_{ij}^i) connect even spin modes to odd spin modes. The cross-correlation between dipolar interactions between spins ij and ik (represented here by δ_i which for three spin system is a sufficient notation but in general should be called δ_{ijik}), connects even order modes to even order modes and odd order modes to odd order modes (Werbelow and Grant, 1977; Canet, 1989).

It may be appropriate to point out at this place that in general the magnitudes of cross-correlations are comparable to auto-correlations. For example the cross-correlations between dipolar interactions of spin i with j and k depend on the angle between the dipolar vectors ij and ik, in addition to the distances between them. The ratios of dipolar cross-correlations to the auto-correlations are -1/8, -1/2 and 1 for this angle being $60°, 90°$ and $180°$ for $r_{ij} = r_{ik}$. Thus dipolar cross-correlations are largest for a linear geometry. The cross term between chemical shift anisotropy relaxation and dipolar relaxation is proportional to the product of these two interactions and much larger than the auto contribution (square) of the smaller interaction.

In a NOE experiment (1D or 2D) one generally inverts or saturates non-selectively all the transitions of a spin and observes the migration of this disturbance to other spins. In other word, one creates at $\tau=0$ a single spin order, say $A_z(0) = -A_z^o$, with all other modes being in equilibrium. In the initial rate approximation the time development of various modes is then obtained as

$$M_z(\tau) = 2\sigma_{AM}\tau A_z^o$$

$$X_z(\tau) = 2\sigma_{AX}\tau A_z^o$$

$$2A_z M_z(\tau) = 2\Delta_{AM}^A \tau A_z^o$$

$$2A_z X_z(\tau) = 2\Delta_{AX}^A \tau A_z^o \qquad [7]$$

$$4A_z M_z X_z(\tau) = 2\delta_A \tau A_z^o$$

$$2M_z X_z(\tau) = 0$$

$$A_z(\tau) = -2(1 - \rho_A \tau)A_z^o$$

$A_z(\tau)$ describes the self relaxation of spin A and $M_z(\tau)$ and $X_z(\tau)$ give the conventional NOE on spins M and X. The multispin modes are created by cross-correlations. Thus cross-correlations create a multiplet effect which is directly proportional to the cross-correlation spectral densities and the mixing time in the initial rate approximation.

If the multiplet effect is suppressed in an experiment (due to absence of J splittings or use of $90°$ detection pulse or integrating the intensities of a multiplet) then one can use, in the initial rate approximation, an equation of motion of only the single spin modes given by

$$-\frac{d}{dt}\begin{pmatrix} A_z \\ M_z \\ X_z \end{pmatrix} = \begin{pmatrix} \rho_A & \sigma_{AM} & \sigma_{AX} \\ \sigma_{AM} & \rho_M & \sigma_{MX} \\ \sigma_{AX} & \sigma_{MX} & \rho_X \end{pmatrix}\begin{pmatrix} A_z - A_z^o \\ M_z - M_z^o \\ X_z - X_z^o \end{pmatrix} \qquad [8]$$

This is the generalized Solomons equation used in most biological studies (Kalk and Berendsen, 1976). This equation uses a relaxation matrix of dimension N*N, where N is the number of the coupled spins, while in Eq(6) the dimension of the relaxation matrix is $2^N \star 2^N$.

As soon as one uses mixing times away from the initial rate approximation, for example to account for spin diffusion using full relaxation matrix analysis, use of

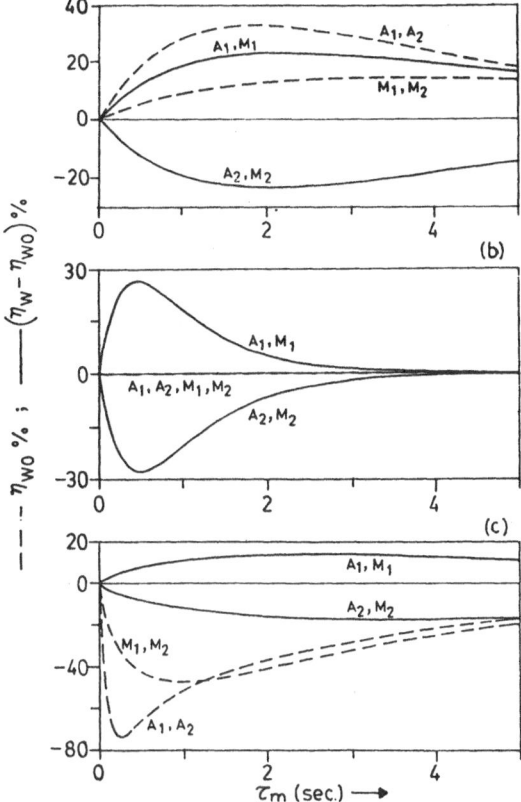

Fig(1): The difference between the calculated transient NOE with and without cross-correlations ($\eta_w - \eta_{wo}$) in percentage (continuous curves), and the NOE without cross-correlations, η_{wo} in percentage (dashed curves), for the A and M multiplets, when the X spin transitions are non-selectively inverted, plotted as a function of the mixing time τ_m for $\omega\tau_c=$(a)0.1, (b)1.118 and (c)10. The parameters used for this simulation are $r_{AM}=2A^o$, $r_{MX}=2.5A^o$, $\angle AMX=\theta=0$, $\omega/2\pi=270$ MHz.

Eq(8) is no more valid since in general there are cross-correlations and one must use Eq(6).

We will calculate in this work and examine the error due to such a neglect, but first we examine the multiplet effect.

MULTIPLET EFFECT

The largest effect of cross-correlations is a multiplet effect. The presence of the multiplet effect in relaxation studies requires design of experiments in which the multispin orders can be detected. For example a non-selective 90° pulse over spin

A and M will not convert $2A_zM_z$ into detectable magnetization. A non-selective 90° pulse on the state of the spin system described by Eq(3) yields equal intensity of all the four transitions given by A_z. A small angle non-selective pulse on all spins or a 90° pulse on one of the spins will make all orders including that spin detectable: the single spin order into a net value and higher orders into antiphase magnetization. A 2D NOESY experiment thus requires, for the detection of multiplet effect, a sequence $(90$-t_1-90-τ_m-α-$t_2)$, in which α is small. The 1D transient NOE experiment also yields similar result in which the detection pulse is a small angle pulse. Fig 1 shows the result of calculation of NOE on spin A or X when the middle spin M is selectively inverted at $\tau_m=0$ assuming that M is in the middle of A and X in a

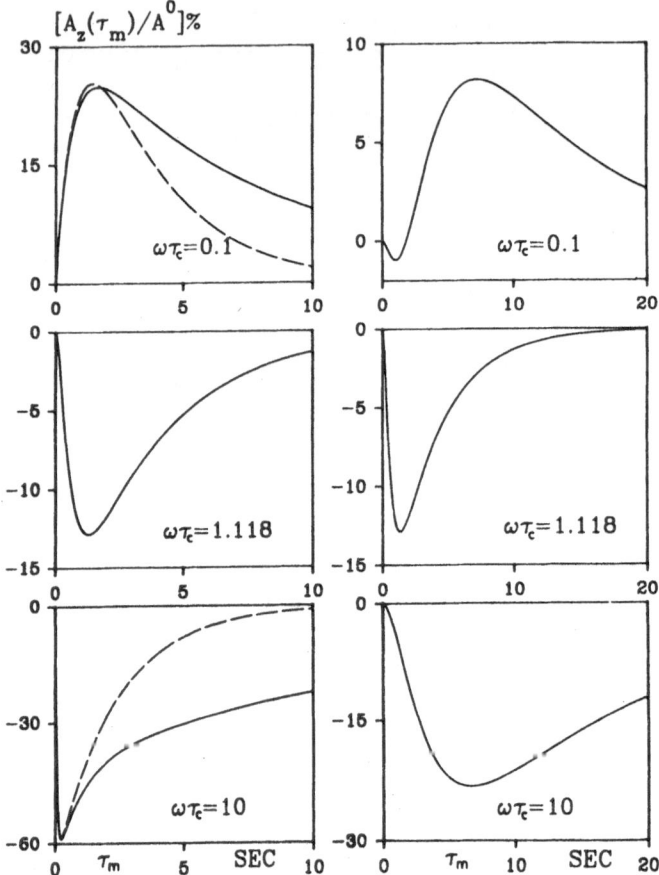

Fig(2): Calculated net NOE in percentage on spin A, after selective inversion of spin M at $\tau_m=0$ is shown as a function of τ_m for the linear geometry ($\theta = 180^o$) of the three spins, for three values of $\omega\tau_c$. In the left hand diagrams the dashed curves represent the calculated net NOE without cross-correlations and the solid curves with cross-correlations and in the right hand diagrams the difference between these two calculated NOE's are shown by solid curves. The parameters used for this simulation are $r_{AM}=r_{MX}=2A^o$ and $\omega/2\pi=270$ MHz.

linear configuration. There is a very large multiplet effect for all values of $\omega\tau_c$ and τ_m. For $\omega\tau_c = 1.118$ the NOE calculated without cross-correlations is zero, but with cross-correlations there is a large multiplet effect (Krishnan and Anil Kumar, 1991).

These multiplet effects are easily detected in NOE experiments in inversion recovery experiments and the multispin order can be spin-locked in experiments utilizing π pulses (Levitt and Di Bari, 1992; Levitt and Di Bari, 1994). The magnitude of the dipole-dipole cross-correlation is geometry dependent and attempts have been made to extract geometric factors from such measurements (Brüschweiler and Ernst, 1992; Ernst and Ernst, 1994).

NET EFFECT

For longer mixing times, the created multispin order reconverts into single spin order using cross-correlation spectral densities. This can be easily seen from curves of Fig 1b, where the positive and negative errors are not exactly equal. Fig 2 summarizes the net NOE in the above spin system calculated with and without cross-correlations. The differences are significant for all correlation times. For $\omega\tau_c=1.118$ the net NOE without cross-correlation is zero and the difference is entirely due to cross-correlations. For $\omega\tau_c=10$, the difference is large but appears much later in time. In the initial rate approximation the error in net NOE is negligible. Similar results have been published by Bull (1987).

An examination of the recovery of inverted spin also shows a large net effect arising due to cross-correlations, confirming the earlier work on non-exponential relaxation of multispins due to cross-correlations (Werbelow and Grant, 1975).

CONCLUSIONS

It is clear from the above analyses that there is a large multiplet effect due to cross-correlations which is first order in time and that there is a significant net effect which is second order in time. The earlier double resonance studies where cross-correlations showed large effects were steady state experiments and included observation of intensities of individual transitions with selective perturbation of selected transitions and had the creation and observation of the multiplet as well as the net effect. In 2D NOE experiments in which the detection pulse is a 90° pulse the multiplet effect is suppressed. The net effect is small for short mixing times but could lead to errors for large mixing times. Inclusion of additional spins and other leakage processes, however, reduce the net NOE as well as the errors arising from cross-correlations (Madhu and Anil Kumar, unpublished results).

Whenever attempts are made to obtain "accurate" distances from NOE measurements utilizing the so called "full" matrix analysis using Eq(6), it may be reasonable to expect that a check is made on the magnitude of cross-correlations. However, since in practice the net effects in multispin systems in presence of leakage processes tend to be small "semi-accurate" distance estimation neglecting cross-correlations should be valid (Wüthrich, 1986).

REFERENCES

Anil Kumar and Nageshwara Rao, B.D., 1968, Relaxation effects in the nuclear magnetic double resonance spectra of a symmetrical three spin system, *Mol.Phys.* 15:377.

Band, M.F., and Hubbard, P.S., 1968, Non-exponential spin-lattice relaxation of protons in solid CH_3CN and solid solutions of CH_3CN in CD_3CN, *Phys.Rev.* 170:384.

Brüschweiler, R., and Ernst, R.R., 1992, Molecular dynamics monitored by cross-correlated cross-relaxation of spins quantized along orthogonal axes, *J.Chem.Phys.* 96:1758 .

Bull, T.E.,1987, Cross-correlation and 2D NOE spectra, *J.Magn.Res.* 72:397.

Canet, D., 1989, Construction, evolution and detection of magnetization modes designed for treating longitudinal relaxation of weakly coupled spin 1/2 systems with magnetic equivalence, *Prog.NMR Spectrosc.* 21:237.

Ernst, R.R., Bodenhausen, G., and Wokaun, A., 1983 "Principles of Nuclear Magnetic Resonance in One and Two Dimensions", Oxford Univ.Press, London.

Ernst, M., and Ernst, R.R.,1994, Heteronuclear dipolar cross-correlated cross relaxation for the investigation of side-chain motions, *J.Magn.Reson.* 110A:202.

Fagerness, P.E., Grant, D.M., Kuhlmann, K.F., Mayne, C.L., and Parry, R.B., 1975, Spin-lattice relaxation in coupled three spin systems of the AIS type, *J.Chem.Phys.* 63:2524.

Hilt, R.L., and Hubbard, P.S., 1964, Nuclear magnetic relaxation of three spin systems undergoing hindered rotations, *Phys.Rev.A.* 134:392.

James, T.L., 1991, *Curr.Opin.Sturct.Biol.* 1:1042.

Kalk, A., and Berendsen, H.J.C., 1976, Proton magnetic relaxation and spin diffusion in proteins, *J.Magn.Reson.* 24:343.

Krishnan, V.V., and Anil Kumar, 1991, Dipolar cross-correlation effects in the nuclear Overhauser experiments of weakly and strongly coupled spins, *J.Magn.Reson.* 92:293.

Levitt, M.H., and Di Bari, L., 1992, The steady state in magnetic resonance pulse experiments, *Phys.Rev.Lett.* 69:3124.

Levitt, M.H., and Di Bari, L., 1994, The homogeneous master equation and the manipulation of relaxation networks, *Bull.Magn.Reson.* 16:94.

Runnels, R.K., 1964, Nuclear spin-lattice relaxation in three spin molecules, *Phys.Rev.A.* 134:28.

Vold, R.L., and Vold, R.R., 1978, Nuclear magnetic relaxation in coupled spin systems, *Prog. NMR Spectrosc.* 12:79.

Werbelow, L.G., and Grant, D.M., 1975, Proton decoupled carbon-13 relaxation in CH_2 and CH_3 spin systems, *J.Chem.Phys.* 63:544.

Werbelow, L.G., and Grant, D.M., 1977, Intramolecular dipolar relaxation in multispin systems, *Adv.Magn.Reson.* 9:189.

Wüthrich, K., 1986, "NMR of Proteins and Nucleic Acids", John Wiley & Sons, New York .

DISCUSSION

David Gorenstein - As we go to higher field, and to larger molecules, and the CSA increases as the square of the field, say if we get a GHz spectrometer, 35-40 kD protein, when do you anticipate the dipole-dipole CSA cross-correlation effects to get worse?

Anil Kumar - My calculations are up to $\omega\tau_c = 10$, but the trend seems to be that as one goes to higher fields, the effect of cross correlation decreases but that of CSA increases. Furthermore as $\omega\tau_c$ increases the magnetization remains within the spin system for a longer time. While the multiplet effect is first order in time and one can certainly use it, the net effect is second order in time (and I am sure that you are worried about the net effect). As one goes to higher $\omega\tau_c$ the net effect will show up at later times and there is less problem at short mixing times.

Thomas James - Your last slide might have left some of us nervous in terms of the possibilities for getting distances accurately. Perhaps you could clarify that and give a little more detail in terms of how careful we need to be in the instance where we are using macromolecules where correlation times might be of the order of nanoseconds.

Kumar - I must say this. What I have taken here is a concocted three spin system without any leakage terms in it. As soon as we add even the fourth spin, the effects tremendously reduce. Add any leakage term to the three spin system, the effects decrease. Basically, it is second order in time; so if you allow the magnetization to remain within the spin system and if you can isolate your spin system, these effects will become significant. If you have a network of spins coupled to many things, my intuition at the moment says that this will not be significant.

B. D. Nageswara Rao - That was the point I was going to ask you about because if there are leakage terms when you have a macromolecule, even the spin system is a very big one and even the concept of keeping the magnetization within the spin system, does mean you distribute it anyway and therefore, it may not be serious if you don't go to too large a mixing time.

Kumar - Absolutely, but there are people who have tried to isolate networks now.

Rao - In those cases you should be careful.

Kumar - That's right.

TOWARDS THE ACCURATE MEASUREMENT OF INTERNUCLEAR DISTANCES IN BIOLOGICAL MACROMOLECULES BY SUPPRESSION OF SPIN DIFFUSION

Sébastien J. F. Vincent, Catherine Zwahlen and Geoffrey Bodenhausen

National High Magnetic Field Laboratory,
1800 E. Paul Dirac Drive, Tallahassee, Florida 32310, USA.

INTRODUCTION

The primary source of information for structure determination of biomolecules in solution is the nuclear Overhauser Effect (nOe) (Overhauser, 1953; Wüthrich, 1986; Neuhaus and Williamson, 1989). Cross-peak intensities a_{ij} obtained by two-dimensional *nuclear Over-hauser effect spectroscopy* (NOESY) (Anil Kumar et al., 1980; Wüthrich, 1986) can be related to proton-proton distances r_{ij}. Cross-relaxation processes stemming from the dipole-dipole interaction between spins can be described by a master equation which for two spins A and X takes the form of two coupled differential equations describing the rate of change of the longitudinal magnetization, the Solomon equations (Solomon, 1955; Ernst et al., 1987):

$$\frac{d\langle I_z^A \rangle}{dt} = -\rho_A \left\{ \langle I_z^A \rangle - \langle I_z^A(0) \rangle \right\} - \sigma_{AX} \left\{ \langle I_z^X \rangle - \langle I_z^X(0) \rangle \right\}$$

$$\frac{d\langle I_z^X \rangle}{dt} = -\rho_X \left\{ \langle I_z^X \rangle - \langle I_z^X(0) \rangle \right\} - \sigma_{XA} \left\{ \langle I_z^A \rangle - \langle I_z^A(0) \rangle \right\}, \tag{1}$$

where I_z^i is the longitudinal magnetization of spin i, $I_z^i(0)$ is the equilibrium longitudinal magnetization of spin i, ρ_i is the *longitudinal relaxation rate* constant of spin i :

$$\rho_i = \rho_i^* + \frac{\kappa}{10} \frac{1}{r_{ij}^6} \left[J_0(0) - 3 J_1(\omega) + 6 J_2(2\omega) \right], \tag{2}$$

with

$$\kappa = \left(\frac{\mu_0}{4\pi}\right)^2 \hbar^2 \gamma^4, \tag{3}$$

NMR As a Structural Tool for Macromolecules: Current Status and Future Directions
Edited by B.D. Nageswara Rao and Marvin D. Kemple, Plenum Press, New York, 1996

and where $J_n(n\omega)$ is the spectral density function at the transition frequency $n\omega$. The second term of Equation (2) describes the dipolar contributions to the relaxation rate, while the first term ρ_i^* represents other pathways for relaxation such as chemical shift anisotropy and external random fields (Bull, 1987; Keeler and Sánchez-Ferrando, 1987; Dalvit and Bodenhausen, 1989). The *cross-relaxation rate* constant σ_{ij} describes the rate of the nOe between spins i and j:

$$\sigma_{ij} = \frac{\kappa}{10} \frac{1}{r_{ij}^6} \left[6 J_2(2\omega) - J_0(0) \right] . \tag{4}$$

Equations (2) and (4) link the relaxation rates ρ and σ with spectral density functions, which in turn can be expressed as a function of molecular parameters once a motional model is defined. Nuclear magnetic resonance allows one to determine molecular structures because the relaxation rates, measurable by NOESY-type experiments, give access to internuclear distances.

For macromolecules, the correlation time τ_c is typically more than 5 ns, and since $J_{n>0}(n\omega) \approx 0$, the cross-relaxation rates (Equation (4)) simplify to:

$$\sigma_{ij} = -\frac{\kappa}{10} \frac{1}{r_{ij}^6} J_0(0) . \tag{5}$$

Longitudinal relaxation is then:

$$\rho_i = \rho_i^* - \sum_{\substack{j=1 \\ j\neq i}}^{N} \sigma_{ij}$$

$$= \rho_i^* + \frac{\kappa}{10} \sum_{\substack{j=1 \\ j\neq i}}^{N} \frac{1}{r_{ij}^6} J_0(0) . \tag{6}$$

For larger spin systems, Equation (1) needs to be extended:

$$\frac{d \, \overrightarrow{\Delta M}(t)}{dt} = - \mathbf{R} \, \overrightarrow{\Delta M}(t) , \tag{7}$$

where the cross-relaxation matrix $\mathbf{R}$ describes the complete dipole-dipole relaxation network and is given by:

$$\mathbf{R} = \begin{pmatrix} \rho_1 & \sigma_{12} & \sigma_{13} & \cdots & \sigma_{1n} \\ \sigma_{21} & \rho_2 & \sigma_{23} & \cdots & \sigma_{2n} \\ \sigma_{31} & \sigma_{32} & \rho_3 & \cdots & \sigma_{3n} \\ \vdots & \vdots & \vdots & \ddots & \vdots \\ \sigma_{n1} & \sigma_{n2} & \sigma_{n3} & \cdots & \rho_n \end{pmatrix} , \tag{8}$$

where n is the number of spins under consideration, and the elements of $\mathbf{R}$ are defined by Equations (5) and (6). The magnetization vector $\overrightarrow{\Delta M}(t)$ describes the deviation from thermal equilibrium of longitudinal magnetization for all spins i, with elements being defined by:

$$\Delta M^i(t) = \left(M_z^i(t) - M_z^i(0) \right)_i .$$ (9)

The system of Equations (7) has the following solution:

$$\overrightarrow{\Delta M}(\tau_m) = \exp\left\{ -\mathbf{R}\,\tau_m \right\} \overrightarrow{\Delta M}_0$$

$$= \mathbf{A}(\tau_m)\,\overrightarrow{\Delta M}_0 .$$ (10)

The two-dimensional NOESY intensities $a_{ij}(\tau_m)$ are proportional to the elements $A_{ij}(\tau_m)$ of the matrix $\mathbf{A}(\tau_m)$, which can be written using a series expansion:

$$\mathbf{A}(\tau_m) = \exp\left\{ -\mathbf{R}\,\tau_m \right\}$$

$$\approx 1 - \mathbf{R}\,\tau_m + \frac{1}{2}\mathbf{R}^2\,\tau_m^2 - \dots + \frac{(-1)^n}{n!}\mathbf{R}^n\,\tau_m^n .$$ (11)

For short mixing times, the higher order terms are negligibly small and the intensity of a cross peak a_{ij} depends only on the direct dipole-dipole interaction between the two spins i and j described by the cross-relaxation rate σ_{ij}. The internuclear distance r_{ij} can therefore be determined directly from the initial slope of a curve displaying the intensity of the cross-peak as a function of the mixing time. This procedure is known as the *initial rate approximation*. To second order in τ_m, however, the amplitudes of cross peaks are also affected by products of cross-relaxation rates $\sigma_{AK}\sigma_{KX}$, $\sigma_{AL}\sigma_{LX}$, etc..., given by the $\mathbf{R}^2$ term in Equation (11). These signal contributions arise from two-step "spin-diffusion" processes such as A $\rightsquigarrow$ K $\rightsquigarrow$ X, A $\rightsquigarrow$ L $\rightsquigarrow$ X, etc... If the duration of the mixing interval is increased, or if the cross-relaxation rates are important as in macromolecules, one also observes n-step processes with contributions proportional to τ_m^n.

The analysis of two-dimensional NOESY spectra can be misleading if spin diffusion effects are not properly taken into account (Kalk and Berendsen, 1976; Anil Kumar et al., 1981; Olejniczak et al., 1986; Lane, 1988; Massefski Jr. and Redfield, 1988). To gain insight into the extent of spin diffusion, one may record "build-up" curves to monitor the cross peak amplitudes a_{ij} as a function of the mixing time τ_m (Anil Kumar et al., 1981). Such curves can be obtained from a series of two-dimensional NOESY spectra, and the resulting build-up curves can be simulated and fitted by considering the simultaneous effects of all cross-relaxation rates σ_{ij} using the "full relaxation matrix" method (Boelens et al., 1988; Borgias and James, 1988; Boelens et al., 1989; Borgias et al., 1990). But while the amplitudes of cross-peaks can easily be determined when the matrix $\mathbf{R}$ is known, the determination of $\mathbf{R}$ from the experimental $\mathbf{A}(\tau_m)$ is not straightforward. Small cross-relaxation rates corresponding to large internuclear distances are particularly prone to errors. The accuracy of the determination of internuclear would be improved if spin diffusion could be quenched (Lane, 1988; Massefski Jr. and Redfield, 1988; Fejzo et al., 1991; Fejzo et al., 1992; Zwahlen et al., 1994), in which case the cross-peak amplitude $a_{ij}(\tau_m)$ would depend *only* on the rate σ_{ij}. An approach consisting in suppressing spin diffusion by selective inversion of the spins undergoing Overhauser effect, known as QUIET-NOESY (Zwahlen et al., 1994), is described in this contribution.

A related feature is the possibility of determining rates of chemical reactions in systems in dynamic equilibrium, where the turnover due to forward- and back-reactions precisely cancels, so that there is no net transfer of material (Jackman and Cotton, 1975). In most forms of spectroscopy, there is little or no evidence that any dynamic processes are taking place, since the concentrations remain time-independent, but magnetic resonance allows one to observe the transfer of magnetization rather than the transfer of material, and it is sufficient that the chemical shifts of the nuclei be affected in the course of the chemical reaction to make the exchange apparent. This can be visualized very effectively by two-dimensional *exchange spectroscopy* (EXSY) (Jeener et al., 1979; Meier and Ernst, 1979), which is essentially equivalent to NOESY (Anil Kumar et al., 1980; Wüthrich, 1986). This effect can also be exploited in line-shape studies (Gutowsky and Saika, 1953; Hoffman, 1970; Stevenson and Binsch, 1979) and in experiments where the longitudinal magnetization of a chosen site is perturbed by selective saturation or inversion (Forsén and Hoffman, 1963; Forsén and Hoffman, 1964; Hoffman and Forsén, 1966) and exchange processes lead to a redistribution of the perturbed magnetization. In a different approach, the behavior of the magnetization in exchanging systems can be modified by selectively inverting suitably chosen magnetization components in the course of the mixing time τ_m, so that the magnetization behaves as if all exchange rates were quenched, except for those leading to the interconversion of a selected pair of sites. The chemical exchange processes proceed unhindered, and the dynamic equilibrium between the chemical reactions is not perturbed. It is only the magnetization which behaves as if the rates were modified. This approach, related to the suppression of spin diffusion by QUIET-NOESY (Zwahlen et al., 1994), is referred to as QUIET-EXSY (Zwahlen et al., 1995), and is presented is this contribution.

SOFT-HOHAHA

Macromolecules usually present crowded one-dimensional spectra. As the sequences presented in this contribution aim at recording information about the dynamic behavior as quickly and efficiently as possible, some means is needed to increase the resolving power, similarly to what can be obtained by resorting to higher-dimensional spectroscopy when a two-dimensional spectrum turns out to be too crowded. The solution consists in using homonuclear selective coherence transfer (soft-HOHAHA) (Konrat et al., 1991; Zwahlen et al., 1993), which allows one to unravel signals buried under many other resonances. Homonuclear coherence transfer arises through a Hartmann-Hahn effect obtained by applying a simultaneous radio-frequency irradiation to two spins of interest, after selective excitation of one of them. The result is an in-phase magnetization transfer from the initially excited spin to its scalar coupling partner. The doubly-selective irradiation shown in Figure 1 is implemented by placing the carrier frequency midway between the two chemical shifts at a frequency $\omega_0 = \frac{1}{2}(\Omega_A + \Omega_X)$, and by modulating a rectangular pulse envelope with $\cos(\omega_a t)$, where $\omega_a = \frac{1}{2}(\Omega_A - \Omega_X)$ (Emsley et al., 1990; Emsley et al., 1991; Konrat et al., 1991; Zwahlen et al., 1993). In order to visualize the effect of amplitude modulation, the cosine function can be expanded:

$$\cos(\omega_a t) = \frac{1}{2}\left\{ \exp(-i\omega_a t) + \exp(+i\omega_a t) \right\} . \tag{12}$$

These two exponential functions correspond to two sidebands rotating in opposite directions at frequencies $\omega_0 \pm \omega_a$, which coincide with the chemical shifts of the two spins of interest, provided the carrier is positioned at $\omega_0 = \frac{1}{2}(\Omega_A + \Omega_X)$. The doubly-selective irradiation needs to have the same amplitude on both spins in order to achieve Hartmann-Hahn transfer, but the matching of the Hartmann-Hahn condition (Hartmann and Hahn, 1962) is always fulfilled in homonuclear experiments as a result of the way the two sidebands

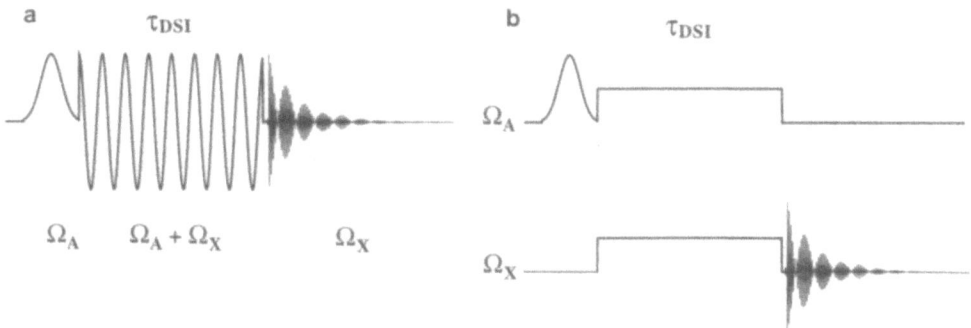

Figure 1: Pulse sequence for soft-HOHAHA. (a) Representation of the actual pulse envelopes. (b) Same pulse sequence, but with one line for every frequency irradiated, similar to heteronuclear pulse sequences. In-phase magnetization I_X^A is initially created by a selective excitation pulse, typically a self-refocusing 270° Gaussian pulse with a duration of 30 ms for a 40 Hz wide multiplet.(Emsley and Bodenhausen, 1989) This is immediately followed by a doubly-selective irradiation of duration τ_{DSI} implemented as described in the text. For optimum transfer, τ_{DSI} is set to $1/J$ (Konrat et al., 1991; Zwahlen et al., 1993). A 180° phase alternation of the initial pulse and the receiver phase is used.

are generated. The pulse sequence shown in Figure 1a represents the envelope of the soft-HOHAHA pulse sequence on one line, where the doubly-selective irradiation is represented as a cosinusoidal oscillation. The location of the two sidebands in frequency domain is indicated below each pulse. Figure 1b is another representation of the same pulse sequence, showing one line for every frequency irradiated. This provides an easy identification of the frequencies and shapes of the various selective pulses and sidebands and gives a direct comparison with non-selective heteronuclear cross-polarization.

In the absence of relaxation, the optimum duration of the doubly-selective irradiation is given by the inverse of the active coupling constant (Konrat et al., 1991; Zwahlen et al., 1993):

$$\tau_{DSI} = \frac{1}{J_{AX}} \cdot \tag{13}$$

In practice, the duration of the doubly-selective irradiation can easily be calibrated empirically by recording a series of one-dimensional soft-HOHAHA experiments with a variable duration τ_{DSI}. This is conveniently achieved by repeating n times an amplitude-modulated pulse of duration:

$$\tau_{DSI} = n\,k\,\frac{2\pi}{\omega_a}, \tag{14}$$

where k is an integer which determines the length of the calibration step and n another integer defining the total length of the pulse (Zwahlen et al., 1994). The amplitude of the doubly-selective irradiation can also be optimized empirically to give the best in-phase transfer (Zwahlen et al., 1992), although this step can in most cases be avoided, if the radio-frequency amplitude is set to around 30 Hz for each sideband. If the amplitude is much weaker, it is impossible to lock all magnetization components within each multiplet simultaneously, whereas if the radio-frequency amplitude is too large, various passive spins, some of which

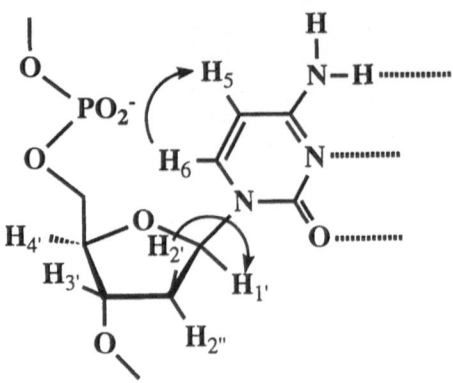

Figure 2: Cytosine-9 residue of the self-complementary duplex dodecamer B-DNA d(CGCGAATTCGCG)$_2$, known as "Dickerson's dodecamer". The two soft-HOHAHA experiments of Figure 3 are represented by arrows starting at the initially excited spin and pointing towards the spin receiving the magnetization (C9-H2' → C9-H1' and C9-H6 → C9-H5).

may have chemical shifts that lie accidentally in the vicinity of the sidebands, will also be perturbed.

Because of its selectivity and thanks to the use of difference spectroscopy, this experiment reveals a multiplet of interest buried under overlapping resonances. The transfer of in-phase magnetization from spin A to (one of) its coupling partners X leads to a single signal in the vicinity of the chemical shift of spin X, the in-phase X-multiplet. This is demonstrated for protons in the self-complementary B-DNA duplex dodecamer d(CGCGAATTCGCG)$_2$ known as "Dickerson's dodecamer" (Drew et al., 1981; Patel et al., 1982; Hare et al., 1983; Nerdal et al., 1989; Ravishankar et al., 1989; Kaluarachchi et al., 1991; Swaminathan et al., 1991; Withka et al., 1991; Withka et al., 1992). The focus was on one of the residues,

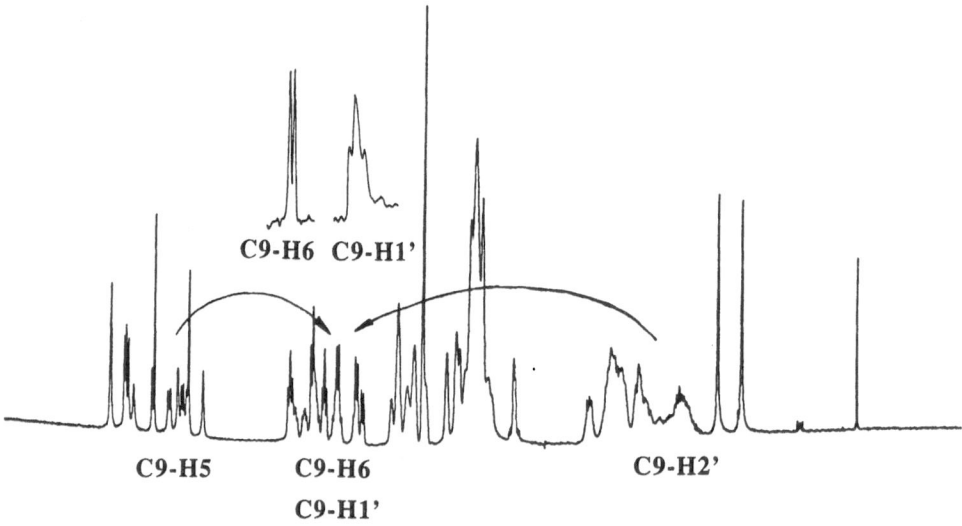

Figure 3: One-dimensional spectrum of the self-complementary duplex dodecamer B-DNA d(CGCGAA-TTCGCG)$_2$. Two soft-HOHAHA transfers are shown, C9-H2' → C9-H1' and C9-H6 → C9-H5, with the resulting multiplets.

Cytosine-9, shown in Figure 2. The spectra of Figure 3 illustrates that with this experiment, two overlapping multiplets (belonging to C9-H1' and C9-H5 in this example) can be separated. There is no limitation concerning the number of overlapping protons.

The soft-HOHAHA experiment has been shown to be very effective in one-dimensional spectroscopy (Konrat et al., 1991), in particular for the selective measurement of relaxation parameters in complicated spectra (Boulat et al., 1992b), and for selective injection of magnetization to separate accidentally overlapping multiplets in two-dimensional spectroscopy (Zwahlen et al., 1992). In two-dimensional experiments, it has been used as a mixing sequence to achieve the selective equivalent of isotropic mixing (Vincent et al., 1992; Vincent et al., 1993).

QUICK-NOESY

With the help of linear prediction and singular value decomposition methods, it has been shown that it is possible to analyze a single build-up curve and decompose the envelope into its constituent exponential components (Malliavin et al., 1992; Reisdorf et al., 1992). This approach requires a good signal-to-noise ratio and a large number of τ_m increments, typically fifty equally spaced mixing times. As extensive series of two-dimensional NOESY experiments are very time consuming, experimental instabilities may affect the cross-peak amplitudes. With the time-efficient one-dimensional QUICK-NOESY method (*q*uantitative *u*nraveling of *i*ntensities for *c*orroborating *k*nowledge in *n*uclear *O*verhauser *e*ffect *s*pectroscopy) (Vincent et al., 1994), the amplitudes of selected individual cross-peaks in a NOESY survey spectrum can be quantitated efficiently and rapidly.

Figure 4 shows the QUICK-NOESY pulse sequence starting with a selective inversion of the longitudinal magnetization of a selected "source" spin A, which is allowed to evolve through cross relaxation during the interval τ_m. A fraction of this magnetization migrates from the source spin A to a "target" spin X. In principle, if there are no overlaps in the NOESY cross-peaks, it should be possible to observe the magnetization of the X spin directly. In practice however, limitations in the accuracy of difference spectroscopy and overlaps typical of macromolecules make this difficult. Therefore, the soft-HOHAHA scheme is added as a second mixing event after τ_m: the longitudinal component of the target spin X is converted into transverse magnetization by a selective excitation pulse, and then transferred to a scalar-coupled "spy" nucleus M. If the transfer efficiency is less than 100%, this affects only the sensitivity of the experiments, but not the accuracy of the build-up curves. The cross-relaxation behavior in this experiment is exactly the same as in NOESY, but unlike the usual NOESY technique, QUICK-NOESY does not suffer from artifacts due to zero-quantum coherences (so-called *J* cross-peaks) (Huang et al., 1981), nor can it be affected by longitudinal two-spin order (Bodenhausen et al., 1984).

The QUICK-NOESY build-up curves can be compared directly with molecular dynamics (MD) simulations. In the example of Figure 5, six build-up curves were measured with the pulse sequence of Figure 4 in the Cytosine-9 residue of the self-complementary B-DNA duplex dodecamer d(CGCGAATTCGCG)$_2$. The relaxation processes towards the aromatic proton C9-H6 were measured from six spatial neighbors on the same residue (Figure 2) in order to compare the experiments with curves predicted by Withka *et al.* (Withka et al., 1991). In all cases a soft-HOHAHA transfer was used from the target spin C9-H6 towards the aromatic coupling partner C9-H5. These experiments lead to an evaluation of the relative merits of the six published models. MD simulations were carried out with two different potentials. A first set was obtained with the standard potential energy function of

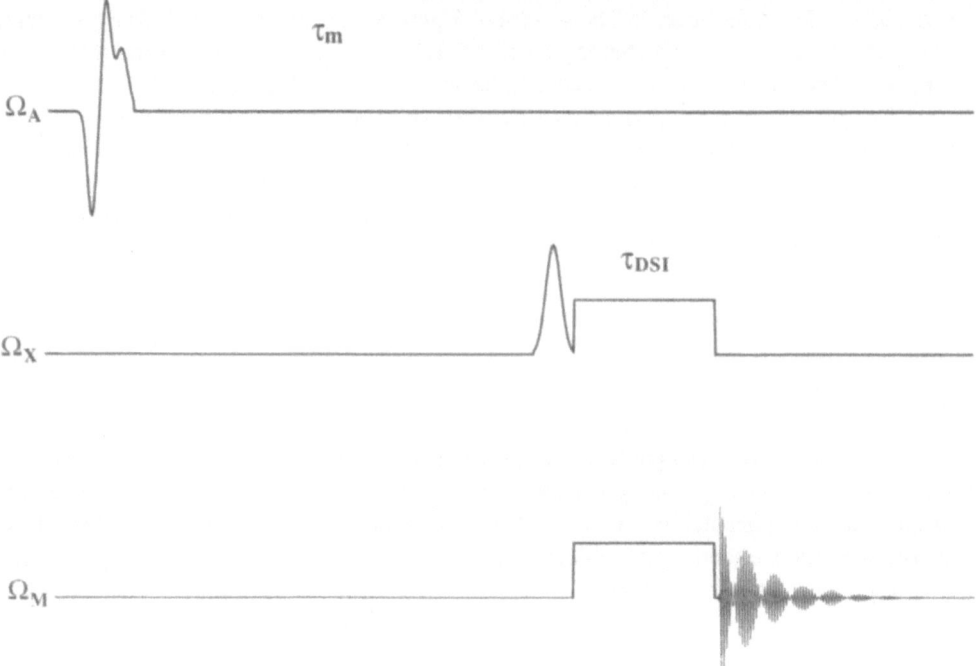

Figure 4. Sequence for selective build-up measurements of transient Overhauser effects (QUICK-NOESY). The longitudinal magnetization of a selected source spin A is first inverted by applying a Q^3 Gaussian cascade (Emsley and Bodenhausen, 1992) at the chemical shift Ω_A. A fraction of the magnetization migrates through cross relaxation during the interval τ_m towards the target spin X. The resulting longitudinal component is converted into transverse magnetization by a 270° Gaussian pulse (Emsley and Bodenhausen, 1989), and transferred to a scalar-coupled partner M through a soft-HOHAHA brought about by a doubly-selective irradiation of duration $\tau_{DSI} \approx 1 / J_{MX}$ (Konrat et al., 1991; Zwahlen et al., 1993). A difference spectrum is obtained by subtracting a signal recorded without the initial inversion so that only the deviations from thermal equilibrium have to be taken into account. For long τ_m values, the signals decays towards zero, like in NOESY.

GROMOS86 (van Gunsteren and Berendsen, 1986; van Gunsteren and Berendsen, 1990), and the second with an improved potential in which harmonic constraints accounting for hydrogen bonds are included (Swaminathan et al., 1991; Withka et al., 1991). For both potentials, three cases of increasing complexity were considered for the calculation of nOe build-up curves (Withka et al., 1991): (i) Isotropic tumbling with a single correlation time of 6 ns. The nOe build-up curves depend only on the interproton distances. (ii) The overall motion was considered to be anisotropic. i.e. two correlation times were considered to account for the barrel-like structure of the duplex B-DNA. (iii) Internal motion was included by considering its effect on the Overhauser build-up rates. In the last two cases, the orientation of each proton-proton vector with respect to the long axis of the duplex B-DNA has been taken into account. The dynamical and geometrical information of the last 80 ps (out of 140 ps) of MD trajectories was used to evaluate nOe build-up curves by integration of the set of coupled equations (Equation (7)) for all 208 protons of the molecule.

The comparison of the calculated build-up curves (Withka et al., 1991) with the corresponding experimental results clearly indicates that the simulations with the GROMOS86

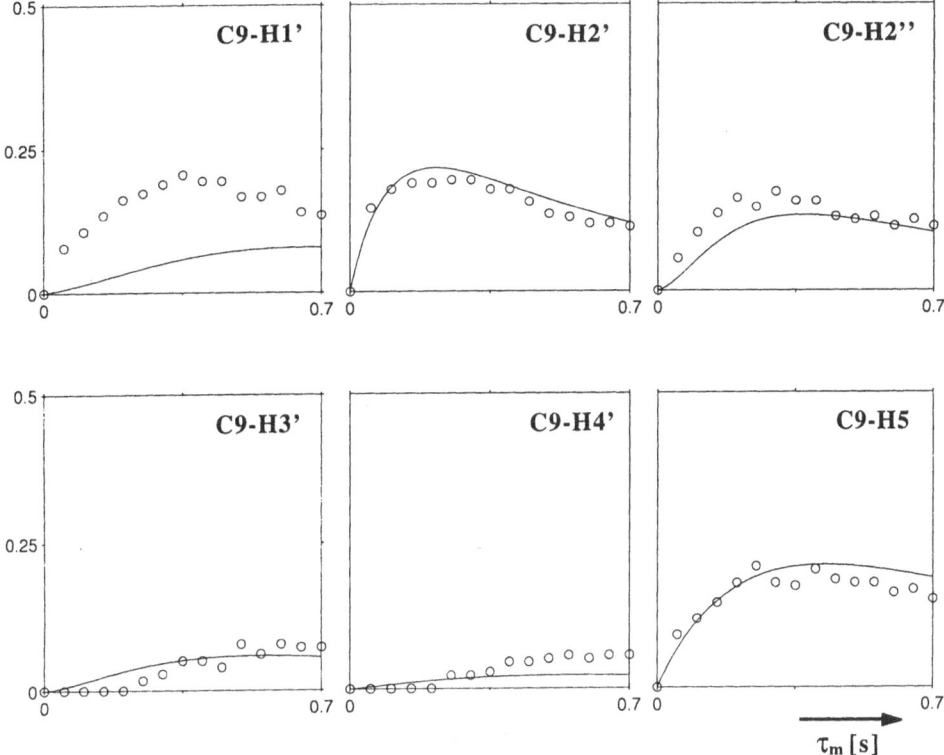

Figure 5. Cross-relaxation processes to the C9-H6 proton of cytosine-9 in the self-complementary B-DNA duplex dodecamer d(CGCGAATTCGCG)$_2$. Comparison of between build-up curves predicted by MD (Withka et al., 1991) and experimental points obtained by QUICK-NOESY (Vincent et al., 1994). The source and target protons are indicated in each graph. Note the discrepancies for the C9-H1' ⇝ C9-H6 process, which cannot be rationalized in the framework of the MD simulations.

potential are not satisfactory, whatever motional model is taken into account. A much better agreement is obtained with the modified potential. By comparing the simulations, which includes internal motion as well as anisotropy, with the experimental results, as shown in Figure 5, one notices that the agreement between experimental observations and calculations is rather good, except for one significant divergence for the process C9-H1' ⇝ C9-H6, which was predicted to be nearly as weak as the C9-H3' ⇝ C9-H6 conversion, but which appears to be much stronger in the experiments. This comparison has been made possible within a very limited time span due to the use of one-dimensional selective QUICK-NOESY experiments.

QUIET-NOESY

The influence of multi-step processes competing with direct single-step cross-relaxation pathways has long been recognized, but methods based on the suppression of this perturbing effect are still in development. The first approach consists in saturating *some* perturbing spins (Massefski Jr. and Redfield, 1988), but this requires the knowledge of the position of the perturbing spins, and a favorable chemical shift dispersion. A more sophisticated approach exploits the fact that cross-relaxation rates in the laboratory and the rotating frames have

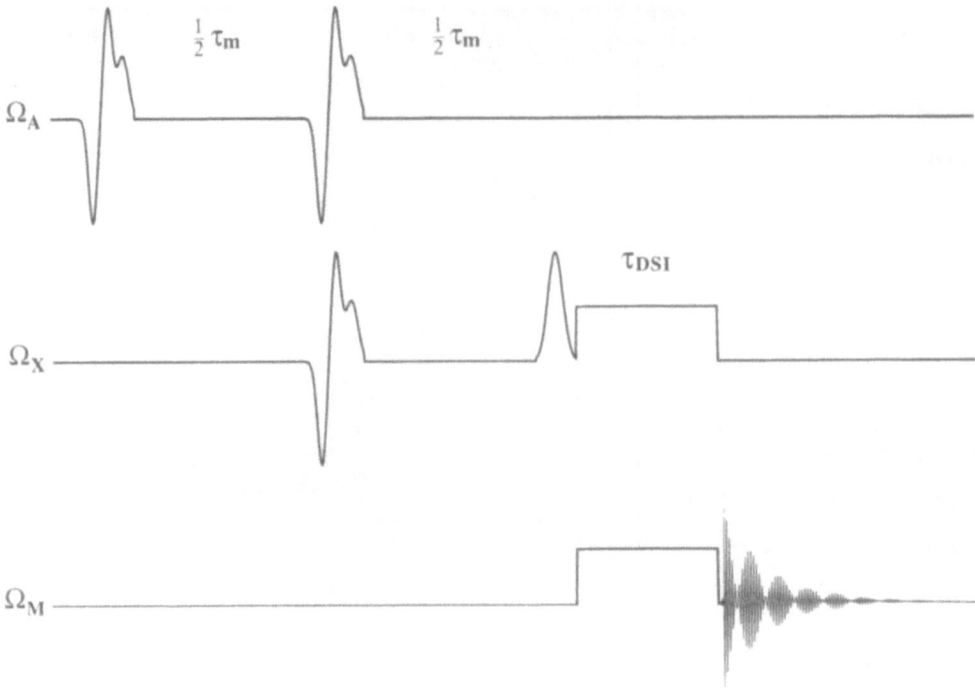

Figure 6. QUIET-NOESY pulse sequence for selective measurements of nOe build-up curves with quenching of spin diffusion (Zwahlen et al., 1994). The longitudinal magnetization of a selected source spin A is first inverted by applying a Q^3 Gaussian cascade (Emsley and Bodenhausen, 1992) at the chemical shift Ω_A. A fraction of the magnetization is then allowed to migrate freely through cross relaxation towards the target spin X during the first interval $\frac{1}{2}\tau_m$, at which point the magnetization vectors of both the source and the target spins A and X are simultaneously inverted using a cosine modulated Q^3 Gaussian cascade (Emsley et al., 1990; Emsley et al., 1991; Zwahlen et al., 1994). The resulting longitudinal component on spin X is converted into transverse magnetization by a 270° Gaussian pulse (Emsley and Bodenhausen, 1989) and transferred to a scalar-coupled partner M through a soft-HOHAHA effect (Konrat et al., 1991; Zwahlen et al., 1993). A difference spectrum is obtained by subtracting a signal recorded without the initial inversion and without the doubly-selective inversion.

opposite signs, so that all Overhauser effects can be canceled, except for those involving a spin or group of spins that are inverted by selective pulses (Fejzo et al., 1991; Fejzo et al., 1992). Another approach (Boulat et al., 1992a; Burghardt et al., 1993; Boulat and Rance, 1994; Zwahlen et al., 1994) consists in focusing on a selected pair of spins and excluding all perturbing non-direct relaxation pathway, without requiring knowledge of the existence or of the shifts of the perturbing spin(s). The QUIET-NOESY method (Zwahlen et al., 1994) shown in Figure 6 makes it possible to study cross relaxation in all motional regimes (i.e. to both small and large molecules) without interference due to spin diffusion.

The mechanism of the QUIET-NOESY can be illustrated by numerical integration of the set of coupled differential equations describing the spin relaxation (Equation (7)). We consider a AXKLMN six spin-system, with a geometry chosen for sake of illustration (Figure 7). With a correlation time of 7 ns, a frequency of 500 MHz, and a contribution from other relaxation mechanisms $\rho_i^* = 1\ \text{s}^{-1}$, the system had the following relaxation matrix:

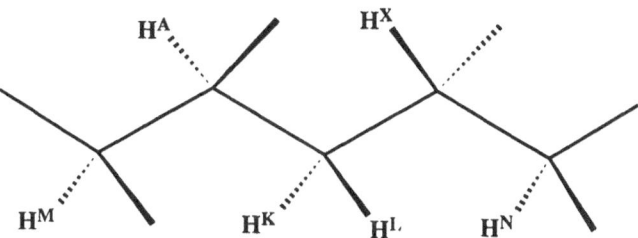

Figure 7. Schematic representation, with some internuclear distances, of the geometry of the six spin system AXKLMN defining the relaxation rates of Equation (15).

$$\mathbf{R} = \begin{array}{cccccc} A & X & K & L & M & N \\ \begin{pmatrix} 3.80 & -0.15 & -1.13 & -0.38 & -1.13 & -0.01 \\ -0.15 & 2.82 & -0.15 & -0.38 & -0.01 & -1.13 \\ -1.13 & -0.15 & 2.39 & -0.07 & -0.03 & -0.01 \\ -0.38 & -0.38 & -0.07 & 1.87 & -0.02 & -0.02 \\ -1.13 & -0.01 & -0.03 & -0.02 & 2.19 & 0.00 \\ -0.01 & -1.13 & -0.01 & -0.02 & 0.00 & 2.17 \end{pmatrix} \end{array}. \qquad (15)$$

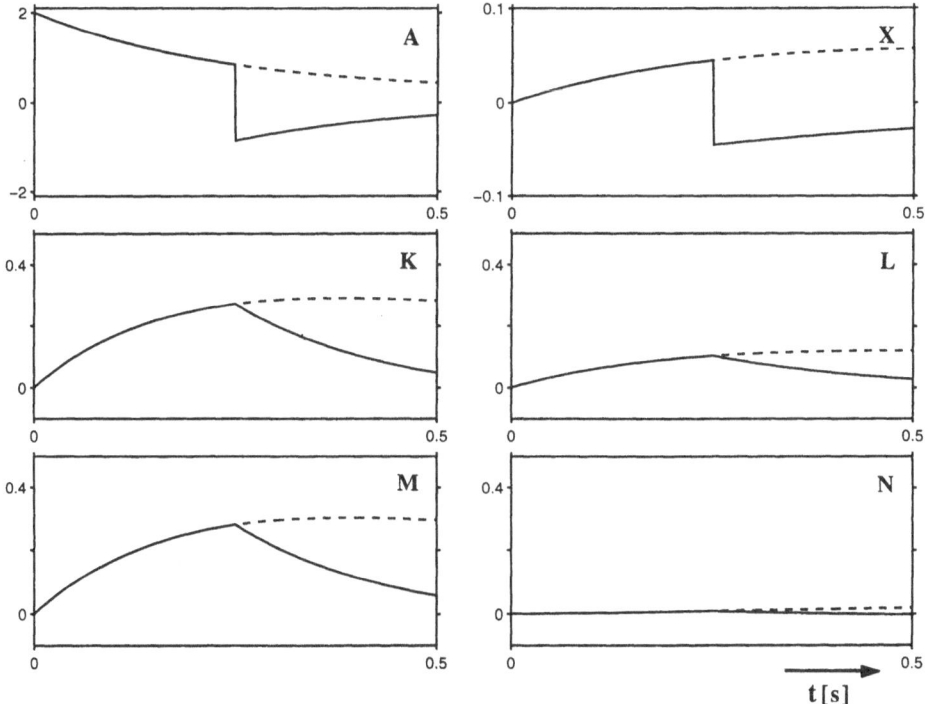

Figure 8. Simulations of the time dependence of the polarizations $\langle I_z^i \rangle$ of the six spins of Figure 7 *in the course* of the relaxation interval τ_m, i.e. for $0 < t < \tau_m$, when the sign of $\langle I_z^A \rangle$ and $\langle I_z^X \rangle$ is changed at $\frac{1}{2}\,\tau_m$ (QUIET-NOESY, —) or unaffected (QUICK-NOESY, – – –). $\mathbf{R}$ is defined in Equation (15).

155

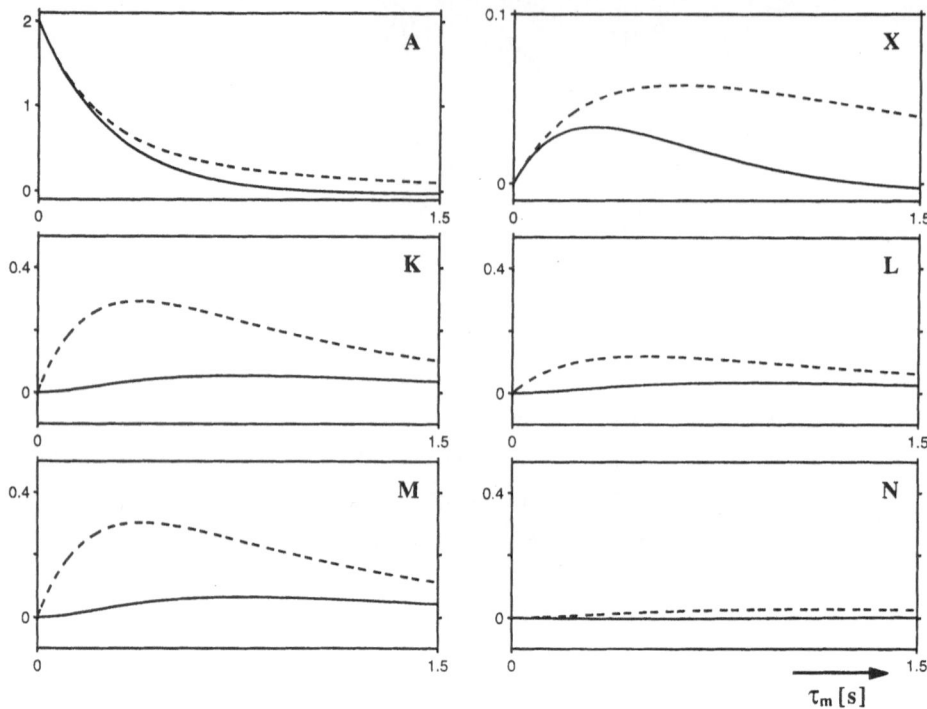

Figure 9. Comparison of the decay of $\langle I_z^A \rangle$ and the build-up of $\langle I_z^i \rangle$ polarizations of the six spins of Figure 7 *as a function* of the relaxation interval τ_m, assuming the relaxation matrix of Equation (15). As in Figure 8, QUICK-NOESY (− − −) is without manipulations during the mixing time, while QUIET-NOESY (—) is with a doubly selective inversion of A and X in the middle of τ_m.

We shall focus attention on the measurement of the internuclear distance $r_{AX} = 3.5 \text{ Å}$ between spins A and X. Figures 8 and 9 show the evolution of longitudinal magnetization $\langle I_z^i \rangle (t)$ for the six spins of Figure (7) for QUICK-NOESY and QUIET-NOESY experiments. Figure 8 was obtained with a single mixing time. In the QUICK-NOESY case, the decay of the source spin A is accompanied by a build-up of the magnetization on the target spin X, but also on all other spins, according to the relaxation rates of Equation (15). If one inverts the two spins of interest to simulate a QUIET-NOESY experiment, the result is that only the magnetization on the target spin X is significant at the end of the mixing time, and its amplitude is due primarily to the first order process A ⤳ X. Figure 9 shows the build-up curves (and the decay for the source spin A) for all six spins, which should be comparable with experimental build-up curves recorded by detecting the longitudinal magnetizations at the end of the variable mixing time τ_m. It is clearly visible that in the QUICK-NOESY experiment, which contains the same information as a series of two-dimensional NOESY spectra, the magnetization flows unrestricted through the whole relaxation network. On the other hand, in the QUIET-NOESY method, only the direct path between the two chosen spins is observed to a significant extent and the build-up curve on the target spin X is largely determined by the internuclear distance between A and X.

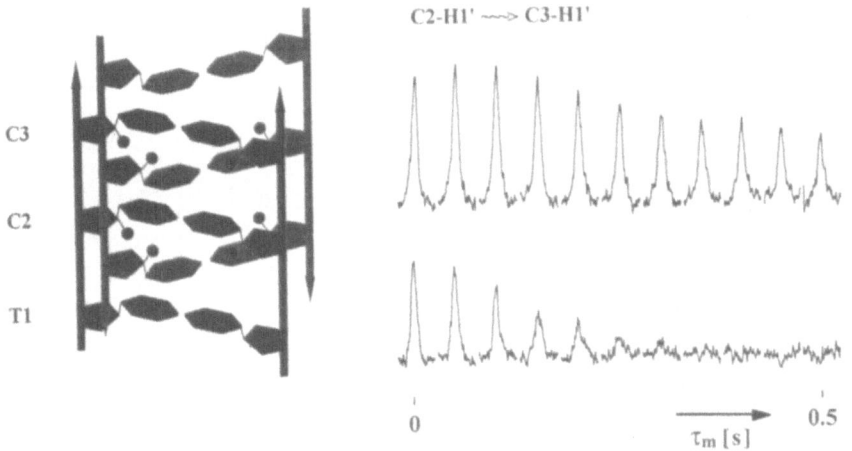

C2-H1' ⤳ C3-H1'

τ_m [s]

Figure 10. (left) Structure of the I-tetrad of d(TCC)₄, as proposed by Guéron and coworkers (Gehring et al., 1993; Guéron, 1995). The C2-H1' and C3-H1' protons are highlighted by spheres. Due to symmetry, the eight protons give rise to only two signals. (top right) QUICK-NOESY and (bottom right) QUIET-NOESY build-up curves for the C2-H1' ⤳ C3-H1' cross-relaxation.

The QUIET-NOESY experiment may be used to confirm the existence of a direct cross-relaxation pathway. In the example of Figure 10, the aim was to determine whether the nOe between two protons in an unusual structure postulated for DNA was genuine or not. The work of Guéron and coworkers (Gehring et al., 1993; Guéron, 1995), based essentially on NMR data, showed that they had characterized a highly unusual DNA tetrameric structure, an I-tetrad, for d(TCCCC)₄ in water solution. In addition to a highly degenerate one-dimensional spectrum which indicates a structure of high symmetry, an unusual Overhauser effect was found between H1' protons on different ribose units. This could not be explained if the strands would lie in the usual double-stranded structure (Gehring et al., 1993; Guéron, 1995). To confirm that this was indeed a direct effect, a series of QUICK-NOESY and QUIET-NOESY spectra were acquired for the sample with the structure shown in Figure 10. The spectra clearly show a conservation of the very strong nOe between C2-H1' and C3-H1' in d(TCC)₄, thus providing additional support for the proposed I-tetrad structure.

QUIET-EXSY

Consider a system in chemical exchange where we wish to focus attention on the conversion of magnetization from a source site A to a target site X. There may be one or several "clandestine" sites K, L, M, ..., so that the magnetization transfer due to the direct conversion A ⤳ X may be contaminated by two-step processes such as A ⤳ K ⤳ X, etc... The misleading contributions to the transfer of magnetization from A to X due to these two-step processes can be removed by selective inversion of longitudinal magnetization in the middle of the mixing time τ_m. Earlier studies (Forsén and Hoffman, 1963; Forsén and Hoffman, 1964; Hoffman and Forsén, 1966) have focused on the selective saturation of unwanted sites, requiring knowledge of their resonance frequencies. The experiment

157

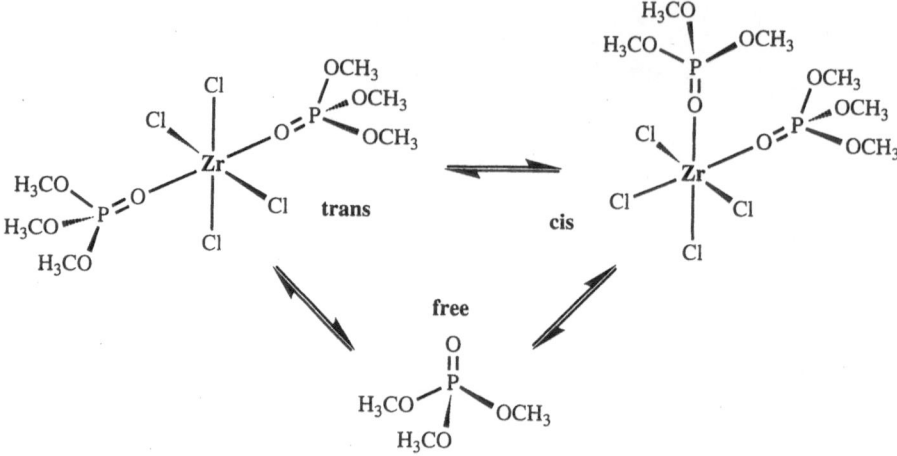

Figure 11. Exchange network of *trans*-ZrCl$_4$L$_2$, *cis*-ZrCl$_4$L$_2$, and *free* ligand L = (CH$_3$O)$_3$PO. The sample solution was prepared as described in the work of Merbach and co-workers (Frey et al., 1990; Turin-Rossier et al., 1990) by mixing appropriate quantities of ZrCl$_4$, trimethyl phosphate (CH$_3$O)$_3$PO, and CDCl$_3$ under dry nitrogen atmosphere.

presented in this contribution, dubbed *q*uenching *u*ndesirable *i*ndirect *e*xternal *t*rouble in *ex*change *s*pectroscopy (QUIET-EXSY) (Zwahlen et al., 1995), relies on the cancellation of unwanted effects, but the only resonance frequencies that must be known are those of the sites of interest and no further assumptions need to be made concerning the number and frequencies of the clandestine sites. In addition, there is no need for saturation required, thus lifting the uncertainties about the efficiency of saturation and long continuous irradiation of the sample can be avoided.

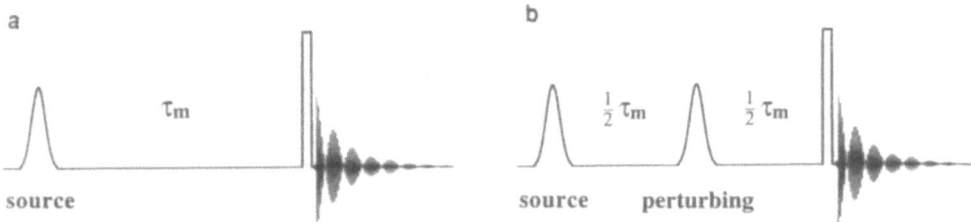

Figure 12. (a) selective one-dimensional EXSY. (b) QUIET-EXSY (Zwahlen et al., 1995). The selective inversion pulses with Gaussian envelopes truncated at 2.5 % have 180° nutation angles. The inversion profile of a Gaussian pulse is very sharp,(James and Roos, 1975) but suitable for singlets. If multiplets with a finite width must be inverted, it is advisable to use shaped pulses with a "top hat" profile such as Q^3 or I-BURP pulses (Geen and Freeman, 1991; Emsley and Bodenhausen, 1992). However, for a given selectivity, more sophisticated shaped pulses tend to have a longer duration than a Gaussian pulse and are therefore less suitable for studying fast exchange processes. In the example of Figure 13, the 180° Gaussian pulse was necessary in order to invert separately the *trans* and the *cis* sites which were only 83 Hz apart. The first 180° Gaussian pulse must be applied to the source site, the second to the perturbing site(s), or alternatively to both the source and the target sites, as described in the text. The final rectangular pulse represents a non-selective 90° pulse. In the difference spectrum obtained by subtracting the resulting spectrum from an ordinary spectrum, the signals decay asymptotically to zero for large τ_m.

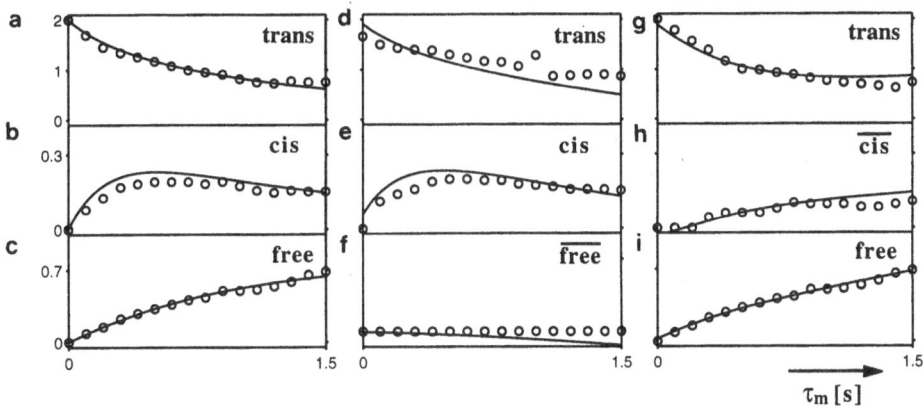

τ_m [s]

Figure 13. Curves corresponding to the matrix of Equation (16). The nine curves were obtained by fitting the experimental data points (o) with a non-linear SIMPLEX algorithm applied to Equation (10) modified with **L** instead of **R** in the MATLAB program (MATLAB,) and using the MINUIT package (James and Roos, 1975). The data points show the time-dependence of the ^{31}P resonances of the zirconium complex of Figure 11. (a) - (c) Conventional difference spectra obtained with the sequence of Figure 12a by inverting the *trans* resonance selectively in odd transients using a 180° Gaussian truncated at 2.5 % of 21 ms duration and subtracting the signals from a normal spectrum. The mixing time was incremented from 0 to 1.5 s. The three resonances are observed after a non-selective 90° pulse of 22 μs, and only three narrow windows (20 Hz wide) are shown for each τ_m value. (d) - (f) QUIET-EXSY experiments obtained by applying a selective inversion pulse to the *free* resonance (inversion symbolized by overbar on $\overline{free}$) in the middle of the τ_m interval. (g) - (i) Analogous QUIET-EXSY experiments, but swapping the roles of the *free* and *cis* sites by using a selective inversion of the *cis* resonance ($\overline{cis}$).

To illustrate the QUIET-EXSY method, we have chosen a three-site exchange network consisting of a zirconium complex present in two forms in solution, *trans*-ZrCl$_4$L$_2$ and *cis*-ZrCl$_4$L$_2$, and an excess of *free* ligand L = (CH$_3$O)$_3$PO, as shown in Figure 11. This equilibrium has been studied by varying the concentration, temperature and pressure in proton NMR by Merbach and co-workers (Frey et al., 1990; Turin-Rossier et al., 1990). The proton-decoupled phosphorus spectrum of this sample consists of three singlets, corresponding to the phosphate resonances of the *trans*, *cis* and *free* ligands respectively. If we consider the *trans* species as the source site A, which may be selectively inverted, the entire spectrum can be observed after a mixing time τ_m and a non-selective 90° monitoring pulse. The sequence shown in Figure 12a allows one to measure the intensity of EXSY cross-peaks in a selective one-dimensional fashion.

As shown in Fig. 13a-c, one can observe the result of selective EXSY: magnetization migrates from the source site *trans* to the target site *cis*, but also to the clandestine site *free*. In the proposed modification of this experiment, the QUIET-EXSY experiment shown in Figure 12b, the longitudinal magnetization of the clandestine site *free* is selectively inverted in the middle of the mixing time. The same result could be achieved by doubly-selective inversion of both the source and target sites, i.e. $\overline{trans}$ and $\overline{cis}$. This alternative approach is also effective if the resonance frequency of the clandestine *free* site were unknown, or indeed if there were an unknown number of invisible clandestine sites. In both cases, the transfer of magnetization from *trans* to *free* changes sign during the second half of τ_m, so that the net transfer is very small at the end of the mixing period. Thus the magnetization behaves, to first order, as if there were no reaction from *trans* to *free*, nor, for that matter, from *free* to *cis*. A different effect, shown in Figure 13g-i, can be achieved if the roles of the target and clandestine site are swapped, i.e. if the target site is the *free* resonance and the clandestine site the *cis* resonance. In this case, only the exchange between the *trans* and *free* sites should be observed, since the *cis* site should be "decoupled". In spite of the inversion of the *cis* resonance ($\overline{cis}$, Figure 13g-i), or equivalently the simultaneous inversion of both the $\overline{trans}$ and $\overline{free}$, the conversion of the magnetization from *trans* to *cis* (Figure 13i) is not completely suppressed because of second order effects.

In cases where exchange has to be taken into account, the set of coupled differential equations (Equation (7)) has to be modified by replacing the relaxation matrix $\mathbf{R}$ by a matrix $\mathbf{L}$ given by the difference of the kinetic matrix $\mathbf{K}$ describing chemical exchange and the relaxation matrix $\mathbf{R}$, describes all incoherent processes (Ernst et al., 1987) taking place during the mixing time τ_m. Numerical simulations can be used to represent the evolution of magnetization during the two halves of the mixing time of the QUIET-EXSY sequence. The effect of the selective pulses can be assessed by including all effects of relaxation and exchange described by the matrix $\mathbf{L}$ in a numerical integration of the Liouville-von Neumann equation in the time-course of the pulses. This shows that the selective EXSY experiment is strongly influenced by the clandestine site, while QUIET-EXSY is very close to the case with continuous saturation. For selective EXSY, the direct determination of the exchange rates is difficult, as the indirect migration of magnetization through the perturbing site is important. From an experimental point of view, saturation is hard to achieve, it is difficult to assess its efficiency and not easy to saturate perfectly $n-2$ sites simultaneously when n is larger than three. On the other hand, the implementation of QUIET-EXSY is straightforward and allows one to isolate a two-site subsystem from an n-site system.

In the $ZrCl_4L_2$ system, further insight can be obtained by initially inverting the magnetization of various source spins other than the *trans* resonance. This would amount to comparing cross-sections taken from two-dimensional exchange matrices at different frequencies in the ω_1 domain. All experimental points, such as the ones shown in Figure 13, can be fitted together in order to retrieve the exchange rates. The results can be represented in matrix form:

$$\mathbf{L} = \begin{pmatrix} L_{trans,\,trans} & L_{trans \leftarrow cis} & L_{trans \leftarrow free} \\ L_{cis \leftarrow trans} & L_{cis,\,cis} & L_{cis \leftarrow free} \\ L_{free \leftarrow trans} & L_{free \leftarrow cis} & L_{free,\,free} \end{pmatrix}$$

$$= \begin{pmatrix} 1.47 \pm 0.02 & -3.32 \pm 0.20 & -0.138 \pm 0.006 \\ -0.71 \pm 0.20 & 3.63 \pm 0.15 & -0.018 \pm 0.005 \\ -0.431 \pm 0.06 & -0.270 \pm 0.005 & 0.22 \pm 0.006 \end{pmatrix}. \qquad (16)$$

The QUIET-EXSY method is related to synchronous nutation (Boulat et al., 1992a; Burghardt et al., 1993; Boulat and Rance, 1994) and to QUIET-NOESY (Zwahlen et al., 1994). Exchange matrices, in contrast to the Solomon matrices describing cross-relaxation, are asymmetrical if the equilibrium populations are unequal. Furthermore, the "kinetic window" of possible exchange rates is far wider than the typical range of cross-relaxation rates. On the other hand, the number of sites is usually more limited in chemically exchanging systems. One of the chief motivations of our technique is the wish to measure slow exchange rates that tend to be overshadowed by faster competing processes. Selective experiments should allow one to determine exchange rates with improved accuracy. The idea of continuous saturation has been proposed in the original work of Hoffman and Forsén (Forsén and Hoffman, 1963; Forsén and Hoffman, 1964; Hoffman and Forsén, 1966), but proper saturation (as opposed to inhomogeneous scrambling of the magnetization) is difficult to achieve quickly, since it can only be obtained on a time-scale on the order of T_2.

In systems with more than three sites, it is generally not possible to "decouple" all undesirable sites with a simple monochromatic inversion pulse, unless they happen to be degenerate, for example because of fast exchange. One has the option between simultaneously inverting all sites K, L, M, ... that should be excluded, or simultaneously inverting the two sites A and X that one wishes to investigate. A doubly-selective inversion can easily be implemented in practice by using cosine-modulated pulses (Emsley et al., 1990; Emsley et al., 1991; Zwahlen et al., 1994). The inversion of A and X allows one to investigate systems where the chemical shifts of the clandestine spins are not known and makes it possible to suppress competing processes regardless of the number of clandestine sites. This stands in contrast to earlier methods for the suppression of unwanted processes (Forsén and Hoffman, 1963; Forsén and Hoffman, 1964; Hoffman, 1970; Massefski Jr. and Redfield, 1988; Fejzo et al., 1991; Fejzo et al., 1992).

ACKNOWLEDGMENTS

We are indebted to M. H. Levitt, L. Di Bari, M. Schwager, J. M. Withka, P. H. Bolton, J.-L. Leroy, M. Guéron, B. Boulat, and L. Helm. This research was supported by the Fonds National Suisse de la Recherche Scientifique (FNRS), by the Commission pour l'Encouragement de la Recherche Scientifique (CERS) of Switzerland, and by the National High Magnetic Field Laboratory (NHMFL), Tallahassee, Florida.

REFERENCES

Anil Kumar, R. R. Ernst and K. Wüthrich, A Two-Dimensional Nuclear Overhauser Enhancement (2D NOE) Experiment for the Elucidation of Complete Proton-Proton Cross-Relaxation Networks in Biological Macromolecules, *Biochem. Biophys. Res. Comm.* 95:1-6 (1980).

Anil Kumar, G. Wagner, R. R. Ernst and K. Wüthrich, Buildup Rates of the Nuclear Overhauser Effect Measured by Two-Dimensional Proton Magnetic Resonance Spectroscopy: Implications for Studies of Protein Conformation, *J. Am. Chem. Soc.* 103:3654-3658 (1981).

G. Bodenhausen, G. Wagner, M. Rance, O. W. Sørensen, K. Wüthrich and R. R. Ernst, Longitudinal Two-Spin Order in 2D Exchange Spectroscopy (NOESY), *J. Magn. Reson.* 59:542-550 (1984).

R. Boelens, T. M. G. Koning and R. Kaptein, Determination of Biomolecular Structures from Proton-Proton NOE's Using a Relaxation Matrix Approach, *J. Mol. Struct.* 173:299-311 (1988).

R. Boelens, T. M. G. Koning, G. A. van der Marel, J. H. van Boom and R. Kaptein, Iterative Procedure for Structure Determination from Proton-Proton NOEs Using a Full Relaxation Matrix Approach. Application to a DNA Octamer, *J. Magn. Reson.* 82:290-308 (1989).

B. A. Borgias, M. Gochin, D. J. Kerwood and T. L. James, Relaxation Matrix Analysis of 2D NMR Data, *Prog. NMR Spectrosc.* 22:83-100 (1990).

B. A. Borgias and T. L. James, COMATOSE, A Method for Constrained Refinement of Macromolecular Structure Based on Two-Dimensional Nuclear Overhauser Effect Spectra, *J. Magn. Reson.* 79:493-512 (1988).

B. Boulat, I. Burghardt and G. Bodenhausen, Measurement of Overhauser Effects in Magnetic Resonance of Proteins by Synchronuous Nutation, *J. Am. Chem. Soc.* 114:10679 (1992a).

B. Boulat, R. Konrat, I. Burghardt and G. Bodenhausen, Measurement of Relaxation Rates in Crowded NMR Spectra by Selective Coherence Transfer, *J. Am. Chem. Soc.* 114:5412-5414 (1992b).

B. Boulat and M. Rance, Monitoring of Slow Conformational Exchange by Doubly-Selective Irradiation in Nuclear Magnetic Resonance Spectroscopy, *J. Chem. Phys.* 101:7273-7282 (1994).

T. E. Bull, Cross Correlation and 2D NOE Spectra, *J. Magn. Reson.* 72:397-413 (1987).

I. Burghardt, R. Konrat, B. Boulat, S. J. F. Vincent and G. Bodenhausen, Measurement of Cross Relaxation between Two Selected Nuclei by Synchronuous Nutation of Magnetization in Nuclear Magnetic Resonance, *J. Chem. Phys.* 98:1721-1736 (1993).

C. Dalvit and G. Bodenhausen, Proton Chemical Shift Anisotropy: Detection of Cross-Correlation with Dipole-Dipole Interations by Double-Quantum Filtered Two-Dimensional NMR Exchange Spectroscopy, *Chem. Phys. Lett.* 161:555-560 (1989).

H. R. Drew, R. M. Wing, T. Takano, C. Broka, S. Tanaka, K. Itakura and R. E. Dickerson, Structure of a B-DNA Dodecamer: Conformation and Dynamics, *Proc. Natl. Acad. Sci. USA* 78:2179-2183 (1981).

L. Emsley and G. Bodenhausen, Self-Refocusing Effect of 270° Gaussian Pulses. Applications to Selective Two-Dimensional Exchange Spectroscopy, *J. Magn. Reson.* 82:211-221 (1989).

L. Emsley and G. Bodenhausen, Optimization of Shaped Selective Pulses for NMR Using a Quaternion Description of Their Overall Propagators, *J. Magn. Reson.* 97:135-148 (1992).

L. Emsley, I. Burghardt and G. Bodenhausen, Double Selective Inversion in NMR and Multiple Quantum Effects in Coupled Spin Systems, *J. Magn. Reson.* 90:214-220 (1990).

L. Emsley, I. Burghardt and G. Bodenhausen, Correction to: Double Selective Inversion in NMR and Multiple Quantum Effects in Coupled Spin Systems, *J. Magn. Reson.* 94:448 (1991).

R. R. Ernst, G. Bodenhausen and A. Wokaun, "Principles of Nuclear Magnetic Resonance in One and Two Dimensions", Clarendon Press, Oxford (1987).

J. Fejzo, A. M. Krezel, W. M. Westler, S. Macura and J. L. Markley, Direct Cross-Relaxation NOESY (D.NOESY). A Method for Removing Spin-Diffusion Cross Peaks from Two-Dimensional NOE Spectra of Macromolecules, *J. Magn. Reson.* 92:651-657 (1991).

J. Fejzo, W. M. Westler, J. L. Markley and S. Macura, Complete Elimination of Spin Diffusion from Selected Resonances in Two-Dimensional Cross-Relaxation Spectra of Macromolecules by a Novel Pulse Sequence (S.NOESY), *J. Am. Chem. Soc.* 114:1523-1524 (1992).

S. Forsén and R. A. Hoffman, Study of Moderately Rapid Chemical Exchange Reactions by Means of Nuclear Magnetic Double Resonance, *J. Chem. Phys.* 39:2892-2901 (1963).

S. Forsén and R. A. Hoffman, Exchange Rates by Nuclear Magnetic Multiple Resonance. III. Exchange Reactions in Systems with Several Nonequivalent Sites, *J. Chem. Phys.* 40:1189-1196 (1964).

U. Frey, L. Helm and A. E. Merbach, 22. A Variable-Pressure 2D 1H-NMR Study of the Mechanism of Trimethyl Phosphate Intermolecular Exchange and cis/trans-Isomerisation of Tetrachlorobis(trimethyl phosphate)zirconium(IV), *Helvetica Chimica Acta* 73:199-202 (1990).

H. Geen and R. Freeman, Band-Selective Radiofrequency Pulses, *J. Magn. Reson.* 93:93-141 (1991).

K. Gehring, J.-L. Leroy and M. Guéron, A Tetrameric DNA Structure with Protonated Cytosine•Cytosine Base Pairs, *Nature* 363:561-565 (1993).

M. Guéron, *NMR of Symmetrical Assemblies of Self-Recognizing Oligonucleotides;* B. D. N. Rao and M. D. Kemple, Eds.; Plenum Press: New York, 1995; see next contribution in this Volume.

H. S. Gutowsky and A. Saika, Dissociation, Chemical Exchange, and the Proton Magnetic Resonance in Some Aqueous Electrolytes, *J. Chem. Phys.* 21:1688-1694 (1953).

D. R. Hare, D. E. Wemmer, S.-H. Chou, G. P. Drobny and B. R. Reid, Assignment of the Non-Exchangeable Proton Resonances of d(CGCGAATTCGCG) Using Two-Dimensional Nuclear Magnetic Resonance Methods, *J. Mol. Biol.* 171:319-336 (1983).

162

S. R. Hartmann and E. L. Hahn, Nuclear Double Resonance in the Rotating Frame, *Phys. Rev.* 128:2042-2053 (1962).

R. A. Hoffman, Line Shapes in High-Resolution NMR, *Adv. Magn. Reson.* 4:87-200 (1970).

R. A. Hoffman and S. Forsén, Transient and Steady-State Overhauser Experiments in the Investigation of Relaxation Processes. Analogies between Chemical Exchange and Relaxation., *J. Chem. Phys.* 45:2049-2060 (1966).

Y. Huang, S. Macura and R. R. Ernst, Carbon-13 Exchange Maps for the Elucidation of Chemical Exchange Networks, *J. Am. Chem. Soc.* 103:5327-5333 (1981).

L. M. Jackman and F. A. Cotton, "Dynamic NMR Spectroscopy", Academic Press, New York (1975).

F. James and M. Roos, MINUIT - A System for Function Minimization and Analysis of the Parameter Errors and Correlation, *Computer Phys. Commun.* 10:343-367 (1975).

J. Jeener, B. H. Meier, P. Bachmann and R. R. Ernst, Investigation of Exchange Processes by Two-Dimensional NMR Spectroscopy, *J. Chem. Phys.* 11:4546-4553 (1979).

A. Kalk and H. J. C. Berendsen, Proton Magnetic Relaxation and Spin-Diffusion in Proteins, *J. Magn. Reson.* 24:343-366 (1976).

K. Kaluarachchi, R. P. Meadows and D. G. Gorenstein, How Accurately Can Oligonucleotide Structures Be Determined from the Hybrid Relaxation Rate Matrix / NOESY Distance Restrained Molecular Dynamics Approach ?, *Biochem.* 30:8785-8797 (1991).

J. Keeler and F. Sánchez-Ferrando, The Influence of Cross Correlation on Multiplet Patterns in Nuclear Overhauser Effect Spectra, *J. Magn. Reson.* 75:96-109 (1987).

R. Konrat, I. Burghardt and G. Bodenhausen, Coherence Transfer in Nuclear Magnetic Resonance by Selective Homonuclear Hartmann-Hahn Correlation Spectroscopy, *J. Am. Chem. Soc.* 113:9135-9140 (1991).

A. N. Lane, The Influence of Spin Diffusion and Internal Motions on NOE Intensities in Proteins, *J. Magn. Reson.* 78:425-439 (1988).

T. E. Malliavin, M. A. Delsuc and J. Y. Lallemand, Computaton of Relaxation Matrix Elements from Incomplete NOESY Data Sets, *J. Biomol. NMR* 2:349-360 (1992).

W. Massefski Jr. and A. G. Redfield, Elimination of Multiple-Step Spin Diffusion Effects in Two-Dimensional NOE Spectroscopy of Nucleic Acids, *J. Magn. Reson.* 78:150-155 (1988).

MATLAB, (c) Copyright 1984-1994 The Mathworks Company, Inc., Version 4.2a

B. H. Meier and R. R. Ernst, Elucidation of Chemical Exchange Networks by Two-Dimensional NMR Spectroscopy: The Heptamethylbenzeonium Ion, *J. Am. Chem. Soc.* 101:6441-6442 (1979).

W. Nerdal, D. R. Hare and B. R. Reid, Solution Structure of the EcoRI DNA Sequence: Refinement of NMR-Derived Distance Geometry Structures by NOESY Spectrum Back-Calculations, *Biochemistry* 28:10008-10021 (1989).

D. Neuhaus and M. Williamson, "The Nuclear Overhauser Effect in Structural and Conformational Analysis", VCH, Weinheim (1989).

E. T. Olejniczak, R. T. Gampe Jr. and S. W. Fesik, Accounting for Spin Diffusion in the Analysis of 2D NOE Data, *J. Magn. Reson.* 67:28-41 (1986).

A. W. Overhauser, Polarization of Nuclei in Metals, *Phys. Rev.* 92:411-415 (1953).

D. J. Patel, A. Pardi and K. Itakura, DNA Conformation, Dynamics, and Interactions in Solution, *Science* 216:581-590 (1982).

G. Ravishankar, S. Swaminathan, D. L. Beveridge, E. Lavery and H. Sklenar, Conformational and helicoidal Analysis of 30 ps of Molecular Dynamics on the d(CGCGAATTCGCG) Double Helix: "Curves", Dials, and Windows, *J. Biomol. Struct. Dyn.* 6:669-699 (1989).

C. Reisdorf, T. E. Malliavin and M. A. Delsuc, Accurate Estimation of Inter-Atomic distances in Large Proteins by NMR, *Biochimie* 74:809-813 (1992).

I. Solomon, Relaxation Processes in a System of Two Spins, *Phys. Rev.* 99:559-565 (1955).

D. S. Stevenson and G. Binsch, Iterative Computer Analysis of Conplex Exchange-Broadened NMR Bandshapes, *J. Magn. Reson.* 32:145-152 (1979).

S. Swaminathan, G. Ravishankar and D. L. Beveridge, Molecular Dynamics of B-DNA Including Water and Counterions: A 140ps Trajectory for d(CGCGAATTCGCG) Based on the GROMOS Force Field, *J. Am. Chem. Soc.* 113:5027-5040 (1991).

M. Turin-Rossier, D. Hugi-Cleary, U. Frey and A. E. Merbach, Adducts of Zirconium and Hafnium Tetrachlorides with Neutral Lewis Bases. 2. Kinetics and Mechanisms: A Variable-Temperature and -Pressure 1H NMR Study, *Inorg. Chem.* 29:1374-1379 (1990).

W. F. van Gunsteren and H. J. C. Berendsen, *Groningen Molecular Simulation System;* ; University of Groningen: 1986;

W. F. van Gunsteren and H. J. C. Berendsen, Computer Simulation of Molecular Dynamics: Methodology, Applications, and Perspectives in Chemistry, *Angew. Chem. Int. Ed. Engl.* 29:992-1023 (1990).

S. J. F. Vincent, C. Zwahlen and G. Bodenhausen, High-Resolution Two-Dimensional In-Phase Multiplets in Nuclear Magnetic Resonance Correlation Spectroscopy, *J. Am. Chem. Soc.* 114:10989-10990 (1992).

S. J. F. Vincent, C. Zwahlen and G. Bodenhausen, Selective Magnetic Resonance Correlation Spectroscopy with In-Phase Multiplets, *J. Am. Chem. Soc.* 115:9202-9209 (1993).

S. J. F. Vincent, C. Zwahlen and G. Bodenhausen, Selective Measurement of the Time-Dependence of Transient Overhauser Effects in Magnetic Resonance. Applications to Oligonucleotides, *Angew. Chem. Int. Ed. Engl.* 33:343-346 (1994).

J. M. Withka, S. Swaminathan, D. L. Beveridge and P. H. Bolton, Time Dependence of Nuclear Overhauser Effects of Duplex DNA from Molecular Dynamics Trajectories, *J. Am. Chem. Soc.* 113:5041-5049 (1991).

J. M. Withka, S. Swaminathan, J. Srinivasan, D. L. Beveridge and P. H. Bolton, Towards a Dynamical Structure of DNA: Comparison of Theoretical and Experimental NOE Intensities, *Science* 255:597-599 (1992).

K. Wüthrich, "NMR of Proteins and Nucleic Acids", Wiley, Chichester (1986).

C. Zwahlen, S. J. F. Vincent and G. Bodenhausen, Separation of Overlapping Multiplets in Two-Dimensional NMR Spectra by Selective "Injection" of Magnetization, *Angew. Chem. Int. Ed. Engl.* 31:1248-1251 (1992).

C. Zwahlen, S. J. F. Vincent and G. Bodenhausen, *Proceedings of the International School of Physics Enrico Fermi;* B. Maraviglia, Eds.; North-Holland Publishers: Amsterdam, 1993; 397-412.

C. Zwahlen, S. J. F. Vincent, L. Di Bari, M. H. Levitt and G. Bodenhausen, Quenching Spin Diffusion in Selective Measurements of Transient Overhauser Effects in Nuclear Magnetic Resonance. Applications to Oligonucleotides, *J. Am. Chem. Soc.* 116:362-368 (1994).

C. Zwahlen, S. J. F. Vincent, M. Schwager and G. Bodenhausen, Isolation of Selected Exchange Processes in Nuclear Magnetic Resonance, *work in preparation* (1995).

C. Zwahlen, S. J. F. Vincent, A. Ziegler and G. Bodenhausen, Characteristic Patterns of Metabolites from Selective Two-Dimensional Proton NMR, *J. Magn. Reson. Ser. B* 103:299-302 (1994).

DISCUSSION

Marc Adler - It is about the exchange studies. When you showed that there was not much difference between a single inversion and a string of inversions, that was based on the assumption that the rates of exchange were on the same timescale. In terms of an exchange, it seems to me that if you have a rapid equilibrium between free and *cis*, that you could still have a fast way to transfer to *cis* completely before the first inversion.

Sébastien Vincent - As Anil Kumar mentioned in his talk and as S. Macura pointed out yesterday, when the processes are getting faster you have to do more inversions; in homonuclear systems, the pulses have to be pretty long and to stay practical, we have to limit ourselves to less inversion. So as far as we can work with one, we will do it. Then if we need to do more, we'll have to do it. That's the way we look at it.

Anil Kumar - Your isolation of the spin system is an ideal situation where I think a cross correlation net effect should be seen. Have you calculated those?

Vincent - We are getting interested in these effects now. That's all I can say. No, we haven't calculated them but we are getting into that.

Philip Hajduk - I just had a question about the relaxation during the pulses. I noticed that when you went to ^{31}P NMR, that you went to the Gaussians because they had better relaxation properties.

Vincent - No, because they are shorter.

Hajduk - Was that done to neglect the cross relaxation effects?

Vincent - Yes.

Hajduk - But I was wondering if there are any non-obvious reasons why you didn't go to other pulses for the Q^3 because there are other pulses that have better effects with respect to relaxation or are there things in the experiment that will preclude this?

Vincent - To be accurate, I should have said that some of the curves recorded were not recorded using only Q^3. In fact, we used the pulses which had better properties at the beginning points in the sequence. We have found in studies of relaxation for the inversion pulse in the beginning of the sequence and in the middle that the two problems are different. In this case it turned out that Q^3 are better in the middle, while 1-BURP-2 are more efficient for the initial inversion. I think we are also working on that seriously to compare different pulses and have a recipe that we can propose like starting with that first and then have one in the middle and so on.

B. D. Nageswara Rao - Does this method of selective double inversion work quite well as $\omega \tau_c$ goes up? Are there any experimental limitations or practical limitations?

Vincent - You mean in the generation of the pulse?

Rao - Well, the mixing times you have used very often are in the second range. One doesn't have that long a time for mixing.

Vincent - If your measurements will be in the 20 ms range, of course these experiments will not be applicable. I don't see how at the present time. The two pulses we use for inversion are now on the order of 30 ms.

Rao - The pulses themselves are about 30 ms?

Vincent - Yes.

Rao - So you have to interpolate it between your half mixing times?

Vincent - Yes.

Rao - So, if your mixing times are shorter than about 100-200 ms, it's not very practical.

Vincent - Smaller than 100 ms, I'd say it would be very difficult to do although even if the process is very fast, you can still see whether the process or the relaxation process that you are looking at is a genuine or direct one. If you still have something, you then know that this comes from a direct process. Now making that in a quantitative fashion is the thing we are looking at now as far as recording and analyzing the relaxation during the pulses are concerned.

Rao - Another thing which prompted me to ask this question is that the numerical relaxation matrix that you showed for the three spin system (one of the later ones) you had some numbers that don't add up to zero.

Vincent - Oh, yes. We added some leakage terms.

Rao - Thank you.

Lev Gorenstein - I didn't catch that. You derived new sequences, right? You compared QUICK, QUIET and TIMES. Did you compare these to build-up curves as in normal NOE?

Vincent - Well, QUICK is providing the same as normal NOESY. If you ask whether we did that experiment, the answer is 'no'. The answer is 'no' because we never ran a series of NOESY sequences with the same sample that we are running in our selective experiment.

NMR OF SYMMETRICAL ASSEMBLIES OF SELF-RECOGNIZING OLIGONUCLEOTIDES

Maurice Guéron, Kalle Gehring[1] and Jean-Louis Leroy

Groupe de Biophysique de l'Ecole Polytechnique et de l'URA D1254 du CNRS, 91128 Palaiseau, France.
[1]Present address: Dept. of Chemistry, McGill University, 3655 Drummond St., Montreal, Quebec H3G 176, Canada.

ABSTRACT OF THE TALK

At slightly acid or even neutral pH, oligodeoxynucleotides carrying a stretch of cytidines form a tetramer in which two parallel-stranded duplexes are intimately associated, with their hemi-protonated $C \cdot C^+$ base pairs face-to-face and fully intercalated, in a so-called "*i-motif*". This structure, first observed in the tetramer of d(TCCCCC)[1], is also formed by other sequences, including the very short d(TCC). In all cases, the four strands appear as identical in the NMR spectrum. Therefore, the spectrum does not reveal the stochiometry, cross-peaks between non-corresponding protons are 4-times ambiguous, and cross-peaks between corresponding protons of different strands are unobservable without isotope substitution. One can nevertheless detect the *i-motif* and solve structures containing it. Detection relies on the determination of stoichiometry, on the recognition of $C \cdot C^+$ pairs, of 3'-endo conformation and of very slow proton exchange, and on the observation of three types of NOE cross-peaks between protons of intercalating strands. The first type is a direct H1'-H1' cross-peak. The second is a direct cross-peak between an amino proton and an H2' ribose proton. The third is a rectangular pattern of indirect NOEs created by the head-to-head stacking of bases. The amino-ribose NOEs determine the intercalation topology and provide distance constraints which are precious for structure resolution because they involve the *bases*: even if all other NOE constraints are ignored, they force the formation of the intercalated structure. A high-resolution structure of the tetramer of d(TCC) will be presented, and its topology will be compared to those of analogous oligonucleotides. We shall also discuss the symmetry requirements for the formation of the NOE rectangular pattern.

COMMENTED REFERENCES ON THE i-MOTIF

1. The article by Gehring *et al.* (in reference 1) reports the discovery of the i-motif structure of the tetramer of d(TC$_5$) by 2D NMR, the identity of the four strands on the NMR time scale, the inter-residue sugar-sugar NOEs such as H1'-H1', and the conjugate NOE cross-peaks for heterologous inter-residue connections such as H1'-H6.

NMR As a Structural Tool for Macromolecules: Current Status and Future Directions
Edited by B.D. Nageswara Rao and Marvin D. Kemple, Plenum Press, New York, 1996

167

2. In the article by Leroy *et al.* (in reference 1), the stoichiometry of other oligomers, some containing as few as three cytidines, is studied, and tetrameric stoichiometry is generally observed. Amino and imino proton spectra are used to establish the structure and symmetry of the base pairs. The amino proton spectrum gives the best lower limit $(8 \cdot 10^4 \ s^{-1})$ on the rate of imino proton jumping across the base pair. An NMR study of the tetramer/single-strand equilibrium of $d(TC_3)$ gives a free energy of -7.6 kJ/mol per cytidine base pair.

It is shown that imino proton exchange is limited by base-pair opening, thanks to efficient intrinsic exchange catalysis. The base-pair lifetime is hundreds of times longer than in any DNA duplex, presumably due to the base-pair intercalation geometry. The internal amino proton exchanges from the open state of the $C \cdot C^+$ pair, at a rate compatible with a pK of 9, as for protonated cytidine. But for the external amino proton, which exchanges from the closed state, the pK is 17! Comparable pK shifts occur upon transfer of organic acids from an aqueous to an organic solvent.

3. A high-definition NMR structure of the d(TCC) tetramer is presented in reference 2. The structure is computed on the basis of 21 inter-residue NOESY cross-peaks, measured at short mixing times and free of spin-diffusion contributions. The spectrum is exceptionally well resolved, due to the small size of the molecule and to the four-fold symmetry. The characteristic (amino proton-sugar proton) cross-peaks entail by themselves the intercalation topology (Figure 1).

The tetramer has two wide and two narrow grooves, across which the shortest inter-phosphorus distances are respectively 1.43 and 0.60 nm. Each duplex is right-handed with a small helical twist, ranging from 10 to 25 degrees. The thymines are on the outside, and appear unpaired. The cytosine base pairs show no systematic propeller twist. The nucleosides are anti. The sugar puckers are C3'-endo/C4'-exo, in the same range as for some complexes of DNA with intercalating drugs. The structure could accomodate a hydrogen-bonded water molecule bridge between amino proton and phosphate, for which decisive NMR evidence is lacking.

The structure of $d(5mCCT)_4$, solved by the same procedures, is similar to that of $d(TCC)_4$. But in the case of d(T5mCC), the NOESY spectrum is that of two separate tetramers in comparable proportions. The intercalation topology is read off the (amino proton)-(sugar proton) cross-peak pattern: in one tetramer, it is similar to $d(TCC)_4$; in the other, meshing of the intercalated strands is shifted by one base, and this avoids the steric hindrance between the methyl groups of the m5C pairs of the two duplexes.

4. The systematic intercalation found in the i-motif may surprise, since intercalating drugs are reported to create excluded sites next to the intercalation site. However, when the electric charge of the intercalating drug is taken into account, the binding isotherm corresponds to only one excluded site, namely that occupied by the drug itself.[3]

There is no obvious reason why intercalation should be restricted to cytosine or to parallel-stranded duplexes. Models of multiple-strand DNAs which respect atomic volumes and covalent bond geometry have been proposed.[4,5]

5. One could expect that the i-motif, which forms by tetrameric assembly of a strand containing a cytosine stretch such as d(TCC), could also form by intramolecular folding of one strand carrying four such stretches. This is of interest since telomeres and centromeres contain multiple repeats of sequences containing cytosine stretches, complementrary to the guanosine stretches which can fold into the so-called G-tetrads. The fragments of the C-rich telomeric strand of vertebrates, $d[CCCTAA]_3CCC$, and of *Tetrahymena*, $d[CCCCAA]_3CCCC$, are studied in reference 6. In that work, the i-motif is recognized by the characteristic Overhauser cross-peaks of the proton NMR spectrum, reflecting short H1'-H1' distances across the narrow groove, and short internucleotide amino-proton - H2'/H2" across the wide groove. The spectra also demonstrate that the

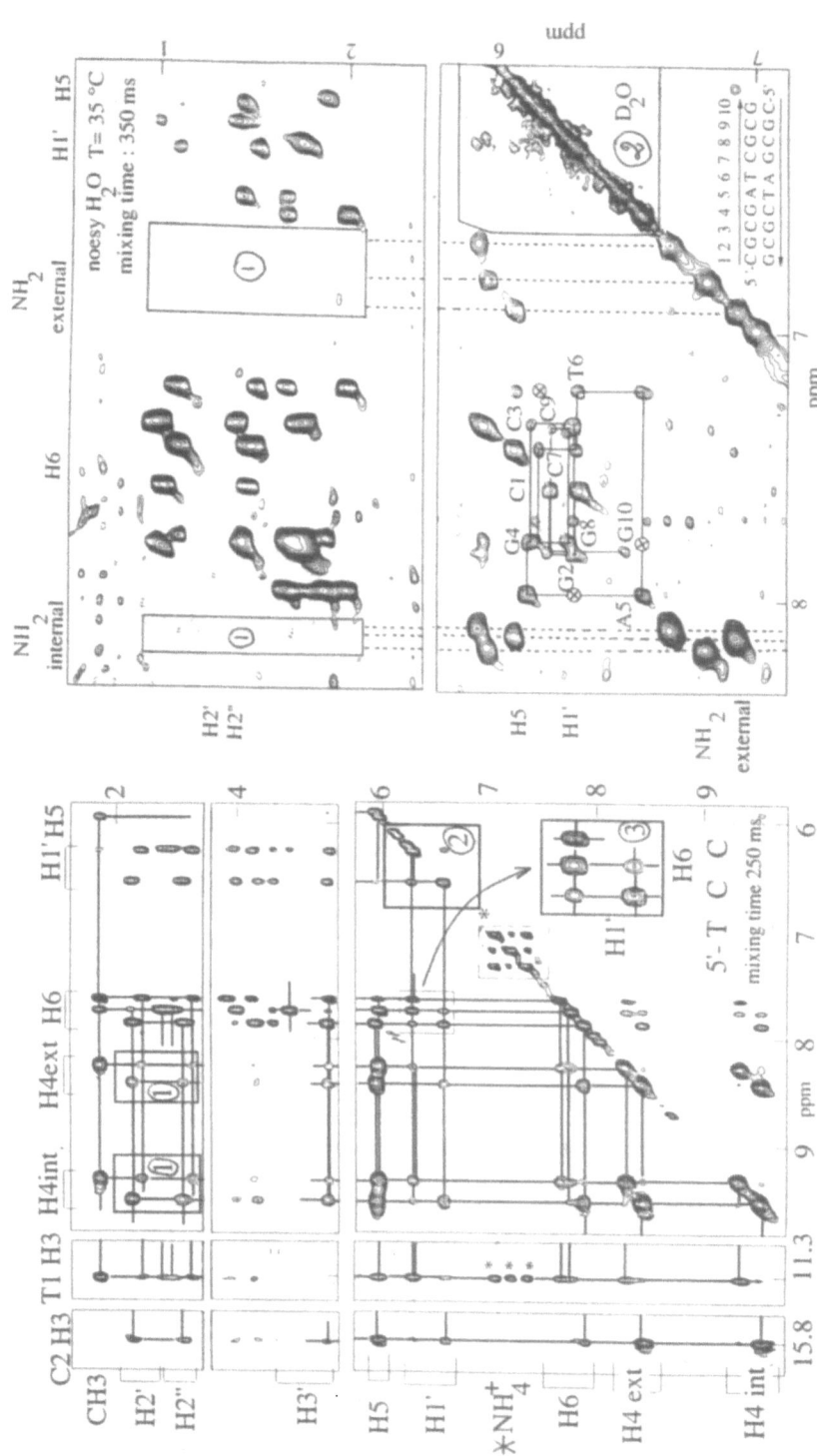

Figure 1. The characteristic inter-nucleotide NOESY cross-peaks of the i-motif, as they appear in a spectrum of the tetramer of 5′-d(TCC), at left. Boxes emphasize (1) the amino proton / sugar proton and (2) H1′-H1′ cross-peaks. In the spectrum of the B-DNA duplex of d(CGCGATCGCG), at right, the corresponding regions are empty, even at long mixing times. The region of the H1′-H6 cross-peaks (expanded, box 3) provides an example of the rectangular pattern of cross-peaks which originates in face-to-face stacking. This does not occur in B-DNA spectra. For instance, the A5H1′-T6H6 cross-peak is conspicuous, but the location of the A5H6-T6H1′ cross-peak, marked by a circled cross, is empty. Conditions: i-motif, pH 4.2, -5°C; B-DNA, pH 7, 35°C.

169

cytosines are base-paired and that proton exchange is very slow, as in other occurrences of the *i-motif*. It is shown by UV absorbance measurement of melting and by gel chromatography measurements of stoichiometry that the i-motif is formed by <u>intramolecular</u> folding. Hypothetically, this could occur *in vivo*. Evidence for the folding of telomere sequences into an i-motif is also given by Ahmed et al.[7]

6. Tetrameric i-motif structures have now been observed by X-ray diffraction of the crystals formed by $d(C_4)$, $d(C_3T)$ and $d(TAAC_3)$.[8-10] The crystal structure of $d(C_4)$ solved at 0.23 nm resolution reveals an i-motif tetramer. The molecule, with wide grooves and narrow grooves, is flat and ribbon-shaped. The ribose conformations are similar to those of the NMR structure, as are the short H1'-H1' distances, and the short amino-proton to sugar proton distances. An interesting difference with the solution structure is that the four strands are not equivalent.

The crystal structure of $d(C_3T)$, solved at 0.14 nm resolution, reveals a similar i-motif tetramer. Bridging water molecules bind a cytosine amino proton to a phosphate group of the other duplex. The i-motif crystal forms even at pH 7.

The crystal structure of d(TAACCC) also involves a tetrameric i-motif. In this case, the meshing of the two duplexes is offset from the maximum, so that A3 of one duplex does not stack on the other duplex. In fact, TAA of the first strand forms a loop, with a Hoogsteen T1·A3 pair stacked on the last C·C pair of the same duplex. The TAA end of the second strand of the same duplex is stretched out, and its T1 is Hoogsteen-paired to A2 of the first strand.

DYNAMIC NUCLEAR POLARIZATION

In response to the mention of Future Directions in the title of the Symposium, a speculation on technical developments in NMR was included in the talk. It was pointed out that poor sensitivity is still an important limitation of NMR, and that only modest progress is to be expected from increases in the magnetic field, which are difficult, slow and costly. It was speculated that solvent polarization could be considerably enhanced by a suitable photo-CIDNP (chemically induced dynamic nuclear polarization) process and that the required light energy dissipated in the sample might be tolerable. The solvent polarization should transfer, via proton exchange and spin diffusion, to most protons of a solute macromolecule.[11] This is addressed by R. Kaptein in the discussion period (see transcript below).

REFERENCES

1. K. Gehring, J.-L. Leroy, and M. Guéron, A tetrameric DNA structure with protonated cytosine-cytosine base pairs, *Nature* 363:561 (1993); and J.-L. Leroy, K. Gehring, A. Kettani, and M. Guéron, Acid multimers of oligo-deoxy-cytidine strands: stoichiometry, base-pair characterization and proton exchange properties, *Biochem.* 32:6019-31 (1993).
2. J.-L. Leroy and M. Guéron, Solution structures of the i-motif tetramers of d(TCC), d(5mCCT) and d(T5mCC). Novel NOE connections between amino protons and sugar protons, *Structure* 3:101 (1995).
3. R. A. G. Friedman and G. S. Manning, Polyelectrolyte effects on site-binding equilibria with application to the intercalation of drugs into DNA, *Bioploymers* 23:2671-2714 (1984).
4. S. McGavin, Intercalated nucleic acid double helices: a stereochemical possibility, *J. Mol. Biol.* 22:187-191 (1966).
5. T. T. Wu, Secondary structures of DNA *Proc. Natl. Acad. US* 63:400-405 (1969).
6. J.-L. Leroy, M. Guéron, J.-L. Mergny, and C. Hélène, Intramolecular folding of a fragment of the cytosine-rich strand of telomeric DNA into an i-motif, *Nucleic Acids Research* 22:1600-1606 (1994).

7. S. Ahmed, A. Kintanar, and E. Henderson, Human telomeric C-strand tetraplexes, *Nature Structural Biology* 1:83-88 (1994).

8. L. Chen, L. Cai, X. Zhang, and A. Rich, Crystal structure of a four-stranded intercalated DNA: d(C4), *Biochemistry* 33:13540 (1994).

9. C. H. Kang, I. Berger, C. Lockshin, R. Ratliff, R. Moyzis, and A. Rich, Crystal structure of intercalated four-stranded d(C3T) at 1.4 A resolution, *Proc. Natl. Acad. Sci. USA* 91:11636 (1994).

10. C. H. Kang *et al.*, *Proc. Natl. Acad. Sci. USA* 92:in the press (1995).

11. J. D. Stoesz, A. G. Redfield, and D. Malinowski, Cross relaxation and spin diffusion effects of the proton NMR of biopolymers in water. Solvent saturation and chemical exchange in superoxide dismutase, *FEBS Letters* 91:320 (1978).

DISCUSSION

Robert Kaptein - Since you mentioned my name, I feel I should discuss this point on using 'CIDNP' to polarize water and also I know that if you come up with an original idea, there are always people that say it can't be done, that shouldn't discourage you really. But let me tell you about our experiences on water polarization. In some of the reactions we studied, we could see a polarized water signal that arose via exchange from a product with polarized OH or NH groups. So it's not impossible but the intensity of the polarized water signal was about a factor 1000 less than the Boltzmann water intensity. So if you want to increase that by, say, a factor of 10 you still have a factor of 10^4 to combat. And I don't really see how I can do that given the limitations of that CIDNP phenomena. Nevertheless, it should not discourage you.

Maurice Guéron - Thank you.

Bill Gmeiner - I have a couple of questions on the idea of i-DNA structures. In particular, I am curious if anybody has looked for proteins analogous to binding proteins that target these i-DNA regions and also I was curious if you had done work with RNA sequences to determine whether analogous sequences can occur in RNA, whether they might be important in folding higher order RNA structure.

Guéron - It's certainly interesting to look for proteins which bind to the i-motif. This is mentioned for instance in a preprint from A. Rich's laboratory on the resolution of an i-motif structure by X-ray crystallography [as noted above, MG]. We have started looking for proteins which bind to such sequences, and we have promising retarded gels, but it's too early to say.

We have not been able to form the i-motif in RNA, for instance in $r(UC_5)$. There are indications that the i-motif is compatible with substitution of one cytidine for deoxycytidine in $d(TC_3)$. We don't understand the intercalation. I mean, there's no obvious reason why you couldn't take a B-DNA duplex, stretch it out and start intercalating bases or base pairs. We've tried that very crudely by adding adenosine and thymidine to a solution of the d(CGCGATCGCG) duplex, and it didn't work. At this point, intercalation is observed only in stretches of deoxycytidine.

David Gorenstein - On the d(T5mCC) i-motif. I am surprised that you find equal contributions of two forms. The second form which is not fully intercalated should be destabilized because you have less stacking-type interactions.

Guéron - The second form of the d(5mCC) tetramer is destabilized by the reduced degree of intercalation, but it does not suffer, as does the first, from a collision between the methyl groups. This explains the comparable stabilities of the two forms. The argument is supported by the observation that the fully intercalated structure is the only one observed for other methylated sequences (e.g. d(5mCCT)), in which the methyls do not fall on contiguous pairs of this structure.

In this interpretation, it is a bit surprising that the concentrations of the two forms of the d(5mCC) tetramer are so close. We think they are not exactly equal, although this is difficult to measure. And there are no cross-peaks whatsoever between the spectra assigned to the two species.

Marc Adler - A quick question on your last spectrum which was quite messy. You said that it was because of the multiple conformations. Also could it be that there is ongoing conformational exchange between those different structures? Have you tried studying different temperatures to see if possibly you could sharpen the lines?

Guéron - My guess is that exchange between different intramolecular foldings of the single-strand telomeric sequence would be very slow. A first indication is the well-resolved aromatic peak with an intensity of approximately 0.5 in the 1D spectrum. Other indications come from comparisons with i-motif tetramers. Thus, the NMR spectrum of the d(TCC) tetramer is in slow exchange with that of the monomer. And tetramers of longer strands migrate on chromatography columns with no indication of dissociation during the run.

PROTEIN-DNA INTERACTION FROM NMR AND MONTE CARLO DOCKING

R. Kaptein, M. Slijper, V.P. Chuprina, J.A.C. Rullmann,
R.M.A. Knegtel and R. Boelens

Bijvoet Center for Biomolecular Research
Utrecht University
Padualaan 8, 3584 CH Utrecht
The Netherlands

INTRODUCTION

An essential step in the regulation of gene expression is the binding of a regulatory protein to a specific DNA sequence in the promotor region of the gene. The understanding of protein-DNA recognition is, therefore, a major theme in structural biology. Much progress has been made since the early '80s when the first structures of bacterial DNA-binding proteins and protein-DNA complexes were solved by X-ray crystallography (for reviews see Steitz (1990), Pabo and Sauer (1992) and Travers (1993)). NMR started to contribute around 1985 with the structure elucidation of the lac repressor headpiece (Kaptein et al., 1985) and a low resolution structure of the headpiece-operator complex (Boelens et al., 1987). These first prokaryotic DNA-binding proteins all contained the helix-turn-helix motif as the DNA-binding subdomain. However, subsequently a large number of other structural motifs has been characterized including zinc-fingers, leucine zippers, helix-loop-helix proteins and β-sheet DNA binding proteins.

The growing importance of NMR as a structural tool is exemplified by the fact that the structural information on several of these motifs came first from NMR spectroscopy. Notably, this was the case for the TFIIIA-type zinc-fingers (Lee et al., 1989; Klevit et al., 1990) and for the nuclear hormone receptors (Härd et al., 1990; Schwabe et al., 1990).

Often a combination of various biochemical and genetic methods with structural work is most fruitful for understanding protein-DNA recognition on a molecular level. X-ray or NMR structures of protein-DNA complexes allow an interpretation of the results of mutagenesis, which would be very difficult in the absence of structural data. Conversely, binding affinities of mutant proteins or DNA variants give insight in the importance of the various interactions seen in X-ray or NMR structures in terms of the free energy of binding.

The lac repressor is a case in point. For this system over the years an enormous body of biochemical and genetic data has been generated, which can now be interpreted in structural terms. Mostly by the work of Miller (1984) and Müller-Hill and his coworkers (Lehming et al., 1987) a staggering number (thousands) of mutants has been characterized. In particular mutant proteins with altered DNA-binding specificity are informative in this respect. With recent structure refinement of the lac headpiece-operator complex by NMR, this work can be put on a firm structural footing (Chuprina et al., 1993).

In this chapter two topics will be discussed. First, our results on the lac repressor-operator system will be presented as an illustration of the NMR approach to the problem of protein-DNA recognition. The second part deals with a novel computational method to model protein-DNA interactions. Since the number of structures of DNA-binding proteins still vastly exceeds that of protein-DNA complexes, it would be useful to have available a predictive method that generates a DNA-complex based on the protein structure. The Monte Carlo modelling procedure implemented in the program MONTY is a step in that direction.

THE LAC OPERON

Lac Repressor Headpiece

The lac repressor of *E. coli* is a tetrameric protein of molecular weight 154 000. The native repressor is too large for high-resolution NMR studies. However, each subunit has a separate DNA-binding domain (headpiece) that can be cleaved off by proteolytic enzymes (Geisler and Weber, 1977). Depending on the proteolytic enzyme used, headpieces can be prepared containing 51, 56 or 59 amino acid residues (HP 51, HP 56 or HP 59). These headpieces retain their original three-dimensional structure and their ability to recognize the lac operator specifically (Ogata and Gilbert, 1979). The trypsin-resistant core is involved in the subunit interaction and contains the inducer binding site. The sequence of the natural lac operator reveals an approximate two-fold symmetry (Gilbert and Maxam, 1973) and in agreement with that, two subunits of lac repressor suffice to recognize lac operator (Kania and Brown, 1976). Therefore lac repressor should bind

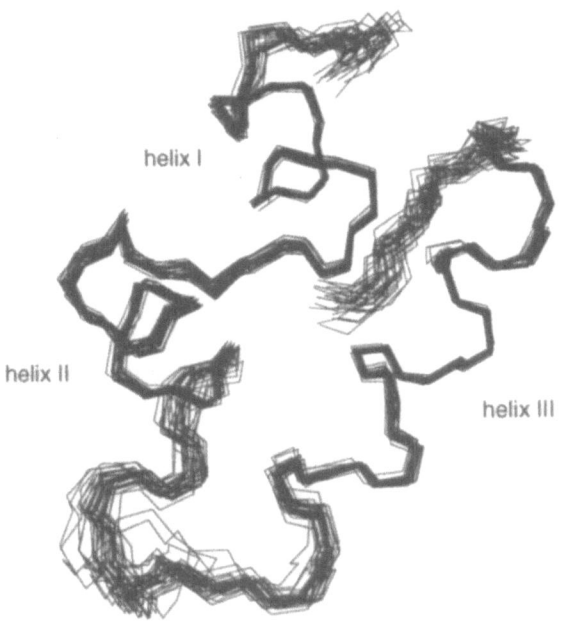

Fig. 1. Superposition of 30 structures for lac headpiece (residues 3-48). The three α-helices are indicated. The structures were calculated using distance geometry and restrained Molecular Dynamics procedures based on ca. 950 distance constraints from NOE's and 46 dihedral angle constraints.

with two headpieces to each half of the operator. Lac headpiece (HP 51) was one of the first proteins for which the three-dimensional structure was determined by NMR (Kaptein et al., 1985; de Vlieg et al., 1986). Recently, the structure has been refined using a more extensive set of constraints from NOE's and J-couplings (Slijper et al., 1995). The family of structures is shown in Fig. 1. Lac headpiece consists of three helices, the first two of which constituting the helix-turn-helix motif, while the third packs against this subdomain forming a hydrophobic core. It can be clearly seen that the loop between helix II and III shows a larger conformational variability than the helical core of the protein. From ^{15}N

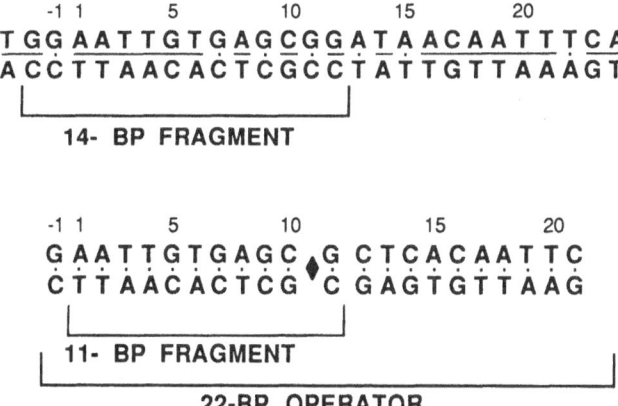

Fig. 2. Sequences of native lac operator (a) and 'ideal' symmetric lac operator (b). Synthetic operator fragments of 11, 14 and 22 bp used in the NMR studies are indicated.

relaxation data measured for bacterially expressed HP 56 it is clear that this corresponds to a larger backbone flexibility for the loop region. As is discussed below this flexibility is functional as it allows the protein to adjust to the DNA upon complex formation.

Lac Headpiece-Operator Complex

Lac operator of E. *coli* is defined genetically as the control region in the *lac* operon, where operator constitutive mutants occur. The region protected by lac repressor is 20-25 bp long, with a pseudo-dyad axis going through GC 11 (Gilbert and Maxam, 1973). It was found by Sadler *et al.* (1983) and Simons *et al.* (1984) that symmetrical *lac* operators lacking the central GC base-pair bind lac repressor up to an order of magnitude stronger than the native one. The sequences of the operators and the fragments discussed here are shown in Figure 2. The initial NMR studies were made with the 14 bp fragment, which turned out to be the stronger binding half-operator as it also occurs in the symmetrical operator. The binding affinity of the isolated headpiece with a half-operator is not extremely high ($K_D \approx 10^{-6}M$), so that the free and DNA-bound forms are in fast exchange on the NMR time-scale. This greatly helps in assigning the resonances in the NMR spectra of the protein-DNA complex, since the resonances of the free species can be followed in titrations, either by adding, for instance, protein to DNA or by titrating the complex with increasing amounts of salt, by which it will gradually dissociate.

An early low-resolution structure of the HP 56-14 bp operator complex was obtained by docking the protein to DNA in standard B conformation guided by 11 NOEs observed between protein and DNA (Boelens et al., 1987). This model was similar to other complexes of helix-turn-helix proteins in that helix II of the headpiece, the "recognition helix" of the helix-turn-helix motif, is inserted in the major groove of DNA and makes the majority of the protein-DNA interactions. However, a surprising result was that the orientation of the recognition helix was opposite to that found in all other known complexes. This means that the first helix points towards the dyad axis of the operator, while for other proteins, such as λ and cro repressors and CAP, it points away from it. Genetic experiments by Lehming *et al.* (1987) confirmed this different orientation. These authors constructed a lac repressor mutant with the first two amino acids of the recognition helix replaced by those of gal repressor (Tyr 17 → Val, Gln 18 → Ala). This mutant repressor had high affinity for the gal operator, which differs from lac operator at positions 7 and 9. Although this already gives some clue as to the orientation of the recognition helix, a more definitive result was their finding of a repressor mutant with Arg 22 replaced by Asn, which now had specificity for a lac operator with GC5 replaced by TA. This fixes unambiguously the orientation of the recognition helix as the opposite of that of cro and λ repressors.

Studies of a complex of two HP 56 molecules with a symmetric 22 bp lac operator (Lamerichs et al., 1989) showed that the two headpieces bind independently and in essentially the same binding mode as they do in half-operator complexes. Therefore, a more detailed study was made of a complex with an 11-bp half-operator (Lamerichs et al., 1990). Based on a larger number of NOEs, among which 39 between protein and DNA, a restrained molecular dynamics refinement of the headpiece operator complex was carried out (Chuprina et al., 1993). Some statistics of the calculation are shown in Table 1. We note that the protein- DNA complex was put in a box containing over 3 000 water molecules to which salt ions were added, and periodic boundery conditions were applied. Inclusion of solvent was felt necessary for a reliable simulation and also because water molecules may play an important role in the interface between protein and DNA. A fairly long equilibration of 60 ps was necessary primarily to equilibrate the ion distribution around the complex. Then a 85 ps trajectory was used for analysis. In addition an annealing procedure was undertaken in order to assess the precision of the structure determination. This consisted of heating the complex to 1000 K and letting it cool down to 300 K in a 5 ps RMD calculation followed by energy minimization. Six structures were obtained by this procedure using different snapshots from the trajectory as starting points.

Table 1. Restrained Molecular Dynamics Refinement of Lac Headpiece-11 bp operator complex

Start:	optimized vacuum structure	
Size:	989 atoms complex, 28 Na+, 10 Cl- 3346 waters	
Constraints:	intra protein	260
	intra DNA	241
	protein-DNA	39
Forcefield:	GROMOS	
Trajectory	60 ps equilibration	
	85 ps analysis	
Annealing:	(6x) 1000 K $\rightarrow$ 300 K	
	5 ps	

The RMD runs yielded a satisfactory structure for the complex with a low total energy (- 175 x 10^3 kJmol^{-1}) while the restraint energy did not exceed the average thermal energy. There were on average 42 bounds (out of 980 total), which were violated by 0.5 Å or more, mostly within the DNA (none involving protein-DNA contacts).

Among a set of 15 structures, 9 from the trajectory and 6 from the annealing procedure, the r.m.s.d. for backbone atoms for both DNA and protein (residues 4-48 for the headpiece) with respect to the mean was found to be 0.9 Å. This value can be

Fig. 3. Backbone trace of lac repressor headpiece for the free protein (grey line) and the protein in complex with operator DNA (dark line).

considered as a measure of the precision of the structure determination. Fig. 3 shows the average backbone confirmation of the headpiece in the complex compared with that of the free protein. It is clear that the loop between helix II and helix III has undergone a considerable conformational change. In particular Asn 25 changes its φ/ψ angles such that the exit from helix II is quite different in the free and complexed protein.

From the analysis of the 85 ps trajectory a picture emerged of the interactions by which lac operator is recognized by the headpiece. Apart from electrostatic interactions, which are probably non-specific, a number of hydrogen bonds are observed between protein and DNA, both to the sugar-phosphate backbone and to the bases. In addition an extensive network of apolar interactions is observed involving the side-chains of Tyr 17, Gln 18 and Ser 21, and the methyl groups of Thy 6 and Thy 8. These interactions are schematically shown in Figure 4. Two direct hydrogen bonds between amino acid side-chains and bases are seen, which appear to be essential for specific recognition. These are the ones between Gln 18 and Cyt 7 and between Arg 22 and Gua 5. Important anchoring hydrogen bonds with DNA phosphates include those of the amide NH of Leu 6 and the side-chain amide of Asn 25. These are conserved among the whole family of helix-turn-helix proteins. Histidine 29, present in the loop between helices II and III is involved in a

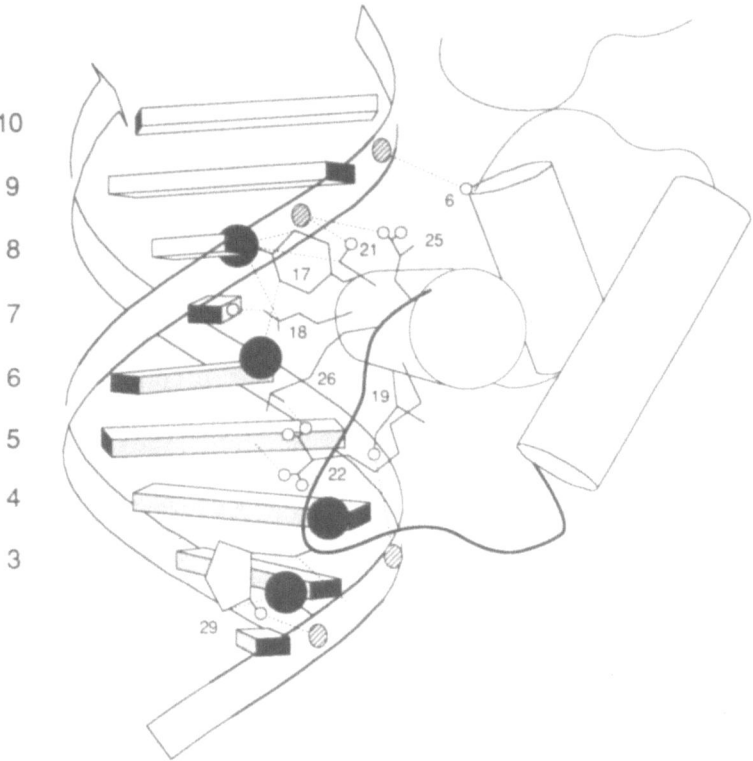

Fig. 4. Schematic view of the interactions between lac headpiece and lac operator. Shown are the hydrogen bonds between protein and DNA and the apolar interactions in the hydrophobic cluster formed by side-chains of Tyr 17, Ser 21, Gln 18 and the methyl groups of Thy 6 and Thy 8. Thymine methyl groups are shown as black balls, phosphate groups are striped circles and protons in H-bonds as small circles.

multitude of interactions: electrostatic interactions because of its positive charge, a hydrogen bond with a phosphate group and van der Waals interactions with the methyl group and sugar ring of Thy 3. For a more complete description of these interactions we refer to the paper by Chuprina *et al.* (1993).

Dynamics of free and complexed lac headpiece

Using ^{15}N-labeled HP 56 the backbone dynamics of the headpiece has been studied for the free protein and in the complex with the operator (Slijper et al., 1995). Fig. 5 shows ^{15}N-$T_{1\rho}$ data for the backbone amide nitrogens. The N-terminal and C-terminal region show a larger mobility than the core of the protein, both in the free and in the bound state, as evidenced by larger values of $T_{1\rho}$. Comparing the relaxation data of Fig. 5A and B it is clear that in general the $T_{1\rho}$ values in the complex are lower than in the free state due to the higher molecular weight. However, more significantly, the higher mobility in the loop region around residue 30 observed for the free protein disappears in the complex. This suggests that the flexible loop adapts itself to the DNA and becomes more

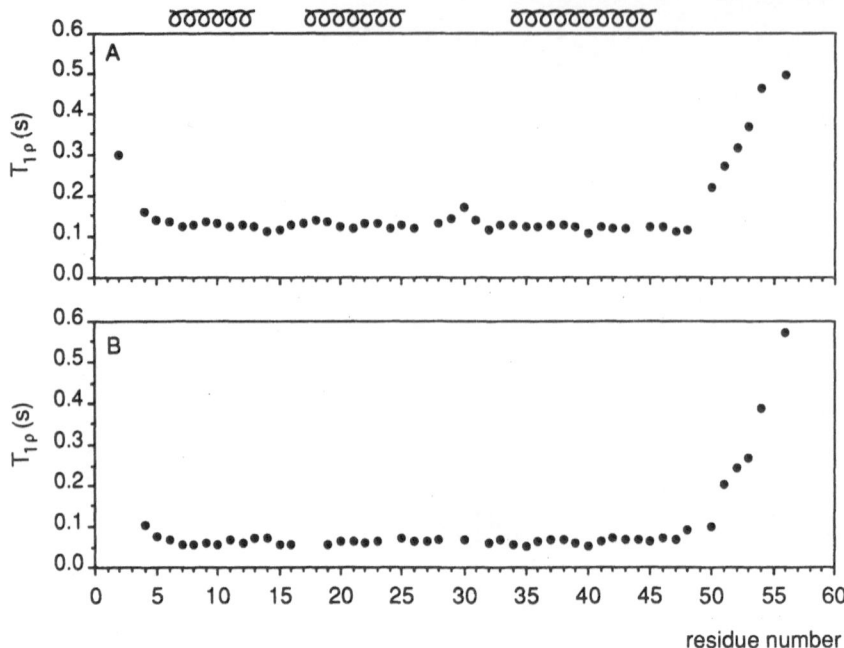

Fig. 5. ^{15}N $T_{1\rho}$'s of the amide nitrogens *vs* residue number for free HP 56 (A) and for HP 56 complexed to lac operator (B). The three α-helices of the headpiece are indicated on top.

rigid. This behaviour is consistent with a negative change in heat capacity upon complex formation, which was interpreted by Spolar and Record (1994) as a (partial) folding transition in the repressor. Based on thermodynamic data this adaptability accompanied by partial folding has been suggested to be a more general phenomenon in protein-DNA recognition (Spolar and Record, 1994). For *lac* headpiece the present structural and relaxation data support this view.

Although part of the protein backbone adopts a more rigid conformation, there is also evidence for residual flexibility in the protein-DNA interface. This comes from an analysis of the restrained MD trajectory in terms of hydrogen-bonds between protein and DNA showing that some of these hydrogen-bonds are not rigid but fluctuate in time. Thus, while anchoring H-bonds involving Leu 6 and Asn 25 are formed close to 100% of the time, the one between Arg 22 and Gua 5, for instance, breaks up during the trajectory and is replaced by an intra-protein H-bond with Glu 26. Apparently, alternative hydrogen bonding schemes are possible with very similar energies. We believe that indeed protein-DNA complexes are rather dynamic in nature and that the current more rigid picture of these complexes needs correction.

PROTEIN-DNA DOCKING: THE MONTY PROGRAM

It is often easier to determine the structure of a free DNA-binding protein than that of a DNA complex. If only the structure of the free protein is known it would be very useful to have a reliable modeling procedure available for docking the protein to its target DNA site. In our hands standard molecular dynamics procedures will not do the job. The conformational space is just too large, so that reliable protein-DNA complexes cannot be generated within a reasonable computer time. For this reason we developed a Monte Carlo docking program called MONTY, in which time savings are obtained by, first, restricting the conformational search problem and, secondly, using a much simplified energy function or "force field" (Knegtel et al., 1994a,b). The procedure is outlined in Fig. 6. After randomization of the dihedral angles of user-selected am side chains the program searches the surface of the major groove of DNA for the optimal binding site by rotating and translating the protein and rotating the selected side chains. A simple intermolecular square well potential is evaluated which takes the form:

$$V_{pot} = V_{H\text{-}bond} + V_{VdW} + V_{rep} + V_{exp}$$

where $V_{H\text{-}bond}$ and V_{VdW} are potentials for hydrogen bonds and van der Waals interactions, respectively; V_{rep} is a large positive repulsion energy and V_{exp} represents various possible potential functions, which allow for the inclusion of experimental biochemical results in the simulation.

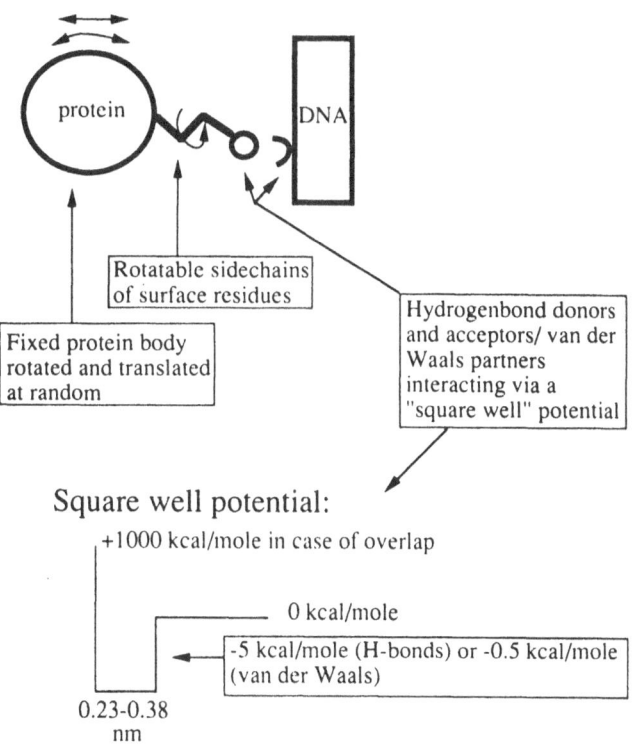

Fig.6. Overview of the MONTY procedure.

Monte Carlo sampling has the great advantage that a variety of interactions can be represented in V_{exp} for which no gradients can be formulated (as would be required for MD based procedures). For instance, if from phosphate ethylation interference experiments it is known that a certain phosphate on the DNA is contacted by the protein, an energy bonus can be given if any side chain of the protein contacts this phosphate in the model. As there is often a great deal of information on protein-DNA interactions known from contact analysis or mutagenesis this feature is of considerable importance. Typically, in a run 200000 structures are evaluated, for which the Monte Carlo acceptance rate is usually rather low (a few percent). Rather then aiming for a single lowest energy structure, the procedure is fast enough so that a family of structures can be generated, not unlike the way protein structures are generated from NMR data. This allows for a statistical description of DNA interactions, which in fact may be more realistic than a description in terms of a single structure. Further details are given in Knegtel et al. (1994 a,b).

To test the method a number of docking calculations were carried out for systems where the structure of the protein-DNA complex is known, either from X-ray crystallography (434 cro repressor, Mondragón and Harrison, 1991) or NMR (lac headpiece, Chruprina et al., 1993). For 434 cro repressor MONTY runs were carried out starting with the X-ray complex, and with complexes where the protein is shifted up and down one base pair or even two base pairs. The results for the two-base pair shift calculations is shown in Fig. 7. Considering that the starting structures are really quite far from the optimal binding sequence of the 434 operator, it can be seen that the convergence of the method is very good. Many of the correct hydrogen bonds and van der Waals contacts are retrieved in the final complexes.

In an attempt to see whether the program can determine the correct orientation of HTH domain proeins in the major groove of DNA calculations were performed on the lac headpiece system starting with the headpiece both in the correct and reversed orientation. Usually this orientation is not *a priori* known, unless it is obvious from the quaternary structure of the protein. As shown in Fig. 8, including only H-bonds and van der Waals terms in the calculations the convergence is rather similar for the two orientations although

crystal complex   starting   six lowest
structure   structures   energy structures
after MONTY

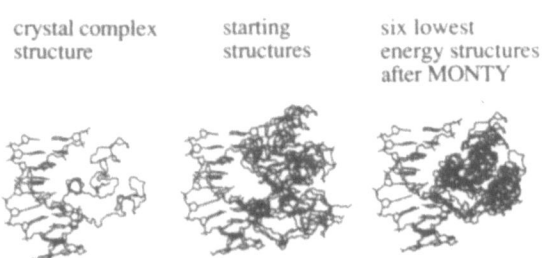

Fig. 7. Results of MONTY simulations of 434 cro protein-DNA complexes shifted by two basepairs. Shown are the native X-ray structure, 3 up- and 3 down-shifted starting structures, and the final 6 lowest energy structures.

Correct   Reversed

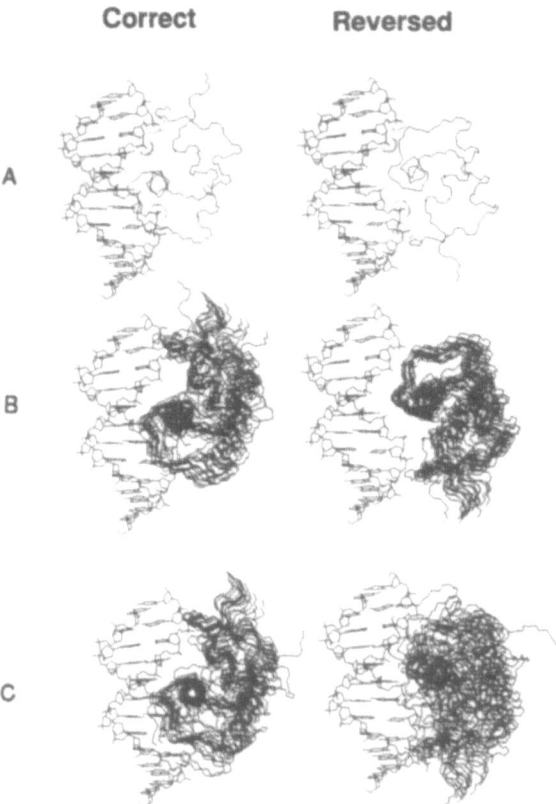

A

B

C

Fig. 8. MONTY simulations on lac headpiece-operator complexes in the correct and reverse orientations. (A) starting structures, B (complexes) generated by MONTY without the use of experimental data, (C) same but with the application of phosphate ethylation interference data.

the energy for the native-like structures is somewhat lower (Knegtel *et al.* 1994b). However, when phosphate ethylation interference data are incorporated in the potential energy function it can be seen (Fig. 7C) that complexes with the wrong orientation in fact diverge during the calculation while good convergence is observed for the correctly oriented headpiece.

Application to the LexA repressor-DNA complex

Recently the solution structure of the DNA-binding domain (residues 1-72) of *E.coli* LexA repressor was solved (Fogh et al., 1994). It was found that the protein fold resembled that of CAP putting LexA repressor in the family of winged HTH proteins also including HNF3/forkhead protein, histone H5 and the heat shock transcription factor. No structural data from NMR or X-ray is known on the interaction with DNA and therefore the complex was modeled using the MONTY procedure (Knegtel et al., 1995). LexA repressor binds to several SOS operators, which are palindromic 16 bp sequences with a CTGTNNNN consensus half site. Monte Carlo docking showed again improved

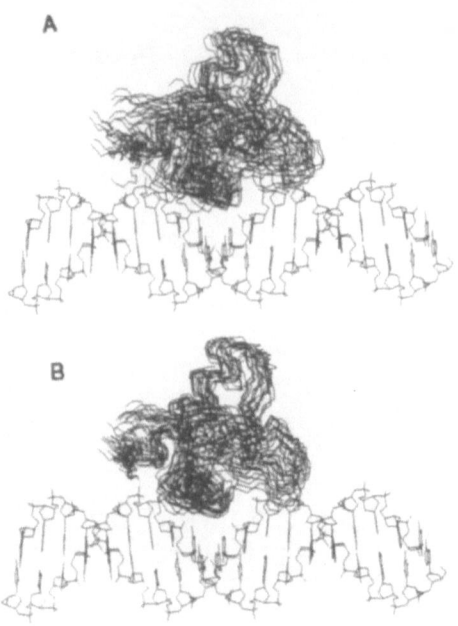

Fig. 9. MONTY derived complexes of LexA repressor with consensus SOS operator DNA (A) without, and (B) with inclusion of data from phosphate ethylation interference experiments.

convergence when apart from H-bonds and van der Waals terms also energy terms representing phosphate ethylation interference data were taken into account (compare Fig. 9A and B). In the resulting complexes helix III of the LexA repressor is located in the major groove of DNA and many of the observed interactions can account for the CTGT consensus sequence specificity. Thus, Asn 41, Glu 44, and Glu 45 form specific hydrogen bonds with bases of the CTGT sequence, while Ser 39, Ala 42 and Asn 41 are involved in hydrophobic interactions with the methyl group of the first thymine base. Furthermore, many non-specific interactions are observed. Residues in the loop region connecting the two β strands have contacts near the dyad axis of the operator. The structures are in agreement with other biochemical and genetic data and in fact allow the results from mutagenesis experiments to be interpreted in structural terms. With the docking calculations in the LexA repressor system the MONTY program has shown its usefulness as it suggested several new mutagenesis experiments to test the model for the complex. Further extensions to include water molecules in the protein-DNA interface are in progress.

REFERENCES

Boelens, R., Scheek, R.M., Boom, J.H. van and Kaptein, R., 1987, *J. Mol. Biol.* 193:213.

Chuprina, V.P., Rullmann, J.A.C., Lamerichs, R.M.J.N., Boom, J.H. van , Boelens, R. and Kaptein, R., 1993, *J. Mol. Biol.* 234:446.

Fogh, R., Ottleben, G., Rüterjans, H., Schnarr, M., Boelens, R., Kaptein, R., 1994, *EMBO J.* 13:3936.

Geisler, N. and Weber, K., 1977, *Biochemistry* 16:938.

Gilbert, W. and Maxam, A., 1973, *Proc. Natl. Acad. Sci. USA* 70:3581.

Härd, T., Kellenbach, E., Boelens, R., Maler, B.A., Dahlman, K., Freedman, L.P., Carlstedt-Duke, J., Yamamoto, K.R., Gustafsson, J.-Å., and Kaptein, R., 1990, *Science* 249:157.

Kania, J. and Brown, D.T., 1976, *Proc. Natl. Acad. Sci. USA* 73:3529.

Kaptein, R., Zuiderweg, E.R.P., Scheek, R.M., Boelens, R. and Gunsteren, W.F. van , 1985, *J. Mol. Biol.* 182:179.

Klevit, R.E., Herriott, J.R., Horvath, S.J., 1990, *Proteins* 7:215.

Knegtel, R.M.A., Rullmann, A.J.C., Boelens, R. and Kaptein, R., 1994a, *J. Mol. Biol.* 235:318.

Knegtel, R.M.A., Boelens, R. and Kaptein, R., 1994b, *Protein Engineering* 7:761.

Knegtel, R.M.A., Fogh, R.H., Ottleben, G., Rüterjans, H., Dumoulin, P., Schnarr, M., Boelens, R. and Kaptein, R., 1995, *Proteins* 21:226.

Lamerichs, R.M.J.N., Boelens, R., Marel, G.A. van der, Boom, J.H. van, Kaptein, R., Buck, F., Fera, B. and Rüterjans, H., 1989, *Biochemistry* 28:2985.

Lamerichs, R.M.J.N., Boelens, R., Marel, G.A. van der, Boom, J.H. van and Kaptein, R., 1990, *Eur. J. Biochem.* 194:629.

Lee, M.S., Gippert, G.P., Soman, K.V., Case, D.A., Wright, P.E., 1989, *Science* 245:635.

Lehming, N., Sartorius, J., Niemöller, M., Genenger, G., Wilcken-Bergmann, B. von and Müller-Hill, B., 1987, *EMBO J* 6:3145.

Miller, J.H., 1984, *J. Mol. Biol.* 180:205.

Mondragón, A. and Harrison, S.C., 1991, *J. Mol. Biol.* 219:321.

Ogata, R.T. and Gilbert, W. , 1979, *J. Mol. Biol* 132:709.

Pabo, C.O. and Sauer, R.T., 1992, *Annu. Rev. Biochem.* 61:1053.

Sadler, J.R., Sasmor, H. and Betz, J.L., 1983, *Proc. Natl. Acad. Sci. USA* 80:6785.

Schwabe, J.W.R., Neuhaus, D. and Rhodes, D., 1990, *Nature* 238:458.

Simons, A., Tils, D., Wilcken-Bergmann, B. von and Müller-Hill, B., 1984, *Proc. Natl. Acad. Sci. USA* 81:1624.

Slijper, M., Boelens, R. and Kaptein, R., 1995, to be published.

Spolar, R.S. and Record, M.T., 1994, *Science* 263:777.

Steitz, T.A., 1990, *Quart. Rev. Biophys.* 23:205.

Travers, A., 1993, DNA-Protein Interactions, Chapman & Hall, London.

Vlieg, J. de, Boelens, R., Scheek, R.M., Kaptein, R. and Gunsteren, W.F. van (1986) *Israel J. Chem.* 27:181.

DISCUSSION

Carol Post - Rob, when you put in those additional terms for the Monte Carlo potential field, how do you give an effective force constant.

Robert Kaptein - This has to be calibrated somehow. There's always noise in these calculations. So you have to put a number there that is significant in terms of the noise that you generate in the structures. That is just by trial and error. There's no a priori way one could do that.

Mike Mossing - The other class of mutagenesis data would be the sequence of the DNA. Have you cross checked these things by changing the sequence of the operator to something that it wouldn't recognize?

Kaptein - We haven't done that yet. Well, in a way we have by shifting it, out of register. That's a complete different sequence.

Mossing - Is it something it didn't recognize? Let's say a single base pair change in the operator.

Kaptein - No, we haven't done that.

Marc Adler - In your structures of protein and DNA, there seems to be a fair amount of wobble between the packing of the protein against the DNA. I was wondering if some of the changes are concerted. This could be shown by cluster analysis where you might be giving concerted change in sidechain packing.

Kaptein - We haven't analyzed it really that way. Of course, what one expects is that there is a change in conformation disrupting a hydrogen bond that it will be forming another hydrogen bond, but whether there are concerted changes other than that, I don't know.

Mengli Cai - When you do the dimer protein you collected a set of NMR data. What's a good indication or evidence to say this protein is a dimer?

Kaptein - This was known before we started work on this system. It was just by some type of molecular weight analysis that it was known that this is a dimer. So we did not do that. Sometimes of course, you will have to establish that first.

Cai - Second question, how much time is needed to run restrained molecular dynamics for 80 ps. How long does it take to complete that?

Kaptein - That was a long run. That took, in fact, several months for one structure done three years ago. That was a big calculation.

DYNAMIC STRUCTURE OF NUCLEIC ACID DUPLEXES

Thomas L. James, Carlos González, He Liu, Uli Schmitz,
and Nikolai B. Ulyanov

Department of Pharmaceutical Chemistry
University of California
San Francisco, CA 94143-0446, U.S.A.

INTRODUCTION

Sequence-dependent structural variations in nucleic acids are important for the protein-nucleic acid and RNA-DNA interactions essential for life. Furthermore, structural variations are important factors in the binding of extrinsic agents, such as mutagens or drugs. While an understanding of these subtle structural variations may have biological significance, the nucleic acid double helix also provides a challenging problem. Since we can fairly readily determine that a double helix is right- (probably) or left-handed, we know *a priori* the structure nearly as well as one often can determine a protein structure. It is therefore incumbent upon us to determine the structure accurately and to high-resolution if we wish to distinguish the three-dimensional, sequence-dependent structural subtleties. The constraints in constructing a double helix from complementary nucleic acid strands via base-stacking with two or three interbase hydrogen bonds in each base pair also suggests that the double helix should assume a fairly stable structure; modeling indicates that some degree of conformational flexibility may still be present though.

To date, there have been no structures available for a free DNA duplex, a protein and the complex formed between the protein and the DNA to ascertain precisely what geometric and energetic features in the three-dimensional structure of the DNA contribute to site-specific recognition by the protein. Crystal structures of protein/DNA complexes indicate that the bound DNA can be modified significantly from canonical B-DNA (Burley, 1994; Pabo & Sauer, 1992; Wright, 1994), but it is not known whether the particular DNA sequence had some tendency for these structural deformations in the absence of protein. This lack of knowledge is chiefly due to the lack of high-resolution structures for free DNA sequences which are specifically recognized by proteins. Analysis of x-ray diffraction from oligonucleotides in single crystals has demonstrated significant sequence-dependent heterogeneity in DNA structures. However, single crystal x-ray diffraction has serious problems.

Perhaps the most readily acknowledged problem has been that crystallizing DNA fragments does not generally leave them in the B-form, i.e., the form in which they are nearly always found in solution. Even in those cases where the B-form is obtained for a crystal, there is concern regarding the influence of crystal packing forces on the DNA structure. Crystallization of a B-DNA duplex into two different crystalline types (space groups) has had a substantial impact upon many local structure parameters (Lipanov *et al.*, 1993).

Within the past few years, the quality of DNA structure determination via NMR has improved sufficiently to provide an option. The methodology for determination of structures with sufficient resolution via NMR has not been easily achieved, but the ability to determine an accurate, high-precision structure of nearly any DNA double helix of length less than 15 base pairs is now possible if sufficient care and effort are expended. We have now obtained enough DNA duplex structures that it is even possible to discern a few tentative generalities about sequence-dependent structure (Ulyanov & James, 1994).

The structure of a molecule can be determined with a sufficient number of structural restraints, e.g., internuclear distances and bond torsion angles, in conjunction with the holonomic constraints of bond lengths and bond angles. Multi-dimensional NMR spectra enable us to obtain many structural restraints. The resolution of NMR structures can be improved by utilizing a greater number and greater accuracy of structural restraints. We will briefly address the methodology for achieving this in the present chapter. However, NMR-derived structural restraints are time-averaged, e.g., a value for the distance between two protons in a molecule will be a complicated average (not arithmetic) which depends on the distance between the two protons in each conformation in which the molecule exists and on the rate of exchange between the conformations relative to the overall molecular tumbling rate. Consequently, a rigid structure determined assuming that there is only a single conformation could be misleading. We will also address this troubling prospect in the current chapter, but note here that this remains an unsolved problem.

STRUCTURE GENERATION USING RESTRAINT BOUNDS

The computational method of restrained molecular dynamics (rMD) attempts to reconcile experimental structural restraints with energetic considerations to define a "structure" or envelope of closely related molecular structures (van Gunsteren, 1993). As noted above, the ability to determine solution structures of biomolecules by NMR spectroscopy is limited by the quality and quantity of distance and torsion angle restraints that can be extracted from the NMR data.

Distance Restraints

Cross-relaxation between two neighboring protons in a molecule during the mixing time period τ_m of the 2D NOE experiment modifies the cross-peak intensities in the spectrum and depends on the distance between the two protons and upon the rate of molecular motion. So the cross-peak intensities contain structural information, i.e., distances. To reliably extract distances from 2D NOE intensities, the spectra should be collected under conditions in which even the slowest proton has had time to nearly completely relax. For biomolecules, the two protons giving rise to a particular cross-peak are not alone. They belong to an array of all protons in the molecular structure which, in principle, experience dipole-dipole interactions with all the others. So cross-relaxation between the two protons is part of a coupled relaxation network. The matrix of 2D NOE intensities $\mathbf{a}$ is related to $\mathbf{R}$, the matrix describing the complete dipole-dipole relaxation network, and τ_m by (Macura & Ernst, 1980):

$$\mathbf{a}(\tau_m) = e^{-\mathbf{R}\tau_m} \approx 1 - \mathbf{R}\tau_m + \tfrac{1}{2}\mathbf{R}^2\tau_m^2 - \cdots + \frac{(-1)^n}{n!}\mathbf{R}^n\tau_m^n + \cdots \qquad (1)$$

where the elements of the relaxation matrix $\mathbf{R}$ depend on spin state transition probabilities, written below as zero-, single- and double-quantum transition probabilities W_m^{ij}, as follows:

$$R_{ii} = 2(n_i - 1)(W_1^{ii} + W_2^{ii}) + \sum_{j \neq i} n_j(W_0^{ij} + 2W_1^{ij} + W_2^{ij}) + R_{1i} \; ; \qquad (2a)$$

$$R_{ij} = n_i(W_2^{ij} - W_0^{ij}). \qquad (2b)$$

where n_i is the number of equivalent spins in a group such as a methyl rotor. For a rigid molecule undergoing isotropic random reorientation with correlation time τ_c, the following equations for transition probabilities hold:

$$W_0^{ij} = \frac{q\tau_c}{r_{ij}^6}; \quad W_1^{ij} = 1.5 \frac{q\tau_c}{r_{ij}^6} \frac{1}{1 + (\omega\tau_c)^2}; \quad W_2^{ij} = 6 \frac{q\tau_c}{r_{ij}^6} \frac{1}{1 + 4(\omega\tau_c)^2} \qquad (3)$$

where ω is the Larmor frequency of the protons, $q = 0.1\gamma^4\hbar^2$ and γ is the proton gyromagnetic ratio. The term R_{1i} represents external sources of relaxation such as paramagnetic impurities. Truncation of the series expansion of equation 1 to the first linear term in $\mathbf{R}$ is valid if τ_m is sufficiently short, which is equivalent to assuming that each cross-peak intensity depends only on the cross-relaxation rate between the two protons of primary interest — the isolated spin pair approximation (ISPA). This assumption is commonly made in distance estimations from 2D NOE intensities; however, this is not a good assumption (Borgias & James, 1988; Borgias & James, 1989; Post et al., 1990). For example, for mixing times generally accepted as being sufficiently short (i.e., 50 to 100 ms) and not including internal motions, ISPA can result in systematic errors of 45 - 80% in distances over 3.5 Å, a range which is quite important in defining molecular structure.

We have demonstrated that complete relaxation matrix analysis of proton homonuclear 2D NOE spectra enables numerous accurate interproton distances to be calculated by accounting for all proton dipole-dipole interactions (Borgias & James, 1988; Borgias & James, 1990; James, 1991; Keepers & James, 1984; Liu et al., 1992). The most efficient techniques for calculating accurate distances entail an iterative approach (Boelens et al., 1989; Borgias & James, 1988; Borgias & James, 1989; Borgias & James, 1990). In particular, the MARDIGRAS algorithm enables the determination of a large number of accurate distance restraints and aids in individually setting bounds (i.e., error bars) for those distances. We can also account for the presence of molecular motions(Liu et al., 1992) and incorporation of exchange with bulk water for exchangeable protons (Liu et al., 1993).

To assess the extent of conformational space consistent with experimental data, we need an estimate of the accuracy of our structural restraints. Explicitly, they are needed for setting bounds in distance geometry or flat-well size in restrained molecular dynamics calculations. Most structural studies have utilized ISPA to estimate interproton distances, but estimates of the error, reflected by the upper and lower bounds assigned, have varied widely throughout the literature. Tighter distance bounds (smaller error bars) lead to a higher resolution structure. But distance bounds made tighter than warranted by experimental accuracy mislead to a highly defined (small atomic RMSD) but incorrect structure (Thomas *et al.*, 1991). So we need to make the bounds as tight as possible but not so tight that the real distance can lie outside the bounds. It is possible with MARDIGRAS to obtain the distance bounds or "error bars" in a logical fashion individually for each proton pair from (James, 1994): (a) distances calculated from spectra at different mixing times; (b) distances calculated using different starting structures; (c) distances calculated using different correlation times centered around best estimate of experimental correlation time; (d) the error estimated from the fit of an element of the converged MARDIGRAS matrix with the corresponding experimental cross-peak intensity, with the calculation repeated several (~30) times using random variations of the experimental intensities within the experimental noise level; (e) in the case of motional or overlap averaging (i.e., with aromatic rings, methylenes, or methyls), using worst case geometries to enable calculation of maximum and minimum distances; and (f) for distances involving exchangeable protons (imino, amino, amide), an upper limit on the exchange rate with bulk water modifies the lower bound.

Torsion Angle Restraints

While NOE-derived distance restraints largely determine duplex structure, the sugar pucker is not really well-defined by distances alone (*vide infra*). Only two intrasugar proton-proton distances exhibit much sensitivity to the conformation of the sugar ring: the distance between H1′ and H4′ and, to a lesser extent, that between H2″ and H4′. Although there is some variation in torsion angle about the glycosidic bond connecting the sugar with its base (angle χ) to complicate matters, the intraresidue distances from base-H6 or base-H8 proton to the H3′ and H2′ protons are dependent on sugar pucker. We do have an independent means to define the sugar structure: with a sufficient number of three-bond coupling constants, the conformation of the sugar rings in nucleic acid fragments can be defined. There are a few multidimensional NMR experiments useful for obtaining scalar coupling constants quantitatively. Two which offer the possibility are the double-quantum-filtered COSY (2QF-COSY) and exclusive COSY (ECOSY) experiments. We emphasize that most proton-proton coupling constants in a sugar ring need to be determined to establish the pucker well — one or two will generally not suffice.

Sometimes coupling constants can be measured directly from ECOSY spectra; and the coupling constants can be extracted from 2QF-COSY spectra via simulation of those spectra utilizing the SPHINX-LINSHA program when the spectral linewidth becomes comparable to or slightly larger than the coupling constants (Celda *et al.*, 1989; Gonzalez *et al.*, 1994; Schmitz *et al.*, 1990; Widmer & Wüthrich, 1987). Torsion angle restraints in the sugar ring may subsequently be derived by use of Karplus-type relationships correlating measured coupling constants to dihedral angles (Rinkel & Altona, 1987). One should consider that with increased correlation times for biological macromolecules, dipolar relaxation effects can lead to altered, generally diminished, coupling constants (Harbison, 1993). However, this effect appears to be negligible for correlation times below 5 ns.

In more detail, the sugar conformation is analyzed as follows. The vicinal proton coupling constants J_j, obtained by fitting the experimental spectra using SPHINX, are related to the respective H-H torsion angles by the modified Karplus equation (Rinkel & Altona, 1987):

$$J_j = a \cos^2 \phi_j + b \cos \phi_j + c + \sum_k \Delta\chi_k [d + e \cos^2 (\xi_k \phi_j + f|\Delta\chi_k|)] \tag{4}$$

where constants a, b...f have been parameterized using 178 x-ray structures, $\Delta\chi_j$ is the difference in Huggins' electronegativity between substituent S_k and hydrogen, and ξ_k is +1 or -1, depending on orientation of S_k relative to the geminal protons. Individual endocyclic torsion angles v_i in a particular sugar conformation can be related to the pseudorotation angle P and the amplitude of pucker θ_m, which describe sugar pucker (Altona & Sundaralingam, 1972):

$$v_i = \theta_m \cos (P + 2(i-2)(360°)/5), \quad i = 0 - 4 \tag{5}$$

Often we have found that a single conformation with characteristic P cannot account for all coupling constants measured for a particular sugar. There may be rapid exchange between more than one sugar conformation with

$$J_i(obs) = \sum_n f_n J_i(n) \tag{6}$$

where f_n is the fraction of conformer n.

Altona and colleagues assume a two-state model with minor N-conformer (P = 9°, i.e., 3'-endo) and major S-conformer (typically P = 171°, i.e., 2'-endo). There are not enough independent variables to describe all parameters, so it is usually assumed that the minor conformer has a fixed pseudorotation angle of 9° and an amplitude equal to that of the major conformer; we have found that variations of this are generally not too important. The three other parameters can be varied within certain ranges: fractional population of S conformer f_s from 50 to 100%, pseudorotation phase angle P_s from 0 to 360°, and pucker amplitude θ_s from 20 to 40°. A (f_s, P_s) map can be constructed for each sugar pucker with contour lines corresponding to the upper and lower bounds of the experimental coupling constants. Some examples are shown in Figure 1. The intersection of the allowed regions for all the experimental J values defines the range of sugar pucker parameters consistent with the experimental data. This establishes the torsion angle restraint bounds to be used in subsequent structure generation. It will be noted in Figure 1 that the fraction of minor conformer is not always 0. However, in the case of DNA duplexes, for nonterminal residues, the minor conformer amounts to no more than 25%, and the conformational exchange is essentially ignored in structure refinement. However, we have considered it in advanced stages of refinement (Gonzalez et al., 1995; Schmitz et al., 1993).

Structure Refinement

Several popular molecular dynamics programs can accommodate NMR restraints. These programs are quite similar in that the potential energy is calculated for a set of atomic coordinates using a force field:

$$V_{total} = V_{bondlengths} + V_{bondangles} + V_{dihedrals} + V_{electrostatics} + V_{NOE} + V_{J\,coupling} \tag{7}$$

195

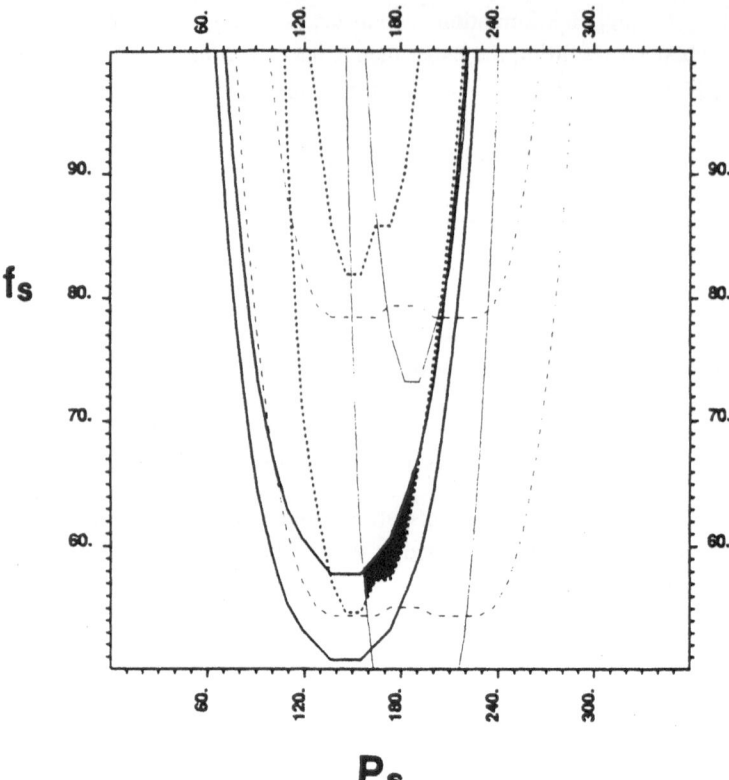

Figure 1. Isocontour plots of the estimated three-bond coupling constant values for some sugars in the DNA strand of [d(GCTATAA$_{ps}$TGG)·r(CCAUUAUAGC)] as a function of pseudorotation phase angle P_s and fraction f_s of the major S conformer for a sugar pucker amplitude θ_s of 35°. Solid bold lines represent $J_{1'2'}$, dashed bold lines $J_{1'2''}$, solid thin lines $J_{2'3'}$, and dashed thin lines $J_{2''3'}$. The region consistent with all the measured coupling constants is blackened. For the other deoxyriboses in the hybrid (data not shown), the minor conformer amounted to about 30-40%.

The first four terms on the right monitor the classical potential energy of the molecule. The last two pseudo-energy terms serve to penalize any molecular structures not fitting the NOE-derived distance restraints and either scalar coupling constants themselves or scalar coupling-derived torsion angle restraints. We utilize a flat-well potential for our penalty terms, with the size of the flat-well individually established by the experimental "error bars".

Using a selected starting structure for the molecule, Newton's equations of motion are solved with the forces generated by taking the derivative of the potential (equation 7), with respect to the coordinates (van Gunsteren, 1993). If all goes well, a global minimum is achieved for the molecular structure. Starting structures should be selected from different parts of conformational space, and different random initial trajectories should be utilized for each starting structure to gain confidence that the global minimum structure has been found. We have also developed a completely alternative method utilizing restrained Monte Carlo (rMC) calculations and internal coordinates, instead of Cartesian coordinates, to insure that the structure found is determined by the experimental data and not by the method of refinement used nor the particular force field employed (Ulyanov *et al.*, 1993). In the initial study, the difference between the structure found via rMD and that via rMC was only 0.5 Å root-mean-square-deviation (RMSD).

We have made sufficient progress that we can now confidently determine the time-averaged solution structure of oligomer duplexes of nearly any chosen sequence <15 bp long (probably longer) using rMD, although improvements are envisioned. We have also examined the effect of the number restraints available on the structure determined in one case where we had on average 20 distance restraints and 5 torsion angle restraints per residue (Weisz *et al.*, 1992). While a structure is fairly well restrained by ~10 distance restraints and 5 torsion angle restraints per nucleotide, we have found that it is better defined by ~15 distance restraints per residue, along with the torsion angle restraints. More than ~15 distance restraints provides redundant information, and the structure determined is little affected by additional restraints. It is assumed, of course, that the restraints are fairly evenly distributed across the molecule. In light of the paucity of protons in the bases, it is also necessary to obtain distances to imino protons; for this, the effects of chemical exchange must be taken into account (Bishop *et al.*, 1994).

NUCLEIC ACID DUPLEX STRUCTURE IN AQUEOUS SOLUTION

We have determined the structure of several nucleic acids in the past several years, as we have been striving to improve the methodology. At this stage, we feel reasonably confident that we can obtain a good quality (high accuracy and high definition) time-averaged structure of duplexes. The sequences we have studied have their own individual structural characteristics, but some generalizations can be tentatively put forward for non-terminal duplex segments containing solely unmodified DNA. Our initial statistical analysis of three DNA oligonucleotide duplex structures has revealed some interesting features (Ulyanov & James, 1994). One of the most remarkable characteristics of these structures is their average helical winding angle of 34.6°. This parameter had been measured to be 34.6° for B-DNA in solution by a very sensitive band-shift method (Wang, 1979), but the vast majority of DNA oligonucleotide crystal structures obtained have an average helical twist of 36.0°.

Helical parameters display a strong dependence on nucleotide sequence. In many cases, this dependence is clearly seen at the level of a sequence of purines (R) and pyrimidines (Y). While average helical twist is 34.6°, the relative sequence-dependent order of helical twist is YR > RR:YY > RY.

Other characteristic features of solution DNA conformations are negative slide, systematically open minor groove (for almost all sequences), and decreased helical rise. The latter, rather unexpected finding, is correlated with a surprisingly strong non-flatness of Watson-Crick base pairs. In essence, the oligonucleotides appear to be shorter and fatter than one would anticipate from canonical B-DNA.

At all three TG steps, local structural variations, including large positive roll ρ, lead to local bending (8-16° from the global helix axis into the major groove of the duplex). Another structure in progress of being refined manifests this same local bend at TG.

While these tentative generalizations are interesting, we must remind that a larger set of solution structures is needed to complete the analysis of the sequence dependence of DNA conformation.

CONFORMATIONAL FLEXIBILITY IN NUCLEIC ACID DUPLEXES

In the process of determining a high-definition structure in solution, we must be cognizant of the possibility of multiple interconverting conformers or conformational flexibility. First, we need to consider the effect of molecular motion on our ability to determine an accurate time-averaged structure — i.e., will conformational jumps or internal motions lead to systematically biased distances and distorted structures. Second, we would like to impose molecular motion information on the time-averaged structure and thus obtain a dynamic picture of the oligonucleotide.

For conformationally flexible molecules in solution, NMR-derived restraints are time-averaged. The time-averaging is complicated. It is dependent upon the rate of interconversion of conformers. For conformational fluctuations slow compared to the reciprocal of the overall molecular tumbling rate but fast compared to the chemical shift difference between the conformational states, the distance measured by the NOE experiment will be subject to $<r^{-6}>$ averaging over individual proton positions, as implied by equation 3, so time-averaging will be heavily weighted toward the shorter distances. There have been various methods of accounting for the 2D NOE intensities when there is exchange between different conformational states (Landy & Nagaswara Rao, 1989). For conformational fluctuations occurring faster than the overall tumbling rate ($>10^8$ s^{-1}), the distance measured is generally slightly weighted toward the shorter distance if the actual motion is not isotropic (Keepers & James, 1984), but the effective distance is identically equal to the distance between the mean positions of nuclear jump states in the case of isotropic jumps (LeMaster et al., 1988).

Linear averaging does not occur in the case of torsion angle restraints either — the well-known Karplus curve is certainly not linear. In short, whenever conformational fluctuations occur, time-averaged torsion angle and distance restraints may be inconsistent, and time-averaged restraints may not correspond to a physically (energetic) realistic structure.

Evidence for Conformational Flexibility in Duplexes

We have generally found that when all proton-proton coupling constants are measured for sugars in nucleic acid duplexes, the results cannot be reconciled with a single sugar conformer. For nonterminal nucleotides in standard DNA duplexes, with few exceptions, we find that the 2'-endo conformer dominates, being populated 75-95% of the time. A slightly more extreme example is shown in Figure 1 for a hybrid RNA·DNA duplex: [d(GCTATAA$_{ps}$TGG)·r(CCAUUAUAGC)], which contains a chiral phosphorothioate in one position of the DNA strand (Gonzalez et al., 1994). In this case, the minor conformer could be present to the extent of ~40% for some of the deoxyribose sugar conformers; the RNA sugars were all clearly 100% 3'-endo. We note that use of the (f_s, P_s) map does not require a two-state equilibrium since, if a single conformer can accommodate all the coupling constant values in a sugar, the points where $f_s = 100\%$ must be contained in the inter-

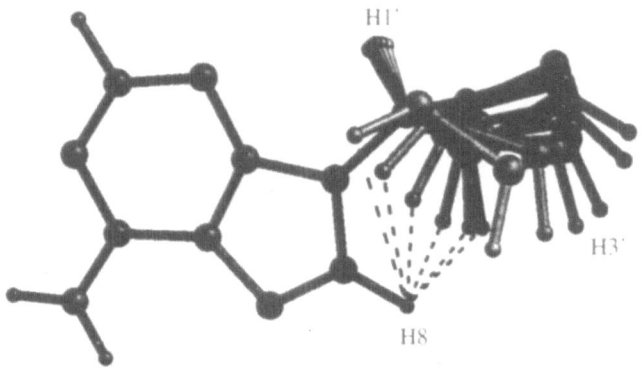

HI′

H3′

H8

Figure 2. Illustration of the change in intraresidue base proton-to-sugar proton distance as the conformation of the deoxyribose changes from 3′-*endo* (shown in light gray) to 2′-*endo* (dark gray). Shown by the dashed line is the H8-H2′ contact. Likewise, distances to other sugar protons vary with sugar pucker. Sugar atoms H2″, H4′, C5′, H5′, and H5″ are omitted for clarity.

section of allowed regions of the (f_s, P_s) map. In the hybrid duplex, this happens only for A6 and G9. It will be noted that the DNA backbone maintains sufficient flexibility via $1-3$ crankshaft-type bond rotations, however, that the disposition of the bases may be altered little by the sugar repuckering.

Careful complete relaxation matrix analysis of the experimental NOE intensities can yield a very good representation of the cross-relaxation rates between proton pairs. Interpretation of those rates in terms of interproton distances can lead to structurally inconsistent distances even with perfect data and perfect analysis, if the distances are calculated assuming a rigid structure but conformational fluctuations are occurring. The scalar coupling constants derived from 2QF-COSY data indicate that sugar puckering occurs. As shown in Figure 2, different sugar conformations will have some different intraresidue distances between the base H8 (purines) or H6 (pyrimidines) protons and sugar protons. If there are conformational jumps, they are anisotropic and will lead to measured distances which are shorter than the simple arithmetic average based on conformer population (Keepers & James, 1984). Consequently, the measured H8-H2′ distance and H8-H3′ distance for a sugar may be internally inconsistent.

Likewise, there may be inconsistencies in measured interresidue distances. A fairly obvious one we have detected is illustrated in Figure 3. As is evident in the figure, the structure of d(GTATAATG)·d(CATTATAC) resulting from rMD has the base ring distorted from planarity in an attempt during the simulation for the force field to simultaneously satisfy conventional potential energy terms (chiefly van der Waals forces) as well as inconsistent NOE-derived distances of 2.69-2.99 Å for H8A5-H8A6 and 3.16-3.39 Å for H2A5-H2A6. However, even this base distortion will not reconcile the experimental distances, as the rMD structure has distances of 3.52 and 3.75 Å, respectively. In rMC structures with idealized base geometries, these two distances are respectively 3.61-3.64 and 3.98-3.99 Å (Ulyanov *et al.*, 1993). For the A5-A6:T11-T12 base pair in d(GTATAATG)·d(CATTATAC) , we had previously found evidence for dynamics via $T_{1\rho}$ measurements (Schmitz *et al.*, 1992).

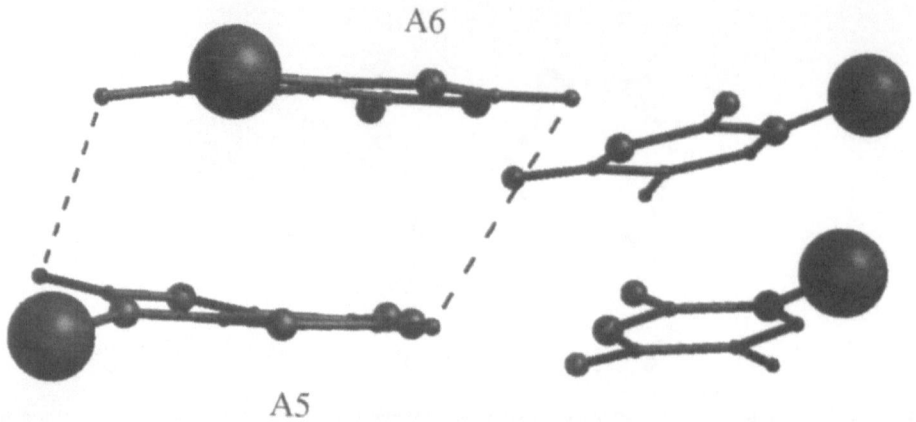

Figure 3. Base pair step A5-A6:T11-T12 in the final structure of d(GTATAATG)·d(CATTATAC) derived from rMD, as viewed from the minor groove. The dashed lines show the contacts H8A5-H8A6 and H2A5-H2A6. Deoxyriboses are depicted schematically by the large spheres.

Molecular Dynamics with Time-averaged Restraints (MD-tar)

We can pose the rhetorical question: how can we deal with the situation of conformational averaging of restraints? Standard rMD requires the NMR-derived restraints to be satisfied at every point of the rMD simulation. Instead, an alternative approach has been proposed, unfortunately termed molecular dynamics with time-averaged restraints, or MD-tar (Torda *et al.*, 1990). The distance between any two protons is monitored as a running average with exponential weighting to emphasize more recent snapshots during an MD trajectory. The running averages are used to calculate the penalty V_{NOE} (equation 7), instead of the distances obtained explicitly at each step in the MD run. This permits sampling of a greater range of conformational space. This approach was applied to experimental 2D NOE distance data, analyzed via complete relaxation matrix analysis, for d(GTATAATG)·d(CATTATAC) (Schmitz *et al.*, 1993). MD-tar produces an ensemble of conformers over the course of an MD trajectory. The resulting structural parameters, as an average over the MD trajectory, are largely the same as with standard rMD. However, MD-tar results yield a more realistic picture of conformational fluctuations in accord with torsion angle determinations made via 2QF-COSY experiments, despite the fact that torsion angle restraints were not used in simulations, as well as a residual index (R-factor), which compares 2D NOE intensities calculated for the MD-tar ensemble of structures with experimental intensities. The resulting conformational ensemble presents a distribution of values for each structural feature, e.g., pseudorotation angle or helix twist, which must be analyzed.

Probability Assessment via Relaxation rates of a Structural Ensemble (PARSE)

Another approach we have recently developed generates an ensemble of conformers and then yields an assessment of the probability of each conformer (Ulyanov *et al.*, 1995). The procedure, termed PARSE, is outlined in Figure 4. A careful analysis of the structure obtained via conventional rMD (or rMC) methods could reveal internal inconsistencies in 2D NOE-derived interproton distances and scalar coupling-based torsion angle restraints, which then are used as the basis for making subsets of distance restraints which are used in

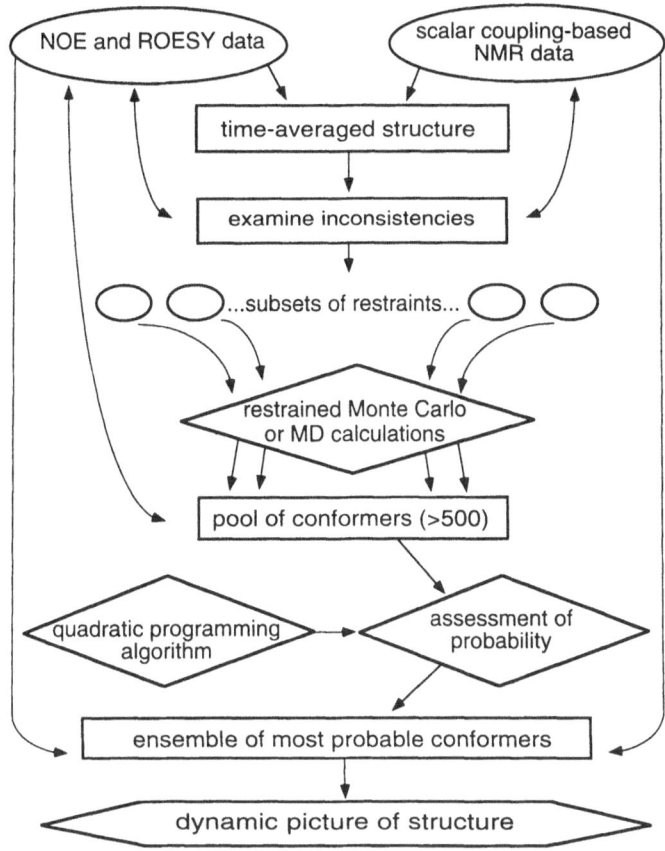

Figure 4. Outline of the PARSE (Probability Assessment via Relaxation rates of a Structural Ensemble) procedure.

restrained Monte Carlo (or feasibly rMD) calculations to generate an ensemble of conformations. Several different subsets of restraints are used, with some remixing of restraints into different subsets. The idea is to generate a large pool of structures each of which should satisfy at least some of the restraints. A quadratic programming algorithm is used to assess the probabilities of the conformers by minimizing a relaxation rate-based residual index, yielding an ensemble of conformers in agreement with all experimental data. The probability assessment is based on fitting experimental values of NMR parameters with theoretical values which can be computed as a linear average over the conformers. Cross-relaxation rates and scalar coupling constants are linearly averaged, so one can set up a quadratic objective function comparing experimental and theoretical values of these NMR parameters, which is subjected to constrained global minimization with the constraints that the sum of probabilities must add to one and the individual probabilities must be greater than or equal to zero. The assessment is carried out using the quadratic programming algorithm (Fletcher, 1981). The result is that we obtain the number of conformers, the identity and the probability of each conformer in the ensemble, which best fits the experimental data. For example, for d(GTATAATG)·d(CATTATAC), only nine of the pool of 500 conformers were found to be necessary to accommodate the experimental data. Of the nine conformers, the most probable (33%) was within 0.5 Å RMSD from the final rMD; the three least probable (<3%) differed the most from the single best structure (~2.5 Å RMSD). The contributions of three conformers with individual probabilities of 14-17%, differing from the most probable by 0.6, 1.3 and 1.6 Å RMSD, are probably quite important in providing the best fit to the experimental data. In our initial analysis of this oligomer, we did not use experimental coupling constant data in generation of the conformational ensemble, but left them as an independent experimental check. Indeed, we found the coupling constants to be much better fit by the nine-conformer ensemble than by the single best rigid structure obtained by either rMD or rMC.

The initial stage of PARSE is quite similar to the MEDUSA procedure developed to investigate conformational equilibria (Brüschweiler *et al.*, 1991). The MEDUSA algorithm has the advantage of doing a random permutation of the restraint list. However, this is not computationally feasible for biomolecules with a large list of restraints. Instead, we used "educated guesses" to design subsets of restraints based on internal conflicts — sixty subsets were used in the example cited above.

While the conformational ensemble constructed using PARSE (or MD-tar for that matter) cannot be declared unique, the members of the ensemble do yield insight about conformationally flexible molecules. The structural characteristics of minor conformers indicate that the molecule must manifest these features some of the time in order to satisfy the experimental data. In this regard, it is therefore necessary that the quality of those restraints be as high as possible to insure the verity of the final conclusions. Future work will need additional experimental data, e.g., heteronuclear relaxation parameters, to elucidate these motional features. The computational framework provided by PARSE is capable of incorporating this additional information.

ACKNOWLEDGMENTS

This work was supported by National Institutes of Health grants GM39247 and RR 01081. CG acknowledges a post-doctoral fellowship from the Spanish Consejo Superior de Investigaciones Científicas (CSIC). We also acknowledge use of the Cray C90 supercomputer, supported by a grant from the Pittsburgh Supercomputing Center through the NIH Division of Research Resources cooperative agreement U41RR04154 and a grant from the National Science Foundation Cooperative Agreement ASC-8500650. Figures 2 and 3 were prepared using the graphics program MIDASplus developed in the UCSF Computer Graphics Laboratory with the support of NIH Grant RR01081.

REFERENCES

Altona, C. & Sundaralingam, M. (1972). *J. Am. Chem. Soc.* **94**, 8205-8212.

Bishop, K. D., Blocker, F. J. H., Egan, W. & James, T. L. (1994). *Biochemistry* **33**, 427-438.

Boelens, R., Koning, T. M. G., van der Marel, G. A., van Boom, J. H. & Kaptein, R. (1989). *J. Magn. Reson.* **82**, 290-308.

Borgias, B. A. & James, T. L. (1988). *J. Magn. Reson.* **79**, 493-512.

Borgias, B. A. & James, T. L. (1989). In *Methods in Enzymology, Nuclear Magnetic Resonance, Part A: Spectral Techniques and Dynamics* (Oppenheimer, N. J. & James, T. L., eds), vol. 176, pp. 169-183, Academic Press, New York.

Borgias, B. A. & James, T. L. (1990). *J. Magn. Reson.* **87**, 475-487.

Brüschweiler, R., Blackledge, M. & Ernst, R. R. (1991). *J. Biomol. NMR* **1**, 3-11.

Burley, S. K. (1994). *Curr. Opin. Struc. Biol.* **4**, 3-11.

Celda, B., Widmer, H., Leupin, W., Chazin, W. J., Denny, W. A. & Wüthrich, K. (1989). *Biochemistry* **28**, 1462-1470.

Fletcher, R. (1981). *Practical Methods of Optimization. Vol. 2: Constrained Optimization*, p. 220 pages, John Wiley & Sons, New York.

Gonzalez, C., Stec, W., Kobylanska, A., Hogrefe, R., Reynolds, M. & James, T. L. (1994). *Biochemistry* **33**, 11062-11072.

Gonzalez, C., Stec, W., Reynolds, M. & James, T. L. (1995). *Biochemistry* **34**, in press.

Harbison, G. S. (1993). *J. Am. Chem. Soc.* **115**, 3026-3027.

James, T. L. (1991). *Curr. Opin. Struct. Biol.* **1**, 1042-1053.

James, T. L. (1994). *Curr. Opin. Struct. Biol.* **4**, 275-284.

Keepers, J. W. & James, T. L. (1984). *J. Magn. Reson.* **57**, 404-426.

Landy, S. B. & Nagaswara Rao, B. D. (1989). *J. Magn. Reson.* **83**, 29-43.

LeMaster, D. M., Kay, L. E., Brünger, A. T. & Prestegard, J. H. (1988). *FEBS Lett.* **236**, 71-76.

Lipanov, A., Kopka, M. L., Kaczor-Grzeskowiak, M., Quintana, J. & Dickerson, R. E. (1993). *Biochemistry* **32**, 1373-1389.

Liu, H., Thomas, P. D. & James, T. L. (1992). *J. Magn. Reson.* **98**, 163-175.

Liu, H., Kumar, A., Weisz, K., Schmitz, U., Bishop, K. D. & James, T. L. (1993). *J. Am. Chem. Soc.* **115**, 1590-1591.

Macura, S. & Ernst, R. R. (1980). *Mol. Phys.* **41**, 95-117.

Pabo, C. O. & Sauer, R. T. (1992). *Ann. Rev. Biochem.* **61**, 1053-1095.

Post, C. B., Meadows, R. P. & Gorenstein, D. G. (1990). *J. Am. Chem. Soc.* **112**, 6796-6803.

Rinkel, L. J. & Altona, C. (1987). *J. Biomol. Struct. Dyn.* **4**, 621-649.

Schmitz, U., Zon, G. & James, T. L. (1990). *Biochemistry* **29**, 2357-2368.

Schmitz, U., Sethson, I., Egan, W. & James, T. L. (1992). *J. Mol. Biol.* **227**, 510-531.

Schmitz, U., Ulyanov, N. B., Kumar, A. & James, T. L. (1993). *J. Mol. Biol.* **234**, 373-389.

Thomas, P. D., Basus, V. J. & James, T. L. (1991). *Proc. Nat. Acad. Sci. USA* **88**, 1237-1241.

Torda, A. E., Scheek, R. M. & van Gunsteren, W. F. (1990). *J. Mol. Biol.* **214**, 223-235.

Ulyanov, N. B., Schmitz, U. & James, T. L. (1993). *J. Biomolec. NMR* **3**, 547-568.

Ulyanov, N. B. & James, T. L. (1994). *Appl. Magn. Reson.* **7**, 21-42.

Ulyanov, N. B., Schmitz, U., Kumar, A. & James, T. L. (1995). *Biophys. J.* **68**, 13-24.

van Gunsteren, W. F. (1993). *Curr. Opin. Struc. Biol.* **3**, 277-281.

Wang, J. C. (1979). *Proc. Nat. Acad. Sci. USA* **76**, 200-203.

Weisz, K., Shafer, R. H., Egan, W. & James, T. L. (1992). *Biochemistry* **31**, 7477-7487.

Widmer, H. & Wüthrich, K. (1987). *J. Magn. Reson.* **74**, 316-336.

Wright, P. E. (1994). *Curr. Opin. Struc. Biol.* **4**, 22-27.

DISCUSSION

David Cistola - Tom, for the determination of the Tat peptide structure, what were the buffer conditions used.

Thomas James - For that particular structure, in order to avoid aggregation which has been a real bug for Tat proteins, we had to run at a concentration of 350 µM. CD spectra run at neutral pH (pH=7.2) and low pH (pH 3.5) were largely the same. So the structure is not strongly dependent on pH. NMR studies were carried out at pH 3.1, 15^0 C.

Cistola - What assay was used to assess biological activity?

James - This was done by my colleagues at UC San Francisco. They have a perfused cell system where they monitor HIV replication. This particular peptide stimulated HIV replication within the cell system that they had.

Cistola - Thank you.

Peter Wright - I want to follow up on that peptide. It is very rare to see such a helical peptide. Are you sure it's not associated even at 350 µM as a dimer?

James - Actually, no, I am not sure of that. It's occurred to me that there may be some kind of aggregation that might stabilize it. However, the NMR spectra are pretty reasonable in terms of lineshapes and linewidths. So if it is associated, I would say that it's not extensively so. While there may be some interactions that take place, I really have an extremely difficult time imagining why such a basic region would like to get together with another one of these basic regions. That's why I think that this peptide may work a little more readily than the intact protein; several different workers have had problems with the whole protein. There maybe other reasons the Tat protein can aggregate; the intact protein also has a cysteine region which could become oxidized giving water solubility problems as well. One of the items which I did fail to point out is why I think the basic region in this particular peptide is stabilized. That is because of the aspartate. If I look at the structures that actually result from our analyses, that aspartate actually points toward and actually helps to stabilize a lysine and an arginine - essentially a bifurcated interaction between the basic α-helical region and the other helical region. From the final calculated structure, it looks like there is a very standard α- helix at least in the basic region in terms of the Ramachandran ϕ, ψ angles.

Wright - Is there a sign of conformational averaging from coupling constants?

James - I do not know that in detail. However, if you're asking whether I think 100% of the peptide in solution assumes this structure? No, I don't.

David Gorenstein - Tom, how do you define conformation? I mean one could argue that in the last ones that just one sugar with an N or S could be a conformation. You're looking at just the major populations obviously.

James - Well, for example in most of the DNA duplexes, the double quantum filtered COSY analysis and E-COSY analysis would suggest that something like 80% to 90% of the deoxyribose rings are in the major conformation and that is the conformer we ultimately end up with when doing our rMD calculations. We perform those calculations recognizing, of course, that it's not exactly the truth, but at least for time average results, it's not so bad.

We have examined conformational flexibility via molecular mechanics calculations, and probably other people have done so as well, and find that you can actually change the conformation of a deoxyribose ring all the way from 2' to a 3' endo with only a few adjustments in the phosphodiester region - nothing else has to change. You compensate with crankshaft types of motion very readily. So in terms of the major players - in terms of the energetics for the double helix, of course - it is going to be up in the bases. The stacking and the Watson-Crick hydrogen bonding will establish three-dimensional structure and will be influenced to only a small extent by the sugar conformation.

Tom OConnell - I have a general question about relaxation matrix calculations. It seems that there is always a weak coupling approximation in perhaps some of the nucleic acid sugars or certainly in carbohydrates NMR structures that probably doesn't apply. So how would that affect some of the NOE's that you would get between spin systems that are involved in higher order coupling?

James - These are questions which we have not ourselves explicitly looked at, but a few other groups have examined. I think the general conclusions for the kinds of systems that most of us are studying, i.e., small biopolymers, is that the problems are not very severe.

PANEL DISCUSSION

Topic: Extension of Techniques to Larger Molecules

Members: G. Marius Clore, National Institutes of Health

Stephen W. Fesik, Abbott Laboratories

David G. Gorenstein, University of Texas Medical Branch

David M. LeMaster, Northwestern University

John L. Markley (Moderator), University of Wisconsin

John Markley - I was given the task here of moderating a panel discussion on larger biomolecules - you can see by my overhead that I have a certain bias toward proteins. One obvious problem with larger biomolecules is that they contain more atoms that give signals in the same chemical shift region as with smaller molecules; this creates a good deal of overlap. The other problem is that the larger the molecule, the more slowly it tumbles in solution, and that also leads to problems with broader lines and more spin diffusion. Finally, these problems are compounded by the fact that the biomolecules are dissolved in solution at a molar concentration high enough for good signal to noise, and so there the solution has more crowding and higher viscosity. Over the years several methods have been developed to overcome these problems. One is to use higher magnetic fields. Protein NMR started back at 40 MHz and has gone up to 750 MHz now, and there are excellent prospects of going to still higher fields. Other approaches are to use higher dimensional NMR. First two-dimensional NMR, then three and four. One question we might raise is, is it worthwhile going to even higher dimensions? As you've heard in the last two days, pulse sequences can be rather generic as designed for proteins at natural abundance, or they can be tailored to particular situations as for a uniformly labeled protein, very finely tuned to getting information about coupling constants or particular kinds of NOE's. Another approach concerns the use of stable isotope labeling, and once again there is a generic versus tailored way of doing this. A generic way would be to use uniform labeling by ^{13}C, ^{15}N, and possibly ^{2}H. There are many special ways of labeling. We heard already from David LeMaster about chiral labeling and random fractional deuteration. We have heard from Ad Bax about reverse labeling, and others have talked about block labeling and selective dual labeling. In our lab we have used particular labeling patterns for looking at *cistrans* isomerizations in Xaa-Propeptide bonds. Rob Kaptein has talked about special labeling for looking at dimers or DNA-protein interactions. Steve Fesik and others have used labeling protocols for looking at drug-receptor interactions.

One thing that we can do, and Richard Ernst made a similar analysis, is to present the present envelope of the state of the art in terms of dealing with proteins of various sizes. I start with the fact that one can work with proteins as large as 150,000 molecular weight if you're willing to take limited information and use selective labeling. Yoji Arata and others have shown with IGG's that one can work with proteins of this size. On the other end, we know that one can carry out full structural analyses of proteins and other biomolecules up

NMR As a Structural Tool for Macromolecules: Current Status and Future Directions
Edited by B.D. Nageswara Rao and Marvin D. Kemple, Plenum Press, New York, 1996

207

to 10,000 or higher at natural abundance. The upper limit on size depends upon how much effort one wants to put into the analysis. It looks like the dual labeling methods that have been discussed at this meeting (uniform labeling with ^{15}N and ^{13}C) will take structural studies certainly into the 30,000 M.W. range. For backbone assignments, even large proteins appear feasible. In the middle is the gray area and this is probably the area that we are going to be discussing. What is the molecular size limit in terms of determining structures, and, what kinds of useful information, in the absence of structures, can one derive from larger biomolecules?

So what I would like to do is put up a number of topics for discussion. These are the ones that I came up with on the plane coming here. I actually had a lot more detail, but my colleagues here convinced me to simplify things. We also would like to find out from people here if we have missed additional topics that are worth discussing. Individual topics may be of more interest to some people than to others. We propose to discuss these in the context of the various steps that one usually follows in carrying out a study of biological macromolecules. These steps include, first of all, doing some so-called "tuning" on the sample. It is certainly true that some samples are more amenable to NMR analysis than others, and you can play around with the solution conditions, such as pH, salt concentration, and solvent agents to make the macromolecule more soluble or to avoid aggregation. Then one collects the data, and usually the step after that is to make backbone assignments. You may need to use certain kinds of labeling to assist you in doing this or to determine starting points for backbone assignments. An analysis of the secondary structure comes out of the pattern of backbone NOE's and now from the assigned chemical shifts associated with the backbone. Then one assigns sidechains, analyzes the structure, and finally refines and validates the three dimensional structure. If the system is a protein, dimer of a protein-nucleic acid complex, then intermolecular interactions have to be worked into the analysis. Then one can analyze the dynamics, and finally what we're ultimately interested in is to learn how these macromolecules actually work. Usually that takes even more data analysis and data collection and further study.

So this is the way we have set things up for this discussion, and what we will do is first ask if there are any burning questions from the audience, items that we might need to add to this list. If not we will start at the top with the issue of why some proteins or other biomolecules show nicer spectra than others and what can be done about it?

Stephen Fesik - We have looked at different proteins and some of them behave very well such as the serine protease domain of urokinase. For this serine protease, surprisingly, the NMR spectrum looks fairly good for a protein of it's size, of about 27 kD. We've also looked at a pepsin/inhibitor complex, which is 37 kD. In particular, we were able to determine the conformation of the bound peptide. The protein signals looked good, compared to other proteins that we've had in the laboratory. In another system, the cyclosporin/cyclophilin complex, for example, it was interesting that some portions of the protein looked very good; whereas, other portions of the protein looked very bad. In some cases one can try to get around this problem by altering sample conditions. We have typically used light scattering and ultracentrifugation experiments to try and probe if there is some evidence of some larger species in the solution and if we can eliminate that by adjusting the sample conditions. However, we don't know why some proteins behave well empirically. One can change the conditions and try to get the sample to look better. That's our experience with some of the proteins that we have looked at.

Markley - I guess one anecdotal result from our experience, proteins that are bound to some prosthetic group seem to give better spectra than ones that are not.

David Gorenstein - One of the things that we have experienced is that we tend to think, as the molecular biologists have started to think, that a protein is a linear polymer. You can cut off the domain, a chunk, and throw away the other part, but as we saw earlier today sometimes one can't do that. When you're looking at a single domain or something like this, the hydrophobic part or some other portion of a protein may be exposed that tends to cause aggregation. So it's very important, when you're not looking at a native globular protein that expresses the whole protein, that you very carefully consider the nature of the fraction that you're actually dealing with and it's three-dimensional fold.

Markley - If anyone in the audience has anything to add on this, please stand up and move forward.

Marc Adler - Dr. Fesik, you said you've looked at light scattering and also centrifugation. Have you seen any correlation?

Fesik - Light scattering is particularly sensitive to the bigger particles that you see in the mixture compared to centrifugation. We do see in some cases that samples that produce broad signals correspond to samples that aggregate or that there are portions that are aggregates as measured by other techniques. Conversely, in those cases in which we see narrow signals and can get it looking like a single species, the other methods give the same results. However, in some instances where other techniques would suggest smaller components, we get broad linewidths. Maybe the signals are broadened by an exchange phenomena. In other cases where what we would have predicted a bad NMR spectrum by other methods, the NMR spectrum looked pretty good. So there's some correlation, but it doesn't always work that way. Probably, the best way to look at whether or not it's going to be suitable for NMR is just to do the NMR experiment. We've been trying to automate a procedure for sample conditioning through a centrifugation technique, whereby you screen a bunch of different additives and then just take the top portion out and analyze it for it's protein concentration. Using this method, you can rapidly analyze a variety of different sample conditions. However, in the end, I think we're going to end up going back to the NMR experiment.

Markley - Does anyone else have something to add to this topic?

Gorenstein - Could I just add one? That seems to argue that we really should be going to larger sample sizes. Clearly, if we're going to use concentrations of 1 or 2 mM, most larger proteins will tend to aggregate.

Markley - David, sample sizes is on the list here. We'll come back to this in a moment.

Unknown - What you said was that, if you study just one domain of a protein which contains a large content of hydrophobic residues, it maybe insoluble in water. What conditions can we adjust to make the protein soluble and suitable for NMR study? Does an organic solvent help solubilize the protein?

Fesik - We've been exploring a variety of mixed solvents. Ad Bax, of course presented the work he's done with CHAPS and has been able to get narrowing of the signals. Brian Sykes has some experience with that as well. Carolyn used TFE, I believe, to get more solubility and good NMR spectra. Also there are various detergents that one could try.

Carol Post - Can I just ask a general question? What is your success rate? When you see a bad spectrum, how often are you able to find conditions where it's suitable for NMR and is there any correlation between stability and suitability? Marius, can you answer this?

Marius Clore - My general experience is that when you see straight away it's not suitable, it remains unsuitable no matter what the conditions. In other words all the things like adding CHAPS and the like take it from difficult to slightly better but they do not provide dramatic improvements. But, what one can often do is, for example, mutate the protein. I can give you one example for work that we did on HIV RNAse-H. The original wild type stuff had only 137 residues, but was completely hopeless. We then mutated a tryptophan to an alanine on the basis that we couldn't see the tryptophan signals and that improved it enough that we were able to get the secondary structure. Subsequent to that we then looked at a sample of RNAse H from a different strain from Agouron which only had a few substitutions and which gave perfect spectra. Using that sort of approach, I think one can do something but that's extremely time consuming. Basically if you want to go after it, it's usually possible but I think altering conditions is usually not very helpful.

John Markley - I think we can take solace in the fact that crystallographers have very similar problems, or inverse problems in that a lot of their proteins don't crystallize, and maybe a single amino acid substitution will lead to a variant that will crystallize and they can't explain why that happens. Perhaps we should move on to the next topic which is, what are the optimal most economical labeling patterns for dealing with larger proteins? I'll ask David to start on this one, and then maybe Steve can talk about it.

David LeMaster - Well, I guess obviously one of the key questions is when do you make a decision between a uniform labeling pattern, which has been so powerful, particularly for main chain assignment work, and ask the question of what benefits can you gain from switching to potentially selective labeling processes? As added earlier today and as others have noted, one of the key problems, particularly in looking at NOE's between residues, is really the sensitivity loss. T_2 losses resulting from situations in which the protons are being relaxed by directly bonded ^{13}C. This comes into play in the uniform patterns in two ways. One, obviously, is the direct relaxation. Furthermore in general you want to use constant time techniques to gain benefits of the refocussing. So the question is with a protein of, say, around 40,000 daltons, where people have been able to assign the main chain but can't complete the structural analysis, 'How are you going to be able to get via selective labeling NOE's sufficiently strong and large enough numbers in such a larger system where you can make say sidechain assignments in order to turn that into structural information? In the earlier work in deuteration, we and other people have seen that when you have a background where you have a proton site that has at least appreciable deuteration in the rest of the neighborhood, the amide relaxation times lengthen by approximately three fold or so. Deuteration basically higher than 75% doesn't make much of a difference. Basically you peak out for the amide positions anyway. Related to that is the sensitivity benefit at selected sites that we have been able to see. You can get sensitivity enhancements as high as 4 to 6 fold over a natural abundance pattern when you have selective protonation fully enriched in a deuterated background.

Then you obviously have the complementary question of, 'How did you get enough of these so as to get enough constraints for structure analysis?' There clearly is a price for any selective labeling pattern. You're wiping out, at least in a given sample, some subset of the interactions, but I think there is one number that I find interesting that doesn't get kicked around quite so much. We often tend to shoot for say 15 or so constraints per residue in a protein structure analysis. If one asks how many cross peaks would be predicted from known structures within 5 $\overset{0}{A}$, it is on average 50 per residue. Hence in a typical analysis we are utilizing only about 30% of such cross peaks. So perhaps at least we've designed their patterns appropriately. All protons are not equal in the structure analysis, clearly methyls and certain aromatic positions are recognized to be key sites for giving constraints between residues and the internal packing, and so intelligent design in selective labeling can surely give you at least a large set of NOE constraints with relaxation properties that can circumvent what we run into in a uniform labeling pattern.

Markley - Anything to add, Steve or Marius?

Fesik - I think the labeling strategy that you want to use depends on what question you are attempting to answer. I am very impressed with the ^{15}N, ^{13}C and ^{2}H labeling approaches for proteins, at least for making the backbone assignments. There might be some other tricks that one can play for sidechain assignments using these labels. However, it is a more difficult problem to make the assignments of the NOE's. Some of the concepts that David brought up we are trying to use within the laboratory as well, trying to examine new patterns of labeling by which we could extract that information. We tried some selective labeling experiments in the study of the Fli-1/DNA complex that I'll be talking about tomorrow by having a deuterated background and adding in selectively protonated amino acids as others have reported, and have had some success with that. Using this technique, we have a deuterated background and we have some selectively protonated amino acids. This can dramatically simplify the NMR spectra. Other approaches may be to shorten the delays and reduce the dimensionality of the NMR experiments to increase the signal to noise has aided in making the assignments of that protein-DNA complex. This has really been a difficult problem.

Gorenstein - Can I raise an issue we haven't heard about yet: transferred NOE's. I know Steve has been a major player in this too. Theoretically if you're only interested in the confirmation of crystal structure, say you're interested in the conformation of a drug bound to the receptor or something like that, NOE transfer occurs if you're in the right time domain of fast exchange, but there are many inherent problems in that, as we're all familiar with, in extending it to much larger molecular weight proteins. Would you want to comment if it is worthwhile labeling just the drug or the agent at the receptor and measuring transferred NOE's or should we forget about that.

Fesik - Well, I don't use transfer NOE's extensively but it's because of the problems that we're typically studying in the lab. We're studying ligands that are tightly bound complexes, so we don't have the need for these experiments. In one case, we used these experiments to study weakly bound inhibitors to the enzyme CMP-KDO synthetase. Using transferred NOE's, we were able to determine the bound conformation of substrates and weakly bound inhibitors. However, as I mentioned earlier, we don't usually have those sorts of complexes to study. Usually we are studying ligands that are bound tightly.

Nuria Assa Munt - This question is for Steve Fesik. We're all talking about labeling, and there are very few people who have experience on labeling mammalian cells. Since we are talking about costs and you seem to have experience in that, can you comment on the methods and how possible it is to do these things nowadays.

Fesik - Yes, we published a paper at the end of 1992 on the labeling of proteins that are expressed in mammalian cells. We labeled the amino terminal fragment of urokinase that was expressed in mammalian cells. Since that time, we have had more experience by labeling the serine protease domain of urokinase. Based on this experience, I would suggest that if you can avoid at all costs labeling in mammalian cells, do it. It's very costly. Moreover, the big problem is that mammalian cells, unlike bacterial cells which will eat anything, are very sensitive to variations in the growth conditions. In addition, you do not get complete labeling of the protein. We've tried to get around some of the deficiencies reported in our earlier paper through a collaboration with Bob Curley at Ohio State, by preparing ^{13}C and ^{15}N labeled cysteine so that we could get complete labeling of the protein. In the past, we would supplement our growth media with unlabeled amino acids, but now we can isotopically label cysteine and get complete labeling. However, it can still be very costly. Furthermore, in some cases the mammalian cells that express the protein could just die on you for no apparent reason, or you don't get reproducible amounts of the protein. We've had really bad experiences with that and have spent a lot of money trying to label the serine protease domain of urokinase with little success. So, I wouldn't advise it unless it is absolutely necessary.

Markley - Do you have something to add to that Marius?

Clore - We've had very little experience with mammalian cells, but one problem with mammalian cells is that you also get glycosylation which you can't control. So that presents a real problem. We've had some experience with yeast. That can work very well providing you can engineer the protein so that you don't get glycosylation, because there again, you can't control it. Doing labeling in yeast is cheap. It's no more expensive than in *E. coli*.

Markley - Have any of you had experience with baculovirus? Does anybody out there want to talk about insects?

Unknown - The system is also a very kinetic one, i. e., the proteins are made in a very specific time and at a very quick rate. So I was wondering if any of you actually thought about the same sort of things or considered the insect system without just sitting there and saying no we haven't done it. But what have you actually thought about the system?

Fesik - We've never had a protein of interest expressed in the baculovirus until recently and now we are going about trying to use the same sort of system that we used to label mammalian cells. We are going to try it. However, I don't have experience with this as yet but the impression that I am getting in talking to my colleagues is number one, it is less finicky than mammalian cells, and number two, it grows on serum-free media which is a plus. Number three, generally people claim that with baculovirus you get higher levels of expression versus some of the mammalian cells, at least the expression we've seen in mammalian cells.

Unknown - It's a big system, right? I might also say that insect cells do seem to glycosylate at the same sites but it's a different glycosylation. It doesn't continue on a long chain; it seems to be a rather specific glycosylation with ends that are very small, much less glycosylation than in mammalian cells.

Markley - Perhaps we should move on to the next topic which is, 'What pulse sequence modules work and which ones fail with larger proteins?' There was some discussion about this in Ad Bax's talk, but if somebody has some more to add to that, please do.

Clore - Well, I think the ones that fail are the ones that have longer relaxation delays; it's as simple as that. So the simpler the sequence is, the better they work. What I can say is that most of the triple resonance experiments tend to fail when you have a correlation time greater than about 15 nanoseconds. They tend to degrade in quality fairly rapidly. It tends to be a very sharp cutoff. At least that's my experience.

Fesik - Similar to what Ad Bax presented, the 3-dimensional ^{13}C-resolved NOESY experiments on the cyclosporin/cyclophilin complex looked good, but the 4-dimensional ^{13}C - ^{13}C resolved NOE experiment which was collected to resolve some of the ambiguous NOE's was not as sensitive. The quality of the 4D data was such that we could not see many of the NOE's that we could see in the 3-dimensional experiment. So, in general for the highest signal to noise, it is important to reduce the delays in the experiment. If you can combine that approach with labeling to get around the resolution problem and reduce the dimensionality of the experiment to try and make it, as Marius said, as simple as possible, you may be better off.

Markley - In our studies of paramagnetic proteins we've found that four-dimensional experiments don't work with their broader lines and faster T_2 values. If there are no comments from the audience here on pulse sequences, we will move to the topic, 'What further are larger magnets?' Is it worthwhile having a 750 MHz spectrometer? Let's ask someone who actually has one and call upon Rob Kaptein and Peter Wright?

Peter Wright - I'll comment briefly. We have the 750 running routine operations for only about six to eight weeks. There are clearly advantages. There's no question about that. You get extra dispersion of course.

Markley - Are there advantages with all experiments?

Wright - Well, we haven't done enough to know. Basically, what we've been doing so far was looking at Walter Chazin's Holliday junctions, looking at some protein-DNA complexes. We did some filtered experiments which really needed the 750 MHz spectrometer to get the extra dispersion for protons. We haven't done very much in the way of multidimensional heteronuclear experiments at this time. So, advantages are clearly the enormous sensitivity; there is a tremendous increase in sensitivity over our old 600 MHz probes. The increase in dispersion for particular applications if you look at HSQC's on 25,000 M.W. proteins, you certainly start to resolve more peaks than you do with the 500 MHz or 600 MHz spectrometers. So there are advantages. As far as the disadvantages that we've come across so far, other than the cost, it is relaxation times. Looking at protein-DNA complexes, some relaxation times are about seven seconds, and that makes it a real pain collecting NOESY data. So, absolutely we are very happy to have one, but they do cost a lot of money.

Rob Kaptein - We have had a system up from about April or May, 1994. It takes a long time to install this system. But I agree for the protons, of course, it is a tremendous change. Now whether the most, or at least a big part of that is due to the extra effort the manufacturers spend on making good probes, is not clear. Part of it is probably true; you know you would expect better results on the 600 MHz spectrometer with new amplifiers and new probes. But for protons, in our hands, it works tremendously. The two-dimensional NOE for protons but also edited time proton NOE's would work very well. With the heteronuclear multidimensional experiments, we have a little more problems. It would be advisable to have a good 600 MHz spectrometer along with your 750. Whether it is worth it, I cannot say. It's a lot of money, but usually one doesn't have a choice of two 600 MHz or a 750 MHz spectrometer.

Gorenstein - I would like to argue that if one wants to go for the homonuclear 3-dimensional NOE-NOE, a 750 MHz spectrometer clearly is the way to go, because if you want to do very careful refinement using complete relaxation matrix methods, really the only way to go probably is the homonuclear three-dimensional NOE-NOE. In principle this will be the way to get many NOE constraints with very good sensitivity for larger proteins.

Markley - Do you have 750 MHz data of this kind?

Gorenstein - No, not yet. Ours is supposed to arrive next month.

Markley - But in principle....

Gorenstein - Yes, in principle, from the back of the envelope calculation.

Kathleen Hall - What about CSA at 750 MHz and then going on to 1 gigahertz. Rob, what kind of problems do you have with heteronuclei? Is that related to CSA?

Kaptein - It is an anticipated problem, and certainly for the carbonyl carbons it is likely to be a problem, but I don't know who has had enough experience with this to comment on it.

Markley - It may be too early to give an answer, but, in principle, it's going to be a problem. Then of course, fluorine and phosphorus will be worse. Well, that leads us into the next topic, which is larger samples. Both Varian and Bruker have come out with larger sample probes and I think some people here may have had some experience with these. I don't know if Peter has an 8 mm probe on his 750 MHz spectrometer. Is it worth it? Do you get the expected increase in sensitivity? Are there other problems with B_1 field inhomogeneities that counter the advantages of the larger sample size? I don't know that many people have had experience, but if anyone has, I personally would like to hear the answer.

Wright - This is a brief comment. We do have an 8 mm probe on the 750 MHz, and on the 600 MHz and 500 MHz, and the 750 certainly has a significant gain in the sensitivity, but you are suffering in terms of homogeneity. We haven't done any real tests; it looks great on ethylbenzene. Everything looks good on ethylbenzene at 750 MHz. But we haven't had any real samples with the 8 mm probe yet, so we don't know how it'll look, but in principle it's about 70% - 75% more sensitive than the 5 mm; that is comparing triple -resonance probes.

Fesik - Peter, what's your 90° pulse length on carbon for some of the probes you've had more experience with on the 600 MHz or 500 MHz. One of the problems associated with going to larger sample sizes is that some of the experiments might not work as well because they have longer 90° ^{13}C pulses.

Wright - Right. Well, on the 750 MHz for example, the 90° on carbon for 5 mm is about 10 µs and it's about 13 µs on the 8 mm. So there's no big increase. That's roughly the same as seen on the other 5 mm probes, with a slight increase on the 8 mm probe.

Gorenstein - These probes, especially Varian's 10 mm, are all really in development. Has anyone gotten any feedback on the 10 mm Varian? Are they still working or beta testing, their triple resonance probes? They don't have a pulsed field gradient.

Markley - Another development too is narrow bore probes. They potentially have some interesting applications. We work with paramagnetic proteins and one idea may be that with shorter 90° pulses, there may be some advantages. We don't have any experience because we don't have the probes installed yet.

Tim Saarinen - I'll just bring up this question which relates to the pulse sequence modules. What about gradients in terms of homogeneity problems on the large volume probes? Can you get by with poor water lineshape with the gradients? Also in terms of pulse sequence elements, I know that with the amide protons you have higher sensitivity, what's the future in terms of large proteins and gradients?

Markley - This seems to be an interesting question with no ready answers.

Bruce Johnson - I'd like to make a few comments on large volume probes. We have almost no experience but we do have the probes. In terms of the Varian probes, we've had a Varian triple resonance 10 mm probe for probably six months or so now. Our biggest problem with these large volume probes is getting enough sample to fill them. I mean, literally, if you look at the spectra they provide, you need about 3 ml of sample, we simply cannot often generate that much sample. The solution to this is to use things like the Shigemi tubes or zero magnetic susceptibility plugs. We've just got an 8 mm probe, and with the Shigemi tubes it is probably going to be very good. So we can get down to probably 600 μl of sample. So effectively we're able to use 600 μl, close to what we used to use for our 5 mm tubes, into an 8 mm tube. It turns out that the signal-to-noise we're getting on the 8 mm with ethylbenzene is about 1500 to 1 relative to about 600 to 1 on a 5 mm probe. So things look very promising I think for the large volume probes. I think it is key that we get a way to squeeze the sample down to fit into the coil on those. There are many of us who are not going to be able to have enough sample to use them.

Markley - That's a very good point.

Gorenstein - But I still think, as I mentioned earlier, that one of the advantages of the large volume is that you can go down to say 0.5 mM, rather than 1 or 2 mM. In our particular case of a 4-3-4 repressor complex, we can't get above 0.5 mM before the complex precipitates out of solution. This is a real problem with many protein-DNA complexes. They just don't stay in solution at 1 or 2 mM, but if you can bring it down to about 0.5 mM, then the avenues for many more studies are opened. So as long as you can overexpress your protein, and you can clearly synthesize a lot of DNA, one can get away with 0.5 mM sample in 3 ml.

Tom James - Well, I'd actually shown a fairly trivial example of a situation where, on that peptide whose structure we determined, we used a 350 μM concentration. It was in a 10 mm Varian tube, and, basically, the experiment would have been impossible without it. At above 0.5 mM concentration, the aggregation became too much of a problem. At 1 mM or 2 mM, where most of us would like to work, you could just forget about it.

Fesik - I must have one more general question about people's experiences with this. Bruce brought up the idea of the Shigemi tubes wit the advantage of using a lot less sample volume. In our experience, if you take the same concentration of protein, (for example 0.5 mM) and put 200 μl in one of these sample tubes, and compare the sensitivity to a 500 μl sample at the same concentration in a regular tube, the difference in sensitivity is not that great. Does anyone have the same experience with that?

Kaptein - Yes, we have the same experience.

Fesik - Thanks.

Markley - I think we should move on to the final topic which relates to data acquisition with larger proteins. I guess, one thing that I had in mind was to call on someone from Gerhard Wagner's laboratory to comment on the nonlinear sampling approach that they are using. Basically, the idea is that when you have quickly decaying signals, you want to sample more of the front part of the FID than the back part. They are combining non-linear sampling with maximum entropy methods to do the reconstruction.

Peter Schmieder - Yes, we are actually applying this. The method works pretty well in test cases. We did run into problems with high dynamic range. One particular case was a protein that has very flexible parts. So it is very mobile, and we observed very sharp lines. This created a lot of trouble in the reconstruction. The principle itself seems to work fairly well.

216

Markley - Are there any other techniques that people have found particularly useful with larger proteins or other macromolecules?

Brian Sykes - Actually, out of all we heard today, I am not sure we have answers that were yes on anything but higher magnetic fields. One proposal, which you well know, is to give up on the NOE and turn to something like the chemical shift for structure determination. You can resolve and assign spectra long after you've lost the ability to accurately measure NOE's, and we see a large development in future in understanding the chemical shift in structural terms. Secondary structure is very well worked out and maybe some aspects of tertiary structure. Maybe in the future we will obtain protein structures totally from chemical shift information, and then the NOE may provide some specified small measurements to link things together.

Markley - I think it's a good point. Are there any comments on that?

Gorenstein - Eric Oldfield has had some wonderful success with gauge invariant quantum mechanical calculations. Calculations of ^{15}N and ^{13}C chemical shifts were done in which he's able to use the structure to come up with rms differences of within about a ppm over 40 ppm range. So you can attribute all of the effects essentially to structure. I agree with Brian; it is a very important potential effect. It will clearly be better to have an empirical approach without having to do *ab initio* calculations for these types of structures.

Markley - One exciting result from the Oldfield laboratory is that one may be able to predict ϕ, ψ angles better from chemical shifts than from coupling constant measurements.

Clore - For C_α, C_β shifts you can do them empirically. In fact, we have incorporated that into direct refinement. We have a paper coming out in the Journal of Magnetic Resonance but that's only good for ϕ, ψ and that's not very precise. Essentially these are very loose constraints. So that isn't good enough to specify structure. As for proton chemical shifts in the absence of structure to start with, you don't have any specific spatial relationships to determine how the shift is going to go. You might know that something is ring current shifted. You don't know what aromatic for example is responsible for that and so on and so forth. I think the NOE is still pretty important.

Gorenstein - In terms of Eric Oldfield's work as I've read it, obviously he has analyzed through space interactions. But what we are referring to here is primarily, as John referred to, a local dihedral angle dependence and its impacts which is clearly not a sufficient interaction. Even if it gets better, so it can obviously be a tighter constraint, it surely isn't going to pull the protein.

David Case - We have worked on proton shifts and have had some success with heme-proteins and getting structural information from shifts. I've had much less success myself applying quantum chemistry to understanding proton chemical shifts particularly for, say, protons connected to nitrogen where you'd like to know what amide proton shifts are doing for you because they're partly quite sensitive, but they don't make much sense to me. So I am not quite convinced yet at least for protons. Eric Oldfield has done much more work with heteronuclei than I have. Quantum chemistry is really not quite ready for prime-time biochemistry yet but maybe in a year or two, we'll know more. I've been disappointed recently in what you can do in that respect.

Sykes - Maybe we are being too restrictive. John, you also mentioned paramagnetics several times and in the past we've worked with lanthanides where you get 15 $\overset{\circ}{A}$ and 20 $\overset{\circ}{A}$ distances and maybe we use chemical shifts in paramagnetics. I know Marius doesn't want to give up with the NOE, but I don't see up that we have barriers here up to this point. Maybe we have to bring in other things. Have you made structure determinations from the prosthetic groups of your proteins for example?

Markley - Tom mentioned that. I'll talk tomorrow about using ^{15}N relaxation measurements in a paramagnetic protein for getting distances.

Gorenstein - I thought Tack Kuntz some time back had selectively modified lysozyme to introduce paramagnetic centers and had very good correspondence between the relaxation changes in the presence or absence of paramagnetic centers in structural terms. I think one needs to think of the future; these are going to be important techniques.

Clore - The problem is that if you measure a long distance and you have a large error in it, it is not very constraining, but if you have a distance that's less than 5 $\overset{\circ}{A}$ between something that can be a 100 $\overset{\circ}{A}$ apart, it's very constraining, and that's why, maybe it doesn't pay off so much.

Gorenstein - But I think the thing that Rob was talking about, that is using a different type of potential function which throws in everything but the kitchen sink, maybe even the kitchen sink, for a structural refinement, is really a good idea for these larger proteins. Put in genetic data, put in chemical modification protection information, put everything in there, and we may be able to get larger structures.

James - Actually, you can readily put in the additional information into all kinds of objective functions. That's true. I did want to follow up on David's comment regarding some of the work that was done in my colleague Tack Kuntz's laboratory regarding spin labels. A number of people in Tack's laboratory did some follow up work, after that initial very nice paper that Paul Schmidt did with Tack, where they spin labeled lysozyme. Phyllis Kosen in particular, did some work with that, and the nice thing about these spin labels is that you do get these 15-20 $\overset{\circ}{A}$ distances, and they're really nice. There is the problem that Steve pointed out in terms of the error bar that you might want to put on to those. But a real major problem, as it turned out for them, is that in Phyllis' experience in trying to label several different kinds of proteins, impurities on the level of 1 or 2% occurred in wrong sites which was enough to obliterate any meaningful information. I think you know that you can do r^{-6} types of calculations to figure out why those incredibly small levels of impurities would indeed do a great deal of harm.

Clore - Also, it's the linker size that is very important, i.e. if you have a flexible linker and then you can have mobility, then you can't triangulate to a single point, and it makes it very difficult as well to try to use that information even though it is longer distance information.

Joshua Wand - I just wanted to actually emphasize something that Marius pointed out.

Ultimately if you're going to use paramagnetic centers, for example cytochrome c, where you have a nice rigid group in place (you know where it is, it is spin 1/2), and look for pseudocontact shifts, you always end up asking what's the diamagnetic contribution to the change in shift say in going to a diamagnetic, oxidized state. So ultimately, I guess it is the kitchen sink, you really do need to know everything, not just understanding the paramagnetic contributions but also the diamagnetic environment. So ultimately I don't think you could do it without knowing the structural environment, which leads to a circular argument. So I think Marius' point is well taken.

Markley - Any further comments?

LeMaster - I was actually going to refer to an earlier comment in terms of the question of whether one needs to give up on NOE's or not. In several cases such as Opella's work on filamentous proteins in micelles, you can see the useful NOE's in complexes of 30,000 molecular weight or so, which are systems of comparable size to those that we can get maintain assignments on. It's not that they can't be observed; it's a question of how to make sure we get the sensitivity and suppress the relaxation problems. I don't think we need to turn to other techniques when working with larger systems. Rather, we need to find ways to get around the problems facing the NOE measurements.

John Markley - Since we have gone on for more than the allotted hour, now is a good time to draw it to a close and over dinner we can discuss or debate what the upper molecular weight limit really is.

NMR STRUCTURES OF PROTEINS INVOLVED IN SIGNAL TRANSDUCTION

S.W. Fesik, R.P. Meadows, E.T. Olejniczak, A.P. Petros, P.J. Hajduk, H.S. Yoon, J.E. Harlan, T.M. Logan, M.-M. Zhou, D.G. Nettesheim, H. Liang, and L. Yu

Abbott Laboratories, Abbott Park, IL USA 60064

INTRODUCTION

Signal transduction is a complicated process that may involve several steps mediated by specific intermolecular interactions. Recently, three-dimensional structures of proteins involved in signal transduction have been obtained and have greatly aided in our understanding of these processes at the molecular level.[1,2] In this paper, three-dimensional structures of three proteins that are involved in signal transduction will be described, including (1) a protein tyrosine phosphatase, (2) a pleckstrin homology (PH) domain, and (3) the DNA-binding domain of a member of the *ets* family of transcription factors, Fli-1, in the DNA-bound form.

PROTEIN TYROSINE PHOSPHATASE

One of the most important means of regulating signal transduction events is through the level of tyrosine phosphorylation. This regulation occurs through the action of protein tyrosine kinases and protein tyrosine phosphatases.[3] Within the cytosolic family of protein tyrosine phosphatases is a group of low molecular weight (18 kDa) proteins that are found in a variety of tissues, including human placenta and bovine heart.[4,5] Although the overall sequence homology between these enzymes and other protein tyrosine phosphatases is quite low, they all contain a highly conserved stretch of amino acid residues, consisting of CXXXXXR(S/T)(G/P) where three of the residues preceeding the Cys are hydrophobic.

In order to gain a better understanding of the important structural features and enzymatic mechanism of this class of protein tyrosine phosphatases, we have determined the solution structure of the low molecular weight protein tyrosine phosphatase from bovine heart.[6,7] The structure of the protein was determined in the presence of sodium phosphate which acts as an inhibitor of the enzyme. The ^1H, ^{13}C, and ^{15}N signals of bovine heart protein tyrosine phosphatase were asssigned from a variety of double and triple resonance multi-dimensional NMR experiments using uniformly ^{15}N- and ^{15}N-,^{13}C-labeled proteins that were expressed in bacteria. The three-dimensional structure of the protein was calculated from a total of over 1700 NMR-derived restraints using a distance geometry/simulated annealing protocol.[8] Figure 1 depicts a stereoview of the 15 lowest energy structures. The regular elements of secondary structure are well-defined by the NMR data (Figure 1B). When an ill-defined loop (residues 116-139) is excluded, the RMSD for residues 5-157 is 0.81 and 1.33 Å for the backbone and all heavy atoms, respectively.[7]

NMR As a Structural Tool for Macromolecules: Current Status and Future Directions
Edited by B.D. Nageswara Rao and Marvin D. Kemple, Plenum Press, New York, 1996

221

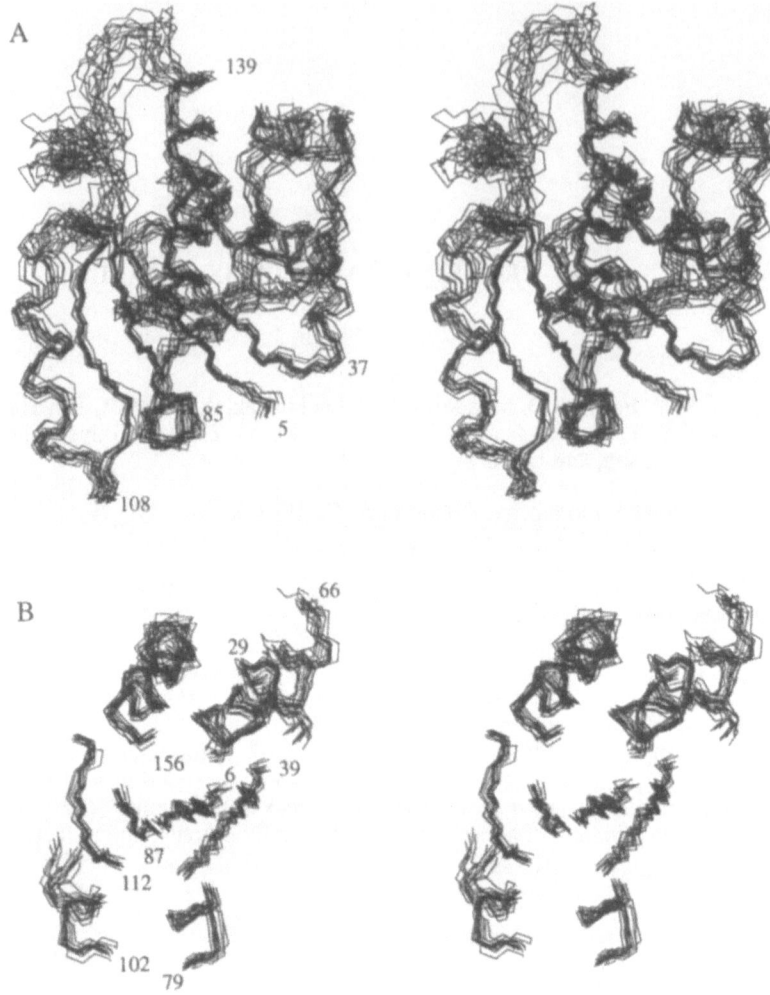

Figure 1. Stereoviews of the 15 lowest energy NMR structures of bovine heart protein tyrosine phosphatase. (A) Superposition of the backbone for residues 5-157 and (B) those residues forming regular elements of secondary structure. Reproduced from Logan et al.,[7] with permission.

As shown in Figure 2, the structure of bovine heart protein tyrosine phosphatase consists of a four-stranded parallel β-sheet that is flanked by five helices with an overall strand topology of +1x, -2x, -1x.[6] β1 and β2 as well as β3 and β4 form a βαβ motif with α1 and α4, respectively. The active site is located at the top of the molecule in the orientation shown in Figure 2. C12 is the active site nucleophile that attacks the phosphotyrosine and forms a thiol-phosphate intermediate.[9,10] R18 which is part of the conserved amino acids found in all protein tyrosine phosphatases is also part of the active site and interacts with the phosphate.

A similar overall topology has been observed in a number of other proteins that are involved in binding to phophorylated (or sulfated) substrates or cofactors. Indeed, the structures of bovine heart protein tyrosine phophatase is closely related to flavodoxin[11] and the bacterial chemotaxis response regulator protein CheY[12] which consist of a 1x, -2x, -1x, -1x folding topology for the central parallel β-sheet.

Recently, several other groups have reported on the three-dimensional structures of protein tyrosine phosphatases by X-ray crystallography.[13-16] A truncated form (321 of 435 residues) of PTP1B[13] and *Yersinia* protein tyrosine phosphatase[14] are structurally similar to each other and consist of an eight-stranded β-sheet flanked by five and two α-helices on each side. In addition to these proteins, two X-ray structures have been reported for low molecular weight protein tyrosine phophatases: one from bovine heart (identical to the protein studied by NMR)[15] and the other with a very similar sequence from bovine liver.[16] The NMR and X-ray structures of the low molecular weight protein tyrosine phophatases were found to be very similar. Although PTP1B and the Yersinia protein tyrosine phosphatase are much larger and not highly homologous to the low molecular weight protein tyrosine phosphatases, there is some local structural similarity near the active site (Figure 3). The hydrophobic core of both classes of proteins consist of a four-stranded β-sheet flanked by α-helices and the active site cysteine which functions as a nucleophile in the enzymatic mechanism is located at the C-terminal end of a central β-strand followed by an α-helix. Thus, despite the differences in sequence and overall structure, the structure of the active site is conserved.

Figure 2. Ribbon diagram of the NMR structure of bovine heart protein tyrosine phosphatase. The side chains of the active site residues, C12 and R18, are shown. Reproduced from Logan et al.,[7] with permission.

PLECKSTRIN HOMOLOGY DOMAIN

Pleckstrin is the major protein kinase C substrate of platelets.[17] This protein contains two homologous domains of approximately 100 amino acids: one at the N-terminus and another at the C-terminus. These conserved sequences of amino acids called pleckstrin homology (PH) domains have been found in a number of proteins, including serine/threonine kinases, GTPase activating proteins, phospholipases, and cytoskeletal proteins.[18-21] Many of the proteins that contain PH domains play an important role in signal transduction pathways and also contain other domains that are known to participate in signal transduction (e.g., SH2 and SH3 domains). Thus, PH domains are thought to play an important role in signal tranduction processes. However, the function of PH domains and the molecules that bind to these modules are still under study. Futhermore, the three-dimensional structure of PH domains was unknown at the time we began our study.

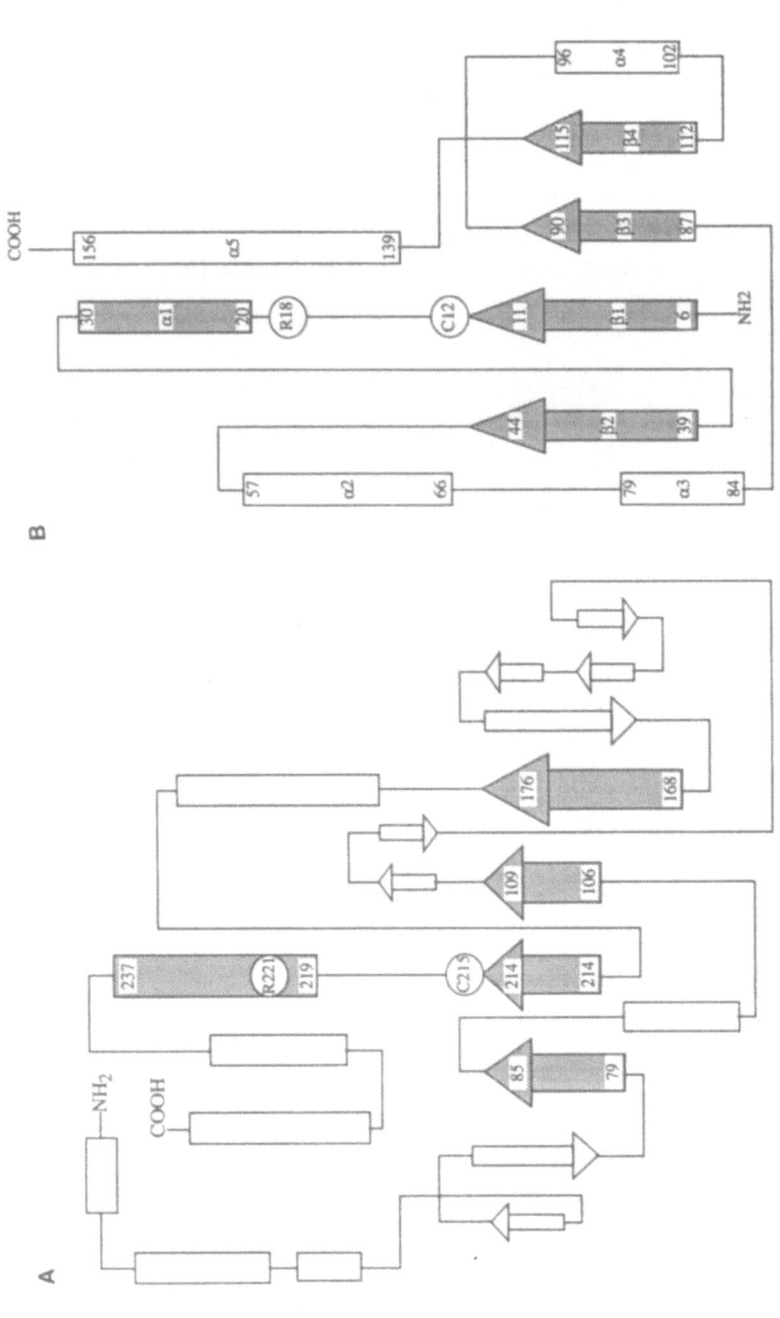

Figure 3. Schematic folding topology diagram comparing the secondary structural elements between (A) PTP1B and (B) bovine heart protein tyrosine phosphatase. Arrows represent β-strands, rectangles represent a-helices, and lines indicate regions of nonregular secondary structure. Reproduced from Logan et al.,[7] with permission.

In order to determine the solution structure of a pleckstrin homology domain, we cloned and expressed the pleckstrin homology domains from five different proteins. The pleckstrin homology domain from the N-terminus of pleckstrin expressed at high levels and gave a well-resolved NMR spectrum. The solution structure of the N-terminal PH domain of pleckstrin, was determined by heteronuclear three-dimensional NMR spectroscopy using uniformly ^{15}N- and ^{15}N-^{13}C-labeled protein.[22] A total of 1376 NMR-derived restraints were used in the structure calculations that employed a distance geometry/simulated annealing protocol.[8] The structure of the backbone was well-defined by the NMR data (Figure 4), except for the first three residues at the N-terminus, the last nine residues at the C-terminus, and two loop regions (residues 15-20 and 59-66). The side chains for the internal hydrophobic residues were also well-defined by the NMR data. The atomic RMSD about the mean coordinate positions (residues 4-104) for the 25 structures shown in Figure 4 is 0.79 +/- 0.1 Å for the backbone atoms and 1.36 +/- 0.1 Å for all heavy atoms.[22]

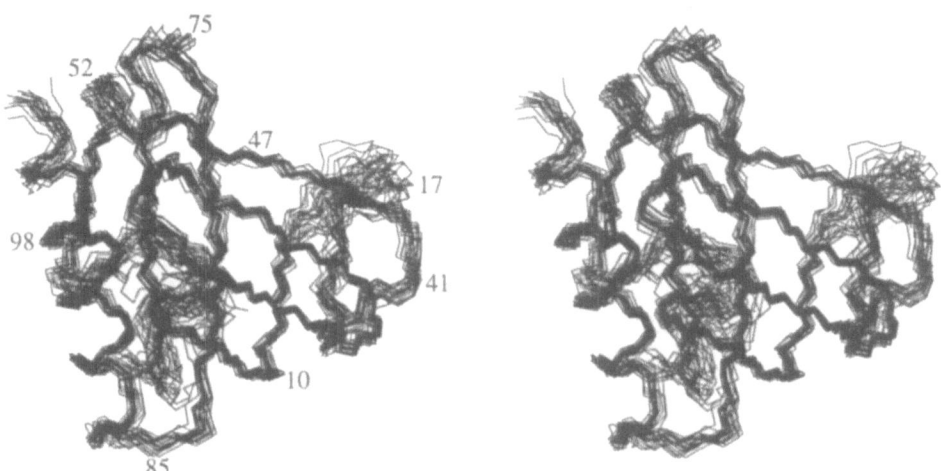

Figure 4. Stereoview of the backbone (N,C^a,C') of 25 superimposed NMR-derived structures (residues 4-104) of the N-terminal PH domain of pleckstrin. Reproduced from Yoon et al.,[22] with permission.

As shown in Figure 5A, the structure of the N-terminal PH domain of pleckstrin consists of two roughly orthogonal β-sheets. In the orientation shown in Figure 5A, the front β-sheet is composed of the first four strands, and the back sheet is defined by the last three strands together with part of the first and second strand. An amphiphilic α-helix (residues 87-103) is present at the C-terminus of the protein. The hydrophobic side chains of W92, I96, and I100 that are part of this helix point towards the interior hydrophobic core of the protein. Thus, W92, which is conserved in all PH domains and points into the hydrophobic interior of the protein, most likely contributes to protein stability rather than mediating binding to other molecules. Although no other amino acid is strictly conserved in all PH domains, many of the PH domains contain positively charged residues in similar positions. It is interesting to note that six of these lysine residues are located on one face of the N-terminal PH domain of pleckstrin (Figure 6).[22]

At the same time that our paper was published on the solution structure of the N-terminal PH domain of pleckstrin,[22] the three-dimensional structure of the PH domain of β-spectrin was reported.[23] The NMR structures of the two different PH domains are very similar. The only significant difference is an additional, short α-helix in the PH domain of β-spectrin. Oschkinat and coworkers[23] also noted the presence of several positively charged residues on one side of the PH domain of β-spectrin. After these initial reports on the PH domain structure were published, several papers appeared on the NMR and X-ray structures of the PH domain of dynamin.[24-27] All of these structures were found to be similar to the PH domains of N-terminal pleckstrin and β-spectrin.

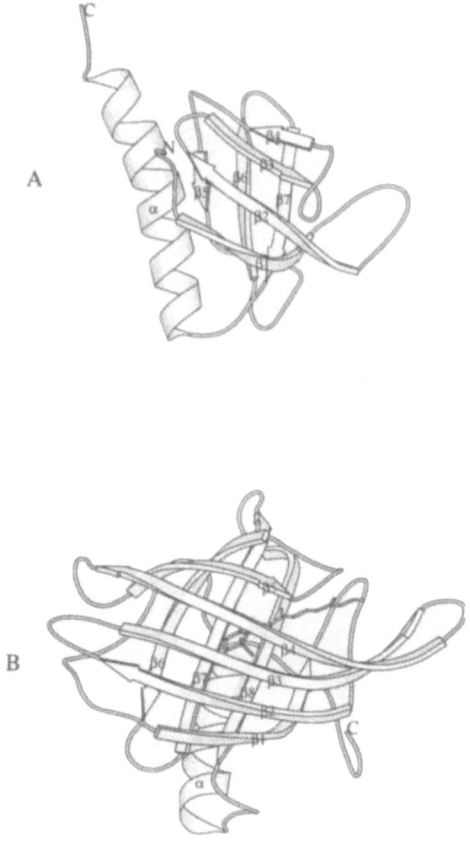

Figure 5. Ribbon plots depicting (A) the averaged minimized NMR structure of the pleckstrin homology domain (residues 4-104) and (B) the X-ray structure of the retinol-binding protein complexed to retinol.[28] Reproduced from Yoon et al.,[22] with permission.

As shown in Figure 5, the overall topology of the PH domain resembles that of retinol-binding protein (RBP).[28] This family of proteins all contain a β-barrel with a hydrophobic core that bind to lipophilic molecules. Based on the structural similarity between the PH domains and the retinol binding protein, we postulated that the PH domains may be binding to lipid molecules rather than to proteins as previously assumed. We also reasoned that the lysine residues located on one side of the molecule (Figure 6) may contribute to binding through interactions with the polar headgroups of the lipid. In order to test this hypothesis, we examined whether the PH domains bind to lipids using a centrifugation assay.[29] The N-terminal PH domain of pleckstrin was unable to bind to many of the phospholipids tested except for the negatively charged lipids, phosphatidylionositol-4-phosphate and phosphatidylionositol-4,5-bisphosphate (PIP$_2$) (Figure 7). Indeed, all five PH domains that we cloned and expressed were found to bind to PIP$_2$ in contrast to the proteins used as a control (Figure 8).[29] Although dynamin was initially reported not to bind to PIP$_2$,[24] binding was observed in subsequent studies (D. Cowburn, personal communication).

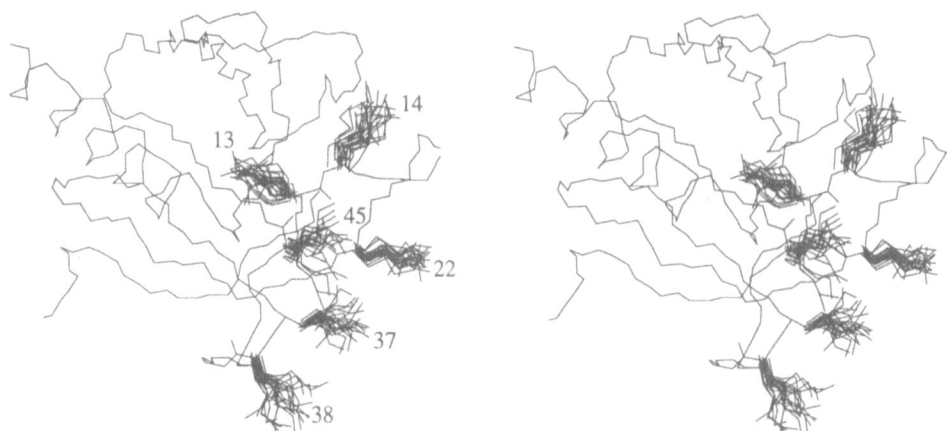

Figure 6. The NMR structure of the N-terminal PH domain of pleckstrin illustrating the six lysines that are located on one side of the molecule. Reproduced from Yoon et al.,[22] with permission.

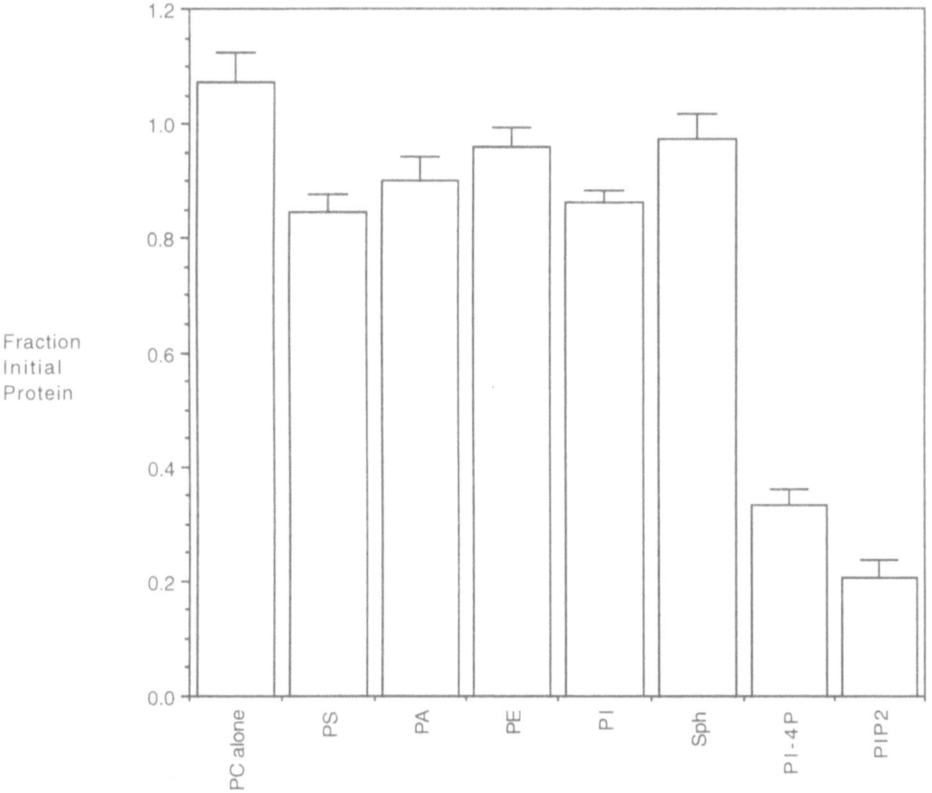

Figure 7. Binding of the N-terminal PH domain of pleckstrin to vesicles containing PC and 5% of other lipids: phosphatidylserine (PS), phosphatidic acid (PA), phosphatidylethanolamine (PE), phosphatidylinositol (PI), sphingomyelin (Sph), phosphatidylinositol 4-phosphate (PI-4P) and phosphatidylinositol 4,5-bisphosphate (PIP2). Binding was determined by measuring the total amount of protein in the supernatant. Reproduced from Harlan et al.,[29] with permission.

In order to determine the location of the PIP$_2$ binding site, the ^{1}H, ^{13}C, and ^{15}N chemical shifts of the N-terminal PH domain of pleckstrin were followed as a function of added lipid. Upon the addition of PIP$_2$, chemical shift changes were observed for K13, K14, S16, V17, N19, T20, W21, K22, F35, Y36, and G46. All of these residues are located in the N-terminal half of the PH domain near the positively charged lysines. The involvement of the PIP$_2$ phosphates in the binding interaction with the PH domain was supported by the changes in the ^{31}P NMR signals of PIP$_2$ upon the addition of the PH domain. These results and the lack of any chemical shift changes for the residues in the hydrophobic core of the protein suggest that the binding of PIP$_2$ to the PH domains is primarily mediated by ionic interactions.[29]

What is the functional relevance of the interaction between the PH domains and PIP$_2$? Most of the proteins that contain PH domains need to be associated with membranes for their function. Although many of these proteins do not contain the classical membrane anchoring groups such as a hydrophobic helix or sites for post-translational addition of lipid molecules, they do contain a PH domain which may play a biologically significant role in the localization

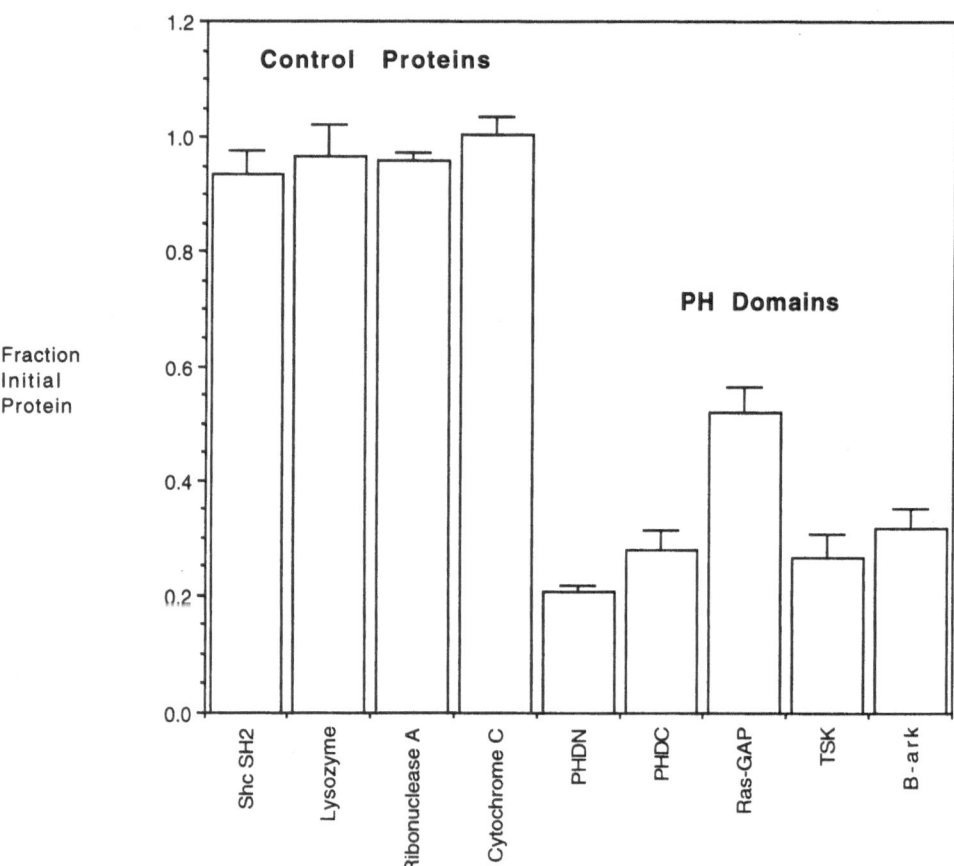

Figure 8. Binding of proteins to PC:PIP$_2$ (95:5) vesicles. The proteins tested include: cytochrome C, lysozyme, ribonuclease A, the Shc SH2 domain containing a (His)$_6$ tag, the N- and C-terminal PH domains of pleckstrin (PHDN and PHDC, respectively), and PH domains of Tsk, ras-GAP and β-Ark. Binding was determined by measuring the total amount of protein in the supernatant. Reproduced from Harlan et al.,[29] with permission.

of proteins to phospholipid membranes.[30-35] The interaction of PH domains with PIP_2 provides a novel mechanism for membrane association and may play a critical role in controlling the biological functions of the many proteins that contain this module. Indeed, the binding of $G\beta\gamma$ and PIP_2 to the PH domain of the β-adrenergic receptor kinase (βARK synergistically enhances agonist-dependent receptor phosphorylation.[36]

FLI-1/DNA COMPLEX

The *ets* family of proteins[37,38] are transcription factors that are involved in the development of a variety of different systems and have been implicated in the regulation of T-cell activation, tissue remodeling, and cancer.[39-48] *Ets* proteins bind DNA as monomers to a GGAA/T core, and all contain a conserved DNA-binding domain of 85 amino acids with three highly conserved tryptophans in the amino terminal half and a large number of conserved basic residues in the carboxyl half of the protein.[38] Since no convincing overall sequence homology was detected between the *ets* domain and the well-characterized DNA-binding motifs,[38] it was thought that the *ets* domain may represent a novel structural motif for binding to DNA.[37,38]

In order to determine the solution structure of the DNA-binding domain of an *ets* protein, we chose to study human Fli-1 when bound to its cognate DNA site.[49] The NMR experiments were conducted on an ^{15}N- and ^{15}N/^{13}C-labeled 98 amino acid fragment of Fli-1 (residues 276-373) complexed to an unlabeled 16 bp DNA fragment containing the consensus binding site for Fli-1.[50] The ^{1}H, ^{13}C, and ^{15}N signals of the Fli-1 *ets* domain in the DNA-bound form were assigned from a series of double and triple resonance NMR experiments,[51] and the three-dimensional structure of the Fli-1 *ets* domain in the DNA-bound form was determined from 810 NMR-derived restraints.[49] Figure 9 depicts the superposition of 30 structures calculated from the NMR data using a distance geometry /simulated annealing protocol.[8] From this set of structures, the RMSD for residues 279-369 to the averaged coordinates is 0.98 ± 0.1 Å for the backbone and 1.51 ± 0.1 Å for all heavy atoms.[49]

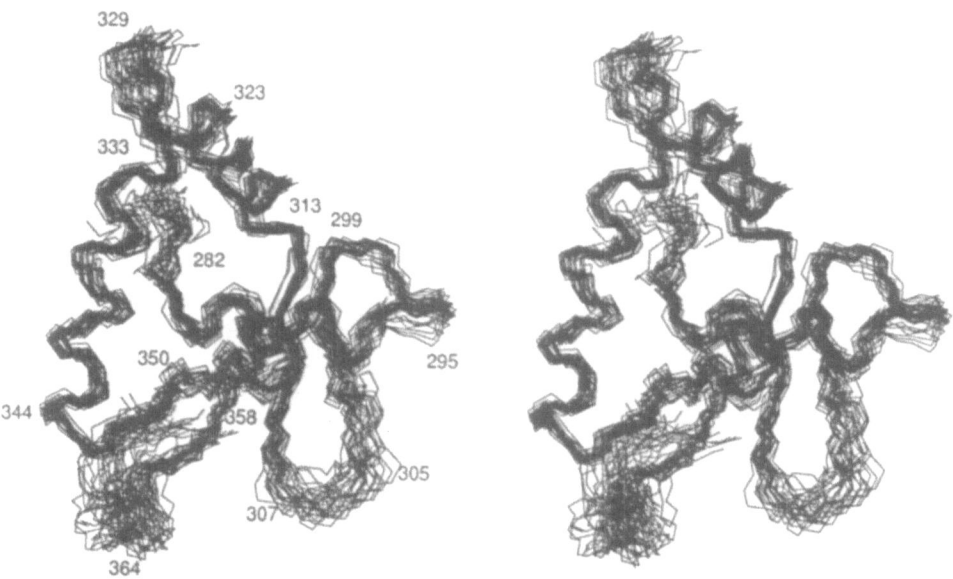

Figure 9. Superposition of 30 NMR-derived structures of the Fli-1 *ets* domain when bound to DNA. A stereoview of the C$^\alpha$ traces of the structures is shown. Reproduced from Liang et al.,[49] with permission.

As shown in Figure 10A, the structure of the Fli-1 *ets* domain consists of three a-helices and a small twisted four-stranded anti-parallel β-sheet.[49,51] Most of the highly conserved hydrophobic residues are buried within the hydrophobic core, including two of the three highly conserved tryptophans (W284 and W321). In contrast to earlier predictions that the ets domain might represent a new DNA-binding motif,[37,38] the tertiary fold of the Fli-1 *ets* domain was found to resemble a class of helix-turn-helix (HTH) proteins[52] represented by the catabolite activator protein (CAP) of *E. coli*[53] (Figure 10B). This family of HTH proteins also includes the biotin operon repressor (BirA),[54] the globular domain of histone H5 (GH5),[55] the HNF-3/*fork head* transcription factor,[56] and the heat shock transcription factor (HSF).[57]

In order to determine the residues of Fli-1 that interact with DNA, the chemical shifts of Fli-1 that change upon binding to DNA were identified.[51] In addition, intermolecular nuclear Overhauser effects (NOEs) between the labeled protein and unlabeled DNA were obtained from a ^{13}C-edited, ^{12}C-filtered NOE spectrum.[49] Site-directed mutagenesis guided by the NMR stucture was also used to probe the residues of Fli-1 important for interacting with DNA.[49]

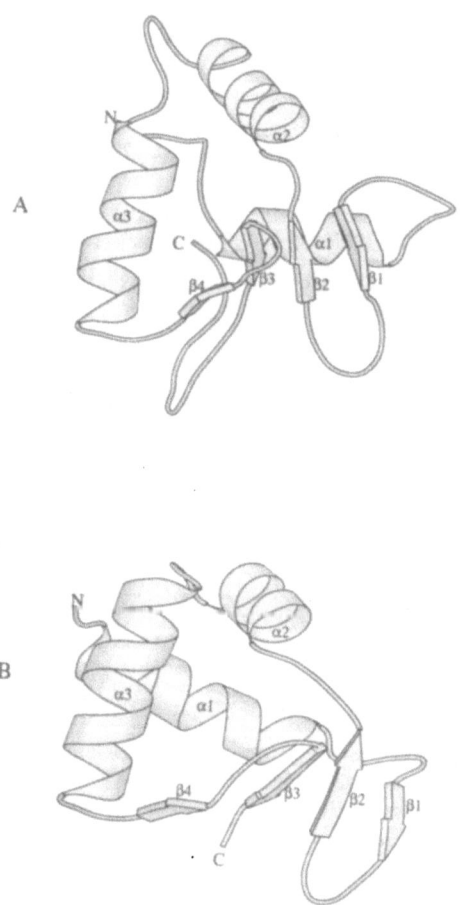

Figure 10. Schematic ribbon representations of (A) the Fli-1 *ets* domain and (B) the DNA-binding domain (residues 136-205) of CAP.[53] The three a-helices, α1 (residues 283-292), α2 (315-323) and α3 (332-344), and four β-strands, β1 (300-303), β2 (308-311), β3 (348-351) and β4 (357-361) of Fli-1 are indicated. Reproduced from Liang et al.,[49] with permission.

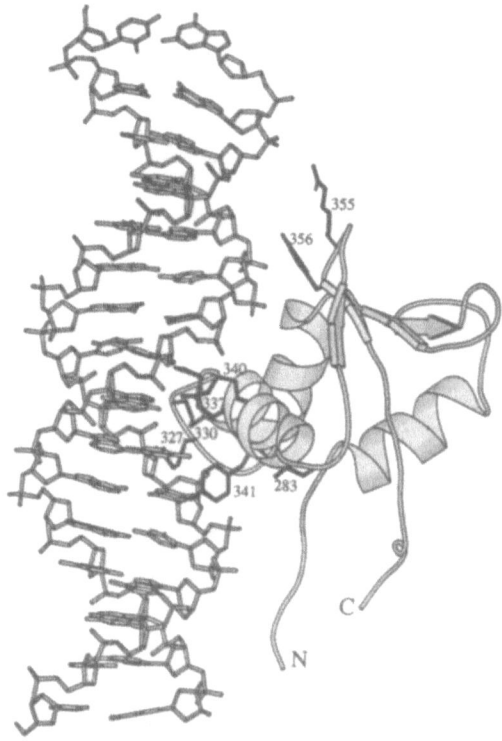

Figure 11. Residues of the Fli-1 *ets* domain suggested by NMR and mutagenesis experiments to be involved in DNA binding. Residues whose backbone resonances showed large chemical shift differences due to DNA binding include: I281, L283-W284, M311, P314, G322, K325, N329-N331, D333, L335-Y341, K350-V351, R355-Y358. The sidechains of residues that showed intermolecular NOEs to DNA L283, K327, M330, A338, Y341, R355, and Y356 and the sidechains of R337 and R340 are displayed. Although canonical B-form DNA is depicted, the DNA may bend as observed previously in protein/DNA complexes of CAP and HNF-3. Reproduced from Liang et al.,[49] with permission.

Figure 11 shows the residues of Fli-1 involved in binding to DNA that were identified from the NMR and mutagenesis experiments in the context of the structure of the *ets* domain. The locations of these residues relative to the geometry of B-form DNA fit well with the results of chemical protection and interference analyses[58] which showed that the *ets* domain of Ets-1 binds to DNA in the major groove as well as to both adjacent minor grooves. On the basis of all of our data, it appears that the residues in the recognition helix (a3) of Fli-1 interact with the bases in the major groove of DNA. In addition, residues in the loop between β3 and β4 and residues in the loop between α2 and α3 may provide contacts with the minor groove and sugar-phosphate backbone.

Other members of the CAP family bind DNA in a somewhat similar fashion. For example, HNF-3 binds to DNA through a recognition helix that interacts with the major groove that is flanked by two extended loops that interact with minor grooves of DNA.[56] This has been referred to as the "winged helix motif". The *ets* domain binds to DNA in analogous manner to the winged helix motif in that *ets* proteins utilize a recognition helix and a loop between β3 and β4 which corresponds to one wing in the winged helix motif. However, Fli-1 lacks the second wing of the winged helix motif, the extended C-terminal loop. Instead, these contacts in Fli-1 with the minor groove are provided by the loop between the helices in the HTH motif. Thus, Fli-1 and the other ets family members may recognize DNA through a variation of the winged helix motif.

CONCLUSIONS

In the last few years, the molecules involved in several signal transduction pathways have been elucidated. In addition, the three-dimensional structures of many of these proteins have been determined by NMR and X-ray crystallography which has greatly aided in our understanding of signal transduction at the molecular level. As new molecules are identified in the future that participate in signal transduction events, the structures of these molecules will be determined, providing important structural information on key cell processes.

ACKNOWLEDGEMENTS

We would like to R.L. Van Etten as well as and X. Mao and C.B. Thompson for fruitful collaborations on the structure determinations of bovine heart protein tyrosine phosphatase and Fli-1, respectively.

REFERENCES

1. G.B. Cohen, R. Ren, and D. Baltimore, Modular binding domains in signal transduction proteins, Cell 80:237 (1995).
2. T. Pawson, Protein modules and signalling networks, Nature 373:573 (1995).
3. T. Hunter, Protein kinases and phosphatases: The yin and yang of protein phosphorylation and signaling, Cell 80:225 (1995).
4. A. Waheed, P.M. Laider, Y.-Y.P. Wo, and R.L. Van Etten, Purification and physicochemical characterization of a human placental acid phosphatase possessing phosphotyrosyl protein phosphatase activity, Biochemistry 27:4265 (1988).
5. Z.-Y. Zhang and R.L. Van Etten, Purification and characterization of a low-molecular-weight acid phosphatase--a phosphotyrosyl-protein phosphatase from bovine heart, Arch. Biochem. Biophys. 228:39 (1990).
6. M.-M. Zhou, T.M. Logan, Y. Thèriault, R.L. Van Etten, and S.W. Fesik, Backbone ^{1}H, ^{13}C, and ^{15}N assignments and secondary structure of bovine low molecular weight phosphotyrosyl protein phosphatase, Biochemistry 33:5221 (1994).
7. T.M. Logan, M.-M. Zhou, D.G. Nettesheim, R.P. Meadows, R.L. Van Etten, and S.W. Fesik, Solution structure of a low molecular weight protein tyrosine phosphatase, Biochemistry 33:11087 (1994).
8. M. Nilges, G.M. Clore, and A.M. Gronenborn, Determination of three-dimensional structures of proteins from interproton distance data by dynamical simulated annealing from a random array of atoms. Circumventing problems associated with folding, FEBS Lett. 239:129 (1988).
9. Z.-Y. Zhang, J.P. Davis, and R.L. Van Etten, Covalent modification and active site-directed inactivation of a low molecular weight phosphotyrosyl protein phosphatase, Biochemistry 31:1701 (1992).
10. Y.-Y.P. Wo, M.-M. Zhou, P. Stevis, J.P. Davis, Z.-Y. Zhong, and R.L. Van Etten, Cloning, expression, and catalytic mechanism of the low molecular weight phosphotyrosyl protein phosphatase from bovine heart, Biochemistry 31:1712 (1992).
11. B.J. Stockman, A.M. Krezel, J.M. Markley, K.G. Leonhardt, and N.A. Strauss, Hydrogen-1, carbon-13, and nitrogen-15 NMR spectroscopy of Anabaena 7120 flavodoxin: assignment of beta-sheet and flavin binding site resonances and analysis of protein-flavin interactionsBiochemistry 29:9600 (1990).
12. A.M. Stock, E. Martinez-Hackert, B.F. Rasmussen, A.M. West, J.B. Stock, D. Ringe, and G.A. Petsko, Structure of the Mg(2+)-bound form of CheY and mechanism of phosphoryl transfer in bacterial chemotaxis, Biochemistry 32:13375 (1993).
13. D. Barford, A.J. Flint, and N.K. Tonks, Crystal structure of human protein tyrosine phosphatase 1B, Science 263:1397 (1994).
14. J.A. Stuckey, H.L. Schubert, E.B. Fauman, Z.-Y. Zhang, J.E. Dixon, and M.A. Saper, Crystal structure of Yersinia protein tyrosine phosphatase at 2.5Å and the complex with tungstate, Nature 370:571 (1994).
15. M. Zhang, R.L. Van Etten, and C.V. Stauffacher, Crystal structure of bovine heart phosphotyrosyl phosphatase at 2.2-Å resolution, Biochemistry 33:11097 (1994).
16. X.-D. Su, N. Taddei, M. Stefani, G. Ramponi, and P. Nordlund, The crystal structure of a low-molecular-weight phosphotyrosine protein phosphatase, Nature 370:575 (1994).
17. M. Tyers, RJ Haslam RA Rachubinski, and CB Harley, Molecular analysis of pleckstrin: the major protein kinase C substrate of platelets.Nature 333:470 (1988).

18. B.J. Mayer, R. Ren, K.L. Clark, and D. Baltimore, A putative modular domain present in diverse signaling proteins [letter], Cell 73:629 (1993).

19. R.J. Haslam, H.B. Koide, and B.A. Hemmings, Pleckstrin domain homology, Nature 363:309 (1993).

20. A. Musacchio, T. Gibson, P. Rice, J. Thompson, and M. Saraste, The PH domain: a common piece in the structural patchwork of signalling proteins, TIBS 18:343 (1993).

21. G. Shaw, Identification of novel pleckstrin homology (PH) domains provides a hypothesis for PH domain function, Biochem. Biophys. Res. Comm. 195:1145 (1993).

22. H.S. Yoon, P.J. Hajduk, A.M. Petros, E.T. Olejniczak, R.P. Meadows, and S.W. Fesik, Solution structure of a pleckstrin-homology domain, Nature 369:672 (1994).

23. M.J. Macias, A. Musacchio, H. Ponstingl, M. Nilges, M. Saraste, and H. Oschkinat, Structure of the pleckstrin homology domain from β-spectrin, Nature 369:675 (1994).

24. K.M. Ferguson, M.A. Lemmon, J. Schlessinger, and P.B. Sigler, Crystal structure at 2.2 Å resolution of the pleckstrin homology domain from human dynamin, Cell 79:199 (1994).

25. A.K. Downing, P.C. Driscoll, I. Gout, K. Salim, M.J. Zvelebil, and M.D. Waterfield, Three-dimensional solution structure of the pleckstrin homology domain from dynamin, Curr. Biol. 4:884 (1994).

26. D. Timm, K. Salim, I. Gout, L. Guruprasad, M. Waterfield, and T. Blundell, Crystal structure of the pleckstrin homology domain from dynamin, Nature Stuct. Biol. 1:782 (1994).

27. D. Fushman, S. Cahill, M.A. Lemmon, J. Schlessinger, and D. Cowburn, Solution structure of pleckstrin homology domain of dynamin by heteronuclear NMR spectroscopy, Proc. Natl. Acad. Sci. USA 92:816 (1995).

28. M.E. Newcomer, T.A. Jones, J. Aqvist, J. Sundelin, U. Eriksson, L. Rask, and P.A. Peterson, The three-dimensional structure of retinol-binding protein, EMBO J. 3:1451 (1984).

29. J.E. Harlan, P.J. Hajduk, H.S. Yoon, and S.W. Fesik, Pleckstrin homology domains bind to phosphatidylinositol-4,5-bisphosphate, Nature 371:168 (1994).

30. L.H. Davis and V. Bennett, Indentification of two regions of β_G-Spectrin that bind to distinct sites in brain membranes, J. Biol. Chem. 269:4409 (1994).

31. M. Rebecchi, A. Peterson, and S. McLaughlin, Phosphoinositide-specific phospholipase C-delta 1 binds with high to phospholipid vesicles containing phosphatidylinositol 4,5-bisphosphate. Biochemistry 31:12742 (1992).

32. H. Mano, K. Mano, B. Tang, M. Koehler, T. Yi, D.J. Gilbert, N.A. Jenkins, N.G. Copeland, and J.N. Ihle, Expression of a novel form of Tec kinase in hematopoietic cells and mapping of the gene to chromosome 5 near Kit, Oncogene 8:417 (1993).

33. S. Gibson, B. Leung, J.A. Squire, M. Hill, N. Arima, P. Goss, D. Hogg, and G.B. Mills, Identification, cloning, and characterization of a novel human T-cell-specific tyrosine kinase located at the hematopoietin complex on chromosome 5q, Blood 5:1561 (1993).

34. D.J. Rawlings, D.C. Saffran, S. Tsukada, D.A. Largaespada, J.C. Grimaldi, L. Cohen, R.N. Mohr, J.F. Bazan, M. Howard, and N.G. Copeland, Mutation of unique region of Bruton's tyrosine kinase in immunodeficient XID mice, Science 261:358 (1993).

35. J.A. Pitcher, J. Inglese, J.B. Higguns, J.L. Arriza, P.J. Casey, C. Kim, J.L.Benovic, M.M. Kwantra, M.G. Caron, and R.J. Lefkowitz, Role of βγ Subunits in targeting the β-adrenergic receptor kinase to membrane-bound receptors, Science, 257:1264 (1992).

36. J.A. Pitcher, K. Touhara, E.S. Payne, and R.J. Lefkowitz, Pleckstrin homology domain-mediated membrane association and activation of the β-adrenergic receptor kinase requires coordinate interaction with Gβγ subunits and lipid, J. Biol. Chem. 270:11707 (1995).

37. F.D. Karim, F.D. Urness, L.S. Thummel, M.J. Llemsz, S.R. McKercher, A. Celada, C. Van Beveren, R.A. Maki, C.V. Gunther, J.A. Nye, and B.J. Graves, The ETS-domain: a new DNA-binding motif that recognizes a purine-rich core DNA sequence, Genes & Dev. 4:1451 (1990).

38. B. Wasylyk, S.L. Hahn, and A. Giovane, The ets family of transcription factors, Eur. J. Biochem. 211:7 (1993).

39. E.W. Scott, M.C. Simon, H. Anastasi, and H. Singh, Requirement of transcription factor PU.1 in the development of multiple hematopoietic lineages, Science 265:1573 (1994).

40. A. Klaes, T.S. Menne, H. Scholz, and C. Klämbt, The ets transcription factors encoded by Drosophila gene pointed direct glial cell differentiation in the embryonic CNS, Cell 78:149 (1994).

41. E.M. O'Neill, I. Rebay, R.Tjian, and G.M. Rubin, The activities of two Ets-related transcription factors required for Drosophila eye development are modulated by the ras/MAPK pathway, Cell 78:137 (1994).

42. J.M. Leiden and C.B. Thompson, Transcriptional regulation of T-cell genes during T-cell development, Curr. Opin. Immunol. 6: 231 (1994).

43. D. Leprince, A. Gegonne, J. Coll, C. de Taisne, A. Schneeberger, C. Lagron, and D. Stehelin, A putative second cell-derived oncogene of the avian leukemia retrovirus E26, Nature 306:395 (1983).

44. F. Moreau-Gachelin, A. Tavitian, and P. Tambourin, Spi-1 is a putative oncogene in virally induced murine erythroleukaemias, Nature 331:277 (1988).

45. Y. Ben-David, E.B. Giddens, K. Letwin, and A. Bernstein, Erythroleukemia induction by Friend murine leukemia virus: insertional activation of a new member of the ets gene family, Fli-1, closely linked to c-ets-1, Genes & Dev. 5:908 (1991).

46. O. Delattre, J. Zucman, B. Plougastel, C. Desmaze, T. Melot, M. Peter, H. Kovar, I. Joubert, P. de Jong, G. Rouleau, A. Aurias, and G. Thomas, Gene fusion with an ETS DNA-binding domain caused by chromosome translocation in human tumours, Nature 359:162 (1992).

47. J. Zucman, T. Melot, C. Desmaze, J. Ghysdael, B. Plougastel, M. Peter, J.M. Zucker, T.J. Triche, D. Sheer, and C. Turc-carel, Combinatorial generation of variable fusion proteins in the Ewing family of tumours, EMBO J. 12:4481 (1993).

48. W.A. May, M.L. Gishizky, S.L. Lessnick, L.B. Lunsford, B.C. Lewis, O. Delattre, J. Zucman, G. Thomas, and C.T. Denny, Ewing sarcoma 11;22 translocation produces a chimeric transcription factor that requires the DNA-binding domain encoded by Fli-1 for transformation, Proc. Natl. Acad. Sci. U.S.A. 90:5752 (1993).

49. H. Liang, X. Mao, E.T. Olejniczak, D.G. Nettesheim, L. Yu, R.P. Meadows, C.B. Thompson, and S.W. Fesik, Solution structure of the ets domain of Fli-1 when bound to DNA, Nature Struct. Biol. 1:871 (1994).

50. X. Mao, S. Miesfeldt, H. Yang, J.M. Leiden, and C.B. Thompson, The Fli-1 and chemimeric EWS-Fli-1 oncoproteins display similar DNA binding specificities, J. Biol. Chem. 269:18216 (1994).

51. H. Liang, E.T. Olejniczak, X. Mao, D.G. Nettesheim, L. Yu, C.B. Thompson, and S.W. Fesik, The secondary structure of the ets domain of human Fli-1 resembles that of the helix-turn-helix DNA-binding motif of the *Escherichia coli* catabolite gene activator protein, Proc. Natl. Acad. Sci. USA 91:11655 (1994).

52. C.O. Pabo and R.T. Sauer, Transcription factors: structural families and principles of DNA recognition, Annu. Rev. Biochem. 61:1053 (1992).

53. S.C. Schultz, G.C. Shields, and T.A. Steitz, Crystal structure and CAP-DNA Complex: the DNA is bent by 90°, Science 253:1001 (1991).

54. K.P. Wilson, L.M. Shewchuk, R.G. Brennan, A.J. Otsuka, and B.W. Matthews, The *E. coli* biotin holoenzyme synthetase/bio repressor crystal structure delineates the biotin and DNA-binding domains, Proc. Natl. Acad. Sci. U.S.A. 89:9257 (1992).

55. V. Ramakrishnan, J.T. Finch, V. Graziano, P.L. Lee, and R.M. Sweet, Crystal structure of globular domain of histone H5 and its implications for nucleosome binding, Nature 362:219 (1993).

56. K.L. Clark, E.D. Halay, E. Lai, and S.K. Burley, Cocrystal structure of the HNF-3/*forkhead* DNA-recognition motif resembles histone H5, Nature 364:412 (1993).

57. C.J. Harrison, A.A. Bohm, and H.C.M. Nelson, Crystal structure of DNA binding domain of the heat shock transcription factor, Science 263:224 (1994).

58 . J.A. Nye, J.M. Petersen, C.V. Gunther, M.D. Jonsen, and B.J. Graves, Interaction of murin Ets-1 with GGA-binding sites establishes the ETS domain as a new DNA-binding motif, Genes Develop. 6:975 (1992).

DISCUSSION

David Cistola - I was particularly intrigued with the pleckstrin homology domain work where you mentioned that the overall topology of the pH domain was a ß-barrel-like protein that resembled serum-retinol binding protein and that family of lipid binding proteins. Of course in those proteins, the lipid binding is in the interior between the two ß-sheets and is usually a single acyl chain or something equivalent to a single acyl chain. I am wondering in the pleckstrin homology domain whether you considered the possibility that phospholipids with a single acyl chain might be able to bind with the acyl chain inside of the ß-barrel such as lysophosphotidylcholine and perhaps acyl coenzyme A. Have you screened those lipids as well?

Fesik - We didn't screen those that you mentioned. However, at first we thought that pH domains may be binding to farnesol or a single lipid, since many proteins that contain pleckstrin homology domains are involved in regulating RAS. For isoprenyl or farnesyl derivatives, however, we didn't see any binding. It is important to note differences between retinol binding proteins and pH domains. If you look at the apo form of retinol binding protein, there's a cavity lined with waters. This is a hydrophobic cavity where the lipid binds. If you look at the structure of the pleckstrin homology domain, that cavity is not preformed. The interior of the protein has a nice hydrophobic packing arrangement. Nonetheless, it is possible that a lipid molecule binds in that cavity. However, the experimental evidence to date suggests that the interaction between pH domains and PIP_2 consists primarily of a charge-charge interaction and that the fatty acid portion of the lipid or fatty acids in general don't greatly contribute to the binding to the pleckstrin homology domains.

Cistola - Thank you.

David Gorenstein - Have you considered that the arginines of Fli-1 are binding to the phosphates? If you looked in your model of the Fli-1/DNA complex, are they far enough away? Is it the dynamics of the phosphates and the arginines that's causing the broadening or something else?

Fesik - In the future, we plan to investigate that further. Today, we can say that we observe the broadening. However, at the present time, we are unsure what is contributing to the line broadening. Since it's a tight complex, you wouldn't think that it is exchange broadening. What we think is going on is that the arginine might be experiencing more than one state and that the motion is on the order of the chemical shift difference between these two states. Now this has been observed before by Kurt Wüthrich in other systems and during this meeting we've heard other examples where this sort of phenomena could be occurring. For example, Julie Forman-Kay spoke about SH2/phosphopeptide interactions that may be a related phenomenon in which an arginine is experiencing two states and that the motion is of the order of chemical shift difference between the two states.

Gorenstein - We've seen that as well in some proteins both from modeling and from phosphorus NMR. I think it is very important. This may be an important component of recognition certainly that the dynamics of the arginine is matching some of the dynamics of the backbone.

Fesik - It may not be as the crystal structures would have you believe in which the side chains were all defined in one location in protein/DNA complexes. In contrast, there might be mobility. In future studies, we plan to try a lot of different sample conditions in an attempt to pin down what's going on with this Fli-1/DNA complex, especially in regards to the recognition helix containing the two arginines that are required for binding to DNA.

Structures of Multimeric Proteins by NMR

G. Marius Clore and Angela M. Gronenborn

Laboratory of Chemical Physics, Building 5, National Institute of Diabetes and Digestive and Kidney Diseases, National Institutes of Health, Bethesda, MD 20892-0520, U.S.A.

Recent advances in NMR methodology, in particular the increase in resolution afforded by the advent of three- and four-dimensional NMR (see Clore & Gronenborn, 1991 for a review) have permitted the structures of proteins in the 15-25 kDa range to be solved in solution. In this paper we discuss the extension of these methods to multimeric proteins and illustrate these with regard to the solution structures of two dimers (interleukin-8; Clore et al., 1990; and human macrophage inflammatory protein 1β; Lodi et al., 1994) and a tetramer (the oligomerization domain of p53; Clore et al., 1994, 1995a,b)

In the case of dimeric proteins, it is usually possible to distinguish intra from intermolecular NOE interactions based on simple considerations of distance, providing the orientation of the two subunits at the interface is approximately anti-parallel. In such situations, it is clear that NOEs corresponding to distances greater than say 8-10 Å in the monomer must arise from intersubunit interactions. When the orientation at the dimer interface is parallel, however, it is necessary to resort to mixed labeling in which one subunit is uniformly $^{13}C/^{15}N$ labeled while the other subunit is at natural isotope abundance (i.e. predominantly ^{12}C and ^{14}N). In this way it is possible to used 4D ^{13}C and ^{15}N-editing experiments to observe interactions only within the $^{13}C/^{15}N$ labeled subunit, and 3D $^{13}C,^{15}N(F_2)/^{13}C,^{15}N(F_3)$-filtered-edited experiments to observe intersubunit NOE interactions solely between protons attached to ^{13}C and/or ^{15}N and protons attached to ^{12}C and/or ^{14}N (Clore & Gronenborn, 1994).

The first dimer to be determined by NMR was interleukin-8 (IL-8), a member of the α-chemokine family of cytokines (Clore et al., 1990), and is illustrated in Fig. 1A. More recently, the structure of the β-chemokine, human macrophage inflammatory protein-1β (hMIP-1β) has been determined (Lodi et al., 1994). While the structure of the monomer is very similar to that of IL-8, as

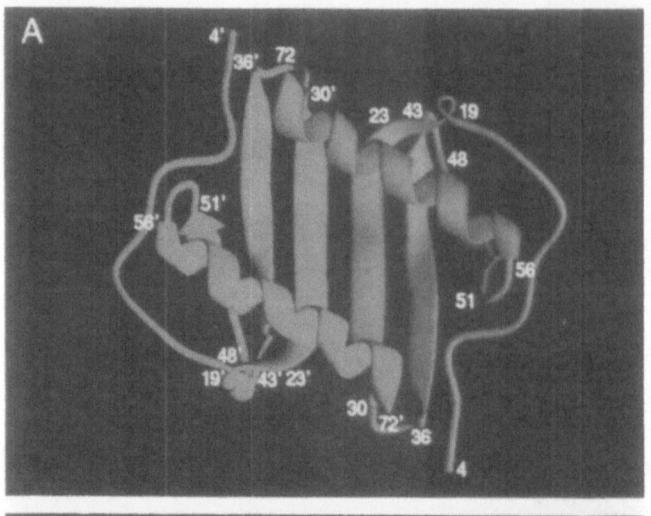

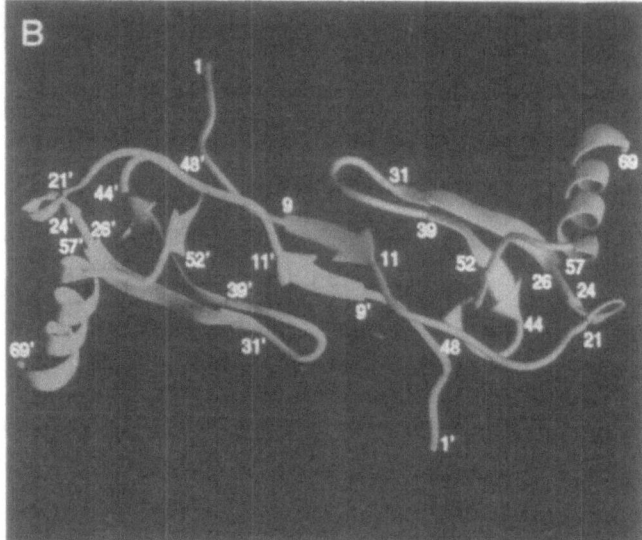

Fig. 1 Ribbon drawings of the IL-8 (A) and hMIP-1β (B) dimers.

expected from the 20% sequence identity, the quaternary structure is entirely different (Fig. 1B). In the case of IL-8, the C_2 axis is located by the CαH protons of Arg26 and Arg26' (equivalent to residue 29 of hMIP1β), and the dimer interface comprises an antiparallel β-sheet in which starnd β1 of one subunit is hydrogen-bonded to strand β1' of the other (Fig. 1A). This results in a structure with a six-stranded antiparallel b-sheet, three from each subunit, on top of which lie two antiparallel a-helices separated by approximately 14 Å. In addition, the C-terminal end of the helix of each subunit interacts with the underlying sheet of the other subunit. In contrast, in the case of hMIP-1β, the C_2 axis is located btween the CαH protons of Ala10 and Ala10'; the two helices are 46 Å apart, located on opposite faces of the molecule, and oriented approximately orthogonal to each other;

strand β1 and β1' are ~30 Å apart and located on the exterior of the protein; and the dimer interface is formed by the N-terminus (residues 2-13), the loop conecting strands β1 and β2 (Leu34 and Cys35), and the loop connecting strands β2 and β3, as well as strand β3 (residues 46-51). The completely different quaternary structures of these two chemokines can be entirely explained on the basis of surface hydrophobicity (Covell et al., 1994). The most hydrophobic surface clusters on the monomer subunits are located in very different regions of the α and β chemokines and comprise in each case the amino acids that are buried at the interafce of their respective dimers. This analysis of hydrophobicity all strongly supports the hypothesis that the distinct dimers observed for IL-8 and hMIP-1β are preserved for all the α and β chemokines, respectively. This provides a rational explanation for the lack of receptor crossbinding and reactivity between the α and β chemokine subfamilies.

Solving the structure of a tetramer is significantly more complex than that of dimer, as not only does one have to distinguish intra- from inter-subunit NOEs, but one also has to ascertain the partners involved in the intersubunit interactions. It is therefore absolutely essential to employ heteronuclear-edited and -filtered experiments to unambiguously assign the intersubunit interactions. The differentiation between the various types of intersubunit interactions can then be made in an

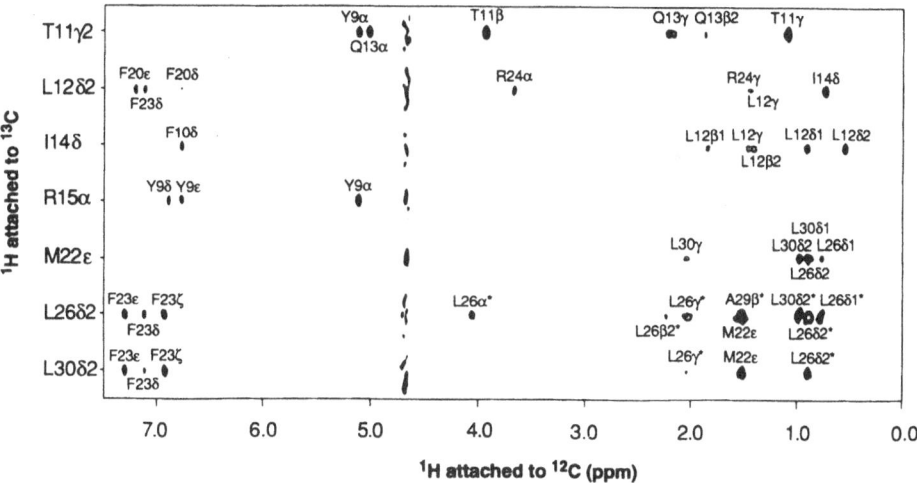

Fig. 2 Composite of ^{13}C-H strips taken from the 3D ^{13}C-edited(F_2)/^{12}C-filtered(F_3) NOE spectrum of the tetrameric oligomerization domain of p53 comprising a 1:1 mixture of unlabeled (^{12}C/^{14}N) and ^{13}C/^{15}N labeled polypeptide. The NOEs labeled with asteriks arise from intersubunit interactions between the A and B subunits (and by symmetry between the C and D subunits), while the remainder arise between the A and C subunits (and by symmetry between the B and D subunits). Note the primary dimer is formed by the A and C subunits (and by symmetry between the B and D subunits as well).

iterative cycle of refinement. Initially, all the intersubunit NOEs can be expressed as an $<r^{-6}>^{-1/6}$ sum (Nilges, 1993), and subsequently, the NOEs can be partitioned between the various subunit-subunit contacts as the refinement proceeds.

The first example of a tetramer solved by NMR is the oligomerization domain (residues 319-360) of the tumour suppressor p53 (Clore et al., 1994; 1995a,b). The overall molecular weight of the oligomerization domain is about 20 kDa and it behaves in solution with a rotational correlation time of about 14.8 ns at 35°C (Clubb et al., 1995). Nevertheless, high quality ^{13}C-edited(F_1)/ ^{12}C-filtered NOE spectra can be obtained as demonstrated in Fig. 2. The strucure was determined on the basis 4472 experimental NMR restraints. The oligomerization domain of p53 comprises a dimer of dimers, oriented in an approximately orthogonal manner. Each dimer (comprising subunits A and C, and subunit B and D) consists of two-antiparallel α-helices and an antiparallel β-sheet. The β-sheets lie on opposite sides of the tetramer and the helices form an unusual, four helix bundle (Fig. 3). The interhelical angles between the helices of subunits A, C, B and D are 155°, 81° and -104°, respectively (Clore et al., 1995). The N- and C-termini (residues 319-323 and 357-360) are completely disordered and highly mobile in solution, and the turn preceeding the core (residues 324-325) and the last quarter turn of helix at the C-terminus (residues 355-356) are poorly ordered (Clore et al., 1995; Clubb et al., 1995). The remainder of the tetramer (residues 326-354) is well defined with a backbone precision of 0.3 Å for the backbone atoms and ~0.4 Å for all ordered atoms. In addition, the rms difference between the NMR and X-ray (Jeffrey et al., 1995) structures for the core of the tetramer is only 0.5 Å for the backbone (N, Cα, C, O) atoms and 0.55 Å for all ordered atoms (Clore et al., 1995b).

The structure of the p53 oligomerization domain readily explains the effects of four point mutations within this domain that have been identified in human cancers (Cariello et al., 1994). The Leu330→His mutation substitutes a completely buried hydrophobic residue by a polar residue in the center of the hydrophobic core of the primary dimer (i.e. the AC or BD dimer). The Gly334→Val mutation is the result of steric hindrance as Val cannot accomodate the unusual φ,ψ angles (86°, 140°) of Gly at this position. The Arg337→Cys mutant removes a salt bridge between the guanidinium group of Arg337(A) and the carboxylate of Asp352(C). Finally, the Glu349→Asp mutation removes a water-bridged hydrogen-bonding interaction between the carboxylate of Glu349(A) and the backbone amide of Arg333(C).

Acknowledgements

This work was supported by the AIDS Targeted Antiviral Program of the Office of the Director of the National Institutes of Health.

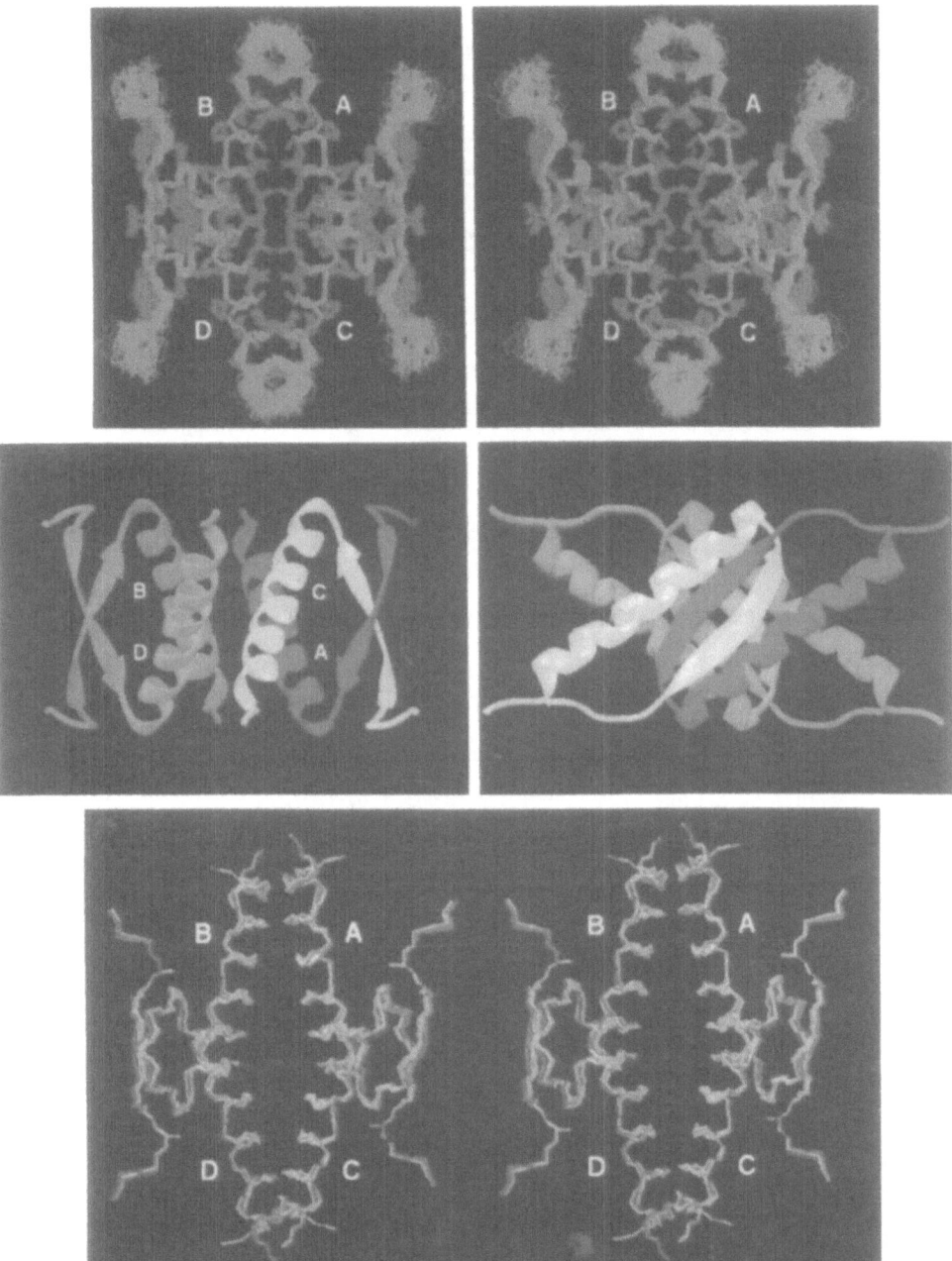

Fig. 3 (A) Best fit superposition of the backbone (blue) and ordered side chains (red) of the ensemble of 76 simulated annealing structures. (B) Two ribbon drawing of the oligomerization domain of p53 (subunits A, C, B and D are colored in red, yellow, green and blue, respectively). (C) Best fit superposition of the backbone of two restrained regularized mean NMR structures (in red and blue) and the X-ray structure (in green). The two restrained regularized mean structures are derived from two different simulated annealing ensembles obtained with slightly different parameters for the effective hard sphere van der Waals radii. Adapted from Clore et al. (1995b).

References

Covell, D.G., Smythers, G.W., Gronenborn, A.M. & Clore, G.M. (1994 *Protein Science 3*, 2064-2072.

Clore, G.M. & Gronenborn, A.M. (1991) *Science 252*, 1390-1399.

Clore, G.M. & Gronenborn, A.M. (1994) *Protein Science 3*, 372-390.

Clore, G.M., Appella, E., Yamada, M., Matsushima, K. & Gronenborn, A.M. (1990) *Biochemistry 29*, 1689-1696.

Clore, G.M., Omichinski, J.G., Sakaguchi, K., Zambrano, N, Sakamoto, H., Appella, E. & Clore, G.M. (1994) *Science 265*, 386-391

Clore, G.M., Omichinski, J.G., Sakaguchi, K., Zambrano, N, Sakamoto, H., Appella, E. & Clore, G.M. (1995a) *Science 267*, 1515-1516

Clore, G.M., Ernst, J., Clubb, R., Omichinski, J.G., Kennedy, W.M.P, Sakaguchi, K., Appella, E. & Gronenborn, A.M. (1995b) *Nature Struct. Biol. 2*, 321-333.

Clubb, R.T., Omichinski, J.G., Sakaguchi, K., Appella, E., Gronenborn, A.M. & Clore, G.M. (1995) *Protein Science 4*, 855-862.

Jeffrey, P.D., Gorina, S. & Pavletich, N.P. (1995) *Science 267*, 1498-1501.

Lodi, P.J., Garrett, D.S., Kuszweski, J., Tsang, M.L.S., Weatherbee, J.A., Leonard, W.J., Gronenborn, A.M. & Clore, G.M. (1994) *Science 263*, 1762-1767.

Nilges, M. (1993) *Proteins Struct. Funct. Genet. 17*, 295-309.

DISCUSSION

Brian Sykes - Marius, coming to your chemokine story, I would have said exactly the same thing when I saw IL-8 to that how could any protein not look like that, not be so beautifully symmetric. While IL-8 is a dimer, we (I and Clark Lewis) have synthesized an analog which is a monomer and fully active *in vitro*. Do you think these dimers and tetramers are important?

Clore - I am glad you brought that up, it's an interesting point. Ian synthesized an interlukin-8 in which he replaced an amide group by an N-methyl group where that N-methyl group was located directly at the ß-sheet interface. In solution that species is monomeric and if you look at the activity *in vitro*, it is active. However, what he didn't show were two aspects. First, he didn't show what the nature of the species was when bound to its receptor, so the dimer isn't excluded. The second thing is that although there's absolutely no question that the form of this chemokine circulating in the blood is monomeric in so far that their association constants range from 10^{-8}-10^{-7} M and their concentrations in the blood are in the picomolar range, it is also quite clear that the chemokines don't act in free solution. They act on solid support and in solid support when they are fixed to vascular endocelium, the local concentration of chemokine is actually very high and under these circumstances, they are definitely going to dimerize or in fact even oligomerize to a higher extent. In the case of hMIP1ß it oligomerizes to a much higher extent. What I think is going on is that, while for just binding to the receptor in a test tube it is fine to be a monomer as far as its *in vitro* action is concerned, the dimeric form is key in so far as it actually has to bind to the surface of the vascular endothelium. I think that these oligomers are a key aspect in fixing the chemokines to the vascular endothelium. If they fail to do that, they just get washed away and then they have no effect whatsoever. So *in vivo* I think it's important.

Carol Post - With regard to the distance restraints involving NOE's between subunits, how were you able to distinguish intersubunit from intrasubunit interactions in the beginning?

Clore - What you can do is allow the NOE's to float. You can make them ambiguous and let them choose whether its intramonomer or intersubunit. In fact, in this particular case, you don't need to go quite to that extent, because it is quite clear from the experiments that we did what is intersubunit and what is intrasubunit. So all we have to do is allow the computer to decide what the particular intersubunit interactions are and that's exactly what we did to start with until we found out roughly what was going on. Once we did we could then bootstrap our way across to figure out all the remaining contacts and the assignments specifically between the subunits.

Carol Post - Marius, I am not real familiar with the p53 story. Could you explain what the oligomerization does? You said p53 interacts by activating a gene for another protein, so what about the oligomerization? Did you say the monomers were active?

Clore - The monomer has weak transcriptional activity.

Post - What about the point mutations which seem to be affecting the oligomerization?

Clore - There are four mutations in the oligomerization domain. There are thousands of mutations in the DNA binding domain. The frequency of mutations in the oligomerization domain is minute compared to that in the DNA binding domain and exactly what role they have is unknown. The problem here is that people have sequenced p53 in plenty of tumors but nobody has sequenced p53 extensively in normal cells. I mean they have sequenced it once but not an infinite number of times. So nobody knows what the variation of the normal p53 sequence is. So I don't even know whether those mutations actually play a role in mutagenesis or not. All I know is that they happen to be involved in the interface and it looks as if they would destabilize the tetramer. What is clear is that the efficiency of the monomer as a transcriptional activator, and for that matter just the DNA binding protein, is much reduced from that in the tetrameric form, and if you look at the DNA binding sites, the DNA binding site comprises four sites in close proximity.

Marc Adler - I have gotten used to seeing the figure in the literature that the rmsd deviation between the X-ray structure of the same protein and different crystal packing forms as 0.4 $\overset{0}{A}$ whereas you had mentioned that you were getting rmsd's of 0.2 $\overset{0}{A}$ for your structure, so I am wondering about the discrepancy. Why do you think your structure has such a small rmsd on the backbone.

Clore - There are two things that I think you are confusing here. First, the rms between crystal structures can be much smaller than 0.4 $\overset{0}{A}$. It can be down to 0.1 $\overset{0}{A}$. It all depends on the quality of the structure, the type of structure, the influence of crystal packing and so on. How well the NMR structure is determined reflects simply the precision of the coordinates. The precision of the crystal structure is even higher than that. Now, why is that structure determined so well. Well, it's determined so well because we happen to have something like 25-30 experimental restraints per residue and, I didn't show this slide because it's rather old, as you increase the number of restraints, the quality of the structure and the precision go up, just as it does in crystallography. If a crystal diffracts only to low resolution it means that there are relatively few experimental data points. As it diffracts to higher resolution, you have larger numbers of experimental data points, and a higher quality structure. It's exactly the same in NMR. This doesn't mean to say there can't be discrepancy between this and the crystal structure of the order of 0.6 or 0.8 $\overset{0}{A}$ and it also doesn't imply that the accuracy of the structure is 0.2 $\overset{0}{A}$. In fact the accuracy of the structure is likely to be a factor of 2 down from that and we have shown that in calculations.

NMR STRUCTURAL STUDIES OF FLEXIBLE MOLECULES

Peter E. Wright and H. Jane Dyson

Department of Molecular Biology,
The Scripps Research Institute,
10666 North Torrey Pines Rd, La Jolla CA 92037

INTRODUCTION

In recent years, NMR has been applied with great success to determine precise three-dimensional solution structures of compact globular proteins and small, inflexible cyclic peptides. NMR also holds promise for providing structural information on flexible biomolecules, although a quantitative description of structure is complicated by the inevitable population-weighted averaging of the key NMR parameters. Here we describe simple procedures for calculation of the dominant structure in the conformational ensemble of a linear peptide and for characterizing the domain motions and overall structural preferences of multidomain proteins.

DETERMINATION OF THE STRUCTURE OF A LINEAR PEPTIDE

Linear peptides exist in aqueous solution as conformational ensembles containing a large number of interconverting conformers. Calculation of three-dimensional structures of short peptides is difficult and generally inadvisable. The use of averaged NMR parameters, which may have contributions from widely differing conformations, as input for structure calculation may well result in a structure that does not resemble any of the conformers actually present in the conformational ensemble. Calculations of 3D structures are not normally necessary for short peptides, since conformational preferences for folded structures can readily be deduced qualitatively from the NMR parameters and can usually be described in terms of common elements of secondary structure such as helices or turns (Dyson & Wright, 1991). In such cases, information on the nature of the structured conformer can be obtained from simple inspection of the NMR data and structural calculations provide no further insights. There are some instances, however, in which structure calculations can be helpful in answering specific questions that cannot be answered directly by inspection of the NMR spectra. An example is the turn conformation inferred for the *cis* form of the peptide SYPFDV in aqueous solution (Dyson et al., 1988; Yao et al., 1994a). Studies of the sequence dependence of the turn population indicate that a highly specific structure is formed that may well differ in detail from the canonical form of the Type VI turn found in proteins (Yao et al., 1994a).

Structures were calculated for the folded conformation of SYPYDV in aqueous solution using distance geometry and restrained molecular dynamics (Yao et al., 1994b). An implicit assumption is that only a single structure or closely related group of structures is populated significantly in the *cis* form of the peptide. This assumption is justified on the basis of the close correlation between the NMR parameters diagnostic of the turn (NOEs, coupling constants, and the mole fraction of *cis* isomer) in a series of related peptides containing varying populations of the folded conformation (Yao et al., 1994a). An important criterion for structure calculation of linear peptides is that the population of the folded conformer in the conformational ensemble be very high and that the relative contributions of the structured and unstructured states to the NMR parameters can be deconvoluted. In the case of SYPFDV, we selected only NOE constraints that are expected to represent only the folded conformation of the peptide for structure calculations, i.e. the NOEs that are not observed in NOESY or ROESY spectra of unfolded peptides (Yao et al., 1994b).

The NOEs that best distinguish folded from unfolded conformers are those that are between non-sequential proton pairs, e.g. $d_{\alpha N}(i,i+2)$, $d_{\alpha N}(i,i+3)$, or $d_{\alpha\beta}(i,i+3)$ NOEs characteristic of turns or helix. These distances are short only in folded conformers; in unfolded states the distances between these proton pairs would be too long (> 5Å) for an NOE to be observable (Billeter et al., 1982). In peptides with a small or negligible population of folded states, medium-range NOEs are not observed (Dyson & Wright, 1991). When the population of folded forms reaches a certain threshold these NOEs are detected, with intensity proportional to the inverse sixth power of interproton distance weighted by population. Unlike medium-range NOEs, sequential backbone-backbone and $C^\beta H$-NH NOEs and intraresidue NOEs have contributions from most states in the conformational ensemble. Since the NOE contributions arising from the different conformers cannot be deconvoluted, these NOEs were not used to generate distance constraints for SYPFDV (Yao et al., 1994b). Omission of such "conformationally ambiguous" NOEs does not result in significant underdetermination of the structures since the covalent geometry of the peptide, the non-sequential inter-residue NOEs, and the dihedral angle constraints obtained from coupling constants are adequate to define the backbone conformation of the folded state.

In calculating the structures of SYPFDV the NOE-derived distance constraints were supplemented by ϕ and χ_1 dihedral angle constraints derived from $^3J_{HN,H\alpha}$, $^3J_{\alpha\beta}$ and $^3J_{N\beta}$ coupling constants. The ϕ dihedral angle was constrained for two residues in which unusually high or low values of the $^3J_{HN,H\alpha}$ coupling constant were observed. For Tyr 2, $^3J_{HN,H\alpha}$ is 3.5 Hz. This value represents a population-weighted average for the conformational ensemble, which includes ~30% unfolded states. The coupling constant for the folded conformer was estimated, from its population determined independently from the *cis/trans* isomer ratio, to be 2.5 Hz and ϕ was therefore constrained to the range -60° to +20° (Yao et al., 1994b). The , $^3J_{HN,H\alpha}$ value for Val 6 is larger than 8 Hz and ϕ was thus constrained to the range -80° to -160°. The $^3J_{\alpha\beta}$ and $^3J_{N\beta}$ (measured with ^{15}N labeled peptide) coupling constants for SYPFDV indicate a high population of of single rotameric forms for Tyr 2 and Phe 4 in the *cis* form of the peptide (Yao et al., 1994b). From the magnitude of these coupling constants, the preferred rotamers were determined and stereospecific assignments of the $C^\beta H$ resonances were made. The χ_1 angle was constrained in the structure calculations on the basis of the most highly populated rotamer.

Distance geometry structures were calculated (Havel & Wüthrich, 1984) using 18 distance constraints and 2 ϕ (Tyr 2 and Val 6) and 2 χ_1 (Tyr 2 and Phe 4) angle constraints derived from the coupling constants. In addition, the ϕ dihedral angle for Phe 4 and Asp 5 was constrained to be negative. All constraints were judiciously selected and reflect the conformation of the folded state; NMR parameters that contain contributions from both folded and unfolded states were not used in structure calculation. Conservative distance constraints were obtained from ROESY cross peak intensities, calibrated for geminal methylene protons (Yao et al., 1994b). While these cross peaks are not ideal as calibrants, the alternative inter-residue sequential NOEs that are routinely used for protein structure calculations are not appropriate for a conformationally averaged linear peptide. Careful calibration of distance constraints is essential; unrealistically tight constraints results in poor sampling of conformational space and physically unreasonable structures.

Initial structures obtained by distance geometry were similar, without severe distance or angle violations (Yao et al., 1994b). The family of structures is shown in Fig. 1. These structures were then refined by restrained molecular dynamics using the AMBER software package and all-atom force field (Weiner et al., 1986; Pearlman et al., 1995). A superposition of 36 structures with no violations of distance constraints greater than 0.1Å or dihedral angle constraints greater than 5° is shown in Fig. 2. The all-atom rms deviations

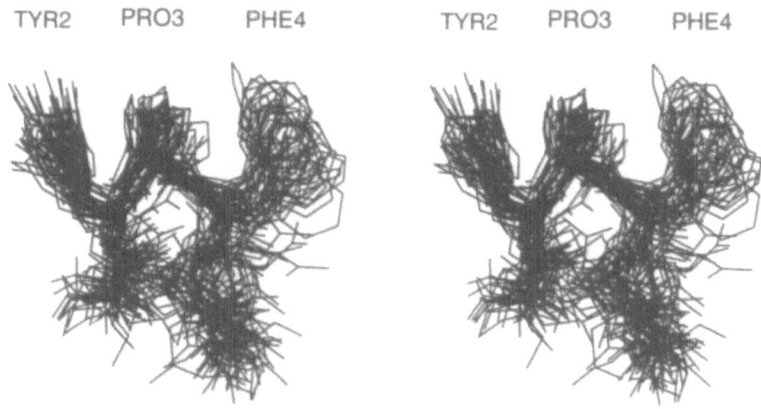

show a significant increase after MD refinement, reflecting the restricted sampling of conformational space that occurs in distance geometry calculations (Metzler et al., 1989). The better sampling properties of restrained MD means that the virtually unrestrained ends of the peptide show a wider range of conformations in the MD structures. By contrast, the central residues of the peptide show significantly lower rms deviations, as the density of restraints constrains them to a few closely related conformations.

The type VI turn structure is well defined by the NMR structures (Yao et al., 1994b). The backbone is well defined throughout the turn region, with a conformation that is similar to type VIa turns found in protein structures (Richardson & Richardson, 1989). The turn is stabilized by packing of the aromatic side chains of Tyr 2 and Phe 4 against the proline side chain (Fig.2). The structure shown in Fig. 2 is consistent with ring current shifts observed for the proline resonances and with measured coupling constants for the proline that were not explicitly included in the structure calculations (Yao et al., 1994b).

The purpose of this structure calculation was to gain insights into the sources of stability of this unusual class of reverse turns, a goal which has clearly been realized. Complete characterization of all the conformations of a linear peptide in aqueous solution is difficult and was not attempted. Instead, NMR parameters were judiciously selected to report on the conformation of the folded state and thus provide a view of the most populated structure in the conformational ensemble.

AVERAGE SOLUTION CONFORMATION OF MULTIDOMAIN PROTEINS

Conformational flexibility is often a characteristic of proteins, as well as peptides, and may well be important for biological function. In modular proteins, for example, characterization of interdomain spatial interactions and relative domain motion is essential

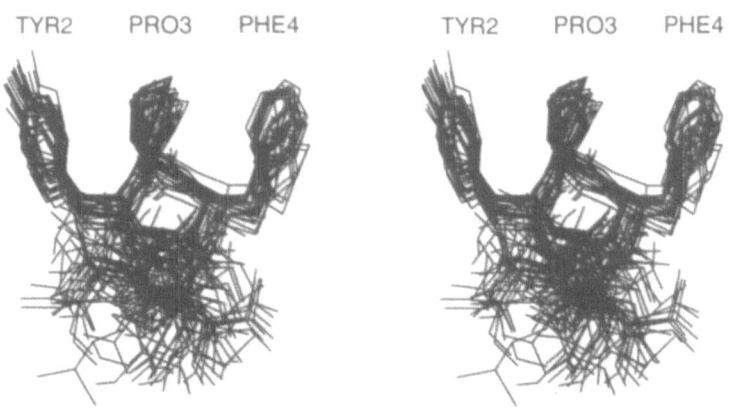

for understanding function and biological activity. Because of the limited distance range of the NOE, information on interdomain interactions can be difficult to obtain in conventional NMR structure determination. However, spin relaxation measurements offer the potential for characterizing the flexibility and providing insights into overall solution structure of multidomain proteins. Interactions between protein domains can result in anisotropic overall molecular tumbling. In such a case, ^{15}N relaxation rates depend on both the average orientations of the ^{15}N-^1H vectors and their fluctuations relative to the molecular diffusion tensor (Woessner, 1962). Thus, by characterizing molecular tumbling using NMR relaxation data, dynamic long-range motional correlations between protein domains can be identified and information of the relative orientation of protein domains can be obtained (Brüschweiler et al., 1995).

The use of NMR relaxation measurements to characterize long-range order in modular proteins has been illustrated for a polypeptide containing three zinc fingers (Brüschweiler et al., 1995). Each zinc finger domain is a compact globular structure containing a β-hairpin and an α-helix (Lee et al., 1989). A key question concerning proteins containing multiple zinc fingers is whether the individual fingers behave as independent beads on a flexible string or whether there are interactions between the finger domains that lead to long-range order in the protein. We have addressed this question by measuring ^{15}N T1 and T2 relaxation times and the ^{15}N-^1H heteronuclear NOE for a protein containing the first three fingers of TFIIIA, termed zf1-3 (Liao et al., 1992). In addition, solution structures for the individual zinc finger domains have been calculated using conventional distance geometry and restrained molecular dynamics methods.

Calculations showed that the ^{15}N relaxation data for zf1-3 could not be adequately fitted using models based on isotropic overall tumbling. Rather, a good fit to the data could only be obtained using an anisotropic rotational diffusion model (Brüschweiler et al., 1995). By utilizing both the relaxation data and the information on the relative orientations of the individual ^{15}N-^1H vectors obtained from the NMR structures, we were able to extract an averaged diffusion tensor for each zinc finger domain (Brüschweiler et al., 1995). The resulting diffusion tensors show substantial anisotropy and asymmetry:

	$(6D_{xx})^{-1}$	$(6D_{yy})^{-1}$	$(6D_{zz})^{-1}$
zf1	10.9	8.5	7.2
zf2	15.1	9.4	6.6
zf3	10.1	7.6	6.3

The overall tumbling of zf1-3 is both anisotropic and much slower than expected on the basis of the correlation time (1.5 ns) measured for a single finger (Palmer et al., 1991), suggesting the presence of interactions between the zinc finger domains and motional restrictions imposed by the linkers. Although restricted rigid-body motions of the individual zinc finger domains do occur, as indicated by the different sizes and shapes of their respective diffusion ellipsoids, the overall structure of zf1-3 in solution is clearly highly elongated on average.

ACKNOWLEDGEMENTS

We are indebted to Drs Rafael Brüschweiler, Xiubei Liao, and Jian Yao whose research is summarized here and to the National Institutes of Health (Grant numbers GM38794 and GM36643) for financial support.

REFERENCES

Billeter, M., Braun, W., & Wüthrich, K. (1982) Sequential resonance assignments in protein ^1H nuclear magnetic resonance spectra. Computation of sterically allowed proton-proton distances and statistical analysis of proton-proton distances in single crystal protein conformations. *J. Mol. Biol.* **155**, 321-346.

Brüschweiler, R., Liao, X., & Wright, P.E. (1995) Long-range motional restrictions in a multidomain zinc-finger protein from anisotropic tumbling. *Science* **268**, 886-889.

Dyson, H.J., Rance, M., Houghten, R.A., Lerner, R.A., & Wright, P.E. (1988) Folding of immunogenic peptide fragments of proteins in water solution. I Sequence requirements for the formation of a reverse turn. *J. Mol. Biol.* **201**, 161-200.

Dyson, H.J. & Wright, P.E. (1991) Defining solution conformations of small linear peptides. *Ann. Rev. Biophys. Biophys. Chem.* **20**, 519-538.

Havel, T.F. & Wüthrich, K. (1984) A distance geometry program for determining the structures of small proteins and other macromolecules from nuclear magnetic resonance measurements of intramolecular ^{1}H-^{1}H proximities in solution. *Bull. Math. Biol.* **46**, 73-698.

Lee, M.S., Gippert, G., Soman, K.Y., Case, D.A., & Wright, P.E. (1989) Three-dimensional solution structure of a single zinc finger binding domain. *Science* **245**, 635-637.

Liao, X., Clemens, K.R., Tennant, L., Wright, P.E., & Gottesfeld, J.M. (1992) Specific interaction of the first three zinc fingers of TFIIIA with the internal control region of the *Xenopus* 5S RNA gene. *J. Mol. Biol.* **223**, 857-871.

Metzler, W.J., Hare, D.R., & Pardi, A. (1989) Limited sampling of conformational space by the distance geometry algorithm: Implications for structures generated from NMR data. *Biochemistry* **28**, 7045-7052.

Palmer, A.G.,III, Rance, M., & Wright, P.E. (1991) Intramolecular motions of a zinc finger DNA-binding domain from Xfin characterized by proton-detected natural abundance ^{13}C NMR spectroscopy. *J. Am. Chem. Soc.* **113**, 4371-4380.

Pearlman, D.A., Case, D.A., Caldwell, J.W., Ross, W.S., Cheatham, T.E.,III, Ferguson, D.M., Seibel, G.L., Singh, U.C., Weiner, P.K., & Kollman, P.A. (1995) *AMBER 4.1*, University of California, San Francisco.

Richardson, J.S. & Richardson, D.C. (1989) Principles and patterns of protein conformation. in *Prediction of Protein Structure and the Principles of Protein Conformation* (Fasman, G.D. Ed.) Plenum Press, New York.. pp 1-99

Weiner, S.J., Kollman, P.A., Nguyen, D.T., & Case, D.A. (1986) An all atom force field for simulations of proteins and nucleic acids. *J. Computat. Chem.* **7**, 230-252.

Woessner, D.E. (1962) Spin relaxation processes in a two-proton system undergoing anisotropic reorientation. *J. Chem. Phys.* **36**, 1-4.

Yao, J., Feher, V.A., Espejo, B.F., Reymond, M.T., Wright, P.E., & Dyson, H.J. (1994a) Stabilization of a Type VI turn in a family of linear peptides in water solution. *J. Mol. Biol.* **243**, 736-753.

Yao, J., Dyson, H.J., & Wright, P.E. (1994b) Three-dimensional structure of a Type VI turn in a linear peptide in water solution: evidence for stacking of aromatic rings as a major stabilizing factor. *J. Mol. Biol.* **243**, 754-766.

IRON-SULFUR PROTEINS: INVESTIGATIONS OF HYPERFINE-SHIFTED HYDROGEN, CARBON, AND NITROGEN RESONANCES

Bin Xia, Hong Cheng,[a] Young Kee Chae, Lars Skjedal,[b] William M. Westler,

and John L. Markley

Department of Biochemistry, University of Wisconsin-Madison,

420 Henry Mall, Madison, WI 53706

Present addresses: [a]Department of Pharmacology, Mayo Clinic and Research Foundation, Rochester, Minnesota 55905; [b]Department of Biochemical Sciences, Agricultural University of Norway, IBF, Kjemi Boks 5036, N-1432 Ås, Norway.

INTRODUCTION

Iron-sulfur proteins are present in almost all living organisms. They are characterized by one or more iron ions ligated to inorganic sulfur and/or cysteine sulfur. Rubredoxin-type clusters contain a single iron ligated to four cysteines and have no inorganic sulfur. All other types of iron-sulfur protein contain two or more irons plus inorganic sulfur. Four types of iron-sulfur centers with cysteine ligands to the protein have been characterized by x-ray crystallography; they can be distinguished by the number of iron and inorganic sulfur atoms as: [1Fe], [2Fe-2S], [4Fe-4S], and [3Fe-4S. An additional type of iron-sulfur center has been shown to have one carboxylic acid ligand, and Rieske centers are believed to contain a [2Fe-2S] cluster ligated to two cysteines and

NMR As a Structural Tool for Macromolecules: Current Status and Future Directions
Edited by B.D. Nageswara Rao and Marvin D. Kemple, Plenum Press, New York, 1996

251

two histidines. Some iron-sulfur proteins contain more than one Fe-S center, and these may be of the same or mixed types. Rubredoxins and all ferredoxins display electron-carrier activity but no classical enzyme function. Some iron-sulfur proteins, such as endonuclease III and aconitase, have been shown to play roles in enzymatic catalysis rather than in redox chemistry (Hentze & Argos, 1991). Other Fe-S proteins are involved in the regulation of transcription (Roualt et al., 1991).

A single iron, or rubredoxin-type, center contains a high-spin Fe(III) (ferric iron) in its oxidized state and a high-spin Fe(II) (ferrous iron) in its reduced state. A [2Fe-2S] cluster has two accessible redox states: an oxidized state, in which both irons are high-spin Fe(III) and antiferromagnetically coupled with each other (EPR silent), and a reduced state in which one of the irons becomes high-spin Fe(II) so that the cluster has a net spin of 1/2. [4Fe-4S] clusters can have three oxidation states: *(A)* 3Fe(III)/1Fe(II), *(B)* 2Fe(III)/2Fe(II), and *(C)* 1Fe(III)/3Fe(II). Typically, only two of these are accessible as the oxidized and reduced states in a given type of [4Fe-4S] protein: states *A* and *B* in high potential iron proteins (HiPIPs) and states *B* and *C* in [4Fe-4S] ferredoxins

Iron-sulfur proteins were among the first proteins studied by nuclear magnetic resonance (NMR) spectroscopy. The hyperfine-shifted signals provided well-resolved markers even in the early days of one-dimensional ^{1}H NMR spectroscopy. However, paramagnetism arising from unpaired electrons in the chromophore complicates the acquisition and analysis of NMR spectra. These unpaired electrons interact with magnetic nuclei either through chemical bonds, by a contact shift mechanism, or through space, by a pseudocontact mechanism. These interactions provide efficient nuclear relaxation pathways that broaden the NMR signals and/or provide new chemical shift mechanisms that give rise to "hyperfine-shifted resonances" located well outside the normal diamagnetic chemical shift range for the particular nucleus (0 ppm to 10 ppm for ^{1}H). The magnitude of the hyperfine shift is defined as the difference between the observed shift and the shift the nucleus would have in the protein in the absence of the paramagnetism. The paramagnetism hinders the determination of sequence-specific assignments, especially to resonances from amino acid residues that are ligated the iron-sulfur cluster. Correlations among resonances affected by the paramagnetism, or between a one affected by the paramagnetism and others that are not, often are not detected by routine NMR methods. Efficient proton relaxation mechanisms mediated by the paramagnetic center may make it difficult or impossible to determine proton-proton nuclear Overhauser enhancements (NOEs) from nuclear spins in the vicinity of the paramagentic center.

This review focuses on NMR studies of the hyperfine-shifted resonances of iron-sulfur proteins and discusses methods for their detection and interpretation. The information that hyperfine-shifted NMR signals provide about the active centers of paramagnetic proteins is not readily accessible by other physical methods. The chemical shifts and relaxation times of these signals are sensitive to the delocalization of unpaired electron density from the metals onto nuclei. The full interpretation of such results is contingent on the availability of assignments of individual signals to specific nuclei in the covalent structure of the iron-sulfur protein. Although resonances from ^1H, ^{13}C, and ^{15}N nuclei in iron-sulfur proteins have been detected and studied for a number of years (Poe et al., 1971; Chan & Markley, 1983; Oh & Markley, 1990b), such assignments have become available only recently and solely for a limited number of signals. A breakthrough came when it was realized that one-dimensional NOE spectroscopy could be used to detect cross relaxation between pairs of hyperfine-shifted resonances and between hyperfine-shifted spins (Dugad et al., 1990). Further application of two-dimensional NOE and magnetization exchange methods, in conjunction with an improved x-ray structure, made it possible to assign most of the cysteinyl ^1H resonances in the spectrum of the reduced [2Fe-2S] ferredoxin from *Anabaena* 7120 (Skjeldal et al., 1991a). Another advance has come from isotopic labeling of recombinant iron-sulfur proteins which has proved to be the most direct and reliable way of determining the origins of hyperfine-shifted resonances by residue type and position (Packer et al., 1977; Cheng et al., 1992; Cheng et al. 1995). Spectra of proteins labeled selectively with ^2H, ^{13}C, or ^{15}N (nuclei that have low natural abundance) are relatively simple and easy to interpret. Finally, the magnetogyric ratios of ^2H, ^{13}C, and ^{15}N are smaller than that of ^1H, and because hyperfine interactions are proportional to the square of the magnetogyric ratio of the nucleus (γ), signals from ^2H, ^{13}C, and ^{15}N nuclei, whose γ values smaller than that of are less affected by paramagnetic centers. We reveiw here recent progress in our NMR investigations of the hyperfine-shifted resonances of four iron-sulfur proteins: *Clostridium pasteurianum* rubredoxin (Rdx), *Anabaena 7120* vegetative [2Fe-2S] ferredoxin (VFd), *Anabaena 7120* heterocyst [2Fe-2S] ferredoxin (HFd), and human [2Fe-2S] ferredoxin (HuFd).

METHODOLOGY

Protein Production   We have developed methods for overproducing all four of these iron-sulfur proteins in *E. coli* under control of the T7 RNA promoter/polymerase system. The DNA fragments containing the coding sequence for the desired protein were cloned into pET3a or pET9a expression vectors of pET system (Novagen), or both. The newly constructed plasmids were then transformed into an appropriate *E. coli* host strain for protein expression.   Four *E. coli* strains (BL21(DE3)/pLysS, JM15/pGp1-2, PA200/pGp1-2, and JM15(DE3)/pLysS) were used for various purposes. BL21(DE3) was used primarily as the host for overexpression of pET system vectors.   The gene encoding T7 RNA polymerase, which is present in its chromosome under control of the *lac* operator, can be induced by IPTG. The plasmid pLysS, which contains a gene encoding T4 lysozyme, a natural inhibitor of T7 RNA polymerase was used to reduce the basal level of protein expression prior to IPTG induction, so as to minimize the toxic effect of the recombinant protein on the host *E. coli* cells.

The following procedure was used f:or protein producton on LB medium. A 5 ml aliquot of an overnight culture of *E. coli* (BL21(DE3)/pLysS) containing the expression plasmid was used to inoculate 1 liter of the medium. The cells were grown at 37 $^\circ$C in an incubator shaker. When the OD_{600} of the culture reached 0.6 (for human ferredoxin) or 1.2 (for the other three proteins), 100 mg IPTG was added to induce the production of T7 RNA polymerase, which in turn induced overexpression of the desired protein.   The culture was incubated at 37 $^\circ$C for an additional 16 h (for human ferredoxin) or 2 h (for the other three proteins). Then, the bacterial cells were harvested by centrifugation, and the cell pellet was resuspended in 20 ml of 50 mM phosphate (or Tris·HCl) buffer (pH 8.0) and frozen at -20 $^\circ$C.   The cells were broken by a freeze-thaw cycle followed by sonication. The metal centers were then incorporated in the apoproteins through an *in vitro* reconstitution process (Cheng et al., 1995).

Isotopic Labeling   For uniform ^{15}N labeling, the host bacteria were grown on M9 medium containing 1 g/L [15]NH$_4$Cl. For selective labeling, the host bacteria were grown on M9 medium plus the isotopically labeled amino acid and a mixture of other unlabeled amino acids (Cheng et al., 1995).   Procedures used for selective isotopic labeling were tailored to the particular site to be labeled. Plasmid pGp1-2, which contains the gene for T7 RNA polymerase, was introduced into different auxotrophic strains of *E. coli*, such as JM15 (Cys auxotroph) or PA200 (Arg auxotroph), which were then used for overexpression of the desired iron-sulfur protein. The temperature-inducible plasmid pGp1-2, which is kanamycin resistant, is not suitable for use as an expression vector with pET 9a which is also kanamycin resistant; only plasmids such as pET3a (ampicillin resistant) that have a different antbiotic resistance can be used with pET 9a. The IPTG-inducible T7 RNA polymerase gene was introduced into the genomic DNA of certain *E. coli* auxotrophic strains by bacterial phage P1 transduction. *E. coli* strains AW608Thr[+]T7 (His auxotroph) and JM15(DE3) (Cys auxotroph), which were constructed by this method, were used as host cells for selective isotopic labeling of the respective amino acids.   In some cases, selective ^{15}N labeling was achieved by using *E. coli* strain BL21(DE3)/pLysS instead of the several auxotrophic strains; in most cases, migration of ^{15}N from the labeled amino acid to other amino acids was negligible (Chae et al., 1994).

VEGETATIVE FERREDOXIN

Anabaena 7120 vegetative ferredoxin is a typical plant-type [2Fe-2S] ferredoxin. It functions as an electron carrier, and transfers electrons from photosystem I to ferredoxin-NADP$^+$ reductase (FNR), which in turn catalyzes the reduction of NADP$^+$ to NADPH (Masaki et al., 1982). The x-ray crystal structure of this protein was first determined at 2.5 Å (Rypniewski et al., 1991) and was refined subsequently to 1.7 Å (Holden et al., 1994). NMR studies of the vegetative ferredoxin have led to assignments of most of the diamagnetic resonances of the oxidized protein (Oh et al., 1990; Oh & Markley, 1990a) and several of the paramagnetic resonances in both oxidation states (Oh & Markley, 1990b; Skjeldal et al, 1991a; Cheng et al., 1995).

In ^{1}H NMR spectra of oxidized VFd, several broad signals were observed in the 34 to 37 ppm range (data not shown). These signals were assigned tentatively to the [1]H$^\beta$

atoms of the cysteines (Skjeldal et al., 1990). In corresponding spectra of the reduced state, eleven well-resolved hyperfine-shifted peaks were detected (Figure 1). Preliminary interpretaton of these signals was in terms of the theoretical predictions of Dunham et al. (1971) who proposed that the hyperfine shifts of the cysteinyl $^1H^\beta$ resonances are larger

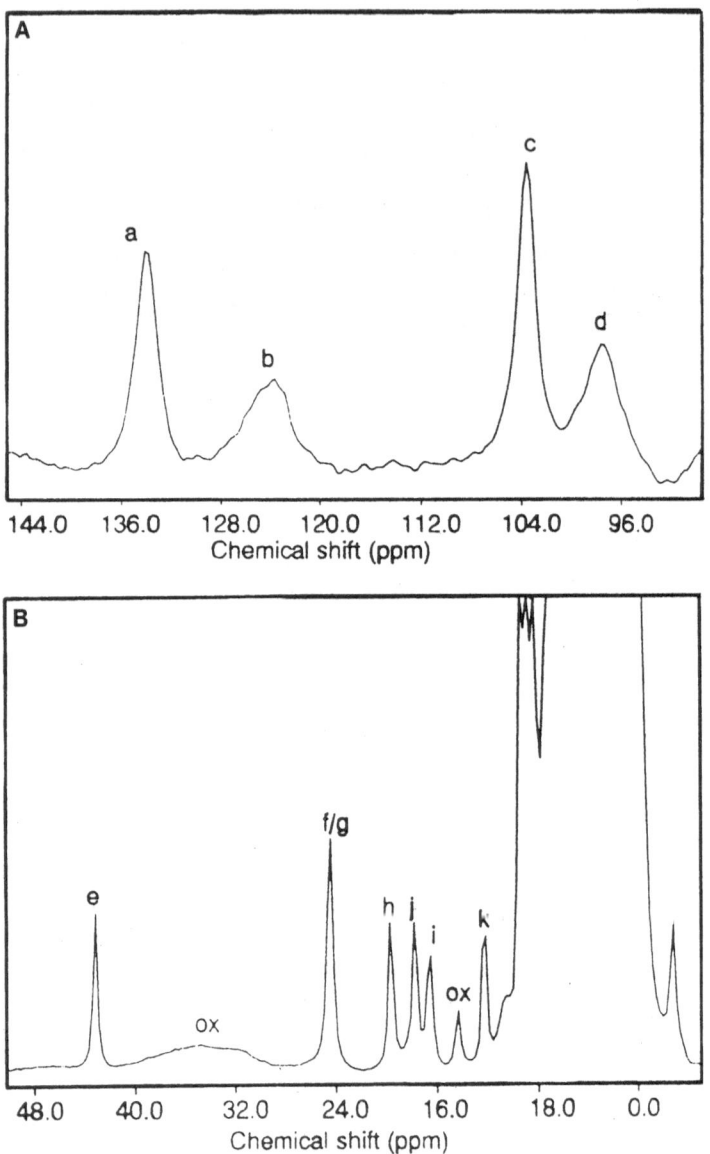

Figure 1. Downfield region of the one-dimensional ^1H NMR spectra of reduced *Anabaena* 7120 vegetative ferredoxin (reproduced with permission from Skjedal et al., 1991a). (A) Four peaks shifted farthest downfield; (B) remainder of the spectrum.

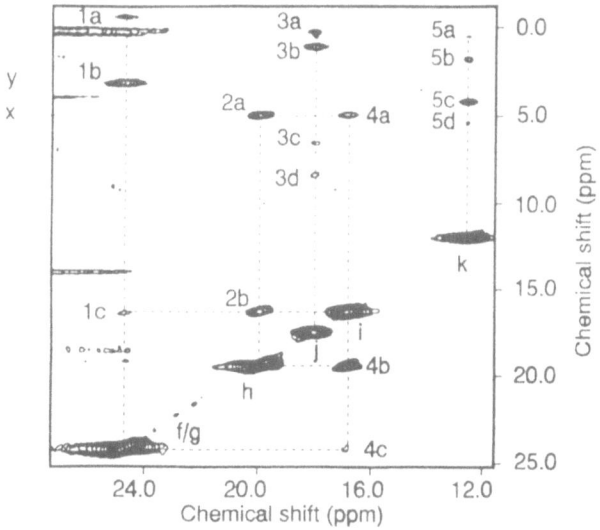

Figure 2. Two-dimensional NOESY spectrum of reduced *Anabaena* 7120 vegetative ferredoxin. Only the region with cross peaks between hyperfine peaks and diamagnetic peaks or other hyperfine peaks is shown (reproduced with permission from Skjedal et al., 1991a).

than those of the $^1H^\alpha$ of cysteines and that signals from cysteines ligated to Fe(III) have Curie-type temperature dependence and that those from cysteines ligated to Fe(II) in the reduced ferredoxin have anti-Curie-type temperature dependence. 1D NOE data (not shown) were collected in analogy with experiments of Dugad et al. (1990) on related plant-type ferredoxins (Skjeldal et al., 1991a). In addition, a 2D NOESY data sets (example shown in Figure 2) were collected with very short mixing times (0.5 to 20 ms), and cross peaks were detected between pairs of hyperfine-shifted resonances or between hyperfine-shifted resonances and diamagnetic resonances. Sequence-specific assignments of some of the paramagnetic resonances were made on the basis of these results in comparison with the x-ray structure of oxidized vegetative ferredoxin (Skjeldal et al., 1991a). Although the NMR data for reduced Vfd could be interpreted in terms of the x-ray structure of oxidized VFd, one would prefer to have an assignment assignment strategy that is independent of the crystal structure and that does not require the untested assumption that the structure of the region around the iron-sulfur cluster is the same in the oxidized and reduced forms of the protein.

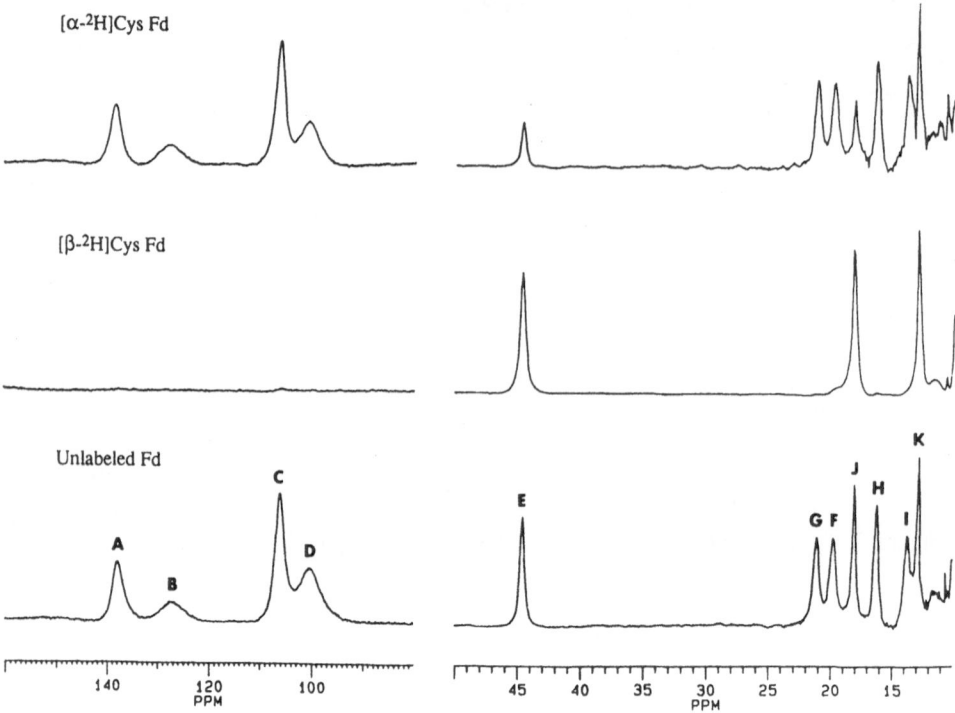

Figure 3. ^{1}H NMR spectra of reduced *Anabaena* 7120 vegetative ferredoxin (reproduced with permission from Cheng et al., 1995). (Top): labeled with [^{2}H$^{\alpha}$]Cys; (middle) labeled with [^{2}H$^{\beta 2\beta 3}$]Cys; (bottom) at natural abundance.

An alternative way of assigning hyperfine-shifted resonances is by selective isotopic labeling; if the labeling pattern is unique, this approach can provide full, unambiguous assignments. In practice, specific labeling by atom type within a given kind of residue is much easier to achieve, generally by biosynthesis, than atom-specific labeling, which may require chemical synthesis of the protein. Biosynthetic incorporation of [^{2}H$^{\alpha}$]Cys and [^{2}H$^{\beta 2\beta 3}$]Cys into VFd, in conjunction with ^{1}H and ^{2}H NMR analysis, was used to test the validiy of the assignments derived from NOE measurements (Cheng et al., 1995). Figure 3 shows 1D ^{1}H NMR spectra of unlabeled, [^{2}H$^{\beta}$]Cys labeled, and [^{2}H$^{\alpha}$]Cys labeled samples of reduced VFd. These spectra, in conjunction with ^{2}H NMR spectra of the same selectively labeled samples, allow the classification of peaks (A-D, F-I) to cysteinyl β-protons and peaks e and j to cysteinyl α-protons.

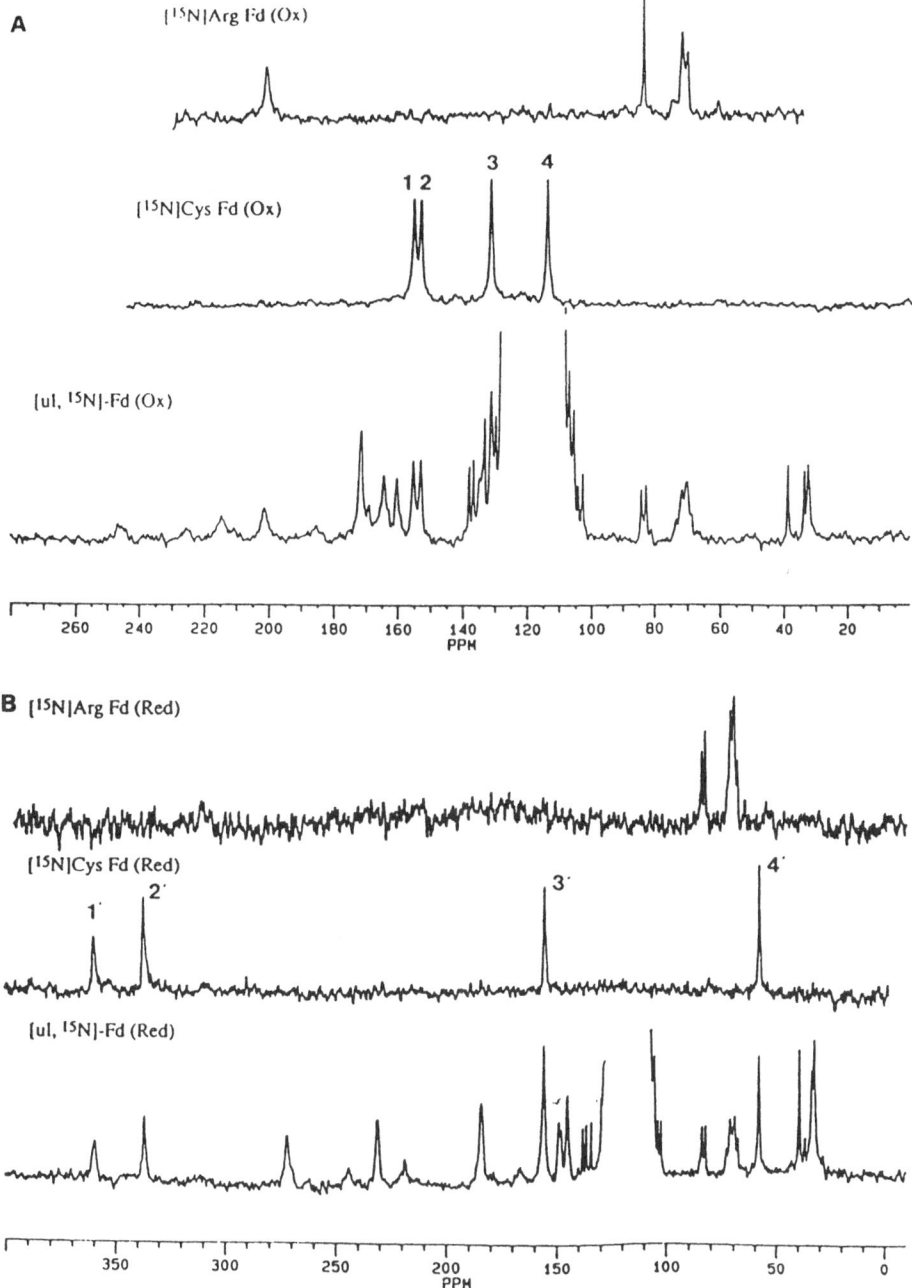

Figure 4. ^{15}N NMR spectra of *Anabaena* 7120 vegetative ferredoxin labeled uniformly with ^{15}N and selectively with [^{15}N]Cys or [^{15}N]Arg ferredoxin (reproduced with permission from Cheng et al., 1995): (A) in the oxidized state; (B) in the reduced state.

The assignment of peak "K" to $^1H^\alpha$ of Arg42 (Skjeldal et al., 1991a) on the basis of NOE results subsequently was shown to be incorrect on the basis of selective isotopic labeling: peak "K" was present at full intensity in the 1D ^1H NMR spectrum of reduced [$^2H^{\alpha\beta2\beta3}$]Arg-VFd; in addition, ^2H NMR spectra of [$^2H^{\alpha\beta2\beta3}$]Arg-VFd showed a broad peak at ~ 2 ppm (Cheng et al., 1995).

[^{15}N-Cys] VFd was prepared, and 1D ^{15}N NMR spectra were collected for this sample in both oxidation states (Figure 4). Four peaks were observed in both oxidized and reduced spectra. The observed cysteinyl ^{15}N hyperfine shifts were larger and more

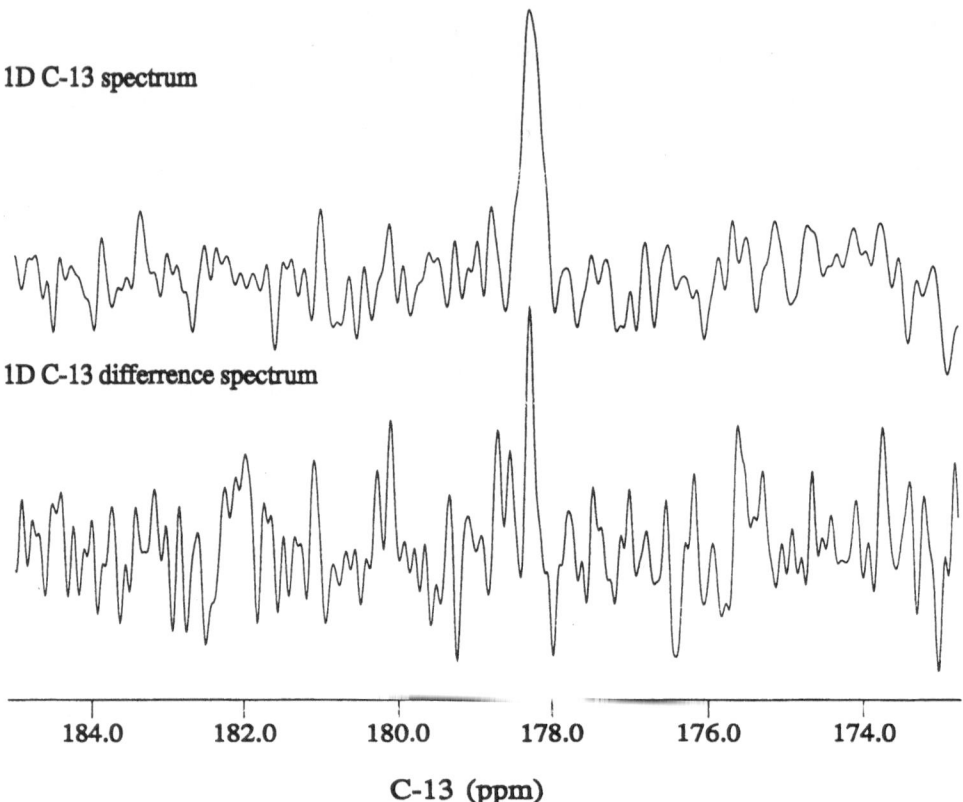

1D C-13 spectrum

1D C-13 differrence spectrum

| 184.0 | 182.0 | 180.0 | 178.0 | 176.0 | 174.0 |

C-13 (ppm)

Figure 5: One-dimensional ^{13}C NMR spectra of oxidized *Anabaena* 7120 vegetative ferredoxin labeled selectively with [^{15}N]Cys and [^{13}C]Ala (reproduced with permission from Cheng et al., 1995): (top) spectrum collected with a rapidly cycled, one-pulse experiment; (bottom) difference-decoupled spectrum with on-resonance irradiation at the ^{15}N frequency of Cys46.

diverse in spectra of reduced Vfd than those of oxidized VFd. In order to make a sequence-specific assignment for one of the cysteinyl ^{15}N signals, a double-labeled sample of Vfd was prepared by incorporating [^{13}C']Ala and [^{15}N]Cys. A ^{13}C{^{15}N} difference-decoupled spectrum of the oxidized form of this sample yielded a carbonyl difference peak at 178.3 ppm (Figure 5) when ^{15}N peak 3 (Figure 4) was irradiated. Because the only Ala-Cys dipeptide in VFd contains Cys46, peak 3 was therefore assigned to this residue. With this assignment, it was reasonable to assign peak 4, whose chemical shift and temperature dependence resembles that of peak 3, to Cys41, which is ligated to the same Fe as Cys46 (Cheng et al., 1995).

According to the crystal structure of Vfd (Rypniewski et al., 1991), the backbone amide proton of Arg42 is hydrogen bonded to a sulfide of the iron-sulfur cluster. The one-dimensional ^{15}N NMR spectrum of oxidized [U-^{15}N]Arg-VFd showed a broad hyperfine-shifted signal at 201.6 ppm (Figure 4), at a hyperfine shift larger than those of the cysteinyl ^{15}N signals from ligand cysteines. This peak, which was assigned to backbone amide nitrogen of Arg42, was not seen in spectra of the reduced [U-^{15}N]Arg-VFd (Figure 4); presumably because it is even broader and/or shifted much farther downfield. These results indicate that unpaired electron density is delocalized into Arg42 and that the hydrogen bond between the backbone amide of Arg42 and a sulfide of the cluster observedin the x-ray structure of the oxidized Vfd is present in both the oxidized and reduced states of the protein in solution (Cheng et al., 1995).

HUMAN FERREDOXIN

Human ferredoxin (HuFd) is a 2Fe-2S ferredoxin found in the mitochondria of human placenta. HuFd is involved in the initial step of progesterone biosynthesis, in which the side-chain of cholesterol is cleaved to yield pregnenolone. It functions to transfer reducing equivalents from a NADPH-dependent ferredoxin oxidoreductase to cytochrome P450scc (Mason et al., 1971; Simpson et al., 1978).

^{1}H NMR studies of hyperfine-shifted resonances of human ferredoxin and other vertebrate ferredoxins suggest that the pattern of electron delocalization in the vertebrate ferredoxin is quite different from that in plant-type ferredoxins (Skjeldal, et al., 1991b). Figure 6 shows the temperature dependence of hyperfine ^{1}H resonances of oxidized and

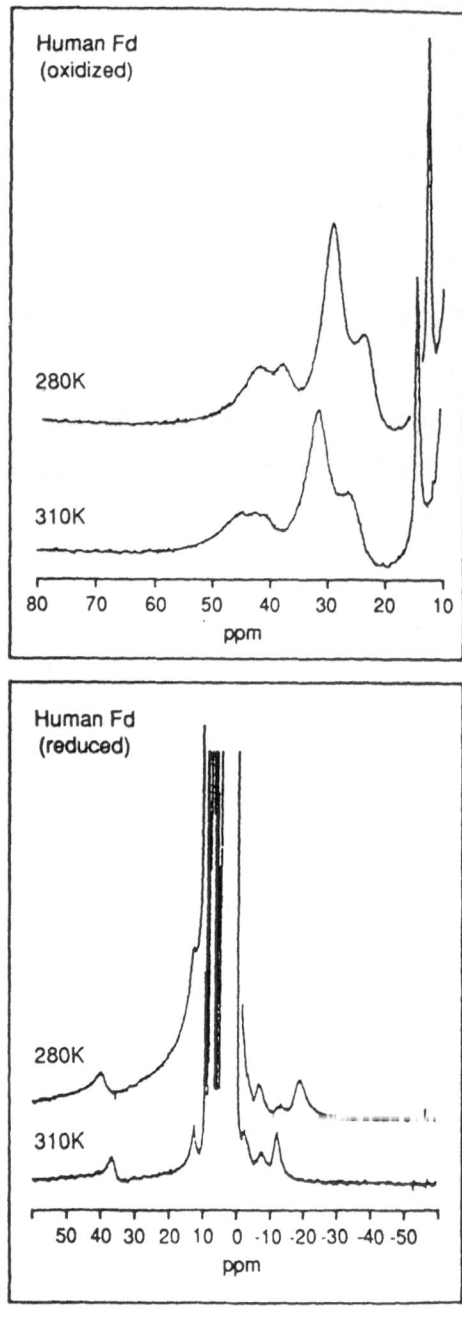

Figure 6. Temperature dependence of ^1H hyperfine-shifted resonaces of human ferredoxin in its oxidized (upper panel) and reduced (lower panel) states (reproduced with permission from Skjedal et al., 1991b).

reduced HuFd. The hyperfine signals from oxidized HuFd resemble those of the plant-type [2Fe-2S] ferredoxins. The signal at 13 ppm corresponds to a single proton intensity, whereas the signals between 23 and 34 ppm appear to arise from overlap of seven to eight proton resonances. Two broad signals were observed at ~ 40 ppm. All these signals showed anti-Curie temperature dependence. A total of five hyperfine-shifted ^{1}H peaks was detected in spectra of the reduced protein. Hyperfine signals from four protons occur upfield, at -6, -12, and -18 (two proton intensity) ppm, and two signals appear downfield at 13 and 40 ppm. All of these hyperfine-shifted peaks have Curie-type temperature dependence. The theoretical model for reduced vertebrate ferredoxins predicts that signals from the $^1H^\beta$ nuclei of cysteines ligated to Fe(II) will occur around -40 ppm and those of cysteines ligated to Fe(III) will occur near 110 ppm (Dunham et al., 1971). It is obvious that the experimental data do not fit this theoretical prediction very well, and suggests that the model needs to be improved.

The redox potential of the vertebrate ferredoxin is pH dependent with a pK$_a$ of 7.2 (Cooper et al., 1973). It was proposed that this pH depencence arises from the interaction of His56 with the iron-sulfur cluster, and that a hydrogen bond is formed between an imidazole NH and a labile sulfide of the cluster (Lambeth et al., 1982; Usanov et al., 1990). Two previous studies of the histidines of bovine ferredoxin reached different

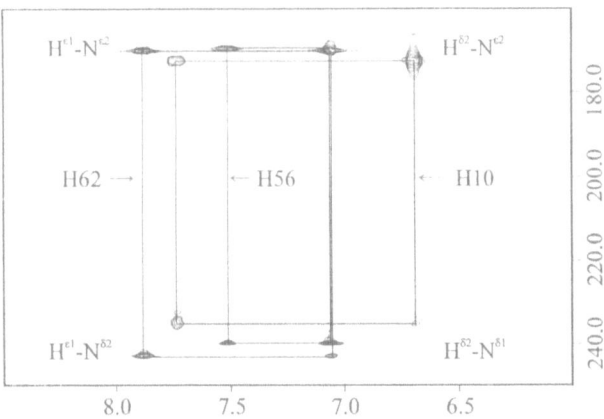

Figure 7: Two-dimensional ^{1}H{^{15}N}MBC spectrum of oxidized uniformly-labeled ^{15}N human ferredoxin. The data were recorded selectively for the His region in the ^{15}N dimension (reproduced with permission from Xia et al., 1995a).

assignments for His56 (Greenfield et al., 1989; Miura & Ichikawa et al., 1991; Murra et al., 1991). One of these studies suggested that signals from $^1H^{\varepsilon 1}$ and $^1H^{\delta 2}$ of His56 may be broadened due to the paramagnetism of the iron-sulfur cluster (Greenfield et al., 1989). In order to assign the histidines unambiguously, the human ferredoxin was labeled uniformly with ^{15}N and selectively with [26% U-^{13}C]histidine (Xia et al., 1995a). The HMBC spectrum of oxidized [U-^{15}N]HuFd clearly revealed the spin systems of each of the histidyl side chains (Figure 7). Also, 1D ^{13}C-edited ^1H NMR spectra of [26% U-^{13}C]His HuFd obtained as a function of pH showed that the ^1H resonances from the His56 side chain do not titrate between pH 6.0 and 8.0; this result suggests that the pKa of His56 is below 5 (Xia et al., 1995a). Moreover, the lack of hyperfine interactions with the cluster, as evidenced by the lack of temperature dependence of its chemical shifts, indicates that His56 does not interact with the iron-sulfur cluster directly. These results demonstrate that His56 is not responsible for the pH dependence of the reduction potential of the cluster.

HETEROCYST FERREDOXIN

The *Anabaena 7120* heterocyst ferredoxin (HFd) is involved in electron transport in nitrogen fixation. It donates one electron to component II (Fe protein) of nitrogenase, which in turn reduces component I (MO-Fe protein) of nitrogenase, which finally reduces N_2 to NH_3 (Schrautemeier and Böhme, 1985; Böhme et al., 1987). Hfd has 98 amino acid residues whose sequence shares a 51% identity with that of the *Anabaena 7120* vegetative ferredoxin (VFd). The crystal structure of oxidized HFd has been solved and refined to a 1.7 Å resolution (Jacobson et al., 1993). Most of the diamagnetic ^1H, ^{15}N, and ^{13}C resonances of the oxidized protein have been assigned, and its secondary structure has been determined (Chae et al., 1994).

Recently, some effort has been made to correlate T_1 relaxation rates of ^{15}N hyperfine-shifted resonances with the distances of these nitrogens from the iron-sulfur cluster (Chae & Markley, 1995). HFd was labeled uniformly with ^{15}N and selectively with some ^{15}N labeled amino acids, and the T_1 relaxation rates of the hyperfine-shifted nitrogen resonances have been determined (Figure 8). The x-ray crystal structure of Hfd shows that the amide nitrogens of five serines, four cysteines, two leucines, two valines, two

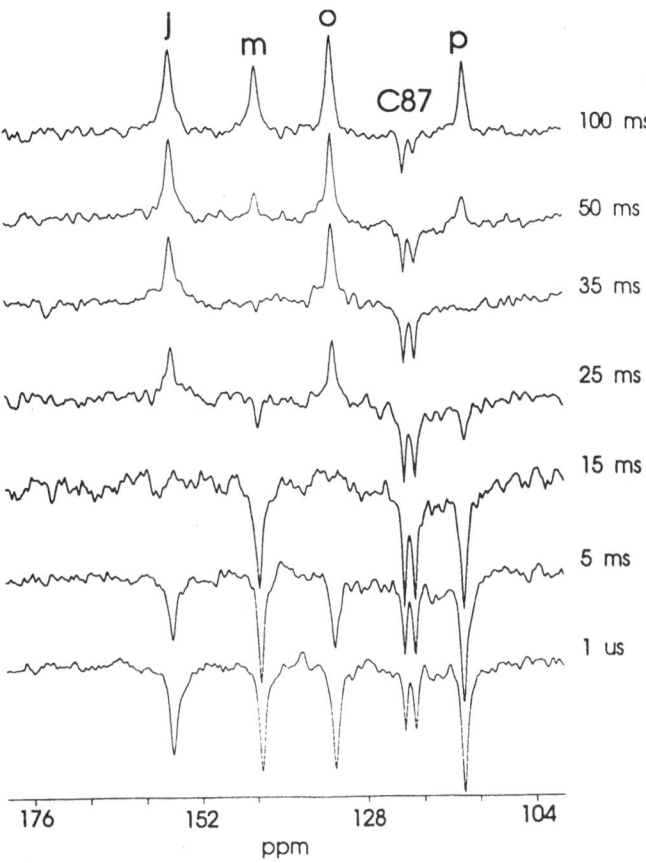

Figure 8: Partially relaxed ^{15}N spectra of oxidized [^{15}N]Cys labeled *Anabaena* 7120 heterocyst ferredoxin obtained with the 180-τ-90 pulse sequence (reproduced with permission from Chae & Markley, 1995). The τ values are given at the end of each spectrum.

phenyalanines, one histidine, and one glycine are within 8.5 Å of both irons in the cluster; it was expected that resonances from these atoms would be affected by the paramagnetic center. HFd samples were labeled selectively with [^{15}N]serine, [^{15}N]cysteine, [^{15}N]leucine and [^{15}N]glycine. Although amino acid auxotrophic strains were not used in the biosynthetic labeling, the only appreciable migration of the label to other amino acid types occurred with added [^{15}N]glycine which was converted to [^{15}N]serine under certain conditions (Chae et al., 1995). One-dimensional ^{15}N NMR spectra of these selectively-labeled samples in their oxidized state have led to the classification by amino acid type of mnay of the ^{15}N hyperfine peaks observed in spectra of [U-^{15}N]HFd. The T_1 relaxation times of these ^{15}N hyperfine-shifted resonances have been measured and compared with their reduced distances d_r (Oh & Markley, 1990b) to the two iron atoms, where

$$d_r = \left[d_1^6 d_2^6 \big/ \left(d_1^6 + d_2^6 \right) \right]^{1/6},$$

and d_1 and d_2 are the distances of a nitrogen from each of the two irons as determined from the x-ray crystal structure (Jacobson et al., 1993).. The nitrogen resonances were tentatively assigned by ordered pairing of the reduced distances of nitrogens and the nitrogen T_1 (Chae & Markley, 1995).

Paramagnetic contributions to T_1 relaxation occur by two major mechanisms (Bertini & Luchinat, 1986; Bertini et al., 1993): contact (through bonds) and dipolar (through space). For nuclei directly bonded to the cluster, the contact contribution may be dominant. For atoms belonging to nonbonded residues, the relaxation can be due only to dipolar mechanism (Bertini et al., 1993). To a first approximation, the dipolar contribution to the T_1 relaxation is proportional to the sixth power of the reduced distance. Thus, the T_1 data obtained for HFd were evaluated by fitting to the equation:

$$1/T_1 \times 10^2 = a / d_r^6 \times 10^4 + b$$

where a and b are adjustable parameters. As shown in Figure 9, data from serine and leucine fit this model precisely, which suggests that these nitrogens are relaxed predominantly by dipolar interaction with the paramagnetic cluster. Remarkably close correlations have been found between distances calculated from ^{15}N T_1 values and those derived from the refined x-ray crystal structure of HFd (Jacobson et al., 1993). However, the points for the nitrogens of one Gly and all four Cys lie well off the fitted curve for the serines and leucine data (Figure 9). These correspond to the nitrogens that are closest the iron atoms, yet they are not relaxed as rapidly, in terms of the expected reciprocal sixth-power dependence on distances, as nitrogens that are farther away. This lack of good agreement for nitrogens at shorter distances from the cluster (within 4.2 Å of either of the iron atoms) probably results either from effects of electron delocalization from the iron atoms onto these atoms or from the breakdown at short distances of the dipolar approximation used in the analysis (Chae & Markley, 1995).

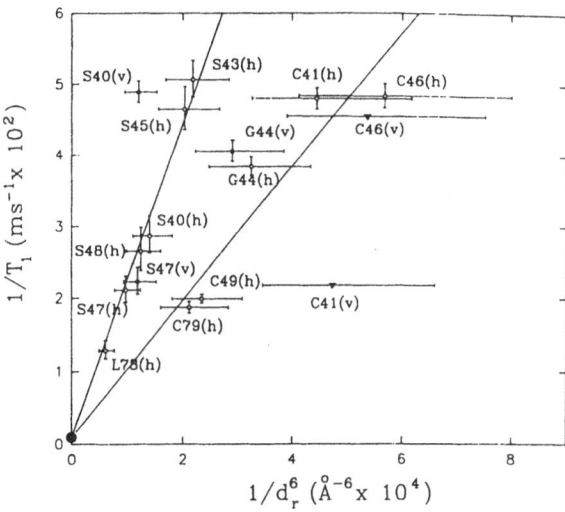

Figure 9: Correlation between the inversed sixth power of the reduced distance of a nitrogen to the [2Fe-2S] cluster and its longitudinal relaxation rate for various residues of oxidized *Anabaena* 7120 heterocyst ferredoxin (reproduced with permission from Chae & Markley, 1995).

The results from this study suggest that the T_1 relaxation for nitrogens that are 4.2 Å or more distant from one or both iron atoms and up to about 8.5 Å from both irons are dominanated by the dipolar mechanism and that relatively accurate distances between these atoms and the metals of the cluster can be calculataed from measured T_1 relaxation times. Thus far these ideas have only been evaluated for two plant-type ferredoxins. It will be interesting to see if this relationship holds with ^{15}N relaxation results from other paramagnetic proteins.

RUBREDOXIN

Rubredoxins belong to the simplest class of iron-sulfur proteins. They contain a single iron coordinated by the sulfurs of four cysteine residues. Rubredoxin was the first non-heme iron protein studied by NMR spectroscopy (Phillips et al., 1970), and it has been the subject of several subsequent NMR investigations (Krishnamoorthi et al., 1986;

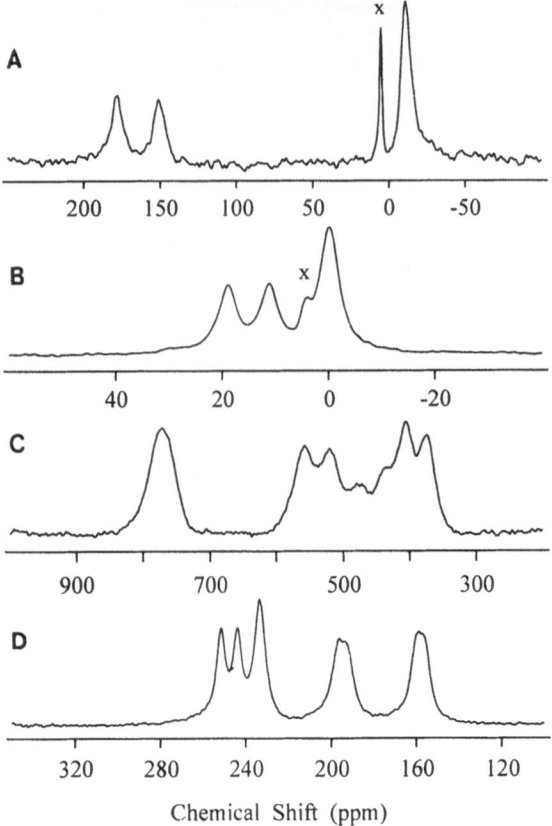

Figure 10: ^{2}H NMR spectra of ^{2}H-cysteine labeled rubredoxin (reproduced with permission from Xia et al., 1995b). (A) oxidized [^{2}H$^\alpha$]Cys labeled rubredoxin; (B) reduced [^{2}H$^\alpha$]Cys labeled rubredoxin; (C) oxidized [^{2}H$^{\beta2,\beta3}$]Cys labeled rubredoxin; (D) reduced [2H$^{\beta2,\beta3}$]Cys labeled rubredoxin. The peak labeled "x" arises from residual ^{1}H^{2}HO.

Blake et al., 1991 & 1992). Recently, the rubredoxin from *Clostridium pateurianum* was cloned and overexpressed in *E. coli* (Mathieu et al., 1992). This has made it possible to isotopically label this protein uniformly and selectively for NMR investigations (Xia et al., 1995b).

Two Rdx samples were labeled selectively by biosynthetic incorporation of [^{2}H$^\alpha$]Cys and [^{2}H$^\beta$]Cys, respectively. One-dimensional ^{2}H NMR spectra of oxidized [^{2}H$^\alpha$-Cys]Rdx (Figure 10A) show that two of the four H$^\alpha$ signals appear downfield at 180 and

150 ppm whereas the other two overlap at -12 ppm in the upfield region. Upon reduction, they all shift toward the diamagnetic region; two signals are seen downfield at 19 and 16 ppm, respectively, and two overlapped peaks occur near 0 ppm (Figure 10B). One-dimensional ^{2}H NMR spectra of oxidized [^{2}H$^\beta$-Cys]Rdx reveal resonances from all eight cysteinyl H$^\beta$ atoms in the extreme downfield region between 900 ppm and 300 ppm (Figure 10C). In the reduced state, the eight H$^\beta$ signals appear in the 280 ppm to 150 ppm region (Figure 10D). All the H$^\alpha$ and H$^\beta$ resonances from cysteines exhibit Curie-type temperature dependence in both the oxidized and reduced states (Xia et al., 1995b).

Thirteen hyperfine-shifted resonances were observed in the ^{15}N NMR spectrum of [U-^{15}N]Rdx in both oxidation states (Fig. 11A & 11B, top spectra). Two of them appear in the upfield region and shift downfield with increasing temperature, whereas the remaining eleven peaks are located in the downfield region and shift upfield with increasing temperature. In order to identify the ^{15}N resonances from the cysteines, an Rdx sample was labeled selectively with [^{15}N]cysteine and studied by ^{15}N NMR at 35 °C (Figure 11). Two hyperfine-shifted peaks from [^{15}N]cysteine appeared in the upfield region (-27 ppm and -71 ppm in oxidized Rdx; 3.5 ppm and -21 ppm in reduced Rdx), and the other two appeared downfield (593 ppm and 570 ppm in oxidized Rdx; 310 ppm and 285 ppm in reduced Rdx).

Apart from the resonances from all eight H$^\beta$ atoms, which appear downfiled in spectra of Rdx in both oxidation states, the signals from all other cysteinyl atoms can be divided into two equal groups: those with upfield hyperfine shifts and those downfield hyperfine shifts. This pattern holds for the four H$^\alpha$ signals, four ^{15}N signals, and four ^{13}C$^\beta$ signals (data not shown); for each group of four, two signals appear downfield whereas the other two lie upfield in both oxidation states. A likely explanation is that dipolar interactions make a significant contribution to the hyperfine shift. The pseudo-contact (dipolar) shift depends upon, among other parameters, the anisotropy of the g-tensor and

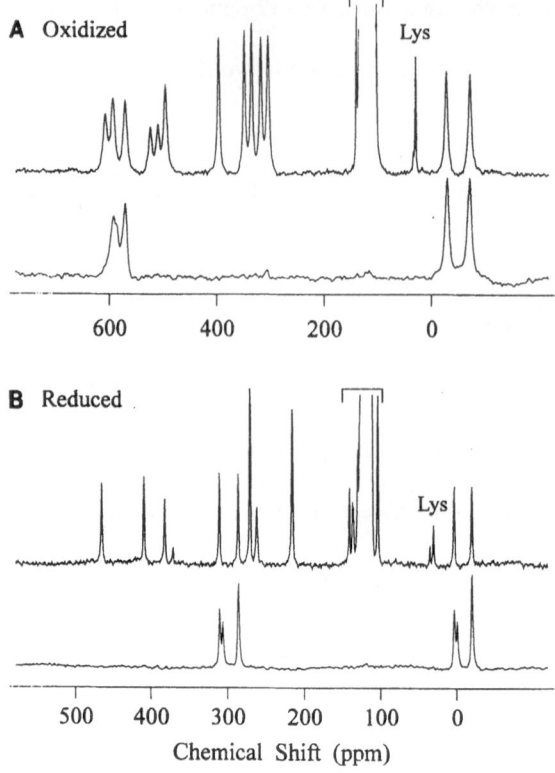

Figure 11: ^{15}N NMR spectra of [UL, ^{15}N] labeled rubredoxin (top) and [^{15}N]Cys labeled rubredoxin (bottom): (A) in the oxidized state; (B) in the reduced state. The brackets enclose the diamagnetic resonances (reproduced with permission from Xia et al., 1995b)

the orientation of the observed nuclei with respect to the paramagnetic center (McConnell & Robertson, 1958; Kurland & McGarvey, 1970). EPR spectroscopy has indeed revealed a highly anisotropic g-tensor for rubredoxin (Lode et al., 1971). Furthermore, the four cysteines of Rdx can be classified into two groups according to the pattern of their hydrogen bonding and the conformation of their side-chains in the x-ray crystal structure (Watenpaugh et al, 1979): Cys6 and Cys39 in one category, and Cys9 and Cys42 in the other. The amide protons of cysteines 9 and 42 are hydrogen-bonded to the sulfur atoms of cysteines 6 and 39, respectively. This structural symmetry probably gives rise to the pattern of observed positive and negative hyperfine shifts. In addition, the signs of the electron-nuclear hyperfine coupling constants may be opposite for the two groups of cysteines.

Although Rdx has the simplest iron ligation pattern of all non-heme iron proteins, its NMR hyperfine-shifted resonances have remained the most poorly characterized. Both oxidation states of Rdx have high electron spin numbers, and leads to extreme line broadening and large isotropic shifts which have made it difficult to detect and categorize its hyperfine-shifted resonances. The recent studies of Xia and coworkers (1995) indicate that selective labeling with low magnetogyric ration nuclei (^{2}H and ^{15}N) provides an effective way to overcome this problem. The next stage in this analysis will be the determination of sequence-specific assignments; these may require the chemical synthesis of Rdx samples with labels at single residue positions.

SUMMARY

Iron-sulfur proteins comprise an important group of proteins. The paramagnetic centers of these proteins very often are involved in their function, as electron carriers, enzymes, or iron sensors within cells. It is thus of importance to understand their three-dimensional structures and, more particularly, the electronic structures of the metal centers. The inherent paramagnetism of iron-sulfur proteins, which contributes additional chemical shift and relaxation mechanisms to nuclear spins near the cluster, carry important information on electron delocalization and conformation in the vicinity of the iron-sulfur cluster. New methodology for labeling proteins with stable isotopes coupled with advances in the collection of NMR data from paramagnetic centers, is providing a wealth of new knowledge about paramagnetic centers in proteins. It is hoped that these data will stimulate continued development of the theoretical basis for structure-function relationships in iron-sulfur proteins. The increasing interest in NMR studies of iron-sulfur proteins forecasts continued rapid development of this exciting field.

REFERENCES

Bertini, I., & Luchinat, C., 1986, "NMR of Paramagnetic Molecules in Biological

  Systems," The Benjamin/Cummings Publishing Company, Inc., Menlo Park, CA.

Bertini, I., Turano, P., & Vila, A. J. 1993, *Chem. Rev.* 93:2833.

Blake, P. R., Park, J. B., Bryant, F. O., Aono, S., Magnuson, J. K., Eccleston, E., Howard, J. B., Summers, M. F., & Adams, M. W. W., 1991, *Biochemistry* 30:10885.

Blake, P. R., Park, J. B., Zhou, Z. H., Hare, D. R., Adams, M. W. W., & Summers, M. F., 1992, *Protein Sci.* 1:1508.

Böhme, H., & Schrautemeier, B., 1987, *Biochim. Biophys. Acta* 891:115.

Chae, Y. K., Abildgaard, F., Mooberry, E. S., & Markley, J. L., 1994, *Biochemistry* 33:3287.

Chae, Y. K., & Markley, J. L., 1994, *Biochemistry* 34:188.

Chan, T.-M., & Markley, J. L., 1983, *Biochemistry* 22:6008.

Cheng, H., Grohmann, K., & Sweeney, W., 1992, *J. Biol. Chem.* 267:8073.

Cheng, H., Westler, W. M., Xia, B., Oh, B.-H., & Markley, J. L., 1995, *Arch. Biochem. Biophys.* 316:619.

Cooper, D. Y., Schleyer, H., Levin, S. S., & Rosenthal, O., 1973, *Ann. NY Acad. Sci.* 212:227.

Dugad, L. B., La Mar, G. N., Banci, L., & Bertini, I., 1990, *Biochemistry* 29:2263.

Dunham, W. R., Palmer, G., Sands, R. H., & Bearden, A. J., 1971, *Biochim. Biophys. Acta* 253:373.

Greenfield, N. J., Wu, X., & Jordan, F., 1989, *Biochim. Biophys. Acta* 995:246.

Hentze, M.W., Argos, P., 1991, *Nucleic Acid Res.* 19:1739.

Holden, H. M., Jacobson, B. L., Hurley, J. K., Tollin, G., Oh, B.-H., Skjeldal, L., Chae, Y. K., Cheng, H., Xia, B., & Markley, J. L., 1993, *J. Bioenerg. Biomemb.* 26:67.

Jacobson, B. L., Chae, Y. K., Markley, J. L., Rayment, I., & Holden, H. M., 1993, *Biochemistry* 32:6788.

Krishnamoorthi, R., Markley, J. L., Cusanovich, M. A., & Przysieki, C. T., 1986, *Biochemistry* 25:50.

Kurland, R. J., & McGarvey, B. R., 1970, *J. Mag. Res.* 2:286.

Lambeth, J. D., Saybert, D. W., Lancaster, Jr. J. R., Salerno, J. C., & Kamin, H., 1982, *Mol. Cell Biochem.* 45:13.

Masaki, R., Yoshikawa, S., & Matsubara, H., 1982, *Biochim. Biophys. Acta* 700:101.

Mason, J. I., & Boyd, G. S., 1971, *Eur. J. Biochem.* 21:308.

Mathieu, I., Meyer, J., & Moulis, J.-M. *Biochem. J.* **1992**, 285, 255.

McConnell, H. M., & Robertson, R. E., 1958, *J. Chem. Phys.* 27:1361.

Miura, S., & Ichikawa, Y., 1991, *J. Biol. Chem.* 266:6252.

Miura, S., Tamita, S., & Ichikawa, Y.,1991, *J. Biol. Chem.* 266:19212.

Oh, B.-H., & Markley, J. L., 1990a, *Biochemistry* 29:3993.

Oh, B.-H., & Markley, J. L., 1990b, *Biochemistry* 29:4012.

Oh, B.-H. Mooberry, E. S., & Markley, J. L., 1990, *Biochemistry* 29:4004.

Packer, E. L., Sweeney, W. V., Rabinowitz, J. C., Sternlicht, H., & Shaw, E. N., 1977, *J. Biol. Chem.* 252:2245.

Phillips, W. D., Poe, M., Weiher, J. F., McDonald, C. C., & Lovenberg, W., 1970, *Nature (London)* 227:574.

Poe, M., Phillips, W. D., Glickson, J. D., & San Pietro, A., 1971, *Proc. Natl. Acad. Sci. USA* 68:68.

Roualt, W. J., Stout, C. D., Kaptain, S., Harford, J. B., & Klausner, R. D., 1991, *Cell* 64:881.

Rypniewski, W. R., Breiter, D. R., Benning, M. M., Wesenberg, G., Oh, B.-H., Markley, J. L., Rayment, I., & Holden, H. M., 1991, *Biochemistry* 30:4126.

Schrautemeier, B., & Böhme, H., 1985, *FEBS Lett.* 184:304.

Simpson, E. R., & Miller, D. A., 1978, *Arch. Biochem. Biophys.* 190:800.

Skjeldal, L., Westler, W. M., & Markley, J. L., 1990, *Arch. Biochem. Biophys.* 278:482.

Skjeldal, L., Westler, W. M., Oh, B.-H., Krezel, A. M., Holden, H. M., Jacobson, B. L., Rayment, I., & Markley, J. L., 1991a, *Biochemistry* 30:7363.

Skjeldal, L., Markley, J. L., Coghlan, V. M., & Vickery, L. E., 1991b, *Biochemistry* 30:9078.

Usanov, S. A., Chashchin, V. L., & Akhrem, A. A., 1990, *Frontiers in Biotransformation* 3:1.

Watenpaugh, K. D., Sieker, L. C., & Jensen, L. H., 1979, *J. Mol. Biol.* 131:509.

Xia, B., Cheng, H., Skjeldal, L., Coghlan, V. M., Vickery, L. E., & Markley, J. L., 1995a, *Biochemistry* 34:180.

Xia, B., Westler, W. M., Cheng, H., Meyer, J., Moulis, J.-M., & Markley, J. L., 1995b, *J. Am. Chem. Soc.* (in press).

ON THE USE OF NMR IN COMPLEX BIOLOGICAL SYSTEMS: NMR STUDIES OF CALCIUM SENSITIVE INTERACTIONS AMONGST MUSCLE PROTEINS

Brian D. Sykes,[1] Carolyn M. Slupsky,[1] David S. Wishart,[2] Frank D. Sönnichsen,[2] and Stéphane M. Gagné[1]

[1]MRC Group in Protein Structure and Function
[2]Protein Engineering Network of Centres of Excellence (PENCE)
Department of Biochemistry
University of Alberta
Edmonton, Alberta T6G 2H7
Canada

INTRODUCTION

The development of the biological NMR field over the last decades have been very rapid and widespread. This has included changes in hardware, software, techniques and applications. In terms of macromolecular applications, it is now possible to determine the complete three dimensional structure of a macromolecule in solution, and to characterize its dynamics both in terms of internal motions and the kinetics of conformational changes and interactions with ligands or other macromolecules. The macromolecules studied include mostly proteins, DNA and their complexes, with increasing applications to other macromolecules such as complex carbohydrates and RNA. For the rest of this article we will refer to proteins only for the sake of simplicity.

As the structural applications of NMR become more widely accepted in the larger biological research community, possibly the most often addressed question relates to how large a protein can one study by these methods. The concomitant question addressed within the NMR community is whether or not the technological developments have reached a plateau. We will not address the latter question, having seen many apparent plateaus disappear in the past to yet another phase of unexpectedly rapid development. The answer to the first question is generally that one can study proteins in the molecular weight range of up to 20,000 - 30,000 daltons. Unstated in this answer is the fact that often protein NMR spectra can be assigned, at least in part, for proteins near 30,000 daltons but complete structure determination is much more difficult at high molecular weights for a variety of reasons. Secondly, as is common for most biophysical applications including NMR, the class of proteins studied is normally the stable, soluble, globular proteins including most serum proteins and enzymes.

NMR As a Structural Tool for Macromolecules: Current Status and Future Directions
Edited by B.D. Nageswara Rao and Marvin D. Kemple, Plenum Press, New York, 1996

The focus of this article is on more complex biological systems. Many biological systems, such as muscle, involve complex arrays of many large and small proteins in a form not so well suited for biophysical studies, especially by NMR. What can be done? How far can one go with NMR? The message we wish to present is that one can obtain biologically significant structural results in these more complex systems but that one has to be inventive in technique, be willing to handle the biological system under a variety of conditions, often be willing to accept partial information, and always be prepared to use as many other biological and biophysical tools as possible in conjunction with NMR.

Many approaches have been used to study biological systems. Considering the various levels of biological complexity, the approaches include working with peptide fragments derived from chemical synthesis or enzymatic fragmentation; domains of proteins produced by enzymatic fragmentation or molecular biological techniques; individual proteins often cloned and expressed outside of the original system; binary and ternary complexes of peptides, domains, and proteins; and the intact system involved. From the technique side, a wide variety of methods have been developed including the transferred NOE to study peptide-protein complexes, multinuclear, multidimensional NMR (specifically including isotope filtered NMR to focus on various components in large complexes), and solid-state NMR techniques. The latter area is most possibly where the unexpected developments will occur in the future.

When one uses any of the above approaches to dissect a complex biological system, there are many questions that are posed and many problems that are encountered. Some of the questions that must be asked are: do the peptides bind in the correct place in the target protein? do the domains represent the intact proteins? do peptide structures provide meaningful information? do the proteins aggregate? and what is the biological relevance of studying isolated proteins from a complex biological system such as muscle? There is no one way to answer all of these questions. One approach, however, is to use as many other approaches as possible. These might include biological assays, other biophysical studies such as cross-linking, binding studies, and other biophysical techniques such as ultracentrifugation studies and spectroscopic methods such as circular dichroism and fluorescence spectroscopy.

Clearly the focus on the biological system brings special problems. One would like to have proteins which are globular, stable, soluble, non-aggregating, and with a variety of secondary structures involved. However, proteins taken from a system such as muscle or a membrane system often are extended, unstable, insoluble or extensively aggregated, and involve extensive regular secondary structure which can make things like NMR assignment more difficult.

Our research has focused largely on the regulation of muscle contraction in skeletal, cardiac, and smooth muscle. In particular, we have focused on the calcium binding proteins from muscle, and addressed the structural question of what conformational changes are induced in these proteins upon binding calcium, and how is this information passed on to other proteins in the muscle and how does it lead to regulation. For the purpose of this manuscript, the main focus will be the protein troponin-C from skeletal muscle. While troponin-C can be isolated from intact muscle by washing the muscle fiber in the presence of a calcium-chelator, and can therefore be studied in isolated form in solution, it does present many of the challenges encountered above including aggregation, extended structure, regular secondary structure, and biological relevance. We have therefore been involved in the development of a variety of NMR/biophysical approaches to study this type of system and will present three of them herein. The first involves novel approaches to the analysis of NOE data with particular focus on the delineation of secondary structures in large proteins. The second involves the use of chemical shift information to delineate secondary structures in proteins. The third involves the use of other solvents, specifically trifluoroethanol, as a denaturant of quaternary structure in aggregating systems. All three of these approaches are focused on aspects of defining structure in larger proteins. Each of these three approaches has been important in defining the calcium-induced structural transition in troponin-C.

RESULTS AND DISCUSSION

Analysis of ^{15}N-Edited NOESY Spectra Using an NOE Ratio

The primary source of structural information in NMR spectra is the distance restraint data derived from NOESY spectra, especially 3D experiments such as the ^{15}N NOESY edited. However, this information can become less easy to interpret and reliable to use as the protein becomes larger because of effects such as cross-relaxation or spin diffusion. We describe herein a useful method for the derivation of distance restraints from NOESY data with particular emphasis on the determination of the ψ angle which helps define the secondary structure of the peptide backbone.

NOE connectivities found in NOESY spectra usually contain, among others, the $d_{\alpha N}(i-1, i)$ and the $d_{N\alpha}(i, i)$ cross peaks. The $d_{N\alpha}(i, i)$ NOE is non-informative by itself, the intra-residue HN-Hα distance being covalently restricted to between 2.2 and 3.1Å, and being realistically found in the 2.7 to 3.1Å range for α-helices and β-sheets. For the $d_{\alpha N}(i-1, i)$ NOE, the allowed HN(i)-Hα(i-1) distance is found in a wider range (2.2 to 3.6Å), and the magnitude of the $d_{\alpha N}(i-1, i)$ NOE could be used to differentiate between a residue found in the right-handed α-helical region (3.1Å < HN(i)-Hα(i-1) < 3.6Å) and one in the β-sheet region (2.2Å < HN(i)-Hα(i-1) < 2.7Å) (see Figure 1). However, the inaccuracy of this measurement (mainly due to the variation of amide exchange rate and differential relaxation along the sequence) makes it a poor criterion for secondary structure identification. This inadequacy can be overcome by using the ratio of these two NOE's, $NOE_{N\alpha}/NOE_{\alpha N}$ even for quite large proteins. As shown in Figure 2, the $NOE_{N\alpha}/NOE_{\alpha N}$ ratio has a strong correlation with secondary structure. The β-sheet region is clearly characterized by a $NOE_{N\alpha}/NOE_{\alpha N} < 1$ whereas the right-handed α-helical region is localized in the $NOE_{N\alpha}/NOE_{\alpha N} > 1$ area.

The distribution of ϕ / ψ angles in proteins is well established (Morris et al., 1992). For this report, however, we were interested in the distribution of the $\phi(i) / \psi(i-1)$ angle distribution. In order to determine the distribution of $\phi(i) / \psi(i-1)$ and also establish the ranges for $d_{N\alpha}$ and $d_{\alpha N}$ distances, we used a representative subset of structures in the PDB-database [the search was restricted to non-homologous high-resolution NMR- and X-ray structures deposited before March 1994 (Hobohm et al., 1992)]. In all, 198 protein chains were extracted from the Brookhaven Data Base. 29263 residues leading to 25379 dipeptide segments containing neither Pro nor Gly residues were obtained and analysed for their $\phi(i) / \psi(i-1)$ angle distribution (Figure 2) and the $d_{N\alpha}$ and $d_{\alpha N}$ distance distributions (Figure 1). The extensive statistical data presented here demonstrated the applicability of the $NOE_{N\alpha}/NOE_{\alpha N}$ ratio. The determination of secondary structure using the $NOE_{N\alpha}/NOE_{\alpha N}$ ratio is easy and reliable.

The use of this ratio has been successfully applied in the secondary structure determination of the regulatory domain of troponin-C (Gagné et al., 1994; Gagné, 1994) and of whole troponin-C (Slupsky et al., 1995b). Figure 2 also shows that the $NOE_{N\alpha}/NOE_{\alpha N}$ ratio has potential in the estimation of ψ angles, even for unusual residues with positive ϕ angle. We are using ψ restraints in our structure calculations based on the $NOE_{N\alpha}/NOE_{\alpha N}$ ratio (Gagné et al., manuscript in preparation). Finally, NOE's in 3D ^{15}N-edited NOESY spectrum are usually difficult to calibrate. However, one can take

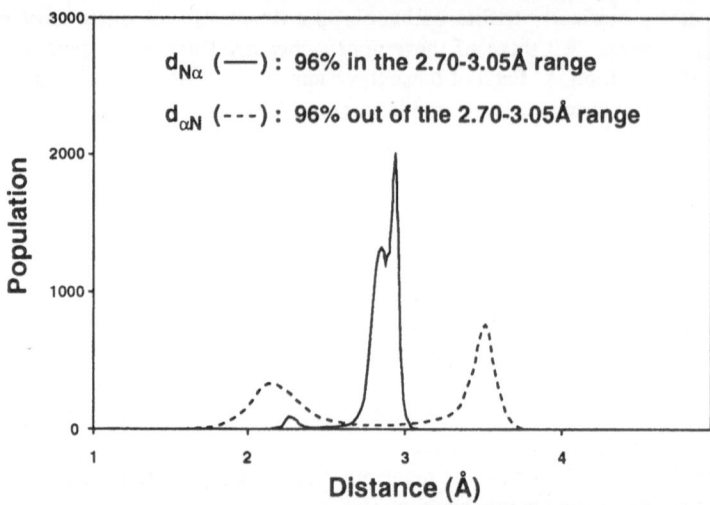

Figure 1. Distribution of the $d_{N\alpha}$ and $d_{\alpha N}$ distances in proteins. There is very little overlap between these two distances. Most of the $d_{N\alpha}$ are found between 2.70 and 3.05 Å (96%). 3% of the $d_{N\alpha}$ are between 2.20 and 2.50 Å, corresponding to residues having positive ϕ angle. The $d_{\alpha N}$'s are found between 1.80 and 2.70 Å (44%) mostly represent the residues in ß-strands, whereas most of the residues in α-helices have their $d_{\alpha N}$ between 3.05 and 3.70 Å (52%).

advantage of the narrow $d_{N\alpha}$ distribution (Figure 1) and calibrate the amide related NOE's as shown in Table 1. This calibration has also been successfully applied in the troponin-C case (Gagné et al., manuscript in preparation). Note that this calibration approach has some similarities with a method proposed by Wagner and coworkers (Hyberts et al., 1992).

Table 1. Calibration method for 3D ^{15}N-edited NOESY.

$$Upper\,bound\,1 \;=\; \sqrt[6]{\dfrac{NOE_{N\alpha} * (3.05)^6}{NOE_{Nx}}} + (correction)$$

$$Upper\,bound\,2 \;=\; \sqrt[6]{\dfrac{NOE_{\alpha N} * (3.70)^6}{NOE_{Nx}}} + (correction)$$

$$Lower\,bound\,1 \;=\; \sqrt[6]{\dfrac{NOE_{N\alpha} * (2.70)^6}{NOE_{Nx}}} - (correction)$$

NOE_{Nx} is the NOE to be calibrated.

Upper bound 2 is used when $NOE_{N\alpha}$ not observed or ambiguous.

Lower bound is used only for residues with negative ϕ angles.

Correction accounts for errors in the measurements.

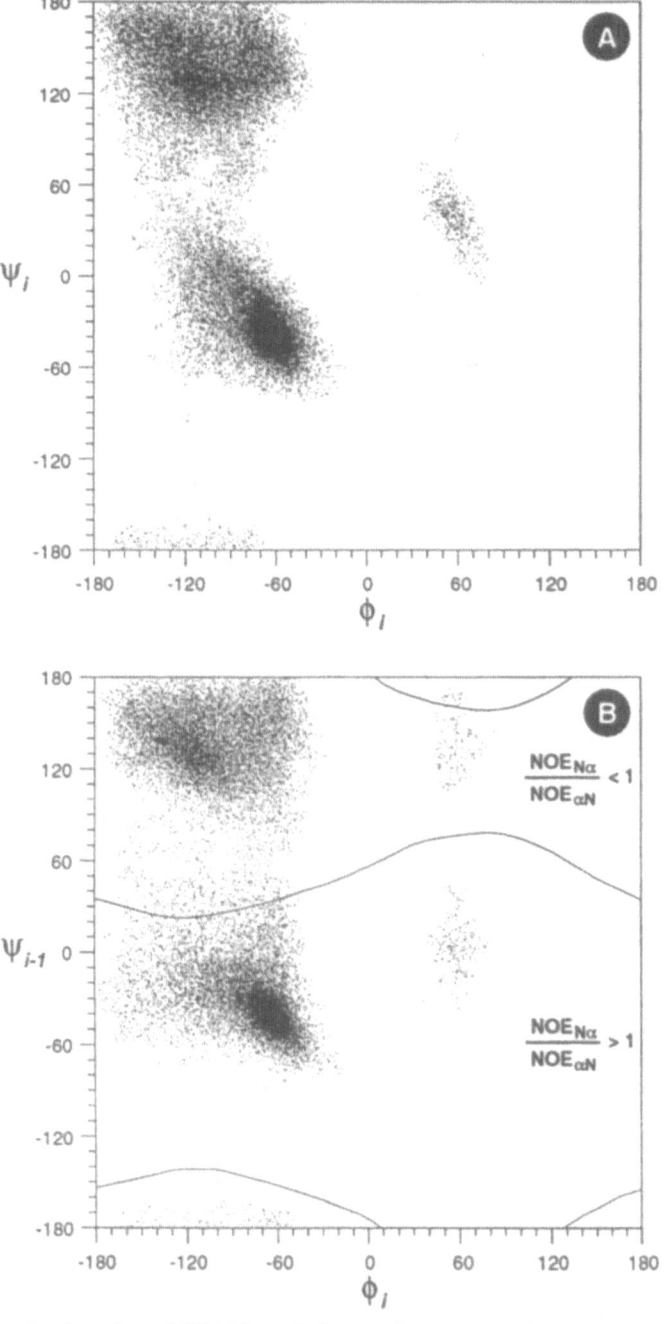

Figure 2. Ramachandran plots of 198 high resolution protein structures. A. Regular $\phi(i)$-$\psi(i)$ plot showing the distribution of backbone dihedral angles in proteins. B. $\phi(i)$-$\psi(i-1)$ plot correlating each ϕ angle with the ψ angle of the previous residue. The $\phi(i)$-$\psi(i-1)$ plot is more relevant to the present discussion than the regular $\phi(i)$-$\psi(i)$ plot. The $\phi(i)$-$\psi(i-1)$ plot is very similar to the $\phi(i)$-$\psi(i)$ plot, except for the positive ϕ half of the diagram. It seems as if residues with positive ϕ angle follow residues with normal (negative) ϕ angle; in other words, residues with positive ϕ angles are not sequential. Points having a population of less than 0.02% within a radius of 5° were not plotted. Diagram B also show the relation between the $NOE_{N\alpha}/NOE_{\alpha N}$ ratio and the major secondary structure regions. The line is the contour for $NOE_{N\alpha}/NOE_{\alpha N} = 1$ from a contour map generated by varying the $\phi 2$ and $\psi 1$ in a dipeptide. $NOE_{N\alpha}/NOE_{\alpha N}$ represent the expected ratio between the $d_{N\alpha}(i, i)$ and $d_{\alpha N}(i-1, i)$ NOE's obtained from the measure of $[r(HN_i\text{-}H\alpha_{i-1}) / r(HN_i\text{-}H\alpha_i)]1/6$, where r is the distance between two protons.

279

Applications of the Chemical Shift Index to Calcium Binding Proteins

Chemical shifts are sensitive measures of molecular structure and dynamics. Over the past four years our group has been attempting to exploit this rich source of information. Based on the well known correlation between ^{1}H, ^{13}C and ^{15}N chemical shifts and polypeptide conformation (Spera and Bax, 1991; Wishart et al., 1991b; De Dios et al., 1993), we have developed several simple techniques for rapidly identifying or quantifying secondary and tertiary structure in peptides and proteins (Wishart et al., 1991a; Wishart and Sykes, 1994a; Wishart and Sykes, 1994b). One such technique is known as the Chemical Shift Index (Wishart et al., 1992). The Chemical Shift Index (CSI) is an empirical technique which exploits the well-known and clearly separable secondary shift trends seen in backbone ^{1}H and ^{13}C nuclei of amino acids in helices, β-strands and unstructured (coil) regions of proteins. Originally developed as a "pencil and paper" technique for α-^{1}H chemical shifts only (Wishart et al., 1992), the Chemical Shift Index has recently been expanded to include α-^{13}C, β-^{13}C and ^{13}C' chemical shifts in a completely automated (computerized) approach (Wishart and Sykes, 1994a). Simply stated, the CSI method takes the raw chemical shift data of a newly assigned peptide and protein and passes it through two filters. The first filter compares the raw chemical shift data to residue-specific, empirically determined, baseline chemical shifts. Those shifts above a certain threshold are given a "1", those below are given a "-1" and those in between are given a "0". This digitally filtered data is then passed through a second pattern recognition or "clustering" filter which permits the identification of stretches of 1's, -1's or 0's as being helices, β-strands or coils. This latter filter also assists in defining the beginning and end of each secondary structural unit. In many respects, the CSI methodology resembles the practice originally developed for analyzing $J_{HNH\alpha}$-coupling constants where $J_{HNH\alpha}$ values greater than 8 Hz indicate β-strands; $J_{HNH\alpha}$ values less than 6 Hz indicate helices and $J_{HNH\alpha}$ values in between are considered "unstructured" (Pardi, 1984). The precise details and necessary tables to implement the CSI method are fully described elsewhere (Wishart and Sykes, 1994a).

When applied to just a single atom type (say α-^{1}H's), the CSI method will correctly identify between 75-85% of the secondary structure in a peptide or protein. However, when the CSI assignments for α-^{1}H, α-^{13}C, β-^{13}C and/or ^{13}C' chemical shifts are combined, the consensus prediction has been found to be more than 90% accurate. The kind of signal averaging used in the consensus CSI approach is important for producing the high quality secondary structure assignments necessary for definitive protein structure identification and refinement. An example of how effective the CSI method can be in identifying the secondary structure of calcium binding proteins is shown in Figure 3. In this particular case, the availability of nearly complete α-^{1}H, α-^{13}C, β-^{13}C and ^{13}C' assignments for the N-domain of troponin-C (Gagné et al., 1994) permitted the nearly 100% correct identification of all of its secondary structural elements long before its NOE data had been completely collected and analyzed. In another example pertaining to calcium binding proteins, the consensus CSI approach was used to identify a functionally important helix in the calcium-activated vision protein called recoverin -- a helix which could not be identified through conventional NOE-based methods (Ames et al., 1994).

There are now more than 100 examples in the literature where the CSI method has been used to identify secondary structures in peptides and proteins ranging in size from 10 to 200 residues. With the push to study protein molecules of ever increasing size and the development of heteronuclear assignment methods which are essentially NOE-independent, there is a strong need to find additional or alternative methods for identifying, delineating and refining protein structures such as the Chemical Shift Index.

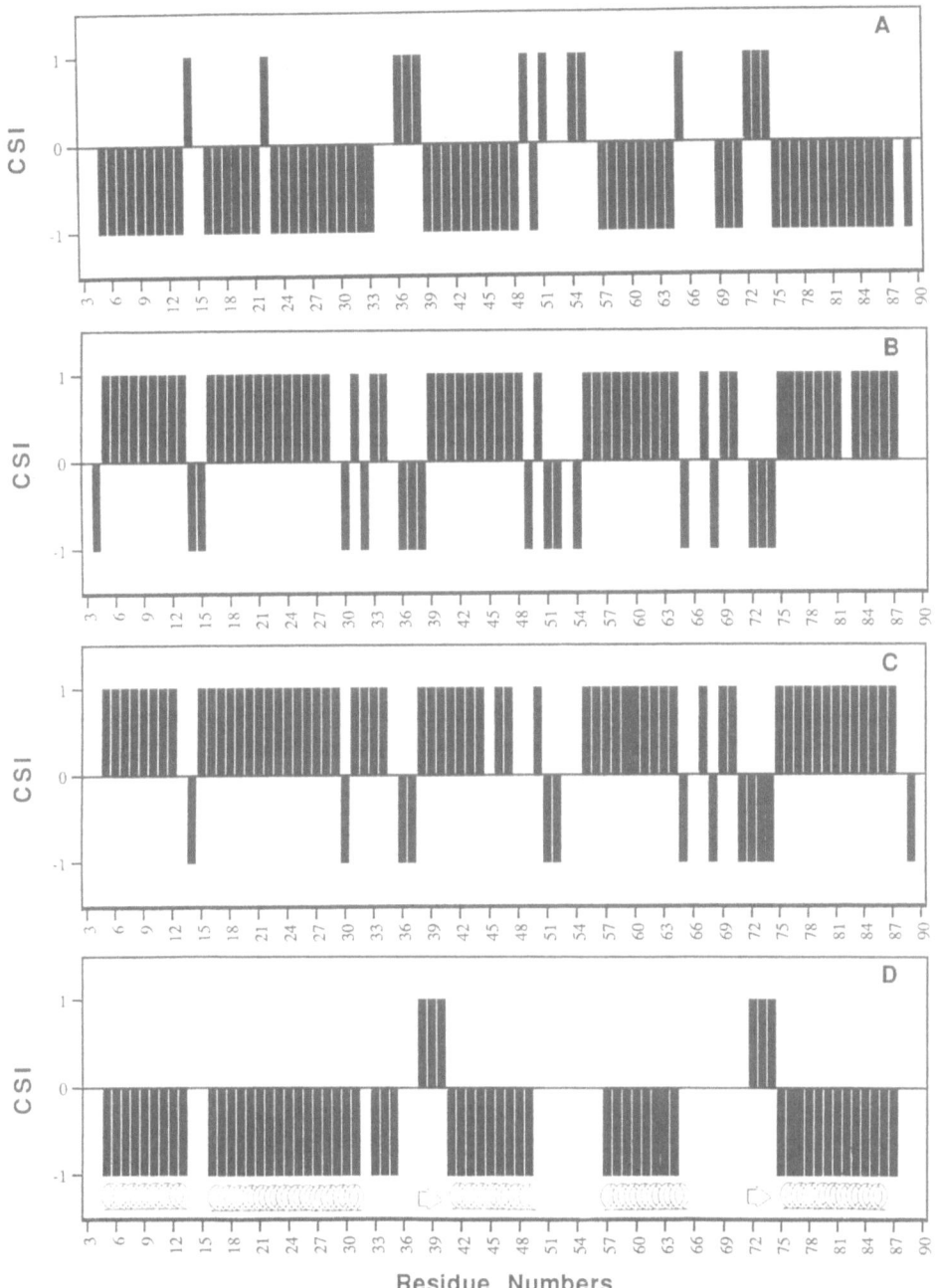

Figure 3. Chemical Shift Index (CSI) graphs derived for the Ca^{++} saturated form of the N-terminal domain of troponin-C (residues 3 - 90). a) CSI graph for α-^{1}H chemical shifts. b) CSI graph for α-^{13}C chemical shifts. c) CSI graph for ^{13}C' chemical shifts. d) Consensus CSI graph determined by combining the information in graphs a - c. The secondary structure, as determined by complete NOE and X-ray analysis is shown using standard icons.

The Use of Trifluoroethanol as a Denaturant of Quaternary Structure

As an increasing number of proteins become available for structural studies using nuclear magnetic resonance spectroscopy as a technique, the problem of protein aggregation arises. Some proteins, such as membrane proteins, are insoluble in water whereas others simply prefer to associate with one another. A number of methods have been used to prevent protein aggregation which include the use of the detergent CHAPS for calcineurin (Anglister et al., 1993; Grzesiek and Bax, 1993), or the use of organic cosolvents such as acetonitrile (Kline and Justice, 1990), acetic acid (Hua and Weiss, 1991a), DMSO (Hua and Weiss, 1991b), or trifluoroethanol (TFE) (Higgens et al., 1990; Craik and Higgens, 1991) for insulin. TFE has been shown to be a denaturant of quaternary structure (Lau et al., 1984a,b) but also retains the distinct characteristic of being a structure-enhancing cosolvent. Most of the work characterizing TFE as a structure enhancing cosolvent, however, has been accomplished on small peptides at relatively high concentrations of TFE (greater than 30% v/v TFE). It has recently been shown that hen egg white lysozyme, a stable protein of 129 amino acid residues containing ß-sheet and α-helix secondary structures, remains relatively unchanged when 15% v/v TFE is added (Buck et al., 1993).

Troponin-C (TnC) is an unusual protein in that it contains two globular domains connected by a long helix. TnC binds four metal ions, two with high affinity in the C-terminal structural domain and two with lower affinity in the N-terminal regulatory domain. Binding of calcium to the N-terminal regulatory domain results in a reversible calcium-induced dimerization which involves the two N-terminal domains interacting (Slupsky et al., 1995a). ^{15}N T_2 studies also illustrate significant mobility in the linker region between the two domains thus allowing the N- and C-terminal domains to act as separate units with separate rotational correlation times for each domain as was shown for calmodulin (Barbato et al., 1992), a protein with similar secondary and tertiary structure. Figure 4 shows 2D ^{1}H-^{15}N HSQC NMR spectra correlating the proton and nitrogen chemical shifts of backbone amide resonances of TnC and the effect that TFE has on the intensity of residues in the N- and C-terminal domains. Linewidths in NMR spectra increase with increasing molecular weights. Thus, crosspeak intensity in nD spectra decrease rapidly as the molecular weight increases (Weiss et al., 1984). In the panel with 0% v/v TFE, the resonances corresponding to residues of the C-domain are visible and strong (isoleucine 149 and phenylalanine 154) whereas the resonances corresponding to residues in the N-terminal domain are either weak (lysine 23 and isoleucine 73) or absent (glutamic acid 16) due to the fact that TnC dimerizes via the N-domain. As the amount of TFE is increased, the intensity (and linewidths) of resonances corresponding to residues in the C-terminal domain remain virtually constant while resonances of the N-terminal domain increase in intensity and decrease in linewidth indicating that the dimer is being dissociated. As the titration is increased from 0% to 13.2% v/v TFE, it is apparent that the chemical shifts of these residues change minimally. Isoleucines 73 and 149 are present in ß-sheet type secondary structures based on an analysis of NOESY spectra (Slupsky et al., 1995b), whereas glutamic acid 16, lysine 23 and phenylalanine 154 are present in α-helical secondary structures. This indicates that TFE does not affect the secondary structure of these residues. Ultracentrifugation studies of TnC in the presence of 15% v/v TFE reveal that the amount of monomer present at 40°C is at least 87% and that there are minimal changes to the backbone chemical shifts upon comparison to TnC in the absence of TFE (Slupsky et al., 1995a). At low concentrations (less than 20% v/v), therefore, TFE is able to act as a denaturant of quaternary structure for stable proteins with a dimerization constant in the millimolar range. Further, at these low concentrations of TFE, there are minimal perturbations of the secondary and tertiary structure.

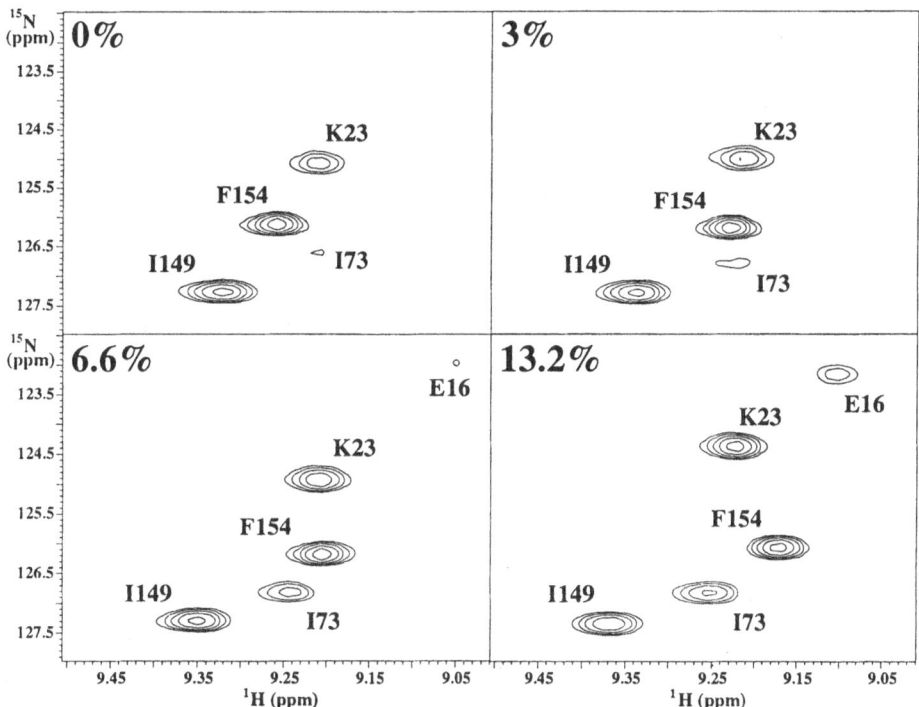

Figure 4. A study of the effect of TFE on the intensity of crosspeaks in a series of ^{1}H-^{15}N HSQC NMR spectra of 1.4 mM uniformly labeled TnC at a temperature of 40°C. The spectra were not scaled to take into account the dilution effects as TFE was added.

REFERENCES

Ames, J.B., Tanaka, T., Styer, L., and Ikura, M., 1994, Secondary structure of myristoylated recoverin determined by three-dimensional heteronuclear NMR: implications for the calcium-myristoyl switch, *Biochemistry* 33:10743.

Anglister, J., Grzesiek, S., Ren, H., Klee, C.B., and Bax, A., 1993, Isotope-edited multidimensional NMR of calcineurin B in the presence of the non-deuterated detergent CHAPS, *J. Biomol. NMR* 3:121.

Barbato, G., Ikura, M., Kay, L.E., Pastor, R.W., and Bax, A., 1992, Backbone dynamics of calmodulin studied by ^{15}N relaxation using inverse detected two-dimensional NMR spectroscopy: The central helix is flexible, *Biochemistry* 31:5269.

Buck, M., Radford, S.E., and Dobson, C.M., 1993, A partially folded state of hen egg white lysozyme in trifluoroethanol: Structural characterization and implications for protein folding, *Biochemistry* 32:669.

Craik, D.J., and Higgins, K.A., 1991, Comparison of ^{1}H NMR chemical shifts of bovine and human insulin's, *Peptide Research* 4:177.

De Dios, A.C., Pearson, J.G., and Oldfield, E., 1993, Secondary and tertiary structural effects on protein NMR chemical shifts: an *ab initio* approach, *Science* 260:1491.

Gagné, S.M., 1994, Calcium-induced structural changes in the regulatory domain of troponin-C by multidimensional NMR spectroscopy, *M.Sc. Thesis*.

Gagné, S.M., Tsuda, S., Li, M.X., Chandra, M., Smillie, L.B., and Sykes, B.D., 1994, Quantification of the calcium-induced secondary structural changes in the regulatory domain of troponin-C, *Protein Science* 3:1961.

Grzesiek, S., and Bax, A., 1993, Measurement of amide proton exchange rates and NOEs with water in $^{13}C/^{15}N$-enriched calcineurin B, *J. Biomol. NMR* 3:627.

Higgins, K.A., Craik, D.J., and Hall, J.G., 1990, ^1H NMR studies of monomeric bovine insulin, *Biochem. Int.* 22:627.

Hobohm, U., Scharf, M., Schneider, R., and Sander, C., 1992, Selection of a representative set of structures from the Brookhaven Protein Data Bank, *Protein Science* 1:409.

Hua, Q., and Weiss, M.A., 1991a, Comparative 2D NMR studies of human insulin and des-pentapeptide insulin: Sequential resonance assignment and implications for protein dynamics and receptor recognition, *Biochemistry* 30:5505.

Hua, Q., and Weiss, M.A., 1991b, Two-dimensional NMR studies of Des-(B26 - B30)-insulin: sequence-specific resonance assignments and effects of solvent composition, *Biochim. Biophys. Acta* 1078:101.

Hyberts, S.G., Goldberg, M.S., Havel, T.F., and Wagner, G., 1992, The solution structure of eglin c based on measurements of many NOEs and coupling constants and its comparison with X-ray structures, *Protein Science* 1:736.

Kline, A.D., and Justice, R.M., 1990, Complete sequence-specific ^1H NMR assignments for human insulin, *Biochemistry* 29:2906.

Lau, S.Y.M., Taneja, A.K., and Hodges, R.S., 1984a, Effects of High-performance liquid chromatographic solvents and hydrophobic matrices on the secondary and quaternary structure of a model protein. Reversed-phase and size-exclusion high-performance liquid chromatography, *J. Chrom.* 317:129.

Lau, S.Y.M., Taneja, A.K., and Hodges, R.S., 1984b, Synthesis of a Model protein of defined secondary and quaternary structure, *J. Biol. Chem.* 259:13253.

Morris, A.M., MacArthur, M.W., Hutchinson, E.G., and Thornton, J.M., 1992, *Proteins* 12:345.

Pardi, A., Billeter, M., and Wüthrich, K., 1984, Calibration of the angular dependence of the amide proton-$C\alpha$ coupling constant JHNα in a globular protein, *J. Mol. Biol.* 180:741.

Slupsky, C.M., Reinach, F.C. and Sykes, B.D., 1995b, Solution secondary structure of calcium-saturated troponin-C monomer determined by multi-dimensional heteronuclear NMR spectroscopy, *Protein Science* submitted.

Slupsky, C.M., Reinach, F.C., Kay, C.M. and Sykes, B.D., 1995a, The calcium-induced dimerization of troponin-C: the mode of interaction and the use of trifluoroethanol as a denaturant of quaternary structure, *Biochemistry* submitted.

Spera, S., and Bax, A., 1991, Empirical correlation between protein backbone conformation and Cα and Cβ ^{13}C nuclear magnetic resonance chemical shifts, *J. Am. Chem. Soc.* 113:5490.

Weiss, M.A., Eliason, J.L., and States, D.J., 1984, Dynamic filtering by two-dimensional ^1H NMR with application to phage 1 repressor, *Proc. Natl. Acad. Sci. USA* 81:6019.

Wishart, D.S., and Sykes, B.D., 1994a, Chemical shifts as a tool for structure determination, *Methods Enzymol.* 239:363.

Wishart, D.S., and Sykes, B.D., 1994b, The ^{13}C chemical shift index: a simple method for the identification of protein secondary structure using ^{13}C chemical shifts, *J. Biomol. NMR* 4:171.

Wishart, D.S., Sykes, B.D., and Richards, F.M., 1991a, Simple techniques for the quantification of protein secondary structure by ^1H NMR spectroscopy, *FEBS Lett.* 293:72.

Wishart, D.S., Sykes, B.D., and Richards, F.M., 1991b, Relationship between nuclear magnetic resonance chemical shift and protein secondary structure, *J. Mol. Biol.* 222:311.

Wishart, D.S., Sykes, B.D., and Richards, F.M., 1992, The chemical shift index: a fast and simple method for the assignment of protein secondary structure through NMR spectroscopy, *Biochemistry* 31:1647.

DISCUSSION

Peter Wright - TFE of course induces helix but it does something else, it weakens hydrophobic interactions and so most proteins in TFE expand dramatically with a large radius of gyration. Can you be sure at the concentrations that you are dealing with, in troponin C, that you are not getting a significant expansion and disordering in the interior.

Brian Sykes - That's obviously a very important question. But it really doesn't look like we are at this stage. You can see at this stage of the structural calculation that we are gathering quite good and quite complete data for the structure in TFE. And we don't see any significantly expanded structure in TFE or anything like that. We do not see any evidence of changes. It would be good to know which residue is responsible for the aggregation or just exactly where is it aggregating. At the time we left we were running a sample where we mixed the ^{12}C-protein and the ^{13}C-protein and tried to look for intermolecular NOE's. But I guess the answer is, we certainly appreciate your concern but we've seen no sort of expansion; we have not gotten anywhere that far in terms of TFE concentration.

Bill Gmeiner - I have a question on the ultracentrifugation techniques and how one goes about doing those tests for dimerization. Do you take the same concentrations that you are working at in the NMR and look for the migration in the ultracentrifuge? Are there any artifactual results that can result from that technique?

Sykes - If you have an older centrifuge, it's easier, and we do. What you do is load the cell at about 1 mg/ml concentration and then it's an equilibrium ultracentrifugation so when the cell has reached equilibrium, it's about a tenth of a mg/ml in one end of the cell and about 2 or 3 mg/ml in the other end of the cell. So it works in a fairly narrowly defined range. All of this corresponds to a range of interest to people like yourself, and all of us, namely 0.1 to 0.5 mM. And then the plot is basically a plot of how the molecule spreads out in the cell and that's proportional to molecular weight. The problems are just a little more subtle. If you looked at the plot I had at the bottom, it was a beautiful plot that went from monomer to dimer but it doesn't fit really well to a classic monomer-dimer equilibrium and that's because there are non-idealities that occur at either end of the cell. Obviously, at the high, centripetal force end you get build up of the protein. I think it gives you a nice result even though there are non idealities that can occur that makes the exact fitting and the exact extrapolation of our constant very difficult. But I think it's a really good way to see if things are monomer or dimer through the range of interest.

THE STRUCTURE OF LENTIVIRAL TAT PROTEINS IN SOLUTION

Paul Rösch[1], Peter Bayer[1], Andrzej Ejchart[1], Rainer Frank[2], Arnona Gazit[3], Franz Herrmann[1], Margot Kraft[2], Rina Rosin-Arbesfeld[3], Heinrich Sticht[1], Dieter Willbold[1], and Abraham Yaniv[3]

[1]Lehrstuhl für Biopolymere, Universität Bayreuth, D-95440 Bayreuth, FRG
[2]Zentrum für Molekularbiologie Heidelberg, D-69120 Heidelberg, FRG
[3]Sackler Faculty of Medicine, Tel-Aviv University, 69978 Tel-Aviv, Israel

INTRODUCTION

Retroviruses are eucaryotic viruses that carry their genetic information as an RNA rather than a DNA sequence. Based on their pathogenic potential, the family of retroviruses can be divided into three different subfamilies, that is the oncoviruses, the lentiviruses, and the spumaviruses. Oncoviruses, such as the Rous sarcoma virus (RSV) and the human T-cell leukemia viruses (HTLV), are cancer causing agents. Lentiviruses, or slow viruses, such as the equine infectious anemia virus (EIAV) and the human immunodeficiency virus (HIV) are non-neoplastic pathogens. Spumaviruses do not have any known pathogenic potential. All lentiviral genomes code for transactivator proteins (Tat) that positively regulate the expression of all viral genes. Tat protein action is absolutely essential for viral replication, and requires the binding to an RNA recognition sequence element (transactivation response element, TAR) located at the 5' end of all viral transcripts. Tat proteins range in size between 75 and 130 amino acids (Jones & Peterlin, 1994). From sequence comparisons of lentiviral Tat proteins it was concluded that immunodeficiency virus Tat protein sequences are in general subdivided into five regions: an NH_2-terminal region, a cysteine-rich region, a core region, a basic region, a glutamine-rich region, and a COOH-terminal region (table 1; (Dorn, 1990; Jones & Peterlin, 1994), and literature therein). The cysteine-rich region and a sequence homologous to the HIV-1 Tat COOH-terminus are not present in the EIAV Tat protein. The highly conserved core region encompasses amino acids Tyr35 through Tyr49 in EIAV Tat protein, and Tyr32 through Tyr47 in HIV-1 Tat protein.

The knowledge of Tat protein structural features is a prerequisite for the design of inhibitors of Tat protein action and thus viral reproduction and, eventually, outbreak of lentivirally caused diseases. Additionally, the monomeric Tat proteins of EIAV (75 amino acids) and HIV-1 (Rice & Chan, 1991) (86 amino acids) and their respective TAR RNAs (25 nucleotides and 29 nucleotides, respectively) are well suited as model compounds for general biophysical studies of protein-RNA interactions (Jones & Peterlin, 1994).

NMR As a Structural Tool for Macromolecules: Current Status and Future Directions
Edited by B.D. Nageswara Rao and Marvin D. Kemple, Plenum Press, New York, 1996

287

Table 1. Multiple sequence alignment of various Tat proteins

```
                10              20              30        40
HV1B1   -MDPVDPNIEP WNH-------PGSQPKT ----ACNRCHCKKC CYHCQVCFIT KGLGISY
HV1Z2   -MEPVDPRLEP WKH-------PGSQPKT ----ACTNCYCKKC CFHCQVCFIT KALGISY
EIAVY   LADRRIPGTAE ENLQKSSGGVPG-QN-T GGQEARPN------ -YHCQLCFLR S-LGIDY
           .     *              ** *    *  *         .***.**.   *** *

                50         60         70          80
HV1B1   G----RK KRRQRRRPSQ GGQTHQDPIP KQPSSQPRGD PTGPKE
HV1Z2   G----RK KRRQRRRPPQ GSQTHQVSLS KQPTSQSRGD PTGPKE
EIAVY   LDASLRK KNKQRLKAIQ QGR------- -QP--QYLL- ------
               ** * .** . *          .       ** *
```

The alignment was done with the CLUSTALW multiple sequence alignment program (Higgins et al. (1992) *CABIOS 8*, 189-191). EIAVY, HV1Z2, HV1B1 are the respective access names of HIV-1 and EIAV isolates in the Swiss-Prot sequence data base. Dots indicate conservative amino acid replacements, asterisks indicate identical amino acids. Starting from the NH_2-terminus, Cys-rich, core, basic, and Gln-rich regions are boxed similar to Dorn et al. (1990) *J. Virol. 64*, 1616-1624.

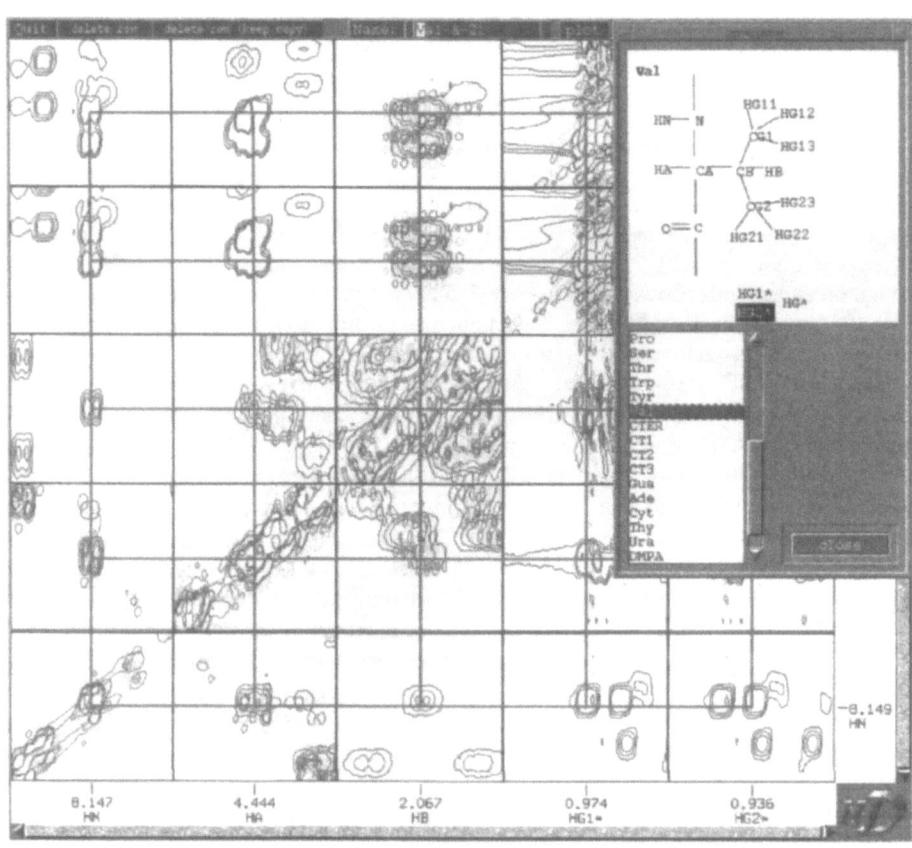

Figure 1. Screen picture of NDEE's "spin system grid" window for simplified recognition of spin systems. An overlay of a COSY- and a TOCSY-spectrum of EIAV-Tat protein reveals all scalar couplings within an amino acid residue. Each "tile" is a detailed view of the surroundings of the crosspeak within a range of ±0.1 ppm. Assignments of individual protons' names are done simply by drag and drop from the molecule sketch window on the top right corner.

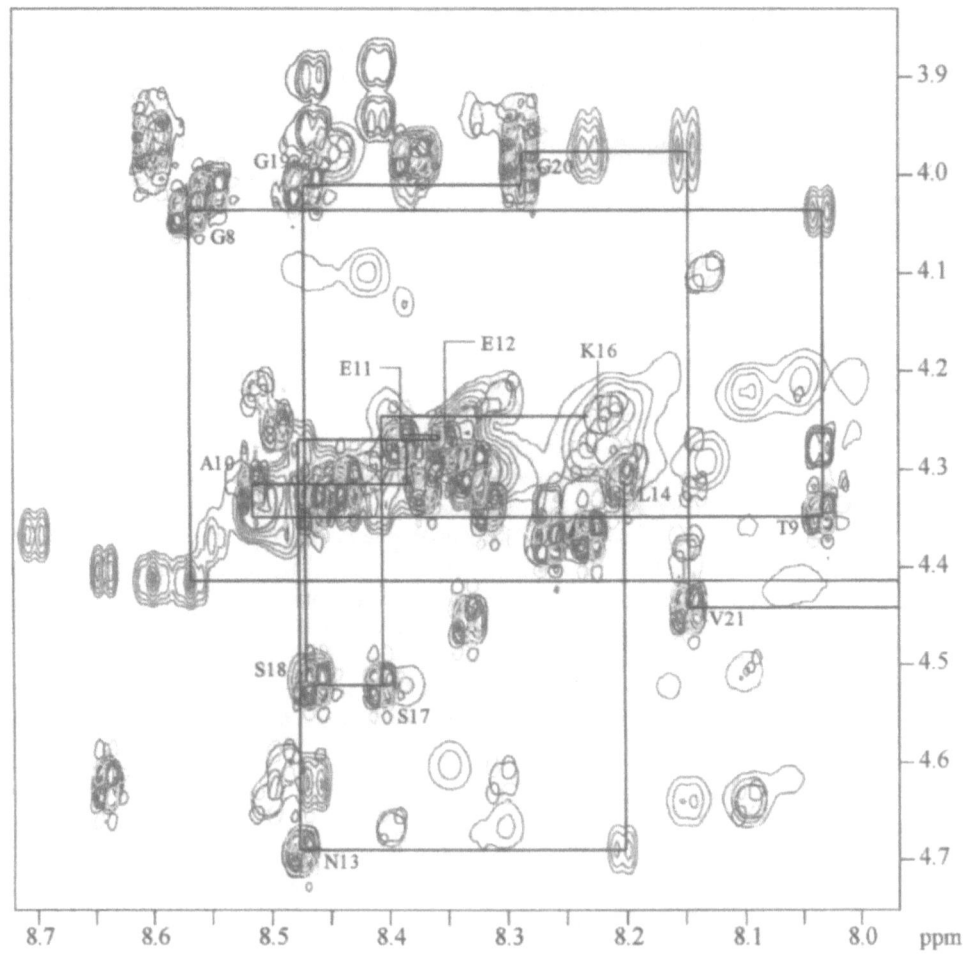

Figure 2. Plot of the H_N-H_α-region of the EIAV-Tat protein shows the backbone trace leading from residue G8 to V21. The overlay of COSY- TOCSY- and NOESY-spectra clearly helps to distinguish between intraresidual H_N-H_α-crosspeaks (found in COSY, TOCSY and NOESY) and interresidual H_N-H_α-crosspeaks (revealed in NOESY only). NDEE assists in connecting an intraresidual H_N-H_α-crosspeak with the sequentially preceding or following interresidual H_N-H_α-crosspeak by supplying an adequate vertical or horizontal line, respectively.

COMPUTER AIDED EVALUATION OF NMR DATA

In the course of structure determination by NMR in general, but particularly in the course of structure determination of biopolymers, evaluation of two- or more-dimensional spectra is a severe bottleneck. Thus, we undertook efforts towards the automation of multi-dimensional spectra evaluation, making use of a variety of hardware platforms such as personal computers (PCs) running under the public domain LINUX operating system and UNIX workstations. The result of our efforts was the development of the 'NDEE' software package that provides an environment for the processing, evaluation and documentation of raw data directly from the spectrometer, thus granting independence of hardware platforms supplied by instrument manufacturers. This platform independence also enables us to use hardware already present in our department. During the evaluation of NMR spectra, NDEE assists with tools such as superposition of different spectra, spin system tracking and documentation, and NOE back calculation. A software interface is integrated to the X-PLOR program package (Brünger, 1993) that we normally use for the subsequent structure calculations. This interface automatically generates X-PLOR input files without manual interference. NDEE provides a completely computer-screen based environment, and facilitates all operations needed to arrive at the input to molecular dynamics calculations from recorded spectra. Processing of raw data starts with weighting by window functions, fast Fourier transformation, phasing and baseline correction. NDEE is able to combine all information from various types of spectra by on-screen superposition of contour plots.

On the basis of superposed spectra, a non-restricted number of assignment elements can be organized in a non-restricted number of transparent 'drawing' planes. In these 'transparencies' spin system networks, text annotations, line segments, quantitative crosspeak information (center of weight, integrals, maximum heights), distance constraints and dihedral constraints can be stored. Single mouse clicks can hide the 'transparency' or make it reappear. The same is true for every contour plot level of the spectra collection.

The intuitive user interface and the strict adherence to the 'what you see is what you get' principle for output on a variety of supported printers (Postscript, PCL, HPGL) drastically decreases the amount of time needed for getting acquainted to the program. NDEE's hardware independence enables use of all accessible standard UNIX/X11-based hardware, such as PCs under public domain LINUX, HP, DEC, SUN, and ESV workstations. NDEE supports background computations on a net independent from the display device. These features drastically decrease the cost of data evaluation.

In our present work, as usual, COSY and TOCSY spectra were used to recognize different spin systems (Wüthrich, 1986). NDEE provided features to focus on spectral regions

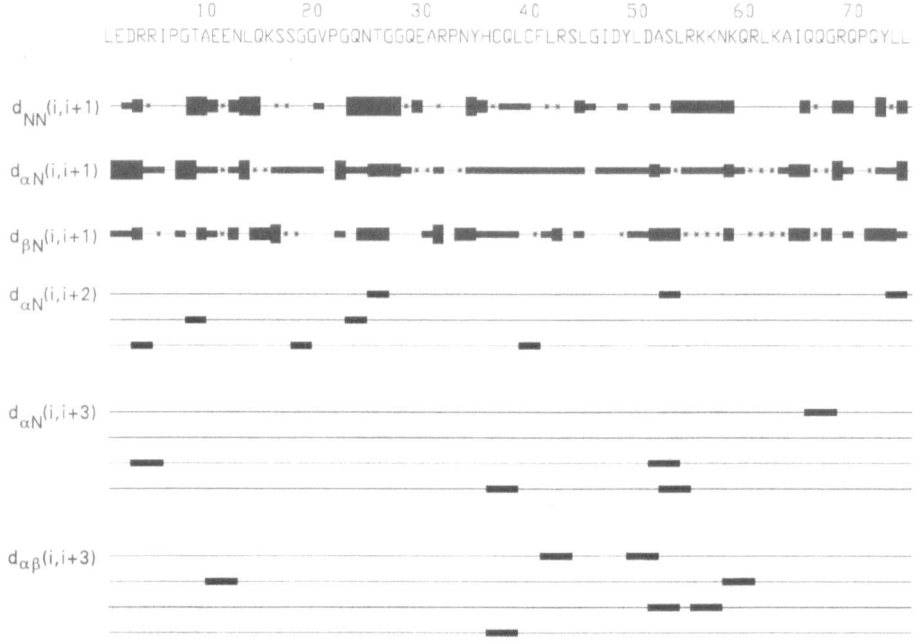

Figure 3. NOESY connectivity table: NOESY connectivities involving backbone protons for amino acids i and j with |i-j| < 4. The height of the bars symbolizes the relative strength (weak, medium, strong) of the cross peaks in a qualitative way. Amino acids are labeled according to the one-letter convention. Horizontal bars connect two residues that are related by the NOE specified to the left. Overlapping and therefore ambiguous cross-peaks are marked by an asterisk.

relevant to specific spin systems, such as, for example, Val21 in EIAV Tat protein (fig. 1). Assignments of individual atom names to the corresponding row or column could be done, and the results could be documented by 'drag and drop' procedures. In fig. 1, the connectivities in the tiled window clearly identify a valine spin system. Fig. 2 gives an example of the chain tracing in EIAV Tat protein COSY, TOCSY and NOESY spectra making use of the superposition capabilities of NDEE.

Once the spin systems were sequentially assigned, the chemical shift of each individual protein proton was automatically documented. Based on unambiguously determined pairs of protons, all information needed by the X-PLOR program package (Brünger, 1986) could be assigned to NOESY cross peaks via NDEE. A continously updated list of actual NOE restraints could be generated as a file suited for direct input into X-PLOR. NDEE suggested peak-volume based distance intervals. Organizing all bookkeeping features such MD restraints, spin systems, and all other kinds of annotation elements in on-screen 'transparencies' simplified the generation of different data files based on a free selection of layers. Results of MD runs for back calculation were used to refine the NOE-restraints.

THE STRUCTURE OF EIAV TAT PROTEIN

Material and Methods

In this study we used both chemically synthesized (Kraft et al., 1994; Willbold et al., 1993) and bacterially expressed EIAV Tat protein (Rosin-Arbesfeld et al., 1994). Both proteins were biologically active and showed the same level of activity in the chloramphenicol-acetyltransferase (CAT) assay. The samples with chemically synthesized EIAV Tat protein were prepared as described earlier (Kraft et al., 1994; Willbold et al., 1993). Except for the early stages of this work, bacterially expressed protein was used for the NMR experiments. E. coli BL21(DE3) were transformed with the pET16b plasmid (Novagen, Madison, Wisconsin, U.S.A.) containing the EIAV Tat sequence with an Ala2Gly replacement and a 10 residue His leader sequence. The mutation did not influence activity. Cells from 6 l fermenter runs were suspended in imidazole, 5 mM, NaCl, 500 mM, Tris/HCl, 20 mM, pH 7.9, guanidinium chloride, 6 M. The extract was passed over a 20 ml metal-chelating sepharose FF (Pharmacia, Freiburg, FRG) column with 50 mM $NiSO_4$ load buffer according to standard procedures. The resulting protein solution was concentrated to a total volume of approximately 5 ml. In order to obtain the Tat protein from the fusion protein a standard bromo cyanide cleavage procedure could be used as the leader sequence ended with a Met residue. The EIAV Tat protein was stored after lyophilization. Activity was tested after this procedure with a CAT assay (Willbold et al., 1993).

The following 600 MHz NMR spectra were employed for the sequence specific assignment of spin systems and the evaluation of NOESY distance constraints: DQF-COSY, TOCSY with mixing times of 80 ms, 100 ms, 150 ms, and 200 ms, respectively, NOESY with mixing times of 150 ms and 300 ms, respectively, and 3D HMQC with ^{15}N labelled protein. Data for long range NOEs ($|i-j| > 5$) were extracted from both NOESY spectra, whereas data for sequential, short and medium range NOEs were extracted from the 150 ms NOESY spectrum only.

Sequential assignments of amino acid spin systems were made using double quantum filtered two-dimensional correlated spectroscopy (COSY), nuclear Overhauser enhancement spectroscopy (NOESY), and total coherence spectroscopy (TOCSY) with the published standard procedures (Wüthrich, 1986). Distance information was extracted from NOESY spectra with mixing times of 150 ms and 300 ms. Sequential ($|i-j| = 1$) to medium range ($1 < |i-j| \leq 5$) NOEs were extracted from the 150 msec spectra to avoid spin diffusion

Table 2. Secondary structure predictions for EIAV Tat protein

```
          10        20        30        40        50        60        70
LEDRRIPGTAEENLQKSSGGVPGQNTGGQEARPNYHCQLCFLRSLGIDYLDASLRKKNKQRLKAIQQGRQPQYLL
1 AA-A------AAAAAB--------------------AA-AB--A---AAAAAAAAAAAAAAAAAAA---------
2 BBTTB----AAAAATTTTT---T---------TTTTTBBBBBTTTTBBBAAAAAAAAAAAAABBBBBBTTT--TBBB
3 ----------AAA-------------------AAAAAA-----AAAAAAAAAAA--AAAAA----------
4 AAAT---T-AAAAAAT--T---TT---TT---T-BBB-----------AAAAAAAAAAAAA-------------
5 ---------------------------BBBBBBBBBBB-----AAAAAAAAAA-------------BBB-
6 --------------------------------aaaaa---aaaaaaaaaaaaaa-------------
7 ---------AAAAA--------------------AAAAAATTTTAAAAAAAAAAAAAAAAAAAAA---------
8 ---------aaaa-------------------aaaaaaaaa---aaaaaaaaaaaaaa-------------
```

Secondary structure predictions were carried out using five algorithms (top to bottom): Bayes statistics (Maxfield & Scheraga, 1976) (1), information theory (Gibrat et al., 1987) (2), neural net (Qian & Sejnowski, 1987) (3) algorithms supplied with the program package "SYBYL 5.4" (Tripos Associates, Inc.), the Chou-Fasman (Chou & Fasman, 1974) algorithm (4) of the FASTA program package (Pearson and Lipman, 1988), and the PHD neural network structure prediction method (Rost & Sander, 1993) (5). The chemical shift indexing strategy (Wishart et al., 1992) was used to estimate the secondary structure for the protein at pH 3.0 (Willbold et al., 1993) (6). The secondary structure found in 40 % (v/v) TFE is shown (7), and compared to the results of the secondary structure for EIAV Tat at pH 6.3 (8). The letters have the following meaning: A, a-helix; B, b-sheet; T, turn; a, helical tendency; -, random coil or no clear decision. Helical tendency means that either the CH chemical shift point to a helical structure according to the procedure of or helical type NOESY cross resonances could be observed for the indicated regions.

effects, long range NOEs ($|i-j| > 5$) were extracted from both NOESY spectra as well as from ^{15}N-^{1}H three-dimensional heteronuclear multiple-quantum coherence (HMQC) spectra. NMR data were evaluated using the NDee program package as described above. 429 intraresidual ($|i-j| = 1$), 250 sequential ($|i-j| = 1$), 146 medium range ($|i-j| = \leq 5$), and 34 long range ($|i-j| > 5$) NOESY cross peaks could be assigned. A summary of the observed backbone crosspeaks is given in Fig.1.

The structures were calculated from the NMR data according to the standard X-PLOR *ab initio* simulated annealing and refinement protocols with minor modifications. NOE crosspeaks were grouped according to their intensity into three categories: strong, 0.18 nm to 0.27 nm; medium, 0.18 nm to 0.4 nm; weak, 0.18 nm to 0.55 nm. Starting from a template structure, 50 minimization steps were employed, followed by 20 ps molecular dynamics at 1000 K with a soft-square NOE energy term (asymptote 0.1), 10 ps molecular dynamics at 1000 K with increased weight on geometry and tilted asymptote (1.0) for the NOE energy term, cooling to 300 K in 50 K steps and 200 steps of minimization. Of the resulting 85 structures the lowest energy structure was used as input structure for a new round of ab initio calculation with identical protocol. The 11 structures from this second round of simulated annealing were subjected to molecular dynamics simulated annealing refinement, the energy function including an electrostatic term with a dielectric constant ε = 1. Six structures were selected on the criteria of smallest number of NOE violations and lowest RMSD values. These structures were used for display in Fig. 2 and for calculation of the parameters in table 3.

Results and Discussion

From the NOESY connectivity table (fig. 3) it was apparent that the protein had a tendency to form weak helices from amino acid Ala10 to Asn13 as well as throughout the core and basic regions, that is amino acids His36 to Arg61. The latter helical part was interrupted by the region from Leu45 to Asp48. It could be seen that the intensity of crosspeaks suggesting helical secondary structures, that is NH_i-$C_\alpha H_{i+3}$ and $C_\alpha H_i$-$C_\beta H_{i+3}$, was very low. Therefore, we deduced that at any given time only a small subset of the total number of protein molecules displayed at least local helix type conformation. These helix type struc-

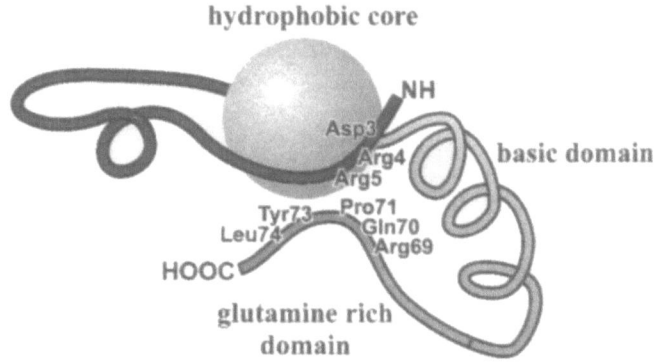

Figure 4. Cartoon of the structure of EIAV Tat. The different domains of Tat are colored coded in the original publication (Willbold et al., 1994): violett, N-terminus and C-terminus; yellow, cysteine rich domain; blue, core domain; magenta, basic domain; green, glutamine rich domain; red, RGD loop.

tures thus formed possible limit structures of the protein (Willbold et al., 1993; Willbold et al., 1994). Interestingly, helix structures at these positions were predicted by most secondary structure prediction methods applied to the EIAV Tat protein sequence (table 2).

The three-dimensional structure of the EIAV Tat protein resulting from the MD calculations was subdivided into domains of different flexibility (table 3, fig.4). Amino acids Tyr35 through Tyr49, which formed a hydrophobic core domain, provided a scaffold for the flexible NH_2-terminal, basic, and glutamine-rich parts of the amino acid sequence. The core region comprised only 20 % of the whole sequence, and the number of intraresidual and sequential NOEs in the core region were also about 20 % of the total number of NOEs in these categories. In contrast, 35 % of the medium range and 63 % of the long range NOEs were observed entirely within the core region. All long range NOEs and 46 % of the medium range NOEs originated from amino acids in the core region (fig. 5). This hydrophobic core domain was well defined by the NMR data according to the MD calculations, with a root mean square deviation (RMSD) of about 0.04 nm for the core region, compared to the corresponding RMSD of more than 0.47 nm for the whole protein (backbone atoms; fig. 6 and table 3). The fold of this core, and thus the formation of the hydrophobic scaffold of protein, was made possible only because the strictly conserved residue Gly46 allowed a structural turn at this position (fig. 6). The basic region wrapped around the core domain, and the NH_2-terminal as well as the COOH-terminal region formed large loop domains that folded back to the hydrophobic core, placing the NH_2- and the COOH-terminus at close distance. The latter, more flexible regions were anchored to the hydrophobic core via the following amino acids (observed NOESY crosspeaks in parentheses): Asp3 ($C_\beta H$ - NH of Tyr49); Gln70 ($C_\gamma H$ - $C_\epsilon H$ of Tyr49); Pro71 ($C_\beta H$ - $C_\epsilon H$ of Tyr49); Tyr73 ($C_\delta H$ to $C_\delta H$ of Tyr49); Leu74 ($C_\alpha H$ - NH of Ile47).

Table 3. Summary statistics for the molecular dynamics calculations

RMSD from ideal Geometry		Average Energies (kJ/mol)		RMSD Protein Backbone (nm)	
NOEs (nm)	0.0140	E_{NOE}	3497.4	whole protein	0.471
angles (deg)	1.445	E_{VDW}	- 688.3	Tyr^{35} to Tyr^{49}	0.042
bonds (nm)	0.0014	E_{total}	- 338.5		
impr (deg)	1.310				

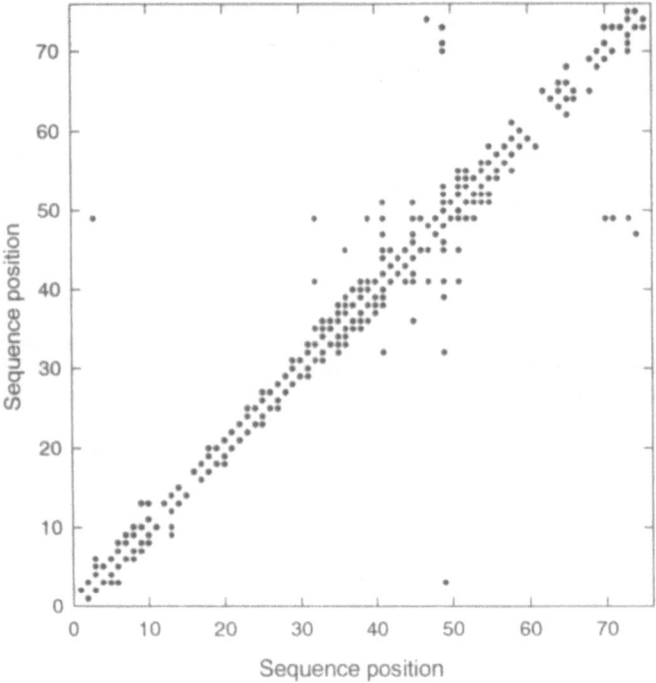

Figure 5. Interresidual NOESY cross peak connectivities: Dots indicate pairs of amino acids connected by a NOESY cross peaks.

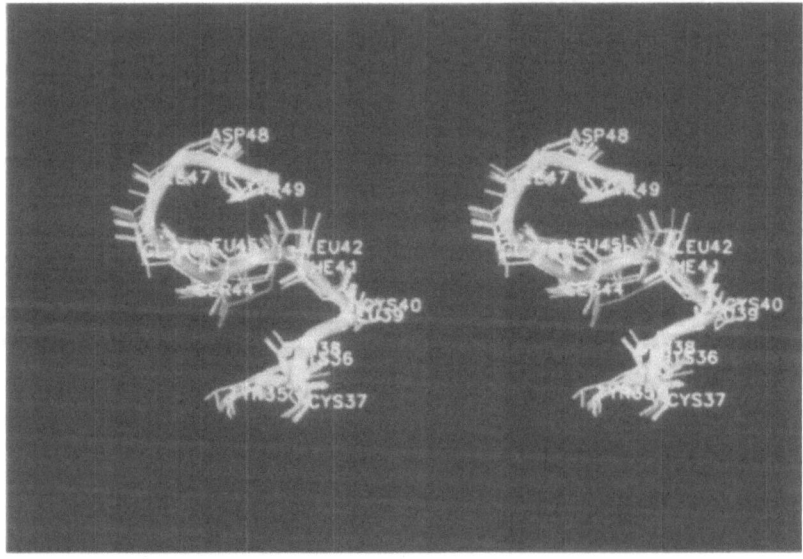

Figure 6. Stereo picture of the core sequence domain: Shown is a superposition of the backbone of the eight solution structures obtained after simulated annealing and refinement procedures. The violet tube indicates the backbone conformation of the average structure.

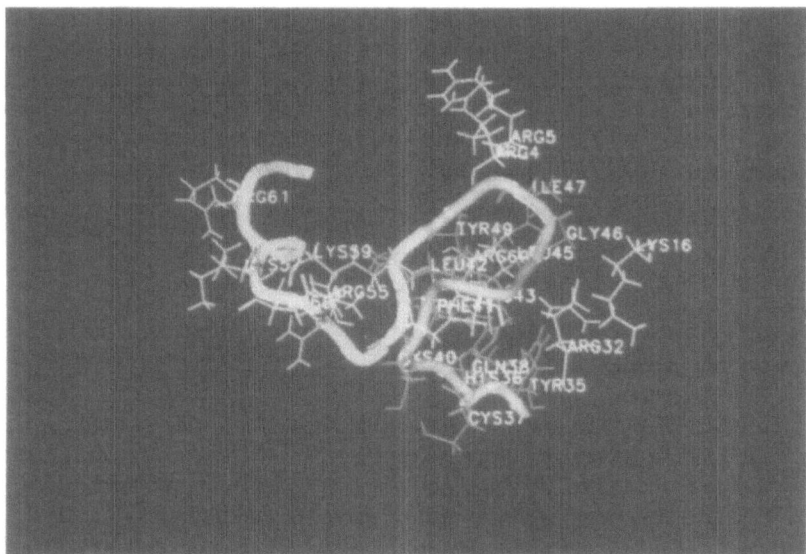

Figure 7. Core and basic domain domain of EIAV Tat protein. A representative structure resulting from the restrained MD calculations is shown.

Most of the basic amino acids are located in the EIAV Tat protein sequence on one side of the core, building a putative RNA binding site. The high density of positively charged amino acids clearly disfavours the formation of a densely packed α-helix in the basic sequence region. Preliminary investigation with EIAV Tat and similar studies with HIV-1 Tat peptides (Loret et al., 1992) indicate that the binding to a polyanion such as TAR enables the formation of a regular α-helix.

In agreement with the lack of stable elements of secondary structure, evidenced for example by the weak intensities of the helix type NOESY cross peaks, all amide protons exchanged rapidly with D_2O in proton-deuteron exchange experiments. Additionally, this observation clearly demonstrated that a time averaged structure is seen in the NMR spectra rather than a mixture of two independent, stable conformations.

Secondary structure prediction algorithms agreed in the prediction of an α-helix in the basic sequence region (table 2). Structural investigations of EIAV Tat at pH 3 and pH 6.3 hint to helical secondary structure in the basic sequence based on C_aH chemical shifts (Willbold et al., 1993; Willbold et al., 1994). We were thus trying to stabilize these transiently present helical regions by addition of 2,2,2-trifluoroethanol (TFE) to the protein solution. TFE is able to stabilize weak helical structures (Dyson et al., 1988; Jasanoff & Fersht, 1994; Sönnichsen et al., 1992). We thus performed a structure determination in the presence of 40 % (v/v) TFE. In this TFE-containing solution, EIAV Tat showed a short regular α-helix from amino acid Ala10 to Leu14, as well as a longer α-helix from Gln18 to Gln76, interrupted by a tight type II turn (Ser44 to Ile47) (fig. 8). The amount of helix present was in agreement with results from CD studies that we performed under identical solution conditions. Thus, the helices observed in TFE-containing solution spanned exactly those regions which show helical propensity in TFE-free solution. This succession of secondary structural motifs is strongly reminiscent of a helix-turn-helix motif as observed for DNA-binding proteins (Sticht et al., 1994). Another secondary structure motif was stabilized in the presence of TFE: In TFE-free solution, a turn type geometry was observed in the hydrophobic core around Gly44. This region formed a classical type II turn in 40 % (v/v) TFE solution. Our studies of EIAV in TFE-containing solution, however, also showed that TFE has a strong influence on the tertiary structure of proteins. The hydrophobic core of EIAV Tat is the only part of the protein with a relatively stable tertiary structure. This hydrophobic core was not present in TFE-containing solution, probably because TFE, with a dielectric constant of 26.7 (Llinas & Klein, 1975), that is one third the dielectric constant of water, weakens hydrophobic interactions. This observation adds a big caveat to studies of structure and folding of proteins performed in the presence of TFE.

THE STRUCTURE OF HIV-1 TAT PROTEIN

The basic region of HIV-1 Tat protein is suggested to form an α-helix from NMR studies of a chimeric EIAV Tat (core region)/HIV-1 Tat (basic region) peptide (Mujeeb et al., 1994). This region is involved in RNA (TAR, trans-action response element) binding (Karn & Graeble, 1992), and Tat proteins thus belong to the family of arginine rich motif (ARM) RNA binding proteins (Burd & Dreyfuss, 1994). The core region of HIV-1 Tat protein is additionally required for sequence specific TAR binding (Churcher et al., 1993). The role of the cysteine rich region is disputed: early reports suggesting that it is a zinc dependent dimerization domain (Frankel et al., 1989) are not universally acknowledged (Churcher et al., 1993), and nature and role of metal ion binding to the HIV-1 Tat protein cysteine rich region is unclear (Jeyapaul et al., 1990). Recent reports doubt the requirement of metal ion binding to the cysteine rich region for transactivation activity altogether (Koken et al., 1994). The glutamine rich region is probably involved in TAR binding (Churcher et al., 1993).

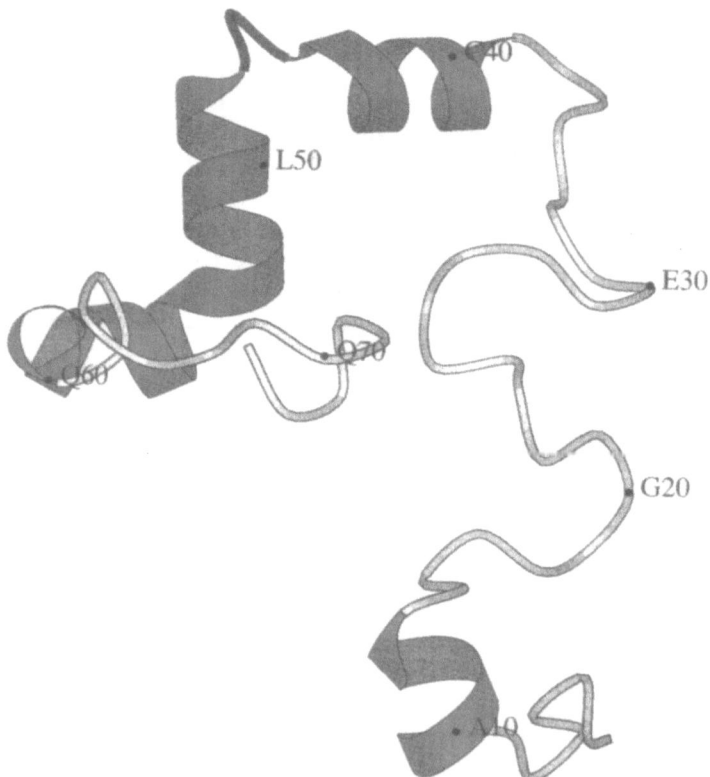

Figure 8. MOLSCRIPT (Kraulis, 1991) sketch of the secondary structure elements found in EIAV Tat protein in TFE-containing solution.

Material and Methods

HIV-1 Tat protein, SwissProt data bank sequence entry HV1Z2, wild type and Thr40Lys mutant, was chemically synthesized, purified, and refolded (Kraft et al., 1994; Slice et al., 1992): 10 ml of a 1 mg/ml solution of Tat protein was unfolded in 1 l buffer solution containing 20 mM potassium phosphat, pH 6.5, $ZnCl_2$, 100 mM, mannitol, 50 mM, ascorbic acid, 10 mM, phenylmethylsulfonyl fluoride (PMSF), 0.5 mM, dithiothreitol (DTT), 0.5 mM, and urea, 6 M. Stepwise refolding was performed in the buffer used for the unfolding procedure, but without $ZnCl_2$, with subsequent concentrations of 4 M, 2 M, and 0 M urea. The protein was finally dialyzed twice against 10 mM ammonium acetate, 10 mM. Helium degassed buffers were used throughout. The protein was lyophilized twice after either procedure to remove ammonium acetate and stored as a lyophilized powder.

110 ml H_2O, NaCl, 30 mM, NaN_3, 0.02 % (v/v), was degassed for two minutes, pressure 1 mbar. The buffer was then bubbled with N_2 to saturation and kept under N_2 atmosphere. $Na_2S_2O_4$, 20 mM, was added, the final pH was approximately 5.5. The lyophilized protein was dissolved rapidly in 900 ml buffer solution, still under N_2 atmosphere. The sample was then concentrated in a Speedvac vacuum concentrator to approximately 450 ml. An NMR sample tube was flushed with N_2 for approximately 10 minutes. The sample was then centrifuged and filled into the NMR tube after pH was determined to be approximately 6.3. N_2 was passed over the sample in the NMR tube and the sample tube sealed. Protein concentrations of up to 2.8 mM without precipitation could be obtained with this procedure. Precipitation of HV1Z2 Tat protein was observed in dialyses in the presence of DTT in excess of 5 mM concentration, and increase in molecular mass of the order of several DTT molecules was detected by matrix assisted laser desorption ionization - time of flight (MALDI-TOF) mass spectrometry under these conditions (Kraft et al., 1994). HIV-1 Tat protein is inactivated by DTT (Koken et al., 1994), in accord with these results. Thus, $Na_2S_2O_4$ was used as a reducing agent.

Resonance assignments were made using DQF-COSY, NOESY, and CLEAN-TOCSY with the published standard procedures (Wüthrich, 1986). The following sets of spectra were recorded on a Bruker AMX600 NMR spectrometer: HV1Z2 wt: COSY-DQF, CLEAN-TOCSY, mixing time 80 ms, NOESY, mixing time 200 and 250 ms, respectively, temperature, 298 K, protein concentration, 1.8 mM, NaCl, 50 mM, NaN_3, 0.04 % (v/v), in H_2O/D_2O (9:1), dithioerythritol (DTT), 1 mM, pH 6.3 - 6.5. HV1Z2 Thr40Lys mutant: DQF-COSY, CLEAN-TOCSY, mixing time 90 ms, NOESY, mixing time 200 ms, temperature, 278 K, protein concentration, 2.0 mM, NaCl, 50 mM, $Na_2S_2O_4$, 50 µM, and NaN_3, 0.04 % (v/v), in H_2O/D_2O (9.1), pH 6.5; CLEAN-TOCSY, mixing time 90 ms, NOESY, mixing time 200 ms, temperature, 293 K, protein concentration, 2.2 mM, NaCl, 50 mM, $Na_2S_2O_4$, 50 µM, NaN_3, 0.04 % (v/v), in D_2O; NOESY, mixing time 200 ms, DQF-COSY, conditions as before, but pH 6.5; CLEAN-TOCSY, mixing time 90 ms, NOESY, mixing time 100, 200 and 250 ms, respectively, temperature, 298 K, protein concentration, 2.8 mM, NaCl, 50 mM, $Na_2S_2O_4$, 50 mM, NaN_3, 0.04% in H_2O/D_2O (9:1), pH 6.0; NOESY, mixing time 200 ms, as before, but in D_2O. Typical spectral parameters were: frequency width 6024 Hz, time-domain data-size 8 k x 0.5 k or 4 k x 0.5 k data points, frequency domain data-size 8 k x 1 k and 4 k x 1 k data points; sinebell squared filter with $\pi/4$ phase shift. 352 intraresidual ($|i-j| = 0$), 185 sequential ($|i-j| = 1$), 61 medium range ($|i-j| = 2,3,4,5$), and 25 long range ($|i-j| > 5$) NOESY crosspeaks could be extracted from these spectra. Distance information was extracted from NOESY spectra with mixing times of 100, 200 and 250 ms. Sequential and medium range ($|i-j| \le 5$) NOEs were extracted from the 100 and 200 ms NOESY spectra to minimize spin diffusion effects. Long range NOEs ($|i-j| > 5$) were extracted from all three NOESY spectra as well as from NOESY spectra recorded in D_2O.

Table 4. Summary statistics for the molecular dynamics calculations

RMSD from ideal Geometry		Average Energies (kJ/mol)		RMSD Protein Backbone (nm)	
angles (deg)	1.542	E_{NOE}	1610	Phe38 to Tyr47	0.08
bonds (nm)	0.0013	E_{VDW}	- 3956	Tyr32 to Tyr47	0.17
impropers (deg)	1.092	E_{total}	- 2117	Gln63 to Gln72	0.15
NOE (nm)	0.015			Pro76 to Pro80	0.07
				Whole protein	0.42

Molecular dynamics (MD) calculations based on the experimental NOESY data were performed with the X-PLOR 3.1 program package basically with the standard hybrid distance geometry/ simulated annealing (dgsa) protocol (Brünger, 1993). NOE cross peaks were grouped according to their intensity into three categories: strong (0.18 to 0.28 nm), medium (0.18 to 0.40 nm) and weak (0.18 to 0.55 nm).The structures were calculated from the nmr data according to the standard X-PLOR distance geometry and refinement protocols with minor modifications. After bound smoothing, embedding and regularization, a family of 30 substructures was produced. Full embedding and subsequent SA refinement with 3.75 ps heating to 1000 K and a square NOE potential was used. MD calculation (12.5 ps) at high temperature with increased weight on geometry and additional cooling in 50 K steps down to 100 K were followed by 300 steps of energy minimization. After another round of refinement with inclusion of electrostatic interactions ($\varepsilon = 1$) and 600 steps of energy minimization, 10 structures were selected on the criteria of smallest NOE violations as well as energy and RMSD values.

Results and Discussion

The NOESY crosspeak patterns do not indicate presence of stable elements of regular secondary structure. Molecular dynamics calculations based on the experimental parameters showed that HV1Z2 Tat protein consisted of three dimensional domains of highly different flexibilty corresponding to the sequence regions defined earlier (Dorn, 1990). This was evidenced by the fact that all observed long range NOESY crosspeaks originated from

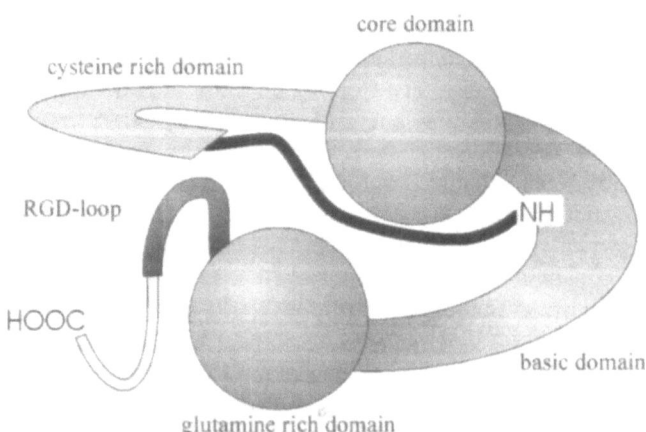

Figure 9. Cartoon of the structure of HIV Tat. The different domains of Tat are colored coded in the original publication (Bayer et al., 1995): violett, N-terminus and C-terminus; yellow, cysteine rich domain; blue, core domain; magenta, basic domain; green, glutamine rich domain; red, RGD loop.

either the glutamine rich or the core sequence region. These regions were clearly defined as three dimensional domains in the final structure as shown by a cartoon of the global fold (fig. 9) and the pairwise root mean square deviations (table 4).

The rigidity of the three dimensional structure made up of the core region amino acids was reflected by the agreement of the final ten backbone structures, Phe38 to Tyr47 and Tyr32 to Tyr47, resulting from the MD calculations (table 4). The complete core domains, Tyr32 to Tyr47, did not superpose well because a flexible hinge, Cys37-Phe38, connected two structurally rigid parts of the core. This hinge is not present in the EIAV Tat protein structure, as a disulfide bond between Cys37 and Cys40 stabilizes the local structure of this protein. A pivotal amino acid in the turn-like core domain structure was Gly44, a residue highly conserved in all known Tat proteins. EIAV Tat protein and HV1Z2 Tat protein are similar in their core domains, Phe41 to Tyr49 and Phe38 to Tyr47, respectively, which can be nearly ideally superposed onto each other (Bayer et al., 1995), in spite of the fact that HV1Z2 Tat protein contains one additional amino acid, Thr40, in this sequence region. This strengthens the view that integrity of the core domain is crucial for activity of Tat proteins, and all conclusions arrived at for the EIAV Tat protein concerning the core domain are equally valid for the HV1Z2 Tat protein. For example, specific RNA binding is only possible with a peptide containing the turn region around Gly44 (Churcher et al., 1993). Lys41, suspected to interact with transcription factor II D (Kashanchi et al., 1994), was solvent exposed and stabilized as part of the rigid hydrophobic core. From its general location, Lys41 could well contribute to RNA binding as suggested recently (Churcher et al., 1993). The glutamine rich sequence region, which formed the second rigid domain in HV1Z2 Tat protein, has no full length counterpart in EIAV Tat protein.

As summarized in the cartoon (fig. 9), core domain and glutamine rich domain both showed NOESY crosspeaks to a stretch of amino acids close to the NH_2-terminus. Val4 to Trp11 were thus sandwiched between core and glutamine rich domains. This relates to observations that HIV-1 Tat protein is inactive in amino acid 2 to 6 (Asp2 to Pro6 in HV1Z2) deletion mutants (Kuppuswamy et al., 1989). The COOH-terminus is also fixed to the glutamine rich domain. This leaves the stretch of amino acids Pro14 to Lys19, the cysteine rich sequence region, and the basic sequence region highly flexible.

Overall, in the case of HV1Z2 Tat protein, one possible interpretation of the experimental data is formation of a stable molecular center made up of core sequence region, glutamine rich sequence region, and N-terminus. An alternative explanation is that the NH_2 terminus attaches to the Gln-rich domain and the core domain for approximately an equal amount of time, with both domains not necessarily being in close contact. This latter interpretation would be justified by the fact that no direct NOESY crosspeak was observed between amino acids from the Gln-rich domain and the core domain. For further studies of Tat and TAR interactions by fluorescence spectroscopy, Trp11 is nearly ideally located in the center of the molecule, sandwiched between the core and basic domains (Fig. 10). This should make Trp11 a sensitive probe for global structural changes imposed on the molecule for example by TAR interaction.

The Arg78-Gly79-Asp80 (RGD loop), supposed to be a general key site for adhesive recognition and receptor interaction (Hynes & Lander, 1992), was also clearly solvent exposed at the tip of a hairpin structure which was experimentally well defined by several NOESY crosspeaks (Ser74-Gly83; Ser75-Glu86; Gln66-Thr82; His65-Asp80; Thr64-Glu86; Gly62-Thr82) as reflected in the low RMSD value of this loop (table 4). Although RGD loop sequences seem to be very flexible in most proteins studied so far (Adler et al., 1991; Klaus et al., 1993; Saudek et al., 1991), the recently determined structure of decorsin (Krezel et al., 1994) shows a rigid RGD loop similar to that found for HV1Z2 Tat protein. The rigidity of the HV1Z2 Tat protein RGD loop structure may well be caused by two proline residues, Pro77 and Pro81, flanking the loop as proline residues have only a very small ϕ torsion angle space. Pro81 is highly conserved in all known HIV-1 Tat se-

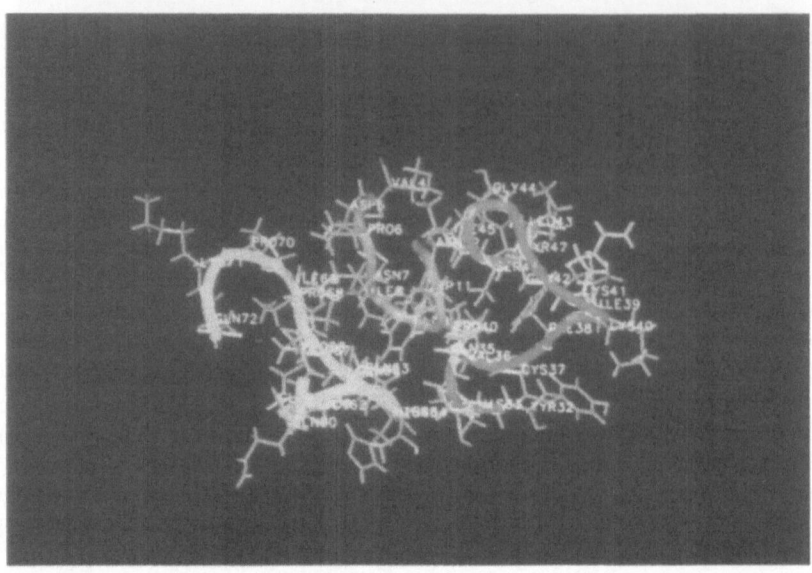

Fig. 10. Part of the core domain - gln-rich region - NH₂-terminal sequence structure. Trp11 is sandwiched between the gln-rich and the core domains.

quences, and Pro77 is only conservatively replaced by Ser77 in some sequences. A similar way of stabilization is suggested for decorsin, where one Pro residue directly flanks the RGD loop, whereas the other proline residue is spaced from the loop by two residues (Krezel et al., 1994).

RELATION TO BIOCHEMICAL FINDINGS

The results of the current structural studies can be related to known features of the Tat protein family:

Mutations in the core region influence the trypsin digestion pattern of HIV-1 Tat protein (Rice & Carlotti, 1990), a result easily derived from the observation that this sequence forms a central hydrophobic core in the protein structure.

HIV-1 Tat protein is inactive in amino acid 2 to 6 (Asp2 to Pro6 in HV1Z2) deletion mutants (Kuppuswamy et al., 1989). These amino acids contribute to the structure formed by the Gln-rich region and the hydrophobic core.

A peptide homologue of the HIV-1 Tat basic sequence region alone does not show full RNA binding activity. This activity is increased when the peptide is extended by amino acids from the core region (Churcher et al., 1993). The ability of the peptides to discriminate between TAR mutants is dramatically reduced if the peptides contain fewer than five amino acids from the core region. From the EIAV Tat structure, it is clear that the hydrophobic core is a necessary structural domain presenting the RNA binding sequence to the TAR RNA at the protein surface. The five COOH-terminal residues of the core region are the necessary components for the formation of the turn around Gly46.

Transcription factor TFIID binds to HIV-1 Tat, and the binding region extends from amino acids Val36 to Gly50 (EIAV Tat: Leu39 to Ala52) with Lys41 (EIAV Tat: Arg43) as a crucial component (Kashanchi et al., 1994). TFIID thus binds to the only basic residue in the most stable domain of the protein. An alternative role of Lys41 could be a contribution to RNA binding.

EIAV and HIV-1 Tat proteins seem to behave somewhat differently with respect to TAR binding (Carroll et al., 1991; Carroll et al., 1992; Derse & Newbold, 1993). Fig. 4 suggests that basic amino acids from the proposed RNA binding region of EIAV Tat, Arg55 to Lys63, together with the NH_2-terminus (Arg4 and Arg5), and the COOH-terminus (Arg69) form a positively charged cleft suitable for RNA binding. This cleft, stabilized by the hydrophobic core domain, was experimentally defined by NOEs from Asp3, Pro71, Tyr73, and Leu74 to the core domain. Indeed, a collapse of the tertiary structure on mutation of COOH-terminal residues, in particular Leu74 and Leu75, is suggested (Derse & Newbold, 1993).

SPECULATIONS

The question as to the conformation of the domain made up of the basic amino acid region in HV1Z2 Tat protein has to go partially unanswered. On one hand, in our experiments this basic region did not show any helix type NOESY cross peaks, which can be observed, although weakly, for the EIAV Tat protein basic region (Willbold et al., 1994). On the other hand, we would not have observed these cross peaks under the present experimental conditions for HV1Z2 Tat protein (2.8 mM concentration) if they were as weak as they were in the EIAV Tat protein spectra (8 mM concentration). Our results thus suggested that the basic domain is not a rigid helix in full length HV1Z2 Tat protein, contrasting the notion of a stable α-helix observed for the chimeric EIAV Tat (core region)/HIV-1 Tat (basic region) peptide (Mujeeb et al., 1994). Induction of α-helical conformation in the basic domain by addition of TFE was possible for EIAV Tat protein (Sticht et al., 1993), but not for HV1Z2 Tat protein.

Interaction of HIV Tat and several cellular factors is reported by various investigators (Jeang et al., 1993; Madore & Cullen, 1993), and direct binding of the TBP subunit of TFIID to Lys41 is reported (Kashanchi et al., 1994). Lys41 is an amino acid located in the rigid core domain of Tat protein. Inspection of the solution structure of HIV Tat discloses that the sidechain of Lys 41 is solvent exposed and thus accessible for docking molecules. Nevertheless, participation in TAR binding could also be a possible role of Lys41.

So far, there is no conclusive evidence that the cysteine rich region contributes to the transcriptional activity of HV1Z2 Tat protein. A different role for the cysteine rich region may be suggested by the fact that HV1Z2 Tat protein, but not EIAV Tat protein, contains the adhesive RGD loop, and that RGD proteins, such as proteins of the disintegrin family, contain six conserved cysteine residues in a total of about 40 amino acids (Klaus et al., 1993; Krezel et al., 1994). Thus, the HV1Z2 Tat protein cysteine rich region may be required for cell adhesion rather than TAR recognition, which would clarify many of the recent ambiguities concerning this sequence. The network of intramolecular disulfide bonds postulated for RGD proteins was not observed for the HV1Z2 Tat. This could be an experimental artifact caused by the reducing conditions under which we had to keep the protein to prevent its aggregation under NMR concentrations. It is interesting to note that a region of high homology to the Cys-rich region in HIV Tat proteins is found in the *Wnt5a* proto-oncogene products, the vertebrate relatives of the developmental *Drosophila melanogaster wingless* gene product (Clark et al., 1993). Wnt proteins are supposed to be cell-cell signalling molecules acting over a short distance:

```
                         22                    37
          HV1Z2          -CNRCHCK--KCCYH-CQVC-
          Wnt5a (mouse)  -TERCHCKFHWCCYVKCKKC-
                         351                   369
```

Also, the requirement of a cofactor which is responsible for the transactivating potential of Tat could therefore bind to the cysteine rich domain (Madore & Cullen, 1993; Zhou & Sharp, 1995). This is currently under investigation. Disulfide bonds between cystein residues in this region were observed by several groups (Koken et al., 1994), but their significance for the physiological role of Tat is still under discussion.

Although both Tat protein structures were very flexible outside the core domain, they may adopt a more ordered structure on RNA binding. The disordered structure of the uncomplexed basic domain is readily understood in light of the high local charge density. Helical structure of the basic domain of HIV-1 Tat protein in the Tat-TAR complex has been suggested (Calnan et al., 1991), but other experimental studies argue to the contrary (Loret et al., 1992; Loret et al., 1991). Formation of helical structure on nucleic acid binding would be reminiscent of the bZIP DNA binding motifs (Patak & Sigler, 1992). If sequence regions with helix forming tendency formed helices simultaneously, the protein would be of the helix-loop-helix-turn-helix type, as is observed in trifluoroethanol solution (Sticht et al., 1994). This motif is known from homeodomain proteins (Laughon, 1991). Indeed, several arguments are in favor of this hypothesis. For example, the third helix is the major groove binding helix in homeodomains and in EIAV Tat protein, and the homeodomain Arg/Lys2, Arg/Lys3 residues, which contribute to DNA minor grove recognition, have their counterpart in Arg4, Arg5 of EIAV Tat protein, which could possibly contribute to TAR minor groove recognition (Weeks & Crothers, 1991).

It is now possible, with determination of basic features of the EIAV Tat protein and the HV1Z2 Tat protein structures, to address more specific questions related to the nature of the Tat/TAR interactions by spectroscopic and biochemical techniques. For a full understanding, however, knowledge of the structure of the Tat/TAR complexes is a necessary prerequisite.

LITERATURE

Adler, M., Lazarus, R. A., Dennis, M. S., & Wagner, G. (1991) *Science 253*, 445-8.

Bayer, P., Kraft, M., Ejchart, A., Westendorp, M., Frank, R., & Rösch, P. (1995) *J. Mol. Biol. 247*, 529-35.

Brünger, A. 1993. *X-PLOR 3.1 Manual.* New Haven: Yale University Press.

Brünger, A. T. C., G. M., Gronenborn, A. M., Karplus, M. (1986) *Proc. Natl. Acad. Sci. U. S. A. 83*, 3801.

Burd, C., & Dreyfuss, G. (1994) *Science 265*, 615-21.

Calnan, B. J., Biancalana, S., Hudson, D., & Frankel, A. D. (1991) *Genes Dev. 5*, 201-10.

Carroll, R., Martarano, L., & Derse, D. (1991) *J. Virol. 65*, 3460-7.

Carroll, R., Peterlin, B. M., & Derse, D. (1992) *J. Virol. 66*, 2000-7.

Chou, P. Y., & Fasman, G. (1974) *Biochemistry 13*, 222-45.

Churcher, M. J., Lamont, C., Hamy, F., Dingwall, C., Green, S. M., Lowe, A. D., Butler, J. G., Gait, M. J., & Karn, J. (1993) *J. Mol. Biol. 230*, 90-110.

Clark, C. C., Cohen, I., Eichstetter, I., Cannizzaro, L. A., McPherson, J. D., Wasmuth, J. J., & Iozzo, R. V. (1993) *Genomics 18*, 249-60.

Derse, D., & Newbold, S. H. (1993) *Virology 194*, 530-6.

Dorn, P., DaSilva, L, Martarano, L Derse, D. (1990) *J. Virol. 64*, 1616-24.

Dyson, J. H., Rance, M., Houghten, R. A., Lerner, R. A., & Wright, P. E. (1988) *J. Mol. Biol. 201*, 161-200.

Frankel, A. D., Biancalana, S., & Hudson, D. (1989) *Proc. Natl. Acad. Sci. U.S.A. 86*, 7397-401.

Gibrat, J. F., Garnier, J., & Robson, B. (1987) *J. Mol. Biol. 198*, 425-43.

Hynes, R. O., & Lander, A. D. (1992) *Cell 68*, 303-22.

Jasanoff, A., & Fersht, A. R. (1994) *Biochemistry 33*, 2129-35.

Jeang, K. T., Chun, R., Lin, N. H., Gatignol, A., Glabe, C. G., & Fan, H. (1993) *J. Virol. 67*, 6224-33.

Jeyapaul, J., Reddy, M. R., & Khan, S. A. (1990) *Proc. Natl. Acad. Sci. U.S.A. 87*, 7030-4.

Jones, K. M., & Peterlin, B. M. (1994) *Annual Review of Biochemistry 63*, 717-44.

Karn, J., & Graeble, M. A. (1992) *Trends Genet. 8*, 365-8.

Kashanchi, F., Piras, G., Radonovich, M. F., Duvall, J. F., Fattaey, A., Chiang, C. M., Roeder, R. G., & Brady, J. N. (1994) *Nature 367*, 295-9.

Klaus, W., Broger, C., Gerber, P., & Senn, H. (1993) *J. Mol. Biol. 232*, 897-906.

Koken, S. E., Greijer, A. E., Verhoef, K., van-Wamel, J., Bukrinskaya, A. G., & Berkhout, B. (1994) *J. Biol. Chem. 269*, 8366-75.

Kraft, M., Westendorp, M., Krammer, P., Bayer, P., Rösch, P., & Frank, R. W. (1994) in *Peptides 1994* (H. L. S. Maja, Ed.) Leiden: Escom.

Kraulis, P. J (1991) *J. Apl. Crystallogr. 24*, 946-50

Krezel, A. M., Wagner, G., Seymour-Ulmer, J., & Lazarus, R. A. (1994) *Science 264*, 1994-7.

Kuppuswamy, M., Subramanian, T., Srinivasan, A., & Chinnadurai, G. (1989) *Nucleic Acids Res. 17*, 3551-61.

Laughon, A. (1991) *Biochemistry 30*, 11357-67.

Llinas, M., & Klein, M. P. (1975) *J. Am. Chem. Soc. 97*, 4731-7.

Loret, E. P., Georgel, P., Johnson, W., Jr., & Ho, P. S. (1992) *Proc. Natl. Acad. Sci. U. S. A. 89*, 9734-8.

Loret, E. P., Vives, E., Ho, P. S., Rochat, H., Van-Rietschoten, J., & Johnson, W., Jr. (1991) *Biochemistry 30*, 6013-23.

Madore, S. J., & Cullen, B. R. (1993) *J Virol 67*, 3703-11.

Maxfield, F. R., & Scheraga, H. A. (1976) *Biochemistry 15*, 5138-53.

Mujeeb, A., Bishop, K., Peterlin, B. M., Turck, C., Parslow, T. G., & James, T. L. (1994) *Proc. Natl. Acad. Sci. U. S. A. 91*, 8248-52.

Pathak, D., & Sigler, P. (1992) *Curr. Opin. Struct. Biol. 2*, 115.

Pearson, W. R., & Lipman, D. J. (1988) *Proc. Natl. Acad. Sci. USA 85*, 2444-8.

Qian, N., & Sejnowski, T. (1988) *J. Mol. Biol. 202*, 865-84.

Rice, A. P., & Carlotti, F. (1990) *J. Virol. 64*, 6018-26.

Rice, A. P., & Chan, F. (1991) *Virology 185*, 451-4.

Rosin-Arbesfeld, R., Mashiah, P., Willbold, D., Rösch, P., Tronick, S. R., Yaniv, A., & Gazit, A. (1994) *Gene 150*, 307-11.

Rost, B., & Sander, C. (1993) *Proc. Natl. Acad. Sci. U. S. A. 90*, 7558-62.

Saudek, V., Atkinson, R. A., & Pelton, J. T. (1991) *Biochemistry 30*, 7369-72.

Slice, L. W., Codner, E., Antelman, D., Holly, M., Wegrzynski, B., Wang, J., Toome, V., Hsu, M. C., & Nalin, C. M. (1992) *Biochemistry 31*, 12062-8.

Sönnichsen, F. D., van Eyk, J. E., Hodges, R. S., & D., S. B. (1992) *Biochemistry 31*, 8790-8.

Sticht, H., Willbold, D., Bayer, P., Ejchart, A., Herrmann, F., Rosin-Arbesfeld, R., Gazit, A., Yaniv, A., Frank, R., & Rösch, P. (1993) *Eur. J. Biochem. 218*, 973-6.

Sticht, H., Willbold, D., Ejchart, A., Rosin-Arbesfeld, R., Yaniv, A., Gazit, A., & Rösch, P. (1994) *Eur J Biochem 225*, 855-61.

Weeks, K. M., & Crothers, D. M. (1991) *Cell 66*, 577-88.

Willbold, D., Krüger, U., Frank, R., Rosin-Arbesfeld, R., Gazit, A., Yaniv, A., & Rösch, P. (1993) *Biochemistry 32*, 8439-45.

Willbold, D., Rosin-Arbesfeld, R., Sticht, H., Frank, R., & Rösch, P. (1994) *Science 264*, 1584-7.

Wüthrich, K. 1986. *NMR of Proteins and Nucleic Acids*. New York: Wiley.

Zhou, Q., & Sharp, P. A. (1995) *EMBO J. 14*, 321-8.

DISCUSSION

Peter Wright - Paul, it is not about TFE. I am intrigued by your suggestions that there might be a helix-turn-helix motif in the Tat protein and I wonder whether you have done any sequence alignments. What is the known helix-turn-helix sequence because there are certain rules that apply there? Certain positions that have small sidechains for example do allow proper packing of the two helices. So it will be very interesting to see whether the sequences indicate the homologies. Even with the detailed ones, you are supposed to see these rules obeyed.

Paul Rösch - We tried to plot this alignment but we were unsuccessful. Maybe we did not try hard enough.

Wright - One has to look at how these helices are formed originally and see how it aligns with the canonical and non-canonical helix-turn-helix motif.

Rösch - It is a very important point. If we could find the helix-turn-helix motif and the RNA binding motif, this would be good.

Julie Forman-Kay - Since Peter Wright did not ask you about TFE, I think I would like to, not to put you on the spot. I have a question. Did you see any difference in the dispersion of the chemical shifts in the aqueous solution from that in TFE and what mole % of TFE were you using?

Rösch - We were using 40% TFE. We did not see any strong alterations in the chemical shift.

Carol Post - Along those lines, have you tried any other kind of additive, or varying pH or ionic strength?

Rösch - We varied pH from 2.8 - 7.2, ionic strength from 0 - 300 mM salt. We tried a variety of conditions. The structure looked like a very flexible structure at the beginning, very unsatisfactory to us, so we tried all the secrets of the trade to induce secondary structure but we did not succeed by these methods, so we used TFE.

Thomas James - Paul, with your EIAV you have done some nitrogen labeling. How did your order parameter correlate with your structural domains that you found?

Rösch - Our calculation of order parameter was not very conclusive. We saw a huge difference between the relaxation times in the hydrophobic corner and the relaxation times in the so called "flexible domain". But we did not come to the point that we calculated the order parameters because we had many problems with peak overlap especially in the flexible regions, so we didn't dare to calculate numbers from that.

A STRUCTURAL BIOLOGIST'S VIEW OF PRECISION AND ACCURACY OF STRUCTURAL MODELS OF PROTEINS BASED ON NMR DATA

A. Joshua Wand

Department of Biochemistry
University of Illinois at Urbana-Champaign
Urbana, IL 61810

INTRODUCTION

Advances in the field of nuclear magnetic resonance (NMR) since its first observation 50 years ago have been staggering and have revolutionized many areas of physics, chemistry and biology. As the technology as a whole has developed there has been an explosion in the number of applications of NMR spectroscopy to studies of proteins and other biopolymers. The dramatic increase in the power, flexibility and efficiency of modern multidimensional and multinuclear NMR spectroscopy now provide the structural biologist with a previously unavailable avenue to high resolution structural information in solution (Clore and Gronenborn, 1991b; Bax and Grzesiek, 1993; Leopold et al., 1994). At the center of the studies of the structure and dynamics of proteins by NMR is the solution of the general problem of assigning individual resonances to specific nuclei within the macromolecule. Recent developments of conceptually new experimental frameworks coupled with technical advancements have had a swift and profound effect on the solution of the resonance assignment problem. The issue of resolving and mapping individual resonances of a protein to specific atomic sites within the molecule has now largely been solved for proteins ranging up to 25 kDa in size. Parallel to these recent advances in resonance assignment methodologies has been an ever increasing level of sophistication in the marshaling of NMR-based structural parameters for the determination of the solution structures of proteins.

In the spirit of this conference, several topics revolving around issues of precision, accuracy and the ultimate utility of NMR-based models of protein structure will be addressed. We have recently solved the structures of reduced and oxidized cytochrome c to relatively high apparent precision and accuracy and this system will be used to illustrate the predictive value of NMR-based models of the solution structures of proteins. Together, these structures provide a template to address a fundamental consideration: How useful are they as predicative tools? To illustrate, results of calculations of various parameters describing electron transfer phenomena will be presented. These structures also

NMR As a Structural Tool for Macromolecules: Current Status and Future Directions
Edited by B.D. Nageswara Rao and Marvin D. Kemple, Plenum Press, New York, 1996

307

serve to show that highly defined models can be obtained using purely homonuclear ^{1}H methods. The second phase of the paper will center on the character of surface and core internal motions in proteins and will be illustrated by published ^{15}N NMR (Schneider et al., 1992) and preliminary ^{13}C relaxation studies of the protein ubiquitin (McEvoy et al., unpublished results) using low pass filtered spectroscopy of randomly fractionally ^{13}C-enriched protein (Wand et al., 1995).

STRUCTURE-FUNCTION RELATIONSHIPS IN MITOCHONDRIAL CYTOCHROME C.

Mitochondrial cytochrome c is a soluble 12.5 kDa protein that mediates single electron transfer between integral membrane protein complexes in the respiratory chain of eukaryotes. Because of its stability, solubility and ease of preparation, cytochrome c has become one of the most thoroughly studied proteins (see Moore and Pettigrew, 1991). Several X-ray crystallographic studies have resulted in a number of high resolution models of the crystal structures of c-type cytochromes (Takano and Dickerson, 1981a,b; Berguis and Brayer, 1992). Although only subtle structural differences between redox states have been observed in these crystal models, a number of NMR-based structural studies suggest the presence of potentially significant structural differences (e.g., Feng et al., 1990). Systems that include cytochrome c and various redox partners have also been used to investigate how the polypeptide chain assists in controlling electron transfer kinetics and thermodynamics (Onuchic et al., 1992; Farid et al., 1993). Experiments involving site-directed mutagenesis, semi-synthesis, and chemical modification of cytochrome c have also been used to probe the role of several key amino acids involved in electron transfer (e.g., Willie et al., 1993). Interpretation of these and numerous other experiments, however, often relies on the availability of applicable high resolution structural models. In an effort to clarify redox dependent structural issues, we have undertaken the determination of the solution structures of horse cytochrome c in its two redox states. Here we will describe our recent determinations of high resolution models for the solution structure of horse heart ferro- and ferricytochrome c using ^{1}H 2D and 3D NMR spectroscopy and hybrid distance geometry-simulated annealing calculations. These detailed structural studies will hopefully provide the basis for a comprehensive re-evaluation of hypotheses concerning the fundamental nature of the electron transfer processes in this protein. They also serve to illustrate that highly defined molecular models of proteins of moderate size can be determined using structural restraints derived solely from ^{1}H NMR experiments.

As mentioned above, the introduction and implementation of heteronuclear-based multidimensional techniques have revolutionized the protein NMR field. Large proteins (>200 residues) are now amenable to detailed NMR studies and structure determination. These techniques, however, necessarily require a scheme by which ^{13}C and ^{15}N isotopes can be incorporated into the protein to yield a uniformly and highly enriched sample. Additional complications, such as extensive covalent post-translational modifications, can seriously limit the ability to efficiently and cost effectively express a protein in isotope enriched media - the c-type cytochromes are an example of such a limitation. In the absence of an effective labeling protocol, one must therefore rely on more traditional proton homonuclear NMR methods. These include two-dimensional (Wüthrich, 1986) and, more recently, three-dimensional ^{1}H experiments (Vuister et al., 1988).

No economic method for uniform ^{13}C and ^{15}N labeling of cytochrome c has been achieved, consequently the use of heteronuclear NMR methods has been precluded. Prior experience has indicated that NOE-based restraint densities on the order of 20 restraints per residue are required to allow for the high precision definition of protein structures. In order to obtain such a density of restraints for cytochrome c in the absence of isotopic enrichment, we have relied heavily on the use of three dimensional NOESY-TOCSY spectra to obtain the necessary library of identified ^{1}H-^{1}H NOEs.

Early ^{1}H resonance assignment studies were led by the pioneering work of Williams, Moore and coworkers and by Wüthrich and coworkers. Much of that work focused on the assignment of ^{1}H resonance of the reduced, or ferrous, form, ferrocytochrome c, in which the heme iron is diamagnetic. For ferrocytochrome c, the ^{1}H chemical shifts are restricted to between -4 and 11 ppm. In the efforts dealing with the ferric form, ferricytochrome c, where the low spin 1/2 iron provides an intrinsic paramagnetic perturbation, the initial efforts concentrated on paramagnetically shifted ^{1}H resonances dispersed in the chemical shift range of -40 to +40 ppm. In the early 1980's the approaches to the resonance assignments became decidedly more aggressive and comprehensive. The work of Williams, Moore and coworkers is particularly noteworthy for their application of a wide variety of analytical NMR techniques to what was then a seemingly impenetrable resonance assignment problem (see Williams et al (1985) and references therein). In their initial ground breaking efforts, these investigators were able, without the benefit of two dimensional NMR techniques or extremely high magnetic fields, to assign a significant fraction of the side chain ^{1}H resonances of both reduced and oxidized cytochrome c from a variety of species. Approaches used included comparison of ^{1}H NMR spectra of homologous cytochromes c, structure dependent interpretations of chemical shifts and steady state and transient nuclear Overhauser effects and spin-decoupling. Wüthrich and coworkers used a variety of NOE-based methods, including two dimensional NOESY spectra, to characterize the role of axial ligand geometry on the setting of the redox potential of c-type cytochromes (see Keller and Wüthrich, 1981 and references therein). Comprehensive ^{1}H resonance assignments were not completed until the work of Wand and coworkers with the essentially complete assignment of the ^{1}H NMR spectrum of the horse protein in both redox states (Wand et al., 1989; Feng et al., 1989). Efforts with the reduced protein led to the formalized NOE-based pattern recognition method for analysis of ^{1}H NMR spectra of proteins subsequently termed the main chain directed resonance assignment strategy (Englander & Wand, 1987) and later demonstrated with the protein ubiquitin (Di Stefano and Wand, 1987). These studies also resulted in the extensive characterization of the backbone conformation of the protein in its two redox states. The resonance assignments of the protein were extensively cross-correlated by use of chemical exchange phenomena in NOESY spectra resulting from interprotein electron transfer in mixtures of the two redox states (Feng et al., 1991).

Using these ^{1}H resonance assignments, a large number of NOE-based restraints can be obtained from the analysis of two dimensional spectra (see Table 1). However, a significant number of additional restraints can be identified by appealing to confirming correlations observed in three dimensional homonuclear NOESY-TOCSY spectra. Though initially designed to provide an avenue to rapid ^{1}H resonance assignments (Vuister et al., 1988), the NOESY-TOCSY spectrum can, in principle, provide support for a proposed assignment of a ^{1}H-^{1}H NOE. It cannot, however, unequivocally confirm the identity of an NOE. This is because the 3D ^{1}H NOESY-TOCSY experiment does not provide simultaneous confirmation of the origin of both NOE-correlated frequencies. This is in distinct contrast to the four dimensional heteronuclear resolved NOESY experiments (Clore et al., 1991). Thus, although confirmation of the origin of both frequencies could be checked by examining both NOESY-TOCSY pathways (i.e., spin A NOE to spin B TOCSY to spin C and spin B NOE to spin A TOCSY to spin D), this does not guarantee that a given crosspeak is entirely due to one spin pair. Therefore initial structures were examined to provide an additional level of confidence on the assignment of a given NOE crosspeak to a given spin pair by rejecting all other possible spin pairs on gross structural grounds. To avoid issues relating to variable transfer efficiencies in the TOCSY component of the three dimensional experiment, all restraints derived from analysis of the three dimensional NOESY-TOCSY spectrum were simply encoded as corresponding to distance upper bounds of 5.0 Å. To illustrate many of the analytical issues involved in taking this approach, a two dimensional TOCSY-plane of a NOESY-TOCSY spectrum of ferricytochrome c is shown in Figure 1. Here, the indicated cross peaks of the three dimensional spectrum confirm the identity of two spins participating in an NOE. Note the congestion of the spectrum which makes this type of analysis extremely cumbersome and often difficult. Nevertheless, the end result is quite satisfying with 16 and 18 NOE-restraints per residue being obtained for oxidized and reduced cytochrome c, respectively (Tables 1 and 2).

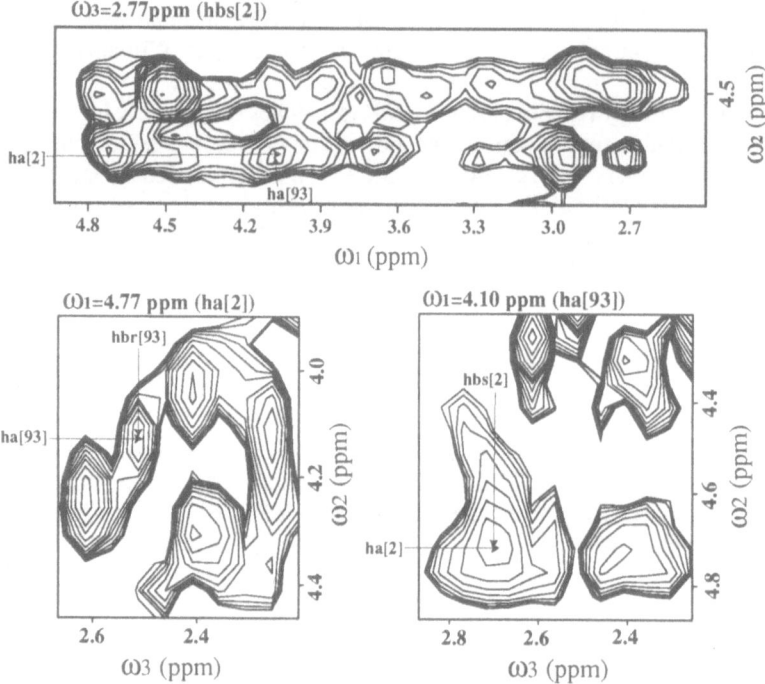

Figure 1. Illustration of the use of the three dimensional NOESY-TOCSY experiment in the identification of the origin of an NOE cross peak. The top panel is an expansion of a two dimensional NOESY slice of a three dimensional NOESY-TOCSY spectrum of horse ferricytochrome c. The indicated cross peak is consistent with an NOE between the CαH of Asp 93 and the CαH of Asp 2. Support for this assignment can derived from the TOCSY planes of the NOESY-TOCSY experiment. Shown in the bottom two panels are the TOCSY planes at the chemical shift of the CαH of Asp 93 (right panel) and the CαH of Asp 2 (left panel). The observed J-correlations support the assignment of the NOE partners to CαH of Asp 93 and the CαH of Asp 2.

The assembled NOE-based restraints were combined with restraints on phi torsion angles obtained from determined alpha-amide scalar coupling constants (Qi et al., 1994a). This restraint set was used in a standard simulated annealing protocol employing no attractive nonbonded potentials and utilized the capabilities of Dspace (Nerdal et al., 1988) and Xplor (Clore et al., 1986). Prochirality was determined using the "floating chirality" method as described by Beckman et al. (1993). No non-bonded attractive potentials were employed during the refinement. In the latter stages of the refinement, long lived structural water detected by use of NOESY-TOCSY spectra were incorporated into the models (Qi et al., 1994b). A relatively large number (44) of structures were refined for each redox state and used to generate the statistics and average structures reported in Tables 1 and 2. The agreement with covalent geometry and the experimentally derived restraints is good for the models of both redox states. For example, no violations greater than 0.5 Å of an NOE-based restraint were observed in either family of structures. Analysis of the families of

Table 1. Structural Statistics and Atomic Root Mean Square Deviations for the Solution Structure of Horse Ferrocytochrome c[a]

	{SA}	$<SA>_{rw}$
Violations of Experimental Restraints (1973)		
distance restraints		
(number ≥ 0.5 Å)	0.00	0
(number ≥ 0.3 Å)	0.25	0
phi torsion angle restraints (85)		
(number ≥ 5°)	0.00	0
(number ≥ 2°)	0.00	0
hydrogen bond distance restraints (52)		
(number ≥ 0.1 Å)	0.00	0
Root Mean Square Deviations		
NOEs (Å)	0.028 ± 0.001	0.029
phi torsion angles (°)	0.121 ± 0.015	0.136
bonds (Å) (1772)	0.005 ± 0.000	0.004
angles (°) (3240)	0.817 ± 0.053	0.795
improper angles (°) (706)	0.464 ± 0.016	0.464

Atomic Root Mean Square Deviations (Å)			
Comparison	Backbone atoms	All heavy atoms	All atoms
{SA} vs <SA>	0.345 ± 0.046	0.842 ± 0.050	1.113 ± 0.052
{SA} vs $<SA>_r$	0.398 ± 0.072	0.975 ± 0.066	1.326 ± 0.067
{SA} vs $<SA>_{rw}$	0.452 ± 0.088	1.012 ± 0.065	1.354 ± 0.064
$<SA>_r$ vs <SA>	0.206	0.493	0.722
$<SA>_r$ vs $<SA>_{rw}$	0.194	0.240	0.262
<SA> vs $<SA>_{rw}$	0.301	0.562	0.773

[a]The notation of the structures is as follows: {SA} is the set of 44 final simulated annealing structures; <SA> is the simple mean structure obtained by averaging the coordinates of the 44 individual SA structures superimposed to each other; $<SA>_r$ is the structure obtained by restrained minimization of <SA>, $<SA>_{rw}$ is the structure obtained by steepest descent restrained minimization of $<SA>_r$ with six structural waters introduced using a total of 34 distance restraints. In all comparisons, noted atom types of all residues were included.

independently refined structures of both redox states indicates that the obtained refined average structures are highly precise with r.m.s. deviations from the mean structure less than 0.5 Å on the main chain (see Tables 1 and 2). The models for both redox states of the protein show highest imprecision in the omega loop region spanning residues 20 to 30 and in a turn region connecting the axial ligand Met-80 to the C-terminal helix.

Table 2. Structural Statistics and Atomic Root Mean Square Deviations for the Solution Structure of Horse Ferricytochrome c[a]

	{SA}	<SA>$_{rw}$
Violations of Experimental Restraints (1973)		
distance restraints		
(number ≥ 0.5 Å)	0.00	0
(number ≥ 0.3 Å)	0.18	0
phi torsion angle restraints (85)		
(number ≥ 5°)	0.00	0
(number ≥ 2°)	0.00	0
hydrogen bond distance restraints (52)		
(number ≥ 0.1 Å)	0.00	0
Root Mean Square Deviations		
NOEs (Å)	0.031 ± 0.001	0.033
phi torsion angles (°)	0.116 ± 0.022	0.090
bonds (Å) (1772)	0.004 ± 0.000	0.005
angles (°) (3240)	0.881 ± 0.125	1.030
improper angles (°) (706)	0.489 ± 0.020	0.480

Atomic Root Mean Square Deviations (Å)

Comparison	Backbone atoms	All heavy atoms	All atoms
{SA} vs <SA>	0.329 ± 0.042	0.832 ± 0.050	1.124 ± 0.054
{SA} vs <SA>$_r$	0.375 ± 0.048	0.956 ± 0.064	1.343 ± 0.068
{SA} vs <SA>$_{rw}$	0.430 ± 0.048	1.014 ± 0.055	1.399 ± 0.060
<SA>$_r$ vs <SA>	0.182	0.473	0.736
<SA>$_r$ vs <SA>$_{rw}$	0.258	0.395	0.485
<SA> vs <SA>$_{rw}$	0.278	0.581	0.833

[a]The notation of the structures is as follows: {SA} is the set of 44 final simulated annealing structures; <SA> is the simple mean structure obtained by averaging the coordinates of the 44 individual SA structures superimposed to each other; <SA>$_r$ is the structure obtained by restrained minimization of <SA>, <SA>$_{rw}$ is the structure obtained by steepest descent restrained minimization of <SA>$_r$ with six structural waters introduced using a total of 38 distance restraints. In all comparisons, noted atom types of all residues were included.

The precision and the general topologies of the protein in its two redox states are illustrated by the superpositions shown Figure 2 and Figure 3. For the most part, the largest variances across the two families of structures are localized to the loop region spanned by residues 21 through 29 and by a turn region encompassing residues 86 to 89. These highly refined structures reveal several internal and surface features that may be important in the function, stability and dynamical behavior of this protein. Some of these features have been noted previously with homologous proteins while others have not. For example, a previously undocummented surface reorganization involving Ile-81 appears to function as a molecular recognition device signalling the current redox state of the protein. On the main chain, the reduced protein displays unusual 3$_{10}$ to alpha helical transitions in the two of the three main helical segments of the protein. This leads to a bifurcated peptide carbonyl in the core of each helical segment. In sum, the determination of the high resolution models of the solution structures of oxidized and reduced cytochrome c by methods based on nuclear magnetic resonance has allowed a detailed comparison between the solution structure of the protein in its two redox states

In an effort to assess the accuracy of the two models, we have employed an analysis of the pseudo contact shifts predicted by the structure of the oxidized protein with the chemical shift changes actually observed upon a change in redox state. This builds upon an earlier application initially by Williams and coworkers (see Turner and Williams, 1993 and references therein) and later taken up by Englander and coworkers (Feng et al., 1990). Earlier efforts used deviations of the observed chemical shift changes from that predicted to occur on the basis of pseudo contact shifts caused by the spin 1/2 heme to detect diamagnetic contributions, i.e. changes in structure upon a change in redox state.

The pseudo contact shift ($\Delta\delta pc$) induced by the spin 1/2 iron of oxidized heme may be conveniently described in the coordinate system defined in Figure 4.

Figure 2. Superpositions of the family of 44 independently refined structures determined for reduced cytochrome c. The orientation has been chosen to maximize the view of the least defined regions of the model.

Figure 3. Superpositions of the family of 44 independently refined structures determined for oxidized cytochrome c. The orientation has been chosen to maximize the view of the least defined regions of the model.

The molecular coordinates must be transformed to this reference coordinate system using a simple translation and three successive rotations about the x, y and z axes of α, β and γ rotation angles. In this coordinate system, the $\Delta\delta pc$ caused by a change in redox state, i.e. from the reduced diamagnetic state to the oxidized spin 1/2 state, is given by:

$$\Delta\delta pc = \frac{\beta^2 S(S+1)}{9kTr^2} \bullet \left[g_{ax}\left(3\cos^2\theta - 1\right) + 1.5 g_{eq}\left(\sin^2\theta\cos 2\phi\right) \right] \qquad (1)$$

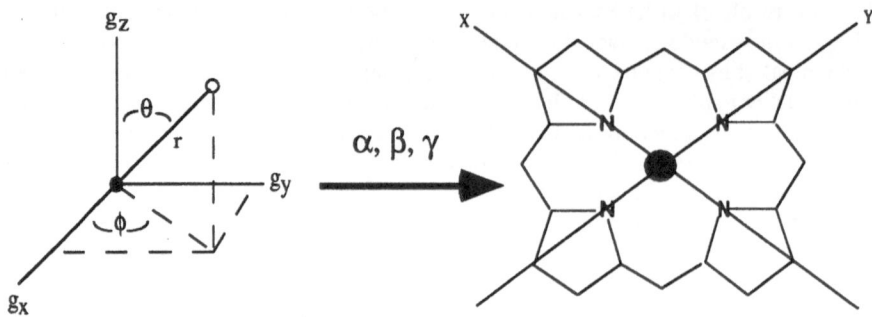

Figure 4. Reference coordinate system for the electronic g-tensor is translated and then successively rotated on the molecular coordinate system defined by the pyrole nitrogens of the heme prosthetic group (see Feng et al., 1990).

where

$$g_{ax} = g_z - 0.5\left(g_x^2 + g_y^2\right)$$
$$g_{eq} = g_x^2 - g_y^2$$

Here β, S, k, and T correspond to the Bohr magneton, the electron spin quantum number of 1/2, the Boltzman constant and the absolute temperature, respectively. Equation 1 effectively assumes that there is no significant change in structure leading to a change in the diamagnetic component of the given hydrogen's chemical shift. Using several different approaches, others have used this assumption to delineate regions of the protein which do not appear to change structure upon a change in redox state. This is generally achieved by identifying those hydrogens whose observed redox-dependent chemical shift changes can be entirely explained by the predicted pseudo contact shift. Equation [1] requires not only definition of the relative geometry of the g-tensor and the interacting hydrogen but also the alignment of the g-tensor within the molecular coordinate system. For the paramagnetic system present in oxidized cytochrome c, the alignment and magnitude of the g-tensor components are adequately described by five parameters: the three rotation angles (α, β, γ) and the axial (g_{ax}) and equatorial (g_{eq}) components of the g-tensor. A reference set of 44 alpha hydrogens was selected on the basis of best agreement between the observed redox-dependent chemical shift change and the pseudo contact shift term predicted by the structure of the oxidized protein. This is a recursive procedure and leads, via a least squares minimization, to the set of rotation angles and g-tensor components best able to describe the observed redox-dependent chemical shift changes of the reference set. The α, β, and γ rotation angles were determined to be 147, 15 and 206 degrees respectively. Values of -2.39 and 3.73 were determined for g_{eq} and g_{ax}, respectively. These values are quite similar to those determined by Williams and coworkers (Turner and Williams, 1993 and references therein).

A plot of the deviation of the predicted pseudo contact shift from the observed chemical shift for all alpha hydrogens in the oxidized protein is shown in Figure 5. Also indicated is the range of the predicted pseudo contact shift over a 0.5 Å variation of the coordinate position of the hydrogen. Those alpha hydrogen sites that show a significant difference between the observed redox-dependent chemical shift change and that predicted for the pseudo contact shift term are predicted to have a significant change in the diamagnetic component of the chemical shift, i.e. they are associated with a significant structure change. Or, these deviations may indicate an inaccuracy in the model determined for oxidized cytochrome c. It is comforting to note, then, that all sites showing such deviations are in fact localized to regions where a significant difference in the structure of the two redox states of the protein is found. These observations provide an independent check on the accuracy of the determined models for the solution structure of reduced and oxidized horse cytochrome c.

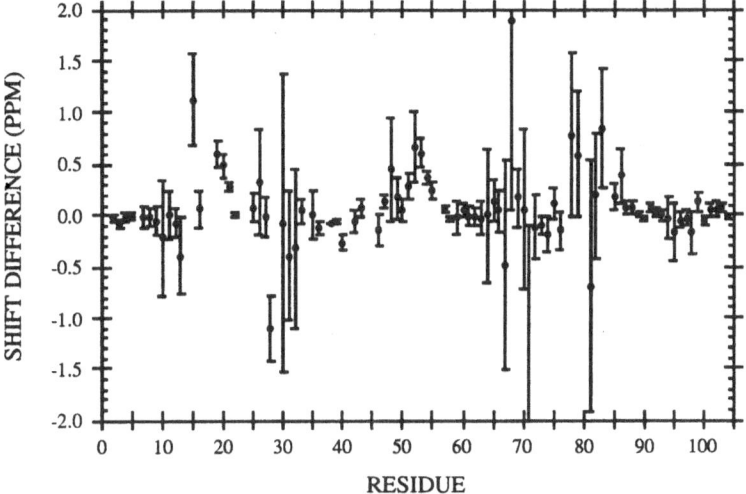

Figure 5. Deviation of predicted pseudo contact shift from the observed redox dependent change in chemical shift for alpha hydrogen resonances of horse cytochrome c. The determined solution structure of oxidized horse cytochrome c was used for the calculation of the pseudo contact shifts. The vertical bars correspond the maximum change in pseudo contact shift predicted to occur with a change of 0.5 Å in any direction by the given atom. Significant deviations from zero indicate a change in the diamagnetic contribution to the chemical shift of a given resonance that arises from a change in structure upon a change in redox state and/or an inaccuracy in the model for the oxidized protein. The major deviations are localized to the 20's loop, the region of the axial methionine ligand to the iron and in the 50's.

A central theme in structural biology is the use of detailed molecular models to use as templates in the description and rationalization of biological phenomena. Cytochrome c has been extensively used as a model system with which to probe the underlying physical parameters governing interprotein electron transfer. Indeed, one of the earliest theoretical applications of Marcus theory (Marcus and Sutin, 1985) to interprotein electron transfer used molecular models of cytochrome c to calculate fundamental parameters governing the electron transfer process (Churg et al., 1983). The relationship between the activation free energy $\Delta g^{\dagger}$ and the reaction free energy ΔG^{0} is central to many aspects of organic and inorganic reaction mechanisms. It is also equally applicable to many processes that occur in biology such as electron transfer. In the context of Marcus theory, the most interesting parameter is perhaps the solvent reorganization energy term which, in the context of interprotein electron transfer, includes not only solvent but also the protein "solvating" the electron transfer centers, in this case, the heme (Churg and Warshel, 1986). The basic elements of the Marcus relationship are shown in Figure 6. According to Marcus theory, which treats the environment as a dielectric continuum, the activation free energy is approximately related to the standard free energy and the solvent reorganization energy by:

$$\Delta g^{\dagger} = \left(\Delta G^{o} + \lambda \right)/4\lambda \qquad (2)$$

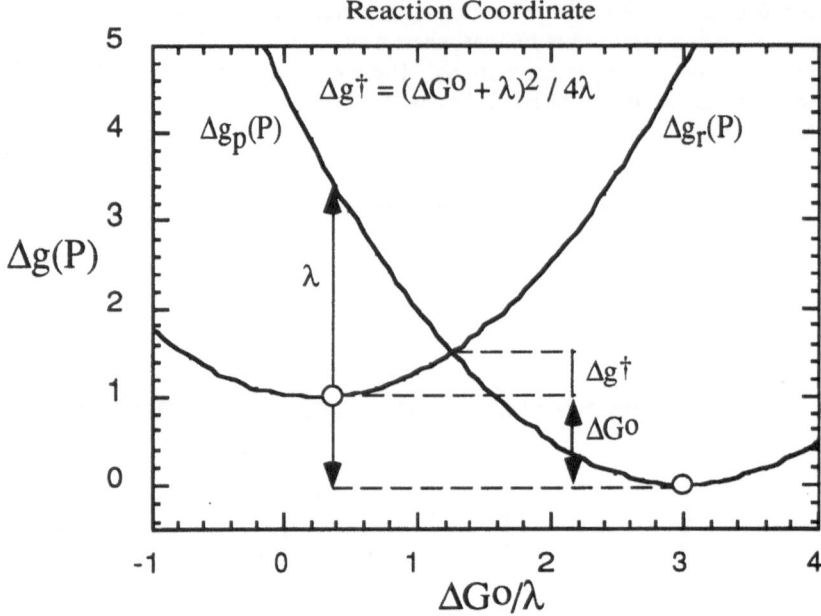

Figure 6. Basic elements of Marcus theory. Here the half-reaction is considered and the two Marcus parabolas are off-set by the redox potential. The two equilibrium structures determined formally rest at the minima of the two curves and are connected by restrained molecular dynamics (see King and Warshel, 1991). The reaction coordinate is given in units of the reorganization energy (λ). The relative positioning of the two Marcus parabolas is governed by the standard free energy (ΔG^O) between the reactant (R) and product (P) states and the activation free energy ($\Delta g^\dagger$).

The development of microscopic simulations employing the equilibrium structures of the product and reactant states of proteins has made it possible to evaluate the solvent reorganization energy using experimentally determined structural information (Churg et al., 1983).

Early treatments of Marcus theory in the context of self-exchange electron transfer between reduced and oxidized cytochrome c led to unexpectedly low values for the solvent reorganization energy (Churg and Warshel, 1983). These calculations were based on crystal structures of tuna cytochrome c in its two redox states which are remarkably isomorphic. We have undertaken a series of calculations of a similar nature using the determined solution structures of the two redox states of the horse protein as a template. The approach has been pioneered by Warshel and coworkers and we have initiated such as calculation in collaboration with Professor Warshel. A preliminary result is shown in Figure 7 where the calculated solvent reorganization energy (λ) is close to the experimental value of ~15 kcal/mole. Marcus parabolas generated by the free energy perturbation method described by King and Warshel (1991) using the program ENZYMIX (Lee et al., 1993) and the determined solution structures of oxidized and reduced horse cytochrome c. These free energy profiles are generated from constrained molecular dynamics trajectories on the experimental structures and keeping track of the energy gap between the reduced protein and the its oxidized (+1 charged) counterpart (see King and Warshel, 1991 for more details). The preliminary effort shown in Figure 7 was carried out without inclusion of induced dipoles but does represent the observed experimental values of the solvent reorganization energy quite well. The flatness of the Marcus curves is also indicative of

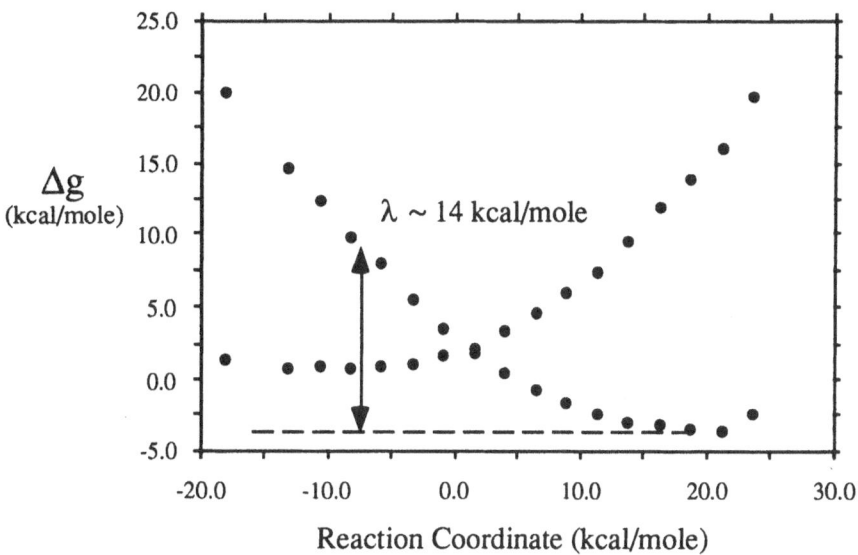

Figure 7. Marcus parabolas generated by the free energy perturbation method described by King and Warshel (1991) using the program ENZYMIX (Lee et al., 1993) using the determined solution structures of oxidized and reduced horse cytochrome c. This preliminary effort was carried out without inclusion of induced dipoles but does represent the observed experimental values of the solvent reorganization energy quite well. The flatness of the Marcus curves is also indicative of undersampling during the molecular dynamics trajectories used to generate these preliminary results.

undersampling during the molecular dynamics trajectories used to generate these preliminary results.

INTERNAL DYNAMICS OF HUMAN UBIQUITIN

I would now like to turn to a related issue which is the role of internal protein dynamics in protein function. Implicit in the calculations just described is the idea that particular motional modes of the cytochrome c molecule may directly promote electron transfer. The physical and functional consequences of internal protein dynamics have long been of interest (see Frauenfelder et al., 1990 and references therein). In many respects, however, the extremely relevant subnanosecond timescale has been difficult to experimentally probe in a comprehensive way. Tremendous strides have been made not only in the methodologies employed in the solution of the resonance assignment problem (Bax and Grzesiek, 1993) and the determination of the solution structures of proteins (Clore and Gronenborn, 1992) but also, most recently, in the characterization of the internal dynamics of proteins by use of NMR relaxation phenomena.

After a rough start where concern was raised about the experimental and analytical strategies of early approaches (Boyd et al., 1990; Peng et al., 1991; Dellwo and Wand, 1991), the application of indirectly detected NMR relaxation techniques to the exploration of the fast internal dynamics of proteins has resulted in the study of two dozen or more systems in the past few years. Most have involved characterization of main chain dynamics by use of ^{15}N relaxation. To date, most analyses of ^{15}N relaxation data have used a particularly simple "model free" form for the spectral density describing internal motion of the ^{15}N-^{1}H heteronuclear vector. The functional form of the spectral density arising from the so-called model-free treatment of Lipari and Szabo is especially appealing on several grounds. This treatment provides for relatively accurate descriptions of the internal motions underlying relaxation in terms of two root parameters: the so-called generalized order parameter (S^2) and the effective correlation time (Lipari and Szabo, 1982):

$$J(\omega) = \frac{2}{5}\left[\frac{S^2\tau_m}{1+\omega^2\tau_m^2} + \frac{(1-S^2)\tau}{1+\omega^2\tau^2}\right]$$ (3)

where $1/\tau = 1/\tau_m + 1/\tau_e$

The effective correlation time (τ_e) and the molecular tumbling time (τ_m) effectively scale the contribution of internal and overall molecular motion to the spectral density $J(\omega)$. The generalized order parameter provides a measure of the spatial amplitude of the motion while the effective correlation time provides an upper limit on the correlation time(s) of the motion(s) contributing to relaxation. There is no requirement to specify the exact physical nature of the internal motion beyond the requirement that they be Markovian or diffusive-like.

Our own work has been with the protein ubiquitin (Schneider et al., 1992). Equation 3 is for the case of isotropic molecular tumbling which allows for the rigorous separation of internal and global motion assuming a suitable separation of time scales (i.e., $\tau_m \gg \tau_e$) (Lipari & Szabo, 1982). As pointed out by Dellwo and Wand (1989), an approximate form for anisotropic tumbling can be used to detect the presence of anisotropic global tumbling and thereby question the applicability of Equation 3. Describing the global tumbling using two correlation times and a mixing parameter followed by a free fit of the global parameters and local model free parameters to the data allows the best fit to point to anisotropic tumbling. In the case of anisotropic tumbling caused by highly asymmetric molecular shape one would not anticipate a fortuitously isotropic distribution of heteronuclear vectors being sampled. Thus, this method would detect anisotropic tumbling by revealing that the best fit to the experimental data is obtained when the two correlation times used are unequal.

In the case of ubiquitin, the freely fitted correlation times were equal at the minimum of the sum of the differences between observed and calculated relaxation parameters. The global correlation time was determined to be 4.2 ns. As one of the first proteins of significant size examined by this method we were surprised and somewhat disappointed by the rather featureless pattern of generalized order parameters obtained for the backbone of the protein (Figure 8). There was no apparent strong correlation with secondary structure beyond a more fundamental correlation with molecular packing and hydrogen bonding. There is also no correlation between the generalized order parameter of a given amide NH and the hydrogen exchange rate of that amide hydrogen or its thermal factor in the crystal structure (unpublished results). Using a chemical shift tensor breadth of 160 ppm, the N-H vectors of peptide linkages participating in one or more hydrogen bonds to the main chain show an average generalized order parameter of 0.80 ± 0.06, while those amide NH of peptide linkages free of hydrogen bonding interactions with the main chain show an average generalized order parameter of 0.69 ± 0.06. The last four residues of the C-terminus display a pattern of generalized order parameters ranging from 0.13 to 0.63 consistent with restricted diffusion about ϕ and ψ rotation axes. These data suggest that molecular packing interactions provide the dominant restriction to internal motion on the subnanosecond time scale and that hydrogen bonding interactions provide a significant but more modest damping of main chain high frequency dynamics (Schneider et al., 1992).

In principle, heteronuclear NMR relaxation phenomena can potentially provide detailed information about the dynamic behavior of internuclear vectors throughout macromolecules such as proteins. Relaxation of ^{13}C nuclei due to dipolar interactions with bonded hydrogen(s) is of particular value because of the potential to report on the dynamic behavior of C-H vectors of the main chain and both buried and surface amino acid side chains of proteins. The interpretation of relaxation for aliphatic methine and methyl carbons in proteins is straightforward owing to the dominant contribution of dipolar interactions to the observed relaxation rates. Unfortunately, however, the application of ^{13}C relaxation measurements to studies of proteins has been hindered by the apparent requirement to selectively and specifically label individual amino acid residues of the protein under study. We have devised a simple two dimensional pulse sequence for the determination of ^{13}C spin lattice relaxation rates that allows for the use of randomly fractionally ^{13}C-enriched proteins (Wand et al., 1995). This offers access to relaxation parameters from methine and methyl carbon sites throughout the protein using a single, easily prepared sample without the need for elaborate chemical synthesis or highly developed bacterial expression systems.

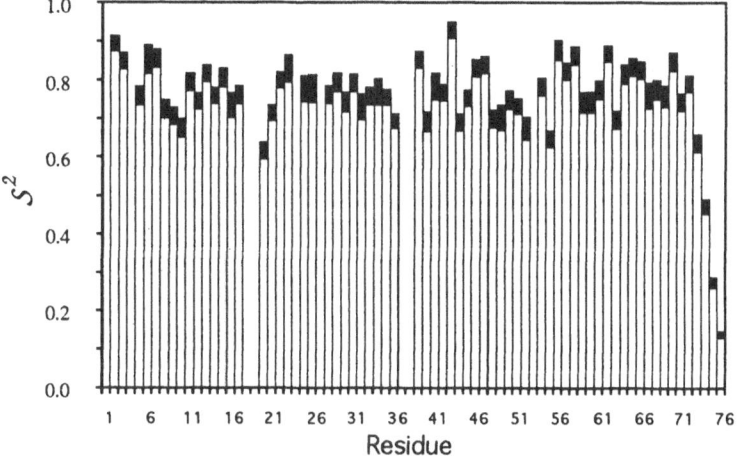

Figure 8. Generalized orders parameters for ^{15}N-^{1}H vectors of human ubiquitin obtained at pH 5 and 30 C. Adapted from Schneider et al. (1992).

A major concern in using uniformly ^{13}C-enriched proteins for relaxation studies of the dynamics of proteins is the contribution of both ^{13}C-^{1}H and ^{13}C-^{13}C dipolar interactions to the observed relaxation behavior. Such a case would lead to multiexponential behavior making analysis of the observed relaxation time courses difficult and unreliable. Randomly and fractionally ^{13}C-enriched protein can be easily prepared by expression during growth on minimal media utilizing a mixture of labeled and unlabeled acetates as the sole carbon source. To suppress ^{13}C-^{13}C pairs, a low pass filter based on ^{13}C-^{13}C J-coupling is inserted into the pulse sequences used to measure the heteronuclear NOE or spin lattice relaxation. This requires use of a constant time period to maintain the fidelity of the filter (Wand et al., 1995). Such a simple extension of the sequence of Skelnár et al. (2) to measure spin lattice relaxation is shown in Figure 9.

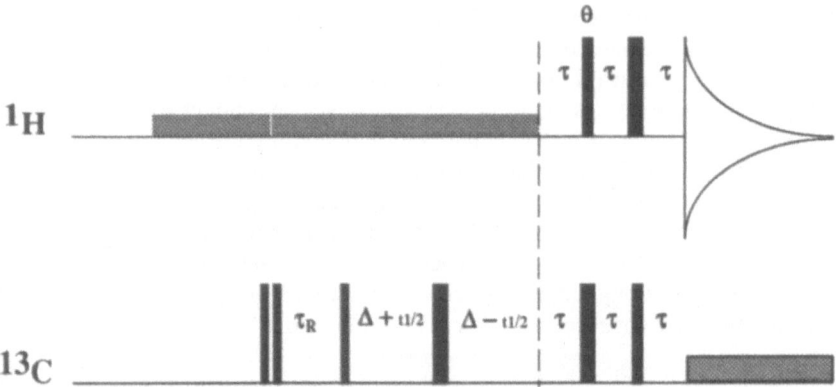

Figure 9. Schematic illustration of the pulse sequence designed to suppress ^{13}C-^{13}C contributions to the observed ^{13}C-spin lattice relaxation in randomly fractionally ^{13}C-enriched proteins. Assuming that τ is set to $1/2J_{CH}$ then 2Δ is set to $1/2J_{CC}$ -$1/J_{CH}$ (Wand et al., 1995).

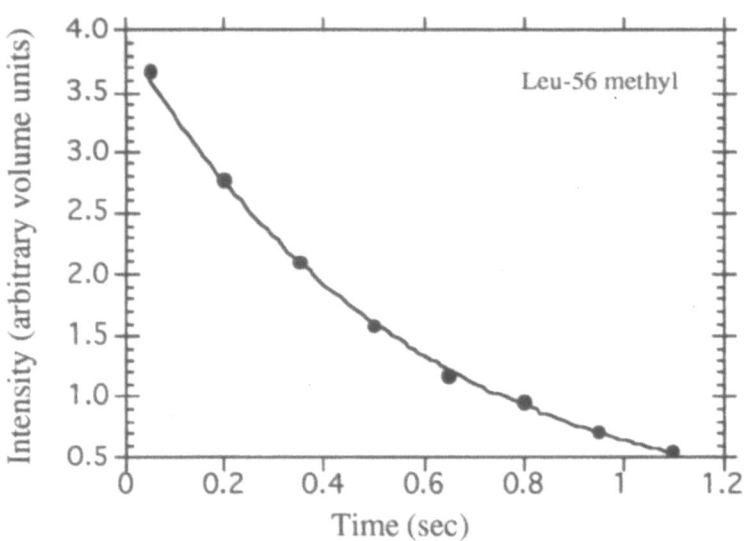

Figure 10. Typical ^{13}C spin lattice relaxation time course measured using 15% randomly ^{13}C-enriched human ubiquitin and employing the low filtered difference T_1 pulse sequence shown in Figure 9. These data were obtained at 11.7 Tesla (125 MHz).

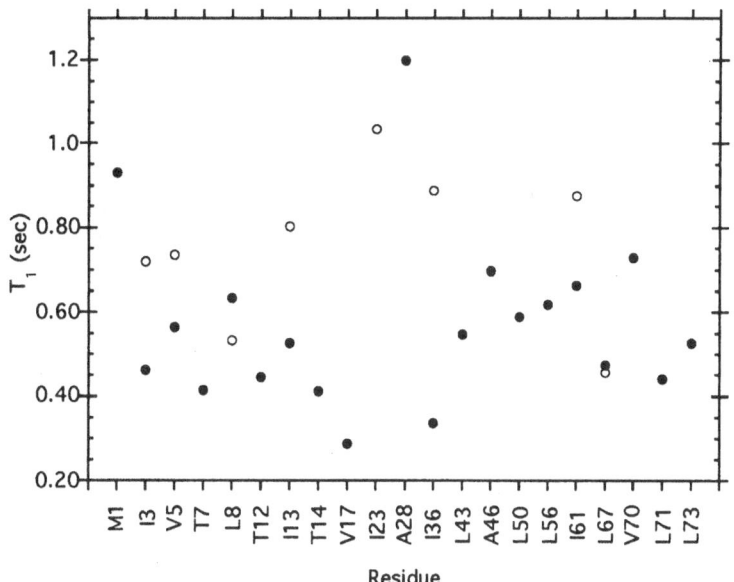

Figure 11. Determined carbon spin lattice relaxation times for methyl groups in human ubiquitin. This data was obtained at 11.7 T (125 MHz).

This sequence is designed to polarize the ^{13}C-nuclei via the NOE , utilizes a difference T_1 time course and increases sensitivity by use of a reverse DEPT (or INEPT) sequence. In the present case, this approach has several distinct advantages over the double DEPT sequences used to study methine carbon or amide nitrogen relaxation (Skelnár et al., 1987). Simple polarization with the NOE avoids creation of unwanted and complicating ^{13}C-^{13}C coherences and allows use of ^{1}H decoupling throughout the low pass filter with similar advantages. It should be noted that the proposed sequence, though suppressing ^{13}C-^{13}C dipolar contributions to the observed relaxation, does not eliminate the problem of cross correlations involved in relaxation at methylene carbon centers. The pulse sequence outlined in Figure 9 is only applicable to methine and methyl carbon centers if the appropriate flip angle (θ) of the reverse DEPT (or delays in the INEPT equivalent) is set to the magic angle as described by Palmer et al. (1991).

The relatively high gyromagnetic ratio of carbon when combined with polarization due to the NOE arising from saturation of bonded hydrogens makes this experiment sufficiently sensitive that avoiding creation of complicating ^{13}C-^{13}C coherences during preparation is a distinct advantage. Illustrations of the very high quality of T_1 relaxation time courses that can be obtained on millimolar protein samples are shown in Figure 10.

A remarkable result that has been obtained is that the methyl groups of ubiquitin show a tremendous range of spin lattice relaxation times which is indicative of the presence of a wide range of motional character for these sites (Figure 11). Assuming that methyl rotation is in the extreme narrowing limit then the difference in spin lattice relaxation rates most likely results from differences in the motion of the methyl symmetry axis. The spin lattice relaxation times for the alpha carbons are much more uniform (not shown). This is consistent with a relatively uniform rigidity along the backbone as indicated by the ^{15}N-relaxation experiments described above (Schneider et al., 1992).

These kinds of data will ultimately allow many important questions to be probed including those central to our understanding of the residual entropy and fundamental dyanmic character of proteins: Can the motions occurring in proteins on the subnanosecond time scale be described by simple Lorenztian spectral densities?

Equivalently, in the language of Frauenfelder and coworkers (1990), are these motions occurring in simple and homogenous energy wells or is the "energy landscape" over which these motions travel much more complex? Ultimately, the broad data base of motional information provided by ^{13}C relaxation studies will present the strongest possible test of the accuracy and validity of the details of molecular dynamics calculations. This is because these data are dominated by "equilibrium fluctuations" that should be consistently manifested in molecular dynamics trajectories of sufficient length. All of these questions and issues have great significance for our understanding of protein structure, dynamics, stability and ultimately function.

ACKNOWLEDGMENTS

I am indebted to the postdoctoral and research associates and graduate students who have contributed centrally to the work discussed here. These include Diane Schneider, Martin Dellwo, Jeffrey Urbauer, Ramona Bieber and Robert McEvoy on studies of the dynamics of ubiquitin and Phoebe Qi, Deena Di Stefano, Ernesto Fuentes and Robert Beckman on the studies of cytochrome c. I am particularly grateful to Professor Arieh Warshel for his collaborative efforts on the calculation of Marcus parameters. This work is supported by NIH grants DK-39806 and GM-35940.

REFERENCES

Bax, A. and Grzesiek, S., 1993, Methodological advances in protein NMR, *Acc. Chem. Res.* 26: 131.

Beckman, R. A., Litwin, S. and Wand, A. J., 1993, Statistical strategy for stereospecific hydrogen NMR assignments: Validation procedures for the floating prochirality method. *J. Biomol. NMR* 3: 675.

Berghuis, A. M. and Brayer, G. D. , 1992, Oxidation state-dependent conformational changes in cytochrome c, *J. Mol. Biol.* 223: 959.

Boyd, J., Hommel, U. and Campbell, I. D., 1990, Influence of cross-correlation between dipolar and anisotropic chemical shift relaxation mechanisms upon longitudinal relaxation rates of ^{15}N in macromolecules, *Chem. Phys.* 175: 477.

Churg, A. K. and Warshel, A., 1986, Control of the redox potential of cytochrome c and microsopic dielectric effects in proteins, *Biochemistry* 25: 1675.

Churg, A. K., Weiss, R. M., Takano, T. and Warshel, A., 1983, On the action of cytochrome c : Correlating geometry changes upon oxidation with activation energies of electron transfer, *J. Phys. Chem.* 87: 1683.

Clore, G. M., Brünger, A. T. and Karplus, M., 1986, Application of molecular dynamics with interproton distance restraints to three-dimensional protein structure determination, J. Mol. Biol. 191: 523

Clore, G. M. and Gronenborn, A. M., 1991a, Structures of larger proteins in solution: Three- and four-dimensional heteronculear NMR spectroscopy, *Science* 252: 1390.

Clore, G. M. and Gronenborn, A. M., 1991b, Two-, three-, and four dimensional NMR methods for obtaining larger and more precise three-dimensional structures of proteins in solution, *Ann. Rev. Biophys. Biophys. Chem.* 20: 29.

Clore, G. M., Kay, L. E., Bax, A. and Gronenborn, A. M., 1991, Four-dimensional ^{13}C/^{13}C-edited nuclear Overhauser enhancement spectroscopy of a protein in solution: Application to interleukin 1β, *Biochemistry* 30: 12.

Dellwo, M. J. and Wand, A. J., 1991, Systematic bias in the model-free analysis of heteronuclear relaxation, *J. Magn. Reson.* 91: 505.

Di Stefano, D. L. and Wand, A. J., 1987, Two-Dimensional ^{1}H NMR study of human ubiquitin: A main chain directed assignment and structure analysis, *Biochemistry* 26: 7272.

Englander, S.W., and Wand, A.J. (1987) Main chain directed strategy for the assignment of ^{1}H NMR spectra of proteins, *Biochemistry* 26: 5953.

Feng, Y., Roder, H., and Englander, S. W., Wand, A. J. and Di Stefano, D. L., 1989, Proton resonance assignments of horse ferrocytochrome c, *Biochemistry* 28: 195.

Feng, Y., Roder, H., and Englander, S. W., 1990, Redox-dependent structure change and hyperfine NMR shifts in cytochrome c, *Biochemistry* 29: 3494.

Feng, Y., Wand, A. J., Roder, H. and Englander, S. W., 1991, Chemical exchange in two dimensions in the ^{1}H NMR assignment of cytochrome c, *Biophys. J.* 59: 323

Frauenfelder, H., Sligar, S. G. and Wolynes, P. G., 1991, The energy landscapes and motions of proteins, *Science* 254: 1598.

Keller, R. and Wüthrich, K., 1981, ^{1}H-NMR studies of structural homologies between the heme environments in horse cytochrome c and in cytochrome c-552 from *Euglena Gracilis. Biochem. Biophys. Acta.* 688: 307

King , G. and Warshel, A., 1991, Investigation of the free energy functions for electron transfer reactions, J. Chem. Phys. 93: 8682

Lee, F. S., Chu, Z. T. and Warshel, A., 1993, Microscopic and semimicroscopic calculations of electrostatic energies in proteins by the POLARIS and ENZYMIX programs, *J. Comp. Chem.* 14: 16.

Leopold, M. F., Urbauer, J. L. and Wand, A. J., 1994, Resonance assignment strategies for the analysis of NMR spectra of proteins, *Molecular Biotechnology* 2: 61.

Lipari, G. and Szabo, A., 1982, Model-free approach to the interpretation of NMR relaxation in macromolecules. 1.Theory and range of validity, *J. Am. Chem. Soc.* 104: 4546.

Marcus, R. A. and Sutin, N., 1985, Electron transfers in chemistry and biology. *Bioch. Biophys. Acta* 811: 265.

Moore, G. R. and Pettigrew, G. W., 'Cytochromes c: Evolutionary, Structural and Physiochemical Aspects', Springer-Verlag, New York, 1990.

Nerdal, W., Hare, D. R. and Reid, B. R., 1988, Three-Dimensional structure of the *wild-type Lac* Pribnow promoter DNA in solution, *J. Mol. Biol.* 201: 717.

Palmer, A. G., III, Rance, M. and Wright, P. E., 1991, Measurement of relaxation time constants for methyl groups by proton-detected NMR spectroscopy, *Chem. Phys.* 185: 41.

Peng, J. W., Thanabal, V. and Wagner, G., 1991, Improved accuracy of heteronuclear transverse relaxation time measurements in macromolecuels. Elimination of antiphase contributions, *J. Magn. Reson.* 95: 421.

Qi, P. X., Di Stefano, D.L., and Wand, A. J., 1994a, The solution structure of horse heart ferrocytochrome c determined by high resolution NMR and restrained simulated annealing. *Biochemistry* 33: 6408.

Qi, P. X., Urbauer, J. L., E. J. Fuentes, Leopold, M. F. and Wand, A. J., 1994b, Structural water in oxidized and reduced horse heart cytochrome c. *Nature Struc. Biol. 1*: 378.

Schneider, D. M., Dellwo, M. J. and Wand, A. J., 1992, Fast internal main-chain dynamics of human ubiquitin, *Biochemistry* 31: 3645.

Sklenár, V., Torchia, D. and Bax, A., 1987 Measurement of carbon-13 longitudinal relaxation using ^{1}H detection, *J. Magn. Reson.* 73: 375.

Takano, T. and Dickerson, R. E., 1981a, Conformation change of cytochrome c 1. Ferrocytochrome c structure refined at 1.5 Å resolution, *J. Mol. Biol.* 153: 79.

Takano, T. and Dickerson, R. E., 1981b, Conformationchange of cytochrome c 2. Ferricytochrome c refinement at 1.8 Å and comparison with the ferrocytochrome structure. *J. Mol. Biol.* 153: 95.

Turner, D. L. and Williams, R. J. P., 1993, H- and C-NMR investigation of redox-state-dependent and temperature-dependent conformation changes in horse cytochrome c, *Eur. J. Biochem.* 211: 563.

Vuister, G. W., Boelens, R. and Kaptein, R., 1988, Nonselective three-dimensional NMR spectroscopy. The 3D NOE-HOHAHA experiment, *J. Magn. Reson.* 80: 176.

Wand, A. J., Di Stefano, D. L., Feng, Y., Roder, H. and Englander, S. W., 1989, Proton resonance assignments of horse ferrocytochrome c, *Biochemistry* 28: 186.

Wand, A. J., Bieber, R. A., Urbauer, J. L., McEvoy, R. P. and Gan, Z., 1995, Carbon relaxation in fractionally randomly ^{13}C-enriched proteins, *J. Magn. Reson. Series B*, in press.

Williams, G., Moore, G. R., Porteous, R., Robinsion, M. N., Soffe, N. and Williams, R. J. P., 1985, Solution structure ofmitochondrial cytochrome c I. [1]H nuclear magnetic resonance of ferricytochrome c, *J. Mol. Biol.* 183:409.

Willie, A., McLean, M., Liu, R-Q, Hilgen-Willis, S., Saunders, A. J., Pielak, G. J., Sligar, S. G., Durham, B. and Millett, F., 1993, Intracomplex electron transfer between ruthenium-65-cytochrome b_5 and position-82 variants of yeast iso-1-cytochrome c, *Biochemistry* 32: 7519.

Wüthrich, K., 1986, NMR of Proteins and Nucleic Acids, Wiley-Interscience, New York.

DISCUSSION

Thomas James - Josh, have you ever thought about calculating a B-factor?

Wand - Yes.

Carol Post - For isoleucine, did the two colors shown represent the two different methyl groups?

Wand - Yes.

Post - Did you look at the adiabatic rotations of those individual methyl groups? Did they correlate at all with the difference there?

Wand - No, we are really not that far along. Our data is very new and hasn't been properly and completely analyzed yet. We haven't put it in the context of a structure.

Post - Is the lower one, the one that's not next to the main chain or anything obvious like that?

Wand - Well, the one that has the lower T_1 is usually the γ, so that's closer to the main chain relative to the δ anyway. That's a qualitative overview, we haven't quantitatively taken apart the data.

PANEL DISCUSSION

Topic: NMR *vis-à-vis* Other Structural Methods

Members: Bernard Brooks, National Institutes of Health

           Mildred Cohn (Moderator), University of Pennsylvania

           Thomas L. James, University of California

           Franklyn G. Prendergast, Mayo Foundation

           Janet L. Smith, Purdue University

Mildred Cohn - The subject of this panel discussion is NMR *vis-à-vis* Other Structural Methods. The way it is going to be run is that each panelist will give a presentation on his or her particular technique, which will include X-ray crystallography by Janet Smith, NMR by Tom James, Fluorescence and Other spectroscopic methods by Frank Prendergast, and finally Computational methods and model building by Bernard Brooks. Speakers will talk about the unique features of their technique, and the limitations and the advantages, and hopefully the contributions that each of them can make to the structures of macromolecules.

One of the questions that I would like to raise is: has there been enough communication among the various practitioners of these techniques. I should like to point out that since this is the closing session of this meeting, and the topic was Current Status and Future Directions, perhaps at this closing we should emphasize future directions. At the opening session Durgu certainly gave us an excellent background of the past, and for three days we have been listening to the current status, and now perhaps it's time to emphasize future directions. Of course, some of the speakers have already done so. We certainly had a glimpse of the future from Al Redfield and some of the other speakers.

I think that it is fitting that we should emphasize that now, and I invite members of the audience to participate in making prophecies not only of the predictable advances that Durgu mentioned but the ones that are really far out, perhaps even in the realm of science fiction. Even those have been known to come to pass in the past.

I will start out myself by raising the possibility of one of the things that we are really concerned about in NMR and that is the sensitivity. I agree with Maurice Guéron that polarization is one way to go but he said he was speculating but I thought he was being very modest indeed. He only asked for a ten-fold increase in sensitivity by polarizing water via a CIDNP mechanism. I should like to point out that in July of this year (1994) there was a paper of a biological experiment with ^{89}Xe by Wishnia and his collaborators where they got a hyper polarization of a factor of 10^5 and this was done by optical pumping and it was done in collaboration with the physicist who described this effect in the 80's. What they did was to let a mouse breathe the Xenon and then showed that they could get a magnetic resonance image of the gas molecules in the lungs. Now if there is any individual sufficiently ingenious to think of how to do this for a nucleus more interesting than Xenon, I think we would be in good shape. By the way, I think it's very nice to bring up a paper by Wishnia at this meeting since more that 30 years ago he was the first one together with

NMR As a Structural Tool for Macromolecules: Current Status and Future Directions
Edited by B.D. Nageswara Rao and Marvin D. Kemple, Plenum Press, New York, 1996

Saunders and Kirkwood to record an NMR spectrum of a protein, ribonuclease, at 40 MHz. They saw two humps. He left the field!! But it is a fact that they were the first to do this. I must say that I have been overwhelmed at this meeting. I am impressed by the ingenuity and the amount of work that has been done on the structure of molecules with NMR. I would however like to emphasize the fact that from my point of view, since one is interested, at least I am interested in the biological implications of structure, I am very pleased to see that many people were interested not only in the anatomy of the macromolecule but it's physiology as well. Not only structure was discussed but function as well. I do think that as the field matures, function will be stressed more. For example X-Ray crystallographers in the early days would show one structure after another but the X-ray crystallographers didn't really care which protein it was or what the particular protein did. Now those days are gone forever. Now it is no longer acceptable. Any respectable X-ray crystallographer who presents something as complicated as the structure of ATP synthase or nitrogenase is very careful to point out the implications for the function. I think NMR is getting there. It isn't quite as mature as X-ray crystallography but I think they are well on the way to having the same perspective. Without further ado I think I will ask Tom James to begin with the NMR.

Thomas James - It's quite a task to be *the* NMR representative here on this kind of a panel. Since everybody here knows an awful lot about NMR, I will cover only a couple of aspects. In his keynote lecture, Prof. Ernst mentioned that as NMR spectroscopists we should be addressing how do we answer a scientific question and leave to the philosophers why we want to study such a question. On the other hand, the approach that we use and exactly how much detail we might want to go into does entail a "why" question, i.e., why do I want to know the structure and possibly the dynamics of a biomolecule. As Mildred just pointed out, probably one of the primary reasons for this is that many people might like to understand the functioning of an important biological process and, as she pointed out, there are many people here in the audience who have moved in this direction, although there are many of us who still have what I might say is a legitimate scientific interest in wanting to be able to describe the true nature of molecules. I think it is completely legitimate if some people want to devote themselves to the question of exactly what are these molecules - what is the true nature of these molecules and how do they really move. What do they really look like - without worrying about the biological implications. That is a very tough problem and an interesting one. It is also legitimate not to be too concerned at all about what the particular function is. In certain cases, if we could understand the structure of a molecule, it might be possible that we could actually even design some kind of a ligand which would bind to that and alter its function, whatever that function was; aside from minor problems such as drug delivery and so forth, it might become a very nice drug which could solve a major medical problem that humanity faces. Not coincidentally, it might even make somebody some money. Of course other people might consider the question "why" from the viewpoint: well, I am an NMR spectroscopist, and I understand that is the sort of thing NMR spectroscopists do; they figure out what structure is, and maybe they might even measure relaxation properties and consider dynamics as well.

An important consideration for NMR spectroscopists is the extent to which understanding of the structure could be enhanced by using information that comes from other methodologies. I think this reluctance to use other data reflects our early days of paranoia. Some of the audience here might remember one meeting in Park City, Utah, in the 1980s in which Kurt Wüthrich got up and donned a ski racing vest; he made the remark that he was the only NMR person on a program, where the rest were crystallographers, to present information about protein structure, but that he was ready to start the race. I think it may have been a good idea that we had very independent-minded individuals who have helped establish NMR as an independent technology. However, rather than being in a downhill race with crystallographers or other people, I think we should view ourselves as perhaps being on a mountain climbing team and work together to achieve the same goals.

Since this is an NMR spectroscopy meeting we might consider from our viewpoint how can we utilize non-NMR data. One possible approach is that we can actually go through the structure determinations - even obtain information about the dynamics - in much the same manner we're presently doing and simply utilize other information that might be available for example, some chemical footprinting. Maybe there is an X-ray structure available, or maybe there is a homologous protein structure that might be available from X-ray crystallography. We can simply use this as a check. We can just make sure that everything we did is in agreement with all of the other known and available experimental data.

Alternatively, we can actually use the non-NMR information explicitly in structure refinement. Rob Kaptein pointed out that they had indeed done this, utilizing some chemical footprinting experiments to aid them with their docking methodology. In fact there is not just chemical footprinting; there are lots of methodologies - mutational analysis, enzymatic footprinting, and other spectroscopies potentially circular dichroism (CD), Raman, and possibly Fourier transform infrared spectroscopy (FTIR). Many experimental parameters could be utilized explicitly with objective or target functions. You could include additional pseudoenergy terms in a force field to help in terms of guiding structural refinement. Of course, in doing these things, one has to take particular care of the balance that we choose between different kinds of methodologies - how we are going to weight different types of information that go into making a structure determination. Nevertheless, I think this is an area which can be and probably should be explored in more detail in a cooperative fashion with our other structural biological colleagues as well as our molecular biology friends.

Valuable information may be available in the form of crystal structures or the NMR structures for homologous molecules. There are several very useful modeling programs available now permitting us to do some homology modeling even prior to doing NMR experiments on a protein. There is a growing emphasis in many labs entailing a whole series of mutants or perhaps a whole series of a protein type which might bear some homologous features. Right now it is possible to calculate NOE intensities of any proposed homology model; you can compare that with the experimental intensities from the structure that you are trying to determine. So you don't have to start from zero. You could find out what is similar and what is different about the structures, simply on the basis of what the computed NOE intensities should be for a particular structural model versus those from a mutant.

Although I don't know how far in the future - maybe it's next week, maybe it's a couple of years from now - but the kinds of things that Brian Sykes' talk brought up yesterday, which David Case and Eric Oldfield have been working on too, of trying to utilize chemical shifts for predicting structures could be employed in very much the same way. So we could increase the throughput for the determination of homologous structures. Since I have taken more than the allotted time, I will stop here.

Cohn - If anyone in the audience has any questions for Tom, please do not hesitate to ask him. Does anybody have any suggestions about - the future of developments in NMR. If not we will go on to the next speaker. We choose the new kid on the block - the computational and modeling technique and see what Bernard Brooks has to say about what it will contribute and what it's future is.

Bernard Brooks - The first thing I would like to point out is that most simulations that are carried out today actually are not done with NMR structures in mind. Throughout the course of the meeting you have been given numerous examples of simulations but all were in conjunction with trying to explain or interpret NMR data. The question as pertains to this panel discussion is: can structures be determined using simulations? But given that as a caveat let me try to answer the question anyway. In order to address this the first thing you have to ask are subquestions. The first subquestion is, "What accuracy is required, i.e., what does it mean to get a "correct" structure?" Also what probability of having a correct structure is required? Whenever you generate a structure using simulation and modeling tools, these structures are generally not unique and you can't say for certain that this is the correct structure. But you can say, this is the correct structure within a certain probability. Sometimes the probability may be as low as 0.1% or less.

One of the factors that affect this is that the global minimization problem is not solved for a multiple minimum problem. It is unlikely to be a solved problem and there are defects in the potential energy surface, there are defects in the modeling, there are a lot of assumptions going into the modeling, and all of these effect the ability of using simulation and modeling to develop structures. In the most general case, can you develop de novo protein structure; can you determine structures using de novo design? The answer I think is generally 'no'. But instead of just stopping at 'no', let's just show what has been done. There has been a lot of effort. There are several groups who are actually trying to build a structure from the sequence, or to determine what sequence fits a given structure. In many cases the overall topology can be found in the correct way. Where the topology is not put into the system, they can get the overall topology. In general, structures are found to within 3 $\overset{0}{A}$ in some cases but I would say that is not routine. If you try to look for details in the structure, they are generally not available. There remains a low probability for finding a correct structure. For every structure found to within 3 or 4 $\overset{0}{A}$, there are many others that are much less accurate. So, this gives perhaps an overly bleak view of the things. But there is much that can be done. If you are trying to model proteins from homology, there are a number of cases in the literature, where, given one sequence in a structure known from either NMR or X-ray, you can then predict what the structure is for a system that is highly homologous. This has been quite successful. The degree of success depends on the degree of homology and it depends on a number of other factors. In general, doing simulations is better than just doing delete and construct methods or minimization methods because you can explore more conformational space.

Another case where there has been a lot of success is in modeling substrate-protein interactions or drug-protein conformations. In this, it is fairly easy to exhaustively enumerate the way in which small compounds fit into proteins and there is a high probability of determining the correct configuration in these cases. Another area where there has been a lot of success is in determining structures of protein loops. Again whether or not you get a protein loop correct rests on what accuracy is required. I have a short example where you can correctly determine the structure of protein loops that I probably will not have time to go into. I also have another example where you want to simulate flexible systems. Molecular dynamics is a very good tool for studying small flexible systems. In terms of the future I think what we are going to see is that potential energy defects are going to be reduced. The effect of having more computer time will make a dent on the global minimum problem, i.e., you will be able to do a better job of modeling because you will have more computer time in which to search for the correct structure and I see that these computational methods will play an increasingly large role both in interpreting data from NMR and crystallography as well as being able to determine structures.

Cohn - Are there any questions one would like to raise about computations and calculations?

Marc Adler - One of the few things we can predict is the spectacular rise in computation time. Do you think that there are fundamental problems in the field like the fact that dipole induced dipoles in an n^3 problem that computational time will not be able to solve?

Brooks - I don't see that as being computationally intractable. Things like the dipole problem can be solved self consistently. So even though it is potentially n^3, you can converge these rather quickly. So the cost of including full polarization models is in the worst case a factor of eight or ten times more expensive than the simulations that we have existing today. So I don't see that as being intractable at all.

Yuan Xu - Computer power might be the limit because right now in the computer industry people are trying to build parallel computers with multiple processors up to several hundred thousand. Probably, it is time to teach the chemistry programmers how to write programs that can be run on parallel computers. Otherwise, if you just run simulations on single processors like a vector machine or on the supercomputers, you have reached the limit.

Brooks - Most simulations we run today are run in parallel across multiple processors. I would say 90% or more of our cycles are run on parallel. Right now most of the parallel computers we use have between 4 and 128 processors. We have been developing the techniques where we can actually exploit machines with as many as 2000 or 3000 processors. I think we are ready for the next generation of machines. I don't see that as being a bottleneck. There are software limitations to what we can do, but I don't think they are in the area of being able to use high speed parallel computers.

Cohn - If there are no further questions in this area, I will ask Frank Prendergast to talk about fluorescence and other spectroscopic methods.

Frank Prendergast - It was very generous of Dr. Cohn to invite me to talk about all optical spectroscopic methods in five minutes! This isn't really possible. What I am going to do is to speak rather generally because I think there are a number of commonalties among the methodologies. I am not going to try to talk about everything, obviously. I trained with Joe Lakowicz as a fluorescence spectroscopist and then spent a lot of time and money building money sinks - mark 1's and 2's - for fluorescence spectroscopy. I don't want to give too bad a picture but I have seen the light and the light says that we need a lot of other bits of spectroscopic information to buttress fluorescence data in order to make meaningful physical interpretations. Accordingly, we now do an awful lot of NMR spectroscopy in collaboration with our colleagues here, Dr. Rao and Dr. Kemple, and also back at Mayo with Dr. Macura and his colleagues. The reason for so doing will become apparent very quickly.

If you take fluorescence as an example, there are very many complexities to its use. Fluorescence spectroscopists have been guilty of hype, and that hype is inimical to the field. The capabilities of fluorescence to provide molecular information is limited. True, we can do very fancy things. We have wonderful picosecond resolution in all of our instruments. In my lab, we have very sophisticated techniques, probably as sophisticated as any in the world. But the problem is that the technique is limited by a large number of complications, not the least of which is the fact that fluorescence reports primarily, but not entirely, on local phenomena within the protein matrix. The second issue is that in time resolved measurements one is faced with the serious problem of heterogeneity of fluorescence lifetimes, and the fluorescence lifetimes are essential for all subsequent calculations of molecular motion.

It is very difficult (not impossible) to sort out the causes of intensity decay heterogeneity. In part they may derive from collisional interactions, an inference which has given rise to the view that heterogeneous lifetimes may reflect the existence of rotational isomers or rotamers. But many other factors cause heterogeneity including, basically, any type of energy transfer, e.g. resonance energy transfer or electron transfer. Heterogeneity may also result from the effects of paramagnetic ions or from dipolar relaxation - a mostly mythical term that everybody talks about but nobody really understands. If such dipolar interactions occur with the excited states of proteins, it could be with dipoles within the protein or through dipole-induced dipole interactions or because of interactions with solvent. Furthermore, limitation in the analysis of the data, all the way from understanding how the convolution integral should be approximated to limitations in the actual type of analysis used, e.g., whether one uses least squares, Levenberg-Marquardt algorithms, maximum likelihood methods, or maximum entropy methods, also cause problems. One gets different results from each of these different techniques, and it is really amusing how often fluorescence spectroscopists tend to worry about how many angels there are on the head of a pin arguing, for example, whether lifetime differences of 4.9 and 4.2 ns are significant and interpreting such differences in ways that are totally spurious. Measurement of motion is even more complicated. Motional parameters determined by fluorescence should be in agreement with those determined by NMR. The second Legendre polynomial is, after all, measured in precisely analogous ways by the two methods. But you don't get the same result from fluorescence and NMR very frequently. There are obviously problems with both methods.

Another of the several problems affecting fluorescence is a simple thing like the determination of the limiting anisotropy, r_0, which can be affected either by motion or by electronic effects. The influence of the local environment on the photophysics of the excited state can actually be profound. In fact you can measure r_0 in large proteins and find limiting anisotropies of close to zero due, apparently, to intramolecular electronic interactions among aromatic moieties. We have also discovered from photoacoustic spectroscopy that when you excite a fluorophore, having pumped a lot of energy into the molecule, there are "local" temperature increases which have consequences for the subsequent behavior of the fluorophore and of the protein matrix in which the fluorphore is embedded. Local volume contraction or volume expansion phenomena occur, which we simply don't understand. We are just beginning to investigate these local signals and the consequences for collective motion. But what is collective motion? Everybody uses the term but usually without explanation. Is it a whole segment moving? Is it merely movement of a domain, or is it reflective of only a loop moving? The point is that all of the above could be collective motion. Even "local motion" is difficult to define and can be very difficult to describe. Moreover, even if some motion is detected, it must be anisotropic given the multiplicity of torsion and dihedral angles to be traversed.

A note, very quickly, on CD. CD can provide information only on average structure. Additionally, Brian Sykes made a very important point this morning, namely, that the distinction between secondary and tertiary structural perturbation of the CD signal is important and cannot be discerned from the CD signal itself. One is, after all, measuring the sum over all possible CD interactions in the protein and therefore symmetry plays an important role in the signal evinced. Some local structural detail is theoretically possible to devine, but until we understand better induced CD effects, it will be impossible to interpret local effects in CD spectra. FTIR, likewise gives structural information averaged over the whole protein. One can get some local structural detail, especially if one employs clever labeling and the reduced mass effect but it is no trivial matter to do so. Raman data suffer from an interesting problem, namely that there is too much information. Resolution of the detail in Raman spectra is going to be extremely difficult. But Raman spectroscopy has a very promising future for protein studies.

What about the future? I think generally that CD and FTIR will continue to be used as adjuncts for other spectroscopies. One is simply not, in the near future, going to be able to decompose a fluorescence spectrum in such a way as to provide the sort of detail possible from NMR or crystallography. Time resolved FT-IR spectroscopy such as that which Dr. Woodruff is doing offers tremendous opportunity for looking at the kinetics of conformational events and chemical change. Likewise, time resolved CD offers some interesting possibilities, especially for looking at ligand protein interactions, once we have a better understanding of induced CD effects. Raman, especially resonance Raman technologies, I think, have very powerful applicability not only for looking for gross structural changes but also for studying local environments.

For fluorescence, better analytic techniques are essential. Ironically, we probably need to develop model-dependent analyses in which, for example, we first get an idea of what patterns of motions might exist from molecular dynamics simulations, and then see if we can iterate towards a meaningful convergence of three sets of information namely that from simulations, from NMR, and from fluorescence. We also need to better understand the photophysics of fluorophores, to be able to interpret how local interactions perturb fluorescence. But the information gathered from fluorescence will always give only indirect information on structure. Photoacoustic spectroscopy has not found its place yet but it's coming. Very exciting work is being done by Dr. Rudzki-Small at Eastern Washington University. I want to end with just one issue. In the final analysis, it is the question which is important not so much the technique.

Cohn - Are there any questions for Dr. Prendergast or comments?

Maurice Guéron - This is not a question. Just a comment. On Sunday, there were questions about the relation between fluorescence and NMR and I made a statement that there was a factor of three between the two correlation times. This was incorrect as has just been stated by you. These times should be equal apparently.

Cohn - Now, we come to X-ray crystallography that Janet Smith is going to talk about. This is the grand daddy of all techniques for studying molecules and of course has been the most successful and the most mature. Just to give you an example, I remember when Perutz first did the structure of hemoglobin, it took him 10 years to do it and now I think it can be done in at least an order of magnitude less time. From Janet we will here about the current status - what it can do and what it can hope to do.

Janet Smith - If Tom felt trepidation up here as the representative of the NMR community, you can imagine how I feel as the sole representative of the crystallographic community. But of course this is 1994 and not 1984 and I think our experimental methods these days are largely complementary rather than competitive. In the last 10 or 12 years, while 3-dimensional structure determination by NMR has been developed for biological macromolecules, there has also been a large change in what we are able to do in crystallography. This goes in three main directions, which can be summarized as faster, better and bigger in that order of importance. Also macromolecular crystallography these days really is driven by biology.

The demand for new structures is constantly growing. Most crystallographers turn collaborators away more frequently than we'd like because there aren't enough hands to do all of the interesting structures. Crystallography remains the most definitive technique for getting very accurate 3-dimensional structures of macromolecules for a very wide range of sizes. The revolution that has happened in our field has come about for many of the same reasons as in NMR. The first is the availability of recombinant materials, which has diminished the crystallization problem. Second, the availability of faster, cheaper computers, has made an enormous difference to the way that we do our research just as it has for NMR. Finally, in crystallography we've had a revolution in the way we collect our diffraction data. We can collect many more data much faster and these are also better data. These things together have led us to solve structures much faster than we were able to do in the past. At the same time, we've had a quantum leap in the quality of structures. Some of this has come from the establishment of a database of structures so that we have much better definition of good quality, but also it comes from getting much better diffraction data. Better data have come from better instruments at home and also through the availability of synchrotron radiation, which is extremely important to biological crystallography. In the future, these trends will continue. Obviously, as we have been able to solve structures faster and more accurately, we have also pushed the frontiers of our end of biology in the direction of size just as people in NMR are doing. For example, in a few years we will have 3-dimensional crystal structures for the subunits of ribosome. This is still a little ways off but definitely within sight. In the future I think we will continue to see structures obtained for larger complexes and structures obtained faster. In favorable cases we're now down to a couple of months for de novo structure determinations where we don't know the folds before we start and, therefore, have to do experimental phasing, and to as short as a week or so for structures that can be solved by molecular replacement from an existing fold.

Where crystallography is going in the future is towards larger structures, and towards better structures from larger macromolecules and smaller crystals. We are pushing the limits of size and determining structures faster. There are a couple of technological developments that will allow us to do this. One of the things going on all around the world right now, which is about to become the standard for crystallography, is to carry out all the experiments at a very low temperature (about -160^0C) by flash freezing crystals. This has made an enormous improvement in the quality of diffraction data that we get, the speed with which we get the diffraction data, and the size of the crystals that we can work on. One of the best things about cryo-crystallography is that one data set comes from one crystal. This has revolutionized how we are doing our experiments. There are some new synchrotron sources just coming online, including the new facility in Grenoble by the ESF and the soon-to-be Advanced Photon Source at Argonne Laboratory outside Chicago. Both of these facilities will have several stations for macromolecular crystallography.

Some of the flashier experiments that are not routine today will become routine. Primary among these is multiwavelength anomalous diffraction (MAD), which is a difficult experiment because of demands on the x-ray optics to rapidly change wavelength, but new beamlines are being built with this in mind. So some of the MAD experiments that you have been hearing about will be applicable to a much larger set of macromolecules. One very clever experiment, which was proposed by Wayne Hendrickson, is a selenomethionine MAD experiment. The cleverness is the genetic incorporation of selenomethionine into proteins in place of methionine. This turns out to be extremely useful in many ways in crystallography, not just for a MAD experiment. For example, it will give us a heavy atom derivative very routinely in many cases. Selenomethionine is going to become more important as people catch onto its versatility.

Finally, I want to say something about where our challenges are and where we are in relation to the NMR community. There are a few things that we don't do well in crystallography, and one of them is to study the dynamics of protein molecules, in particular the string-of-beads motif and the large scale motions between beads that Peter Wright talked about this morning. Obviously, we are never going to see anything in crystal structures unless we compare many different crystal lattices. I hope that a marriage of NMR and crystallography can address some of these problems of large scale motion. We can get very accurate structures for the subset of conformations that we are able to crystallize. There may be many other conformations of biological importance that we are not going to be able to see in the crystalline state. NMR could also possibly be an analytical tool for structural changes in smaller parts of molecules that are important for biological functions. It would also be very useful if, from a set of 3-dimensional coordinates for a macromolecule, one could compute an NMR spectrum or some NMR parameters and bridge the gap between the smaller macromolecules that NMR spectroscopists can work on and the larger ones for which crystallographers can determine structures. The two methodologies may be able to come together there. One of the challenges that remains in crystallography is that we still haven't figured out very well how to crystallize membrane proteins. Now I believe that these systems also cause problems for NMR, but this is a field where our requirements for crystallization have been a big problem. Our knowledge about structures of membrane proteins is limited by our ability to crystallize them. Macromolecular complexes are very important and can be extremely difficult to crystallize. I think there is going to be a growing interaction between crystallographic methodology and structural methods, particularly electron microscopy, that can look at even larger structures and aggregates than we are currently able to crystallize.

Cohn - Are there any questions for Dr. Smith?

B. D. Nageswara Rao - This is not necessarily a question for Dr. Smith. I would like to raise the following question. As Maurice pointed out, two different answers may come from two different techniques like fluorescence and NMR giving two different correlation times. Even that seemingly simple point we have not resolved. I am sure there are differences like that coming from spectroscopies. We should address these things more and more to really understand how to accept certain differences and then inquire into some of the others. With reference to the correlation time problem one could explain it if you bring in anisotropy of motion or question the Debye model of rotational random walk for small angular displacements but it may be forcing you in a direction which is quite wrong. Maybe the Debye model is quite right or quite acceptable for these situations and maybe then it is indicative of anisotropy. Once again it is a can of worms but we have to address such questions. Similarly, between NMR and x-ray crystallography the few times x-ray crystallographers tried to locate the metal ions for example in ATP complexes, they came out with answers which did not make sense at all from the NMR point of view. If the x-ray crystallographers asked somebody to run the NMR spectrum they would have found out that this answer cannot be right. But this has not happened. These are the kinds of issues that we have to address. The question then that I want to generally ask is, "if you do find some differences, are there certain things in your mind of how we want to approach this issue of obtaining different results from the different spectroscopies."

Prendergast - Let me describe the tact that we took a few years ago upon realizing the problem we were having in trying to interpret what is truth when a fluorescence result presents itself and realizing that all fluorescence results we have, whether they be lifetimes or correlation times, are parameters derived from fitting data. You determine that you are going to use a particular mathematical model, and you get the parameter. Then you say "is this the truth"? But if you look at most papers that are published in the field, you'll see that they derive for example a τ_e or a τ_m and they immediately make the assumption that it must be correct because it's there on paper. Then you do an NMR experiment and you find that you get a different answer, i.e., it does not agree. One of the things we decided to do was to go to molecular dynamics simulations using techniques that Bernard Brooks, Charlie Brooks, Martin Karplus, and everybody else has developed. Actually we use Charlie Brooks' technique and ask if we can get any information from that. We set up a triangle. We had three different methodologies, one theoretical and two experimental. Then we made the argument before hand that if the two experiments agree we didn't care what the theory said. On the other hand what if the theory and one of the experimental methods agree? This is what happened and it is the dilemma that we face now. The critical issue in my assessment is to try to find the best molecular models. For example with tryptophan motion, we find that tryptophan likes to lie on surfaces; it does not like to wiggle freely around and, try as we may, we cannot find proteins where the tryptophan likes to wiggle around. The consequence is that when we do NMR, we get order parameters that approach one, and fluorescence anisotropy decays that do not show much decay except with whole protein motion. We then go for a peptide where there are in fact substantial degrees of freedom for the motion and where things do wiggle around. But that is not really simulating a protein. So we are caught in a dilemma. We are right at the stage now of looking for the perfect protein. Let me define it for you. It is a protein of reasonable molecular size with not too atypical surface structure so it is not going to give you abnormal water adsorption to the surface; in other words, it should have reasonable hydration patterns. It must be recombinantly produced in large amounts, and it must be sufficiently soluble for easy NMR measurements. Also the tryptophan or some other residue which we can measure by fluorescence must be wiggling around giving us order parameters of about 0.45. And I must have enough money to label it with ^{13}C and ^{15}N. If any of you have such a protein, I will pay for it.

Cohn - You are asking for quite a bit. If there are no further questions, I would like to thank the panelists for an education and particularly all our thanks to Dr. Rao for arranging this Symposium and for his generous support.

Solution Conformation of Mucin Derived Multiply O-linked Glycopeptides
Thomas A. Gerken, Departments of Pediatrics and Biochemistry, Case Western Reserve University, Cleveland Ohio 44106-4948.

Mucin glycoproteins are high molecular weight heavily O-glycosylated glycoproteins containing 50 to 80% carbohydrate by weight and are responsible for the viscoelastic properties of mucous secretions. Mucins behave as extended random coil peptides with chain dimensions 3 fold greater than those of denatured random coiled peptides. This expansion is solely due to the presence of the O-linked carbohydrate residues (Shogren et al., Biochemistry 28, 5525 (1989)). In order to examine the interactions responsible for the unique conformation of mucins and other highly O-glycosylated glycoproteins we have isolated short O-linked glycopeptides from porcine submaxillary mucin (PSM) for detailed NMR and modeling studies. Several short glycopeptides have been obtained and their structures confirmed by amino acid sequencing and NMR. 600 MHz 2DNMR studies of two glycopeptide, GA $\hat{T}$GA$\acute{S}$ IGQPE $\hat{T}$ $\acute{S}$ R, and $\acute{S}$ IGQPE $\hat{T}$ $\acute{S}$ R (where $\hat{T}$ and $\acute{S}$ represent Thr and Ser residues containing O-linked GalNAc) are being performed and compared to their nonglycosylated peptides. Except for the GalNAc H5 and H6 protons each carbohydrate and amino acid residue's protons have been specifically assigned. The conformational effects of O-glycosylation are being examined by monitoring the peptide and carbohydrate amide proton chemical shift temperature dependencies, coupling constants, and 2D ROESY NOE connectivities. Preliminary results from the ROESY data suggest that both the glycosylated and nonglycosylated peptides exist in extended conformations, although, the glycosylated peptides may be more extended due to an apparent increase in solvent accessibility obtained from the amide temperature dependencies. In contrast, the amide proton of the GalNAc attached to Thr-12 may be involved in a hydrogen bond. With further analysis of the data for these peptides and for others with different sequences, the detailed effects of multiple O-glycosylation on peptide conformation should be revealed. Supported by NIH grant DK 39918.

Solution Structure of Neuropeptides Isolated from the Migratory Locust

G. E. Jackson[†], K. Sunter[†] and [*]G. Gadé
Department of Chemistry[†] and Zoology[*], University of Cape Town, South Africa.

The migratory locust *(locusta migratoria)* adipokinetic hormone I (AKH I) is one of a family of neuropeptides which is involved in energy metabolism during flight. In locusts the hormone is responsible for the release of diacylglycols from fat stores and enhanced lipid oxidation during flight. The family of peptides share several common features, all being octa or decapeptides with blocked N and C termini. It is thought that the reason for the structural similarity between the neuropeptides from different insect species is the necessity for a particular secondary conformational structure for receptor binding.

We have calculated the structure of AKH I in DMSO and aqueous SDS solution using nmr determined distance and torsion angle constraints. Spin diffusion has been taken into account using MORASS.

¹H NMR Derived Molecular Structure of Certain Rhodamine Conformations in Water

P. K. Mishra, S. Macura, P. Ilich, K. Ajtai, T. P. Burghardt
Mayo Foundation, Rochester, MN 55905

Rhodamines are widely used as laser fluorescing dyes. Their ability to bind specifically at some muscle sites make them useful as "markers" in the study of the movement of muscles. In order to draw conclusions about the muscle movement from the laser fluorescence spectroscopy studies it is necessary to understand the geometrical structures of rhodamines in aqueous solution. Optical absorption studies in N, N-alkyl rhodamines have hinted self-association at a concentration of few micromoles in solution. ¹H NMR studies in rhodamine-B, rhodamine-590 and their mixture in water establish the geometrical structure of their homo- and hetero- dimers and provide information about the orientation of their side chains.

The Solution Structure of μ-AGA-I, a Sodium Channel Activator Isolated from the Venom of Agelenopsis aperta. Diana Omecinsky, Michael Adams† and Michael D. Reily, Department of Chemistry, Parke-Davis Pharmaceutical Research, Division of Warner-Lambert Co., 2800 Plymouth Road, Ann Arbor, Michigan 48105. † Departments of Entomology and Neuroscience, University of California, Riverside, California 92521.

We report the solution structure of μ-agatoxin I, a voltage-gated sodium channel activator in insects, isolated from the venom of the funnel web spider, *Agelenopsis aperta*. The structure of this 36 residue polypeptide neurotoxin was determined by NMR, distance geometry and molecular dynamics calculations. 160 NMR-derived distance restraints and 20 angle restraints obtained from vicinal coupling constants were used as input for the structure calculations. The peptide contains eight cysteines of which the pairings of four cysteine residues were uncertain and had to be determined from preliminary structure calculations. The cysteine rich segment containing residues 2-34 has an average rmsd of 0.84 Å for the backbone atoms among 19 converged conformers. The structure consists of a triple-stranded β-sheet involving residues 6-9, 20-24 and 30-34. The μ-agatoxins have been shown to be very effective insecticides in contrast to structurally similar neurotoxins which interact with different ion channels in mammals. Our results together with recent reports on these structurally related, yet functionally diverse, peptides should aid us in understanding the basis of selectivity in these neurotoxic peptides.

Characterization of the Bound Conformation of a Receptor Peptide
Antagonist to Interferon-gamma and Mutants of Interferon-gamma using
Transferred Nuclear Overhauser Ennhancement Spectroscopy.
Gail F. Seelig, Winifred W. Prosise, Julio C. Hawkins, Charles Lunn,
Daniel Lundell, and Mary M. Senior. Schering Plough Research Institute,
2015 Galloping Hill Road, Kenilworth, New Jersey, 07003

   Transferred nuclear Overhauser enhancement (TrNOE) studies on
recombinant interferon-gamma and a twenty-two amino acid receptor
peptide antagonist [120-141(Tyr)acm] were performed at 500MHz. The
peptide was identified using anti-idiotypic antibody mapping
techniques. In addition to native interferon, TrNOE studies were also
performed on the peptide and two mutant forms of interferon-gamma. The
location of these mutations are at the carboxyl terminus of the protein
and in a loop located between helices A and B. These mutant proteins
have been shown to have diminished receptor binding activity
demonstrating the importance of the carboxyl terminus [Lundell et. al,
(1991) Protein Engineering, Vol 4, 335-341] and the AB loop [Lundell et
al., (1994) J. Biol. Chem. Vol. 269, 16159-16162].
   The TrNOE data shows that the peptide antagonist binds to interferon-
gamma. Intramolecular interactions within the bound peptide were
detected at two non-contiguous regions and at a third region comprising
a beta turn formed by the sequence DIRK. When TrNOE studies were
performed on the mutant proteins, the beta turn was observed for the
carboxyl truncated mutant, but intramolecular interactions in other
parts of the peptide were not detected. The peptide in the presence of
the interferon with the modified AB loop showed no TrNOE effect
suggesting that mutation of the AB loop prevents binding to the peptide
antagonist. These studies support biological data which indicates that
the carboxyl terminus and AB loop are important regions for receptor
binding to interferon-gamma.

The FISINOE-2 Program for Determination of Polypeptide Conformations from NMR

Simon A. Sherman* and Igor L. Tomchin.
Eppley Cancer Institute, University of Nebraska Medical Center, Omaha, NE 68198-6805

The FISINOE-2 program, that is a new generation of the FISINOE program[1], allows the
determination of the accurate polypeptide conformations that are consistent with a given set of NMR
data. The new version of the program uses both values of intraresidue and sequential cross-peak
intensities, graded as strong, medium, and weak, and values of coupling constants, $^3J_{\alpha N}$ and $^3J_{\alpha\beta}$,
graded as high, intermediate, low, and negligible, as input data to determine the mathematical
expectations of the ϕ, ψ and χ_1 angles and their standard deviations, σ_ϕ, σ_ψ, and $\sigma_{\chi 1}$. The upper limit
for distance when a strong cross peak is observed is set at 2.5Å, for a medium cross peak - 3.0Å,
and for a weak cross peak - 3.6Å. Values of vicinal coupling constants are graded as high ($> 8Hz$),
intermediate (in the interval from 5 to 8Hz), low (in the interval from 2 to 5Hz), and negligible
($< 2Hz$). FISINOE-2 uses joint density distributions of the three angles, ϕ, ψ, and χ_1, for all 20
types of residues. The joint density distributions were obtained on the basis of statistical analysis of
Brookhaven Protein Data Bank (Sclove & Sherman, unpublished data). FISINOE-2 uses NMR
measurements and the joint angular distributions to estimate the upgraded values of the ϕ, ψ and χ_1
angles and their standard deviations by a Bayesian inferential paradigm. Comprehensive
computational experiments made on the basis of simulated NMR data have shown that FISINOE-2
allows the determination of the backbone angles, ϕ and ψ, with an accuracy of about 20°, and the
χ_1 angles - with an accuracy of about 15°. FISINOE-2 also gives stereospecific assignments of β-
protons as a by-product of the χ_1 determination. The results of applying the FISINOE-2 program to
study conformational features of several proteins and peptides using real NMR data are discussed.

[1] Sherman, S.A. & Johnson, M.E. *J. Magn.Res.*, Vol.96. pp.457-472, 1992.

Pharmacophor Model of Substrates that Receive O-linked Glycosylation

Simon A. Sherman*, William H. Gmeiner, Leonid I. Kirnarskiy, Sam D. Sanderson, and Michael A. Hollingsworth. Eppley Cancer Institute, University of Nebraska Medical Center, Omaha, NE 68198-6805

To better understand the molecular processes that govern the specificity and kinetics of O-linked glycosylation of serine and threonine with N-acetylgalactosamine by the enzyme UDP-GalNAc:polypeptide N-acetylgalactosaminyltransferase (GalNAc transferase), the structure of six nonapeptide substrates that contain glycosylated and non glycosylated residues were investigated by NMR. A probabilistic approach[1] was applied to determine low-energy NMR-matched structures of the peptide substrates. Comparison of spatial structural features of active and inactive substrate analogues allowed us to formulate hypotheses that may explain the observed differences in substrate effectiveness. GalNAc transferase recognizes the stereospecific orientation of the side chain, amide and carboxyl groups of the backbone of L-hydroxyamino acids. For enzymatic recognition, it is hypothesized that the hydroxyl group must be exposed, the backbone of the residue must form an elongated structure, and both amide proton and oxygen must be also exposed and oriented toward the surrounding solvent. The backbone of several residues flanking to the reacting hydroxyamino acid should form an elongated frame, that has restricted flexibility and maintains the amide proton of the glycosylated residue (position g) and oxygens of the residues in positions g-2, g, and g+2 in a parallel orientation. An energy barrier for transition from an elongated to any twisted conformation predisposes the substrate to the elongated conformation that facilitates its initial interaction with GalNAc transferase. It is hypothesized that a net of hydrogen bonding between glycosylated substrate and the GalNAc transferase may be formed in the process of their interaction. The amide proton of the glycosylated residue may serve as a donor for hydrogen bonding with the active site of the enzyme. The backbone oxygens of residues in the g-2, g, and g+2 positions may also be involved in the net of hydrogen bonding with GalNAc transferase. Presumably, residues in the positions g-1, g+1, and g+3 influence the spatial arrangement and the rigidity of the frame, rather than being involved in a direct interaction with GalNAc transferase. This model can be used for the design of new substrates or inhibitors of GalNAc transferase.

[1] Sherman, S.A. & Johnson, M.E. *Progr. Biophys. molec. Biol.*, Vol.59. pp.285-339, 1993.

The NMR Solution Structure of the Amyloid ß-Peptide Provides a Molecular Approach to the Treatment of Alzheimer's Disease. J. Talafous, K. J. Marcinowski, C. Keane, S. Jao, A. Salomon, M. Eisenenberg, D. Goldgaber, M. G. Zagorski, Department of Chemistry, Case Western Reserve University, Cleveland, Ohio 44106-7078 USA.

Patients with Alzheimer's disease have amyloid plaques in their brain tissue. The major proteinaceous component of the amyloid plaques is the ß-peptide. 2D NMR data was collected for residues 1 to 28 of the amyloid ß-peptide [ß-(1-28)], a fragment of the ß-peptide. Restraints were derived from the data which included NOE crosspeaks, vicinal coupling constants, and temperature coefficients of the amide-NH chemical shifts. The three-dimensional solution structure of ß-(1-28) was derived with X-PLOR 3.0 using the DGSA protocol. The resulting structure is a monomeric α-helix and is the first model of any amyloid ß-peptide fragment. We will also present possible pathways of aggregation motivated by our structure. Since plaque formation probably occurs through aggregation, inhibition of these pathways may lead to therapeutic treatments for Alzheimer's disease.

NMR Study of A 36-Residue C-Terminal Fragment From Apolipophorin-III --- An exchangeable Apolipoprotein

Jianjun Wang, Vasanthy Narayanaswami*, Robert O. Ryan* and Brian D. Sykes

Protein Engineering Network of Centres of Excellence and *Lipid and Lipoprotein Research Group, Department of Biochemistry, University of Alberta, Edmonton, Canada T6G 2S2

Apolipophorin-III is a 166 residue exchangeable apolipoprotein from the sphine moth, Manduca sexta. It plays a critical role in stabilizing LDLp particle, therefore transporting lipoprotein-associated diacylglycerol in hemolymph during sustained flight. While the structure of M. sexta apoliphorin is unknown, an analogous apolipophorin-III structure from Locusta migratoria, which bears considerable sequence and functional homology, has been solved and shown to be a five helix bundle connected by loops. A 36 residue peptide fragment from the C-terminal of apoliphorin-III was obtained by cyanogen bromide treatment. The NMR results indicate random structures of the peptide in aqueous solution and helical structures in both TFE and in presence of SDS micelles. These helical structures were generated by molecular modeling based on NMR experimental data. Compared to the model structure of the intact protein generated from the homology of locusta migratoria apolipophorin-III, this 36 residue peptide corresponds to the fifth helix of the intact protein. The present study suggests that although this 36 residue peptide has strong tendency to adopt a helix structure, it fails to form a helix structure in aqueous solution. The environmental changes such as TFE solvent or the presence of SDS micelles stabilize the peptide and induce a protein-like structure. It is shown in the present study that in the intact protein, the interactions between different helices stabilize this 36 residue fragment, therefore induce the formation of the fifth helix. The possible interactions between this 36 residue peptide with the other helix fragments in the intact protein are discussed.

A Variable Target Intensity-Restrained Global Optimization (**VARTIGO**) Procedure for Determining Three Dimensional Structures of Polypeptides from NOESY Data

Yuan Xu, Istvan P. Sugar and N. Rama Krishna. Dept. of Biochemistry and Comprehensive Cancer center, University of Alabama at Birmingham, Birmingham, AL 35294; Dept. of Biomathematical Sciences and Physiology and Biophysics, The Mount Sinai Medical Center, New York, NY 10029, USA

A new global optimization method for intensity-restrained structure refinement based on a variable target function (VTF) analysis has been illustrated using experimental data on a model peptide, gramicidin-S (GS) dissolved in dimethyl sulfoxide (DMSO). The method (referred to as VARTIGO for *variable target intensity-restrained global optimization*) involves minimization of a target function in which the range of NOESY contacts is gradually increased in successive cycles of optimization in dihedral angle space. Several different starting conformations (including all-trans) have been tested to establish the validity of the method. Not all optimizations were successful, but these were readily identifiable from their large NOE R-factors. We have also shown that it is possible to simultaneously optimize the rotational correlation time along with the dihedral angles. The structural features of GS thus obtained from the successful optimizations are in excellent agreement with the available experimental data. A comparison has been made with structures generated from an intensity-restrained single target function (STF) analysis. The results on GS suggest that VARTIGO refinement is capable of yielding better quality structures. Our work also underscores the need for a simultaneous analysis of different NOE R-factors in judging the quality of optimized structures. The NOESY data on GS/DMSO appear to provide evidence for the presence of two orientations for the ornithine sidechain, in fast exchange. The NOESY spectra for this case were computed using a relaxation matrix which is a weighted average of the relaxation matrices for the individual conformations.

Dynamics of Two Model Peptides

Peng Yuan*, Marvin D. Kemple*, and Franklyn G. Prendergast[†], *Physics Department, Indiana University-Purdue University Indianapolis, Indianapolis, IN 46202-3273; [†]Department of Pharmacology, Mayo Foundation, Rochester, MN 55905

The rotational motion of a tryptophan (trp) side chain in a protein or peptide can be characterized by fluorescence anisotropy and NMR relaxation measurements when the trp moiety is appropriately labeled with ^{13}C or ^{15}N. Physically reasonable motional parameters and agreement of the two techniques have been obtained from studies of free trp. For trp as a component of a protein and/or a relatively large peptide physically reasonable motional parameters are obtained, but there are some discrepancies between the two techniques. It then becomes of interest to study the dynamics of trp in small peptides. Presented here are the results of a dynamics investigation of two peptides, KSAWSGK with a ^{13}C enrichment at the W-Cδ_1 and G-Cα positions, and KWK with ^{13}C enrichment at the W-Cδ_1 position. The peptides were dissolved in D_2O at near neutral pH and at concentrations of 2 mM for KSAWSGK and 4.5 mM for KWK. ^{13}C NMR relaxation rates, T_1 and NOE, were measured at 75.4 and 125.7 MHz. The overall rotational correlation times were determined to be 0.39 ns for KSAWSGK, and 0.2 ns for KWK. Proton NMR measurements indicate that KSAWSGK aggregates at a concentration of 2 mM. Additional experiments are underway to determine if concentration-dependent dynamical parameters result and if this peptide possesses any structural features. Surprisingly, trp in both peptides showed quite restricted motion, with S^2 being 0.45 for KSAWSGK and 0.58 for KWK. Fluorescence anisotropy measurements will be performed to provide additional information on the trp dynamics in these peptides. The work was supported in part by NSF Grant DMB-9105885 to MDK and by PHS Grant GM34847 to FGP.

The pKa Values of Melittin in Solution and in Lipid Micelles

Lingyang Zhu*, Marvin D. Kemple*, and Franklyn G. Prendergast[†]
*Department of Physics, Indiana University - Purdue University Indianapolis, Indianapolis, IN 46202 and [†]Department of Pharmacology, Mayo Foundation, Rochester, MN 55905

Melittin (MLT) is a 26-amino acid cytolytic peptide from the *Api mellifera* honey bee. It is known to exist as an α-helical tetramer or as a monomeric random coil depending on solvent conditions. The charge state of MLT is believed to be a major factor in determining its aggregation properties and its interaction with lipids. Several contradictory indirect measurements of the pK_a values of the three lysine groups in MLT have been reported. In the present study, high resolution ^{15}N-NMR at 50.6 MHz was used to directly measure the pK_a values of MLT. The pH dependence of MLT ^{15}N chemical shifts were measured separately for the isotopically labeled backbone α-^{15}N of gly-1 and the side chain ζ-^{15}N's of lys-7, lys-21 and lys-23 at a MLT concentration of 1.2 mM and a temperature of 23°C. Measurements were made for MLT in phosphate buffer, in neat water, and in MMPC lipid micelles. The results show that each lys residue has a pK_a of 10.2 and gly-1 has a pK_a of 8.1 in phosphate buffer. Similar results were obtained for MLT in 48 mM MMPC. The pK_a values were somewhat lower for MLT in water with gly-1 at 7.6 and lys-21 and lys-23 at 9.9 and 9.8 respectively. These pK_a values indicate that the lysine residues are positively charged when MLT forms a tetramer at pH values between 8.5 and 10 contrary to prevailing expectations. This work was supported in part by NSF Grant DMB-9105885 to MDK and by PHS Grant GM34847 to FGP.

Differences in Peptide-Micelle Interactions Observed Under Kinetic and Thermodynamic Conditions

Philip K. Hammen, Henry Weiner and David G. Gorenstein^, Purdue University, West Lafayette, IN 47907 and University of Texas Medical Branch^, Galveston TX 77555

Measurement of polypeptide NMR spectra in the presence of micelles has provided a membrane binding model for many years. While micelles are a membrane mimetic medium, they lack many of the properties of biological membranes. The major deficits are smaller size and lack of an ordered bilayer. Do peptides really bind to micelles in a way that accurately models a peptide-membrane interaction?

To address this question, we have studied synthetic amphiphilic α-helical peptides, which correspond to mitochondrial targeting sequences, in the presence of micelles. The three peptides we have studied become helical in the presence of micelles and display a range of affinity (2 uM to 50 uM) for lipid bilayers. We have measured amide proton exchange rates for each peptide in the presence of micelles, under two conditions. One is with a preformed peptide-micelle complex and the second involves the mixing of peptide and micelle solutions. The latter experiment allows one to observe the first amide protons sequestered from solvent in the interaction with the micelle. In general, the results are consistent with the affinities for lipid bilayers. The peptide with the greatest affinity for the bilayer demonstrates the greatest protection from solvent exchange. The other peptides rank according to their bilayer affinities. However, the protons protected from exchange and the pattern of protection can differ for the the two conditions, demonstrating that the initially formed complex may not resemble the complex at equilibrium.

NMR Study of High Pressure Effects on Lateral Diffusion Coefficients of Sonicated Phospholipid Vesicles.

Bao-Shiang Lee*, Stephanie A. Mabry*, Ana Jonas[†], and Jiri Jonas* *Department of Chemistry, School of Chemical Sciences, and [†]Department of Biochemistry, College of Medicine, University of Illinois, Urbana, IL 61801

High pressure effects on self-diffusion of the sonicated pure dipalmitoylphosphatidylcholine (DPPC) vesicles and sonicated pure 1-palmitoyl-2-oleoyl-phosphatidylcholine (POPC) vesicles (15% by wt) in D_2O were measured at pressures of 1 bar to 5 kbar and at temperatures of 50°C to 70°C for DPPC and 5°C to 20°C for POPC using proton NMR spin-lattice relaxation times in the rotating frame ($T_{1\rho}$). The diffusion coefficients (D) were calculated from the plots of the relaxation rate in the rotating frame ($T_{1\rho}^{-1}$) versus the square root of the spin-locking field ($\omega_1^{1/2}$). The decrease of the lateral diffusion coefficient with increasing pressure in the LC phase was attributed to volumetric effects. Very little pressure dependence of D was observed in the pressure-induced gel I (GI) or interdigitated gel (Gi) phases. As expected the lateral diffusion coefficient sharply decreased at the LC to GI phase transition pressure. The activation volumes of diffusion ($\Delta V^{\neq}$) were calculated for both DPPC (37 ml/mole at 50°C, 34 ml/mole at 60°C, and 25 ml/mole at 70°C) and POPC (16 ml/mole at 5°C, 9 ml/mole at 20°C, and 6 ml/mole at 35°C) sonicated vesicles in the LC phase. The activation energy of diffusion (E_a) was calculated for both DPPC (3.48 kcal/mole) and POPC (3.87 kcal/mole) in the LC phase, and for POPC (4.4 kcal/mole) in the pressure-induced GI phase.

^{2}H NMR Studies of Unsaturated Lipid Bilayers

Stephen R. Wassall, M. Alan McCabe, Cynthia W. Browning, Sherrel D. Harris and Stephen P. Schuh, Department of Physics, Indiana University-Purdue University Indianapolis, Indianapolis, IN 46202

Our current solid state NMR studies investigate the considerable impact that the location of acyl chain unsaturation has on the properties of membranes. Specifically, we have applied broadline ^{2}H NMR techniques to compare [[2]H$_{31}$] 16:0 - 18:1 PC (1 - [[2]H$_{31}$] palmitoyl-2-*cis*-octadecenoylphosphatidylcholine) bilayers in which the double bond is at the $\Delta 6$, $\Delta 9$, $\Delta 12$ or $\Delta 15$ position. Moments reveal dramatic differences in phase behaviour. The temperature of the gel to liquid crystalline transition, which varies by more than 40°C, exhibits a minimum when the double bond is in the centre of the chain. Moment analysis also demonstrates that average order in the liquid crystalline state changes by over 35%. When comparison is made at the same temperature the dependence resembles that seen for the transition temperature, while the variation is inverted and the $\Delta 12$ isomer is most ordered at equal reduced temperature. Order parameter profiles mapped from depaked spectra confirm the trends identified. They indicate, moreover, that the plateau region of approximately constant order in the upper portion of the *sn*-1 chain becomes longer as the unsaturation is moved down the *sn*-2 chain. Determination of the conformation of the double bond is the focus of recently initiated work which attempts to explain the differences observed.

NMR Study of MMPC Micelle Dynamics and Peptide-Micelle Interaction

Peng Yuan*, Marvin D. Kemple*, and Franklyn G. Prendergast[†], *Physics Department, Indiana University-Purdue University Indianapolis, Indianapolis, IN 46202-3273; [†]Department of Pharmacology, Mayo Foundation, Rochester, MN 55905

The lipid micelle has been used as a simplified model to study membrane binding and interacti and related informationr many years. Presented here are the results of studies of monomyristoylphosphocholine (MMPC) micelle interaction and binding with the cytolytic peptide melittin (MLT). The ^{13}C and ^{1}H NMR chemical shifts of the lipid were assigned based on the result of conventional 1D, DQCOSY, and HMQC experiments. The dynamic behavior of the lipid with and without MLT binding at pH=4 and 25 °C in 50 mM phosphate buffer solution was characterized by three motional parameters of the "model-free" formalism, which were extracted from NMR relaxation rates, T_1 and NOE, measured at two different magnetic fields. In the case of the micelle only, the overall correlation time, $\tau_m \approx$ 14 ns, was obtained consistent with the aggregation state of the lipid. It was observed that the fastest and least restricted motions occur for the methyl groups at the end of both the acyl chain (the ω-1 position) and the choline head group. The motions become gradually more restricted toward the middle of the glycerol backbone. Also the motion becomes more restricted for CH_2 groups in the acyl chain near the backbone. With MLT bound to the micelle, much the same trends are observed, but the hydrophobic core of the micelle apparently becomes more packed as indicated by increases in order parameters of a number of groups along the acyl chain. Both the glycerol and choline groups were also affected upon MLT binding. The work was supported in part by NSF Grant DMB-9105885 to MDK and by PHS Grant GM34847 to FGP.

Structure-Function Relationships in <u>Cucurbita maxima</u> Trypsin Inhibitor-V (CMTI-V), as Studied by NMR Spectroscopy, M. Cai[1], Y.X. Gong[1], O. Prakash[1], J. Kao[2], L. Wen[3], S. Han[3], G. Chen[3], J.-K. Huang[3], and R. Krishnamoorthi[1*], [1]Department of Biochemistry, Kansas State University, Manhattan, KS 66506; [2]Department of Chemistry, Washington University, St. Louis, MO 63130; [3]Department of Chemistry, Western Illinois University, Macomb, IL 61455.

<u>Cucurbita maxima</u> Trypsin Inhibitor-V (CMTI-V; M_r 7 Kd) belongs to the Potato I inhibitor family. It has 68 amino acid residues, including a disulfide linkage. It is also a specific inhibitor of a blood coagulation protein, Factor XII_a (Hageman Factor). The inhibitory activity toward Factor XII_a is lost upon hydrolysis of the reactive-site peptide bond between Lys44 and Asp45. In an effort to delineate structure-function relationships in CMTI-V, we have undertaken a multi-dimensional NMR investigation of native CMTI-V, its reactive-site hydrolyzed form (CMTI-V*), recombinant CMTI-V (rCMTI-V), and a set of engineered mutants with dramatically altered functional properties--R50A, R52A, and D45I. As a first step, we have determined an average three-dimensional solution structure of native CMTI-V by the application of standard 2D NMR and computational methodologies, utilizing 213 intra-residue connectivities, 267 inter-residue connectivities, and 25 pairs of hydrogen bonds based on the slow amide NH exchange rates and inter-residue nOe's, and 76 torsional angles. The secondary structural elements of the inhibitor protein are: an α-helix between residues 17 and 28; three pairs of β-sheets, one parallel comprising residues 50 --> 56 and 31 --> 38; and two anti-parallel sheets--one involving residues 67 --> 65 and 7 --> 9, and the other, residues 62 --> 60 and 12 --> 15; and three β-turns--a type II turn from residues 12 - 15, and two type I turns from residues 28 - 31 and from residues 57 - 60.

NMR work with CMTI-V* has allowed a comparison of the chemical shifts of the backbone hydrogen atoms, $C_\alpha H$'s and amide NH's, of the intact and modified forms of the inhibitor, which indicates that the Lys44-Asp45 peptide bond cleavage perturbs only those residues that form the binding loop; a few other residues that interact with the binding loop also exhibit changes in the chemical shifts. The secondary structural elements, which form the scaffold region of the inhibitor, are not affected. A three-dimensional solution structure is being determined to characterize the altered conformation of the binding loop in CMTI-V*.

Also in progress is work involving rCMTI-V and the above-mentioned mutants in order to characterize structural changes due to the amino acid substitutions.

^{1}H NMR Studies of the Three Dimensional Solution Structure of a 6-kD Proteinase Inhibitor Isolated from the Stigmas of *Nicotiana alata*

David J Craik[a], Katherine J Nielsen[a], Robyn L Heath[b] & Marilyn A Anderson[b]

Centre for Drug Design and Development, University of Queensland, St. Lucia 4072, Australia
[b]Plant Cell Biology Research Centre, School of Botany, University of Melbourne, Parkville, 3052, Victoria, Australia

The three-dimensional structure and disulfide connectivities of a 6-kD protein isolated from the stigma of the ornamental tobacco *Nicotiana alata* has been determined by ^{1}H NMR spectroscopy combined with simulated annealing calculations. The protein, termed C1, is a chymotrypsin inhibitor and is one of five homologous proteinase inhibitors that are proteolytically cleaved from a 40.3-kD precursor protein, (Atkinson, A.H., Heath, R.L., Simpson, R.J., Clarke, A.J., & Anderson, M.A, (1993) *The Plant Cell*, 5, 203-213). The other four proteinase inhibitors (T1-T4) contain reactive sites for trypsin. The three dimensional structure of C1 is generally well defined and contains a triple stranded β-sheet as the dominant secondary structural feature. Several turns and a short region of 3_{10} helix are also present. The putative chymotrypsin-reactive site is present on an exposed loop which is less defined than the rest of the protein. The overall shape of C1 is disc-like and the N- and C- termini are exposed, supporting the proposal that this protein results from post-translational processing of the 40.3-kD precursor protein. C1 has 70% sequence identity with the potato chymotrypsin inhibitor (PCI-1) and this is reflected in a structural similarity and identical disulfide connectivities of the proteins.

347

NMR Structural Studies of a Human *src*-Homology 2 Domain Complex

Robert Gampe[*], Robert Xu, Michael Word, Donald Davis, Martin Rink and Derril Willard
Glaxo Research Institute, Five Moore Drive, Research Triangle Park, N.C. 27709

Cytosolic *src* is a nonreceptor kinase that participates in intracellular signal transduction and has been implicated in the development of malignant tumors in the human colon and breast. *Src* is comprised of conserved domains that engage in various functions during intracellular signal transduction and in self regulatory processes. One of the domains, the *src*-homology 2 or SH2 domain, mediates with high selectivity the signal transduction by binding phosphorylated tyrosine motifs located on specific receptor types. In order to elucidate the molecular interactions and conformation in solution, we have determined a family of solution structures of the human *src* SH2 domain in complex with a high affinity phosphorylated pentapeptide, Acetyl-pYEEIE-OH. The structures, generated with distance geometry and a dynamical simulated annealing protocol, satisfied 2072 experimental restraints derived from a variety of multi-frequency/-dimensional and isotope filtered NMR data. Residues 143-246 from the family of 23 structures superimposed upon the mean coordinate set yielded a rmsd of 0.58 ± 0.09Å for the N, Cα, C' atoms and 1.04 ± 0.08Å for all the nonhydrogen atoms. The protein exhibits 3 antiparallel β-sheets that traverse a compact core with an α-helix on each side of the core near the N- and C- termini. Intermolecular nOes from the pY, +1E and +3I sidechains positioned the ligand in an extended conformation across the SH2 domain surface with the pY and +3I sidechains inserted into protein binding pockets. The solution conformation is consistent with previously reported crystal structures and a solution structure of different SH2 domain complexes. A comparison is made to a crystal structure of the same SH2 domain complex, which exhibited crystal contacts at the ligand binding site.

Comparison of the Solution Structures of α- and β-Toxins from the New World Scorpion *Centruroides sculpturatus* Ewing.

Michael J. Jablonsky, Patricia L. Jackson, Dean D. Watt, and N. Rama Krishna. University of Alabama at Birmingham, Birmingham, AL 35294; Creighton University, Omaha, NE 68178.

We have determined the solution structures of a β-toxin (CsE-I) and an α-toxin (CsE-V) isolated from the venom of the scorpion *Centruroides sculpturatus* Ewing (CsE). These toxins interfere with the transport of Na[+] through the cell membrane. The CsE-I affects the activation of, while CsE-V affects the inactivation of the Na[+] current. CsE-I and CsE-V have 43% sequence homology. CsE-V is unique in that it represents a transition protein between the New World and the Old World scorpion toxins. Sequence-specific assignments for both toxins have been made using 2D ^{1}H data at 600 MHz. Data from NOESY (100-400 ms mixing times), phase-sensitive COSY, and amide-exchange experiments were used as constraints for molecular modeling calculations. Distance geometry and dynamical simulated annealing calculations were used to generate families of structures for both toxins. The solution structures of both toxins will be compared with each other, and with other scorpion neurotoxins. These data give insight into the structure-function relationship of the toxins, and provide a basis for the study of toxin-sodium channel interactions.

Solution NMR Studies of a Pheromone Binding Protein

Smita Mohanty and G. Prestwich, Department of Chemistry,
State University of New York at Stony Brook, New York 11794.

Lepidopteran PBP (Pheromone Binding Protein) forms a subfamily of the insect odorant binding proteins (OBPs) in which ligand specificity is presumed to exist for the pheromone components of the female produced sex pheromone blend. The exquisite sensitivity and molecular specificity of moth olfaction offers the optimal model for unraveling the mechanism of signal transduction. The first insect OBP identified was the PBP of the wild silk moth *Antherea polyphemus*. These proteins, located in the sensillum lymph of the pheromone responsive sensory hairs in the male antenna, were supposed to mediate delivery of hydrophobic sex pheromones to specific receptor proteins in the dendrite membrane.

This 142 residue protein has a molecular mass of 14 kDa and is mostly α-helical as observed from the CD studies. Solution NMR studies on this pheromone binding protein from the silk moth *Antherea polyphemus* will be presented. ^{1}H NMR assignment is in progress on a 3 mM solution of PBP in phosphate buffer at pH 6.0, using both homo and heteronuclear NMR experiments.

The Lumazine Protein from *Photobacterium leiognathi*: NMR study of structure and interactions. John Lee, Bruce Gibson, and Jacques Vervoort. Departments of Biochemistry and Molecular Biology, University of Georgia, Athens, GA, and Wageningen Agricultural University, The Netherlands.

Lumazine protein is a 22 kDa-protein functioning in bioluminescent bacteria as an "antenna". The excitation energy is chemically generated at the active site of a second protein, the enzyme luciferase (77 kDa). The excited state of the lumazine is populated by an excitation transfer process within a protein-protein complex with a luciferase reaction intermediate. Fluorescence dynamics results indicate that the donor site on the luciferase and the lumazine acceptor site on its protein, are separated by only 15 Å. Both these proteins have defied attempts at producing crystals suitable for X-ray determination of 3D-structure. The structure of lumazine protein can be determined by NMR methods. To this end the lumazine protein primary sequence was determined and then an *E. coli* expression system established. The expressed protein is identical to the native, both in spectroscopic and bioluminescence functional properties. Homonuclear NOESY and TOCSY results also indicate closely similar 3D structures. A high yielding growth method in minimal media has been developed and the lumazine protein labelled in ^{13}C and [15]N. The results from double and triple resonance experiments will be shown. This work is supported by the NIH and NWO.

Refined Solution Structure and Dynamics of the Glycosylated Adhesion Domain of the Human T-Cell Glycoprotein CD2 and its Proposed Interaction with CD58

Daniel F. Wyss§§, Johnathan S. Choi¶, Kwaku T. Dayie¶, Antonio R.N. Arulanandam§^(a), Alexander Kister^(a), Malcolm J. McGregor§, Ellis L. Reinherz^(b) and Gerhard Wagner¶.

§Procept, Inc., 840 Memorial Drive, Cambridge, MA 02139 ; ¶Department of Biological Chemistry and Molecular Pharmacology, Harvard Medical School, Boston, MA 02115; ^Laboratory of Immunobiology, Dana Farber Cancer Institute and Departments of (a)Pathology and (b)Medicine Harvard Medical School, Boston, MA 02115.

CD2, a 50-55 kDa T-cell specific surface glycoprotein, initiates the adhesion of T lymphocytes to target cells and antigen presenting cells prior to up-regulation of other receptor-mediated adhesion pathways. Additionally, CD2 serves a signal transducing function which may act to stimulate or inhibit the activation of T-cells. In humans, CD2 binds specifically to CD58 expressed on the cell to which the T-cell binds. Human CD2 is a member of the IgSF and consists of a two-domain 185 amino acid residue extracellular segment, a 25 residue hydrophobic transmembrane-spanning region and a 117 residue cytoplasmic tail. All adhesion functions have been mapped to the extracellular amino-terminal 105 residue domain of human CD2 (hu-sCD2$_{105}$) which contains a single consensus N-glycosylation site at Asn^{65} that is occupied with high mannose oligosaccharides [(Man)$_n$GlcNAc$_2$,n=5-8] when a soluble version of the CD2 extracellular segment (hu-sCD2$_{182}$) is expressed in CHO cells. This adhesion domain of human CD2 requires a N-linked carbohydrate to maintain its native conformation and ability to bind CD58.

To better understand the structural aspects that regulate human CD2 adhesion functions we recently determined the solution structure of the protein part of hu-sCD2$_{105}$ using predominantly homonuclear NMR methods and DG calculations [J. M. Withka, D. F. Wyss et al. Structure 1, 69-81 (1993)]. On the basis of this structure, we have made alanine-substitution mutations in human CD2 to probe the binding site for CD58. We have also identified the other binding surface of the CD2-CD58 adhesion pair by mutating charged residues shared among CD2 ligands that are predicted to be solvent exposed on a molecular model of the Ig-like adhesion domain of human CD58. Possible docking orientations for the CD2-CD58 molecular complex will be offered. We recently refined the structure of hu-sCD2$_{105}$ using heteronuclear, multidimensional NMR methods on labeled samples. We also studied the backbone dynamics of the glycoprotein by ^{15}N spin relaxation measurements. Moreover, we achieved sequential assignments for most ^{1}H and ^{13}C resonances of the heterogeneous high mannose carbohydrate in hu-sCD2$_{105}$ on the intact glycoprotein. Protein-glycan contacts were identified and used together with intra-carbohydrate NOEs to determine the conformation of the attached glycan and how it interacts with the protein. To our knowledge, this is the first time that the structure has been obtained by NMR for both the protein and glycan components of an intact glycoprotein of this size.

NMR Investigations of the SH2 domain from the *c-fes* protooncogene

Weixing Zhang, Scott D. Briggs, Thomas Smithgall, and William H. Gmeiner Eppley Cancer Institute, University of Nebraska Medical Center, Omaha, NE 68198-6805

The c-fes protooncogene is the normal human homolog of several transforming oncogenes and encodes a 93 kDa protein tyrosine kinase (PTK) of the cytoplasmic class[1]. FES is important in growth control of myelogenous cells. Present in FES are an SH2 domain and a kinase domain that are homologous to other PTK proteins. The N-terminal domain of FES does not show homology with other PTK family members and is probably responsible for the unique signal-transducing properties of FES. Deletion of the SH2 domain from FES or FPS has a negative impact on PTK activity in-vitro. It has been proposed that the SH2 domain of FES binds to an autophosphorylated Y in the kinase domain as a means of folding the enzyme into its active conformation. To investigate the structural basis for the interactions of the SH2 domain from FES with target Y residues, the 91 amino acid SH2 domain was expressed from E. coli as a recombinant protein using the pET system. 15 mg of purified SH2 domain have been obtained from a 2 L culture using minimal media. The protein was purified by ion exchange chromatography and its purity determined by SDS-PAGE electrophoresis, mass spectrometry, and amino acid analysis. CD spectroscopy reveals this SH2 domain contains much β-sheet (46%) and some α-helix (9%) and thus may be similar structurally to the SH2 domains from c-abl[2] and other proteins for which SH2 domain structures are known. Preliminary 1D and 2D NMR experiments have been performed on this unlabelled sample. ^{15}N enriched and ^{15}N, ^{13}C dually labelled samples of the recombinant SH2 domain from FES are being prepared. 3D ^{1}H-^{15}N HMQC-NOESY and ^{1}H-^{15}N HMQC-TOCSY experiments will be performed on the ^{15}N enriched sample and this data will be presented and discussed.

1 - Hjermstad, S.J. et al. (1993) Oncogene 8, 2283-2292.

2 - Overduin M. et al. (1992) Cell 70, 697-704.

Sequential ^{1}H and ^{15}N Resonance Assignments and Solution Secondary Structure of Oxidized *Desulfovibrio desulfuricans* Flavodoxin. <u>John R. Pollock</u>[1], Brian J. Stockman[1], and Richard P. Swenson[2], [1]Upjohn Laboratories, The Upjohn Company, 301 Henrietta St., Kalamazoo, MI 49007 and [2]Department of Biochemistry, Ohio State University, Columbus, OH 43210

Sequence-specific ^{1}H and ^{15}N resonance assignments have been made for the 146 non-prolyl residues in oxidized *Desulfovibrio desulfuricans* flavodoxin. Assignments were obtained by analysis of heteronuclear three-dimensional ^{1}H-^{15}N NOESY-HMQC and TOCSY-HMQC data sets recorded on uniformly ^{15}N-enriched protein, pH 6.5 at 300 K. Numerous side-chain resonances have also been assigned. Residues with overlapping ^{15}N chemical shifts were resolved by a three-dimensional ^{1}H-^{15}N HMQC-NOESY-HMQC spectrum. Medium- and long-ranged NOEs, $^3J_{NH\alpha}$ coupling constants, and ^{1}H^N exchange data indicate a secondary structure consisting of five parallel β-strands and four α-helices with a topology similar to that of *Desulfovibrio vulgaris* flavodoxin. Two of the β-sheet strands in *D. desulfuricans* appear twisted. In addition, prolines at residues 106 and 134, which are not conserved in *D. vulgaris* flavodoxin, distort the two C-terminal α-helices.

Solution Structure of the N-Terminal Receiver Domain of NTRC Determined by Multidimensional Heteronuclear NMR

Brian F. Volkman[‖], Michael J. Nohaile[#], Sydney Kustu[#], and David E. Wemmer[*‖]
[#]Department of Molecular and Cell Biology, [‖]Department of Chemistry, University of California, and the [*]Structural Biology Division, Lawrence Berkeley Laboratory, 1 Cyclotron Road, Berkeley, California 94720.

Two-component regulatory systems have been widely studied in bacteria and recently identified in yeast. A kinase domain and the receiver domain, which is phosphorylated, are the conserved features of all two-component systems. NTRC is a transcriptional enhancer-binding protein involved in nitrogen regulation. Phosphorylation of the receiver domain of NTRC turns on an ATPase in the central domain of NTRC, which activates transcription of genes controlling nitrogen metabolism. Using 3- and 4-dimensional NMR spectroscopy, we have completed the ^{1}H, ^{15}N, and ^{13}C resonance assignments and used distance geometry methods to determine the solution structure of the N-terminal receiver domain of the NTRC protein. The receiver domain is comprised of the N-terminal 124 amino acids of NTRC, and contains five α-helices and a five-stranded parallel β-sheet in a (β-α)$_5$ topology. NMR spectral changes upon phosphorylation may indicate how the receiver domain interacts with the central domain of NTRC to signal activation of transcription. Implications for further characterization of receiver domain signal transduction will be discussed.

Bioactive Conformation of Stromelysin Inhibitors Determined by Transferred Nuclear Overhauser Effects

Nina C. Gonnella*, Regine Bohacek, Xiaolu Zhang, István Kolossváry, C. Gregory Paris, Richard Melton, Cindy Winter, Shou-Ih Hu, and Vishwas Ganu

Research Department, Pharmaceutical Division, Ciba-Geigy Corporation, Summit, NJ 07901

Stromelysin-1 (SLN) is a zinc metalloendoproteinase (MMP-3) that is secreted by synoviocytes and articular chondrocytes in response to inflammatory mediators such as interleukin-1 (1). This enzyme is believed to cause the destruction of cartilage proteoglycans associated with osteo- and rheumatoid arthritis (2). Because of stromelysin's link to cartilage degradation, it is anticipated that the design of SLN inhibitors could result in the next major class of drugs for the treatment of arthritis.

As part of a rational drug design strategy, it is important to obtain structural knowledge of the enzyme-inhibitor complex. In particular, the shape of the enzyme's binding pocket and/or the conformation of the bound inhibitor can provide valuable structural information that can lead to the design of conformationally restricted analogs with enhanced biological properties.

Since the transferred nuclear Overhauser effect (TR-NOE) can provide information on the conformation of an inhibitor in the bound state, TR-NOE and 2D NMR spectroscopy were used to determine the biologically active conformations of two stromelysin inhibitors. Both inhibitors were hydroxamic acids generated via chemical synthesis. This study established the stromelysin bound conformations of these inhibitors and provided information on the shape and orientation of the S1' and S2' pockets of SLN relative to thermolysin (TLN). Comparisons were made between SLN and TLN inhibitors to critically examine thermolysin as a template for stromelysin inhibitor design. The NMR derived structures were determined for use as a template in conformationally restricted drug design and were successfully used as templates in the design of 3D database search queries.

1. Murphy, G., Hembry, R.M., and Reynolds, J.J. (1986) *Collagen Relat. Res.* **6**, 351-363.
2. Flannery, C., Lark, M.W., and Sandt, J.D. (1992) *J. Biol. Chem.* **267**, 1008-1014.

The Effect of Exchange Rates on TRNOE Buildup Curves

Steven B. Landy and B. D. Nageswara Rao
Department of Physics
Indiana University-Purdue University Indianapolis
Indianapolis, IN 46202-3273

TRNOE experiments are a major tool in the determination of the conformations of substrates which exchange between enzyme-bound and free forms. Often the rate of exchange between conformations is unknown and must be determined as part of the data analysis. Recent investigations have explored the manner in which the exchange rates impact on TRNOE buildup curves. Initial results have indicated that buildup is usually enhanced by elevated exchange rates.[1,2]

We have performed computer simulations on three spin systems undergoing exchange between free and enzyme-bound forms. The relaxation matrix elements of the bound form (for which $\omega\tau_c \gg 1$) were assumed to be much larger than those of the free, and, in most cases, no external leakage was included. In spite of these simplifying assumptions, the dependence of buildup rates on exchange was found to be non-trivial. It was generally found that for short mixing times, elevated exchange does enhance buildups, but at long mixing times, the opposite is the case. Molecular geometry is also a factor. If magnetization transfer is promoted mainly by spin diffusion, elevated exchange tends to reduce buildup rates. At intermediate mixing times buildup rates may optimize at an intermediate exchange rate. The work was supported in part by grants from NIH (GM 43966) and IUPUI

1. F. Ni, *J. Magn. Reson.*, **96**, 651 (1992).
2. R. E. London, M. E. Perlman, and D. G. Davis, *J. Magn. Reson*, **97**, 79, (1992).

Conformations Of Nucleotides At The Active Sites Of ATP - Utilizing Enzymes By Two-Dimensional Transferred Nuclear Overhauser Spectroscopy (TRNOESY)[†]

N. Murali, G. K. Jarori[††], and B. D. Nageswara Rao

Department of Physics, Indiana University Purdue University Indianapolis (IUPUI), 402 North Blackford Street, Indianapolis, IN 46202, U. S. A.

Conformations of nucleotides at the active sites of ATP-utilizing enzymes were determined by using two-dimensional transferred nuclear Overhauser spectroscopy (TRNOESY). Three classes of enzymes were chosen for this study viz., (i) four phosphoryl transfer (kinases) enzymes; creatine kinase, pyruvate kinase, arginine kinase, 3-phospho glycerate kinase, (ii) one nucleotidyl (adenylyl) transfer enzyme; methionyl tRNA synthetase and (iii) one pyrophosphoryl transfer enzyme; phosphoribosylpyrophosphate synthetase. The nucleotides used as substrates were MgATP which is a substrate in complexes with all the enzymes, and MgADP which is a substrate only for the kinases. Contributions to the observed TRNOE from weak non-specific binding of the enzyme were minimized by studying the ligand concentration dependence of the NOE, and by choosing a sample protocol for which the adventitious binding effects are negligible. The TRNOESY experiments were performed at 500 MHz and at 10 ^{0}C for several mixing times in the range 40 ms to 300 ms in the NOESY algorithm. The TRNOE buildup curves were analyzed, using the complete relaxation matrix, to obtain inter-proton distances in the adenosine moiety. The distances obtained from such an analysis were used as constraints in energy-minimization calculations using CHARMm in order to obtain the bound conformations of the nucleotides that fit best the NOE data and the energy considerations. The glycosidic torsion angle (O'_4-$C_{1'}$-N_9-C_8) in the bound nucleotides in all the complexes were found to be clustered around 50 ± 5^0. However, the sugar pucker varied from one enzyme complex to another.

[†] Supported in part by NIH GM 43966 and IUPUI.
[††] Present address: SSP NMR, Tata Institute of Fundamental Research, Colaba, Bombay 400 005, India.

Kinetic, Binding and TRNOE studies of Perdeuterated Yeast Phosphoglycerate Kinase and Its Interactions with Substrates

Celeste G. Shibata, Jay D. Gregory, Brenda S. Gerhardt, and Engin H. Serpersu
The University of Tennessee, Department of Biochemistry, M407 Walters Life Sciences Building, Knoxville, TN 37996-0840.

Perdeuterated yeast phosphoglycerate kinase (^{2}HPGK) was isolated from yeast cells grown in 99.9% ^{2}H$_2$O. Kinetic and binding studies suggested that perdeuterated enzyme was similar to the isotopically normal PGK. The use of ^{2}HPGK prevented not only overlap between the enzyme and substrate NOE cross-peaks but also permitted observation of weak NOE cross-peaks between the substrate protons that are >4 Å apart. Although many spin diffusion pathways are eliminated by protein deuteration, significant spin diffusion was present between the protons of the substrate at mixing times longer than 25 ms. The effect of spin diffusion was reduced by increasing the substrate to enzyme ratio; however, the dynamic range and the magnitude of observed NOEs were also reduced, and consequently weak NOEs became unobservable.

Intestinal Fatty Acid-Binding Protein: Effect of Bound Ligand on Backbone Amide Proton Exchange Rates

Michael E. Hodsdon, James J. Toner and David P. Cistola
Department of Biochemistry and Molecular Biophysics, Washington University School of Medicine, St. Louis, MO 63110

Intestinal fatty acid-binding protein (131 residues, 15.4 kDa) is thought to function in the absorption and intracellular transport of dietary fatty acids. It specifically binds a single molecule of long-chain fatty acid with a dissociation constant of 200 nM. The holo-protein adopts a β-clam structure comprised of two, five-stranded antiparallel β-sheets and the ligand binds in an interior cavity between the two sheets. In the absence of ligand, the protein adopts a similar overall conformation and the ligand-binding cavity is occupied by water molecules. In one sense, the ligand forms a part of the globular core of the protein and may have a significant effect on the protein's stability. To begin to investigate this issue, we have measured and compared amide proton exchange rates for the holo- and apo-proteins. Essentially complete ^1H, ^{13}C and ^{15}N assignments were established for the holo-protein using seven, 3-D triple-resonance experiments. Even at pH 7.2 and 37°C, 118 of the 131 amide protons were observable and could be assigned. Amide proton exchange rates were monitored by dissolving the lyophilized, unenriched protein in D_2O and collecting 2-D TOCSY experiments at 2.5 hr increments. The data were analyzed in two ways: (i) peak volumes were measured and plotted as a function of incubation time and the resulting exponential was analyzed to give exchange rate constants; and (ii) the spectra were combined into a single binary file and then subjected to a 3-D Fourier transform. The resulting linewidths in the "third" dimension provided the amide exchange rate constants. Substantial differences were observed in the exchange rates across the protein backbone, as will be discussed.

Study of Molecular Motion of ATP in Viscous Solutions via ^2H, ^{13}C and ^{31}P NMR Line Shapes[†]

Yan Lin and B. D. Nageswara Rao
Department of Physics, Indiana University Purdue University Indianapolis
402 N. Blackford St., Indianapolis, Indiana, IN 46202

Adenosine triphosphate (ATP) is a rather floppy molecule. An isolated ATP molecule has three categories of internal mobilities: 1. the glycosidic reorientation of the adenine base with respect to the ribose, 2. the motion of the phosphate chain, and 3. the ribose pucker. The first two of these motions are of large amplitude, and the third motion is of low amplitude. It is reasonable to surmise that when MgATP is bound to enzymes for which it is a substrate, the large amplitude motions will be arrested due to the topology of the binding site. In order to gain some insight into these motions, ^{13}C, ^{31}P and ^2H line shape measurements have been carried out on ATP, MgATP and AMP as a function of viscosity using glycerol and sucrose solutions. ^{13}C measurements were made on [2-^{13}C]ATP(or AMP) and ^2H measurements on [8-^2H]ATP(or AMP). ^{13}C results have been analyzed by considering the relaxation mechanism of ^{13}C-^1H dipolar interaction and ^{13}C chemical shift anisotropy(CSA), along with their interference effects. ^2H results were analyzed on the basis of the quadrupole relaxation of ^2H. All the interaction parameters, viz. the ^{13}C CSA tensor elements and the ^2H quadrupole coupling constant have been independently determined by solid-state experiments. The results are clearly indicative of the presence of the large amplitude internal motions superposed on an overall motion. The overall motion is assumed to be isotropic with a linear dependence on viscosity. In comparison, the internal mobilities display a more complex viscosity dependence suggesting that the internal mobilities are weakly attenuated by solvent packing at high viscosities.

† Supported in part by NIH GM 43966 and IUPUI.

Characterization of Differences in Structure and Dynamics between Intact <u>Cucurbita maxima</u> **Trypsin Inhibitor-III (CMTI-III) and Reactive-Site Hydrolyzed CMTI-III (CMTI-III*) by 2D Proton-Detected** [13]**C NMR, J. Liu**[1]**, S. Nemmers**[1]**, Y. Huang**[1]**, B. Tobias**[2]**, and R. Krishnamoorthi**[1*]**,** [1]**Department of Biochemistry, Kansas State University, Manhattan, KS 66506;** [2]**Department of Chemistry, Parke-Davis Pharmaceutical Research Division, Ann Arbor, MI 48105.**

<u>Cucurbita maxima</u> Trypsin Inhibitor-III (CMTI-V; M_r 3 Kd) of the Squash family of inhibitors loses its inhibitory activity toward another serine proteinase, viz., Factor XII_a (Hageman Factor) of the blood coagulation cascade, when its reactive-site peptide bond between Arg5 and Ile6 is hydrolyzed. In order to characterize the attendant structural and dynamic changes, if any, that lead to functional loss, we have followed our ^{1}H NMR work with the intact and modified forms of the inhibitor (CMTI-III and CMTI-III*, respectively) with the application of proton-detected [1]H-^{13}C heteronuclear spectroscopy at natural ^{13}C abundance level. Thus, we have assigned all the 29 [13]C^α-^{1}H units and most of the side-chain [13]C-^{1}H groups of both forms of the inhibitor by single-bond and multiple-bond correlation techniques. Comparison of backbone and side-chain ^{13}C chemical shifts of CMTI-III and CMTI-III* reveals that residues Val2, Cys3, Pro4, Arg5, Ile6, Leu7, Met8, Lys9, Asp15, Cys16, Leu17, Glu19, Cys20, Leu23, His25, Tyr27, and Cys28 all show significant changes, with the N-terminal residues exhibiting the largest changes up to 2.1 ppm. These results confirm our earlier conclusion based on ^{1}H and ^{15}N NMR work that structural perturbations due to cleavage of the Arg5-Ile6 peptide bond are transmitted throughout the protein molecule. In order to characterize changes in molecular dynamics of the two forms, we have measured the T_1 relaxation times of the [13]C^α atoms. The T_1 values range from 180 to 378 ms. Comparison of T_1 values of CMTI-III and CMTI-III* indicates that the relaxation times are significantly increased in CMTI-III* for residues in the reactive-site region--Cys3, Pro4, Arg5, Ile6, and Leu7; only a few of the other residues are affected--Cys16 shows an increase in the T_1 value, while Cys10 and Glu24 register decreases. These preliminary observations suggest, in agreement with our earlier thermodynamic study, that in CMTI-III*, the liberated amino and carboxy termini are less flexible, and hence less mobile, due to interaction with water molecules. Further work is in progress to quantitatively determine changes in the internal mobilities of the various C_α atoms.

Dynamics of Fluorotryptophan Residues in *E. coli* **Glucose/Galactose Receptor.** Linda A. Luck, Tom O'Connell, and <u>Robert E. London</u>, Laboratory of Molecular Biophysics, NIEHS, Research Triangle Park, NC 27709

Fluorinated amino acid residues have found increasing popularity for probing structural and conformational features of larger biologically interesting macromolecules. Although initial studies suggested that relaxation due to chemical shift anisotropy might present a serious problem at higher fields, there are now several examples of well resolved ^{19}F spectra at 11.75 T. Recent studies of the periplasmic glucose/galactose receptor (GGR), MW = 33 kD, grown on a medium containing 5-fluorotryptophan (5-FTrp) yielded five well-resolved resonances which were assigned using site-directed mutagenesis. We have investigated the spin-lattice relaxation behavior of the GGR containing either 5-FT or the ring-deuterated analog, $[2,4,6,7-^2H_4]$-5-fluorotryptophan (5-F-DTrp). As a consequence of the large deuterium-fluorine isotope shifts, resonances of both species can be observed simultaneously. Spin-lattice relaxation rates for the tryptophan residues depend on both intra-residue and inter-residue dipolar interactions, and hence are sensitive to both the motion of the individual fluoroindole rings, and to relative motion of tryptophan and protons on interacting sidechains. Alternatively, the spin lattice relaxation of the 5-F-DTrp fluorine depends only on the relative motion of the tryptophan and interacting sidechains. Spin lattice relaxation rates determined for the individual tryptophan residues showed considerable variation. The observed proportionality of the measured T_1 (5-FTrp) and T_1 (5-F-DTrp) values indicates that both measurements are sensitive to the same types of effects. The contribution of the two ortho protons to the fluorine relaxation was found to be less dominant than might have been expected. Motion of Trp-183 which is located in the binding pocket next to the bound glucose is significantly altered upon formation of the glucose or galactose complex.

Probing the Dynamics of the Human Erythrocyte Glucose Transporter with Fluorodeoxyglucose. Tom O'Connell, Scott A. Gabel, and Robert E. London, Laboratory of Molecular Biophysics, NIEHS, Research Triangle Park, NC 27709

The transport of several n-fluoro-n-deoxy-D-glucose derivatives across the human erythrocyte membrane has been studied under equilibrium exchange conditions using one and two dimensional NMR techniques. This approach is based on the intracellular ^{19}F shift, which was found to depend on the anomeric form and on the F/OH substitution position. Since the transport behavior of both glucose anomers can be followed simultaneously, this approach is particularly sensitive to anomeric differences in permeability. For 2-, 3-, 4-, and 6-fluoro- deoxyglucose analogs, the α anomers permeate more rapidly, and the $P\alpha/P\beta$ ratio is dependent on the position of fluorination, with values of 1.06, 1.3, 2.5, and 1.6, respectively, obtained at 37° C. Although mutarotase activity has been reported for red cells, mutarotation rates were found to be completely negligible on the transport and spin lattice relaxation time scales. Metabolic transformation of the fluorinated glucose analogs, primarily to fluorinated gluconate and sorbitol analogs, is very slow and does not significantly interfere with the transport measurements. These data can be interpreted in terms of a simple alternating carrier model in which the anomeric preference is a function of the carrier conformation.

Structure and Dynamics Studies of Proteins in Non-aqueous Solvents by 2D ^1H NMR Spectroscopy
Jian Wu and David G. Gorenstein, Department of Human Biological Chemistry and Genetics, University of Texas Medical Branch, Galveston, Texas 77555

In this study, the NMR NH-exchange methodology has been extended to probe the role of water in defining the structure and dynamics of proteins in nonaqueous solutions. The NH-exchange rates of oxidized horse cytochrome c suspended in tetrahydrofuran (THF) containing 1% D_2O (vol/vol) at 37 °C, pH 8.9, have been determined indirectly. The relative solvent protection factors for amide exchange of cytochrome c in THF/1% D_2O have been calculated for 35 amide protons and compared with the results obtained at the same pH in aqueous solution. In both aqueous solution and THF/1% D_2O at pH 8.9, the NH-exchange profile for cytochrome c is similar to that in the native state (pD 7.0). The protection factors in THF/1% D_2O at pH 8.9 vary between 10^2 and 10^5, which are in turn 10^2-10^3 -fold smaller than those observed at pH 8.9 in purely aqueous solution. These results suggest that while cytochrome c remains folded and compact at pH 8.9 in both purely aqueous solution and THF/1% D_2O, the flexibility of the protein is clearly enhanced, especially in the organic medium. However, some differences were also observed between the two solvent systems. Most noticeably part of the C-terminal helix (Leu94, Ala96, and Lys99) has the lowest NH-exchange rates in D_2O, whereas in THF/1% D_2O these exchange rates increased dramatically. The decrease in protection factor for those residues of the C-terminal helix facing the N-terminal helix suggests that the interaction between N- and C-terminal helices is destabilized in the hydrated organic solvent, exposing the surface residues Leu94, Ala96, and Lys99. In another study, the structure of the highly hydrophobic protein crambin in various organic solvents have been determined by ^1H NMR spectroscopy. The results showed that crambin remained in the native state in polar organic solvents, whereas in non-polar organic solvents, the folding of the protein is highly dependent on the water content in the solution. The above studies suggest that water does indeed serve to "lubricate" the local dynamics of the protein in the hydrated organic solvents.

NMR STRUCTURE OF A DITHIOPHOSPHATE MODIFIED, NON-PALINDROMIC, OLIGONUCLEOTIDE OCTAMER OF THE SEQUENCE: d(GAGAGAGA) • d(TCTCTpS$_2$⁻CTC)

Jill Nelson Granger, Florence Lebreton, Bruce A. Luxon, Jennifer L. Jordan and David G. Gorenstein
Dept. of Chemistry, Purdue University, W. Lafayette, IN 47907

Abstract: Thiophosphoramidite chemistry has been used to prepare a dithiophosphate analogue of a non-palindromic oligonucleotide octamer d(GAGAGAGA) • d(TCTCTpS$_2$⁻CTC). The ^{1}H and ^{31}P resonances have been assigned by two-dimensional NMR. Using NOESY distance-restrained hybrid-matrix refinement, a family of solution-state structures has been determined. The computational results stem from both A-DNA and B-DNA model structures and converge to a self-consistent family of structures whose helical and conformational parameters have been analyzed using a modified version of NEWHELIX for UNIX. Unlike the non-modified octamer (Rajagopal and Feigon, *Biochemistry* 1989, 28, 7859), the singly dithiophosphate-modified analog does not resemble a standard B-DNA conformation but adopts an intermediate-type structure with unusual helical parameters and transitions. In addition to chemical shift perturbations in the NMR assignments (prominent in the dithiophosphate modified region), there is also a destabilization of the formation of the duplex. Values for local helical and sugar-phosphate backbone parameters for the refined structures have a transition between the typical values for A-DNA and B-DNA at or near the site of the dithiophosphate modification. The pseudorotation phase angles (representative of sugar conformation) for the sugar rings in the refined structures further demonstrate that most of the sugars have adopted an intermediate conformation between that of A- and B-DNA.

Structural Characterization of Carcinogen-Modified DNA Oligomers.

Thomas R. Krugh, Charles W. Bailey, Linda M. Eckel, and Matthew A. Fountain

Department of Chemistry   University of Rochester   Rochester, NY 14627

One and two-dimensional NMR spectroscopy, energy minimization and molecular dynamics simulations have been used to characterize the structure and dynamics of benzo(a)pyrene, acetylaminofluorene, and aminofluorene bound to oligo-deoxynucleotides in solution. A covalent adduct of (+)-trans-*anti*-benzo[*a*]pyrene (BP) attached to a guanine amino group in the BP-modified oligomer shows that the pyrene ring is oriented in the 5' direction of the modified strand and situated near the sugar phosphate backbone of the unmodified strand. N-acetyl-2-aminofluorene adducts bound at the C$_8$ position of guanine in 9-mer and 10-mer duplexes show that the AAF-G[5] base adopts a *syn* conformation, with the flourene moiety stacked within the duplex. A 2-aminofluorene (AF) modified deoxyoligonucleotide duplex reveals the presence of two major conformations. In one conformation of the AF-10-mer, AF[ext], the AF moiety is located in the major groove with the DNA maintaining a conformation similar to B-DNA; in the second conformation, AF[int], the fluorene is stacked within the duplex.

Solution Structure of the Second RNA-Binding Domain of Sex-lethal

Andrew L. Lee[‡], Roland Kanaar[§], Donald C. Rio[§], & David E. Wemmer[‡]
University of California, Berkeley

[‡]Department of Chemistry, University of California, Berkeley, CA 94720-1460
[§]Department of Molecular and Cell Biology, 401 Barker Hall, University of California, Berkeley, CA 94720-1460

The RNA-binding protein, Sex-lethal (Sxl), is a critical regulator of sexual differentiation and dosage compensation in *Drosophila*. This regulatory activity is a consequence of the ability of Sxl to bind uridine-rich RNA tracts involved in pre-mRNA splicing. Sxl contains two RNP consensus-type RNA-binding domains (RBDs). A structural study of a portion of Sxl (amino acids 199-294) containing the second RNA-binding domain (RBD-2) using multidimensional heteronuclear NMR is presented here. Nearly complete ^{1}H, ^{13}C, and ^{15}N resonance assignments have been obtained from ^{15}N- and ^{13}C/^{15}N-uniformly labeled protein. These assignments were used to analyze 3D ^{15}N-separated NOESY and ^{13}C/^{13}C-separated 4D NOESY spectra. Initially, 494 total and 169 long-range NOE-derived distance restraints have been extracted from these data sets. The calculated family of structures to be presented exhibits the $\beta\alpha\beta$-$\beta\alpha\beta$ tertiary fold found in other RBD-containing proteins. The RMSD to the average structure for the backbone atoms of residues 11-93 is 1.55 ± 0.30 Å, while the RMSD for backbone atoms involved in secondary structure is 0.76 ± 0.14 Å. A capping box has been identified at the N-terminus of the first helix and has been characterized by short- and medium-range NOEs, ^{13}C chemical shifts, and amide exchange patterns. The structure calculations reproduce the capping box as a well-defined motif.

Solution Structure of the *ets* Domain of Human Fli-1 in the DNA-Bound Form

H. Liang[†], E. T. Olejniczak[†], D. G. Nettesheim[†], L. Yu[†],

X. Mao[§], C. B. Thompson[§] and S. W. Fesik[†*].

[†]Pharmaceutical Discovery Division, Abbott Laboratories, Abbott Park, IL 60064 and

[§]Howard Hughes Medical Institute, The University of Chicago, IL 60637

The *ets* family of eukaryotic transcription factors have in common a conserved DNA-binding domain, the *ets* domain. By using multidimensional NMR, we have determined the three-dimensional structure of the *ets* domain of human Fli-1 in the DNA-bound form. We find that the *ets* domain is structurally similar to that of the helix-turn-helix motif of CAP and those of several eukaryotic DNA-binding proteins. Residues of the Fli-1 that are involved in DNA-binding were identified from NMR and mutagenesis experiments. We present a model of the complex of the Fli-1 *ets* domain and its cognate DNA site.

Does Form Follow Function? Structure and Function of RNA Binding Domains.

J. Lu & K.B. Hall, Department of Biochemistry and Molecular Biophysics, Washington University School of Medicine, St. Louis, MO 63110

The U1A protein is part of the U1 snRNP, which is one of the small nuclear RiboNucleoproteinParticles that participate in pre-mRNA splicing. Like many proteins that bind to RNA, U1A contains two distinctive RNA Binding Domains (RBD). The RBD is small, about 90 amino acids, and is distinguished by the presence of two sequences, RNP1 and RNP2, an octamer and hexamer sequence, respectively, that are conserved both in sequence and in position within the RBD. The N-terminal 102 amino acid domain associates with a hairpin RNA, while the C-terminal RBD apparently does not bind RNA. The structure of the 102A RBD is known, and consists of a $\beta\alpha\beta$-$\beta\alpha\beta$ fold with a β sheet formed from 4 antiparallel β strands. Two α-helices are positioned on one side of the β sheet. For the U1A 102A RBD, the surface of the protein that binds the RNA has not been defined.

Since the RBD is a very common motif that can associate with different forms of RNA, its properties are of interest for what they might reveal about the mechanism of RNA recognition. Both structural features and amino acid sequence of a given RBD are certain to contribute to its RNA binding properties, and so through a comparison of several RBDs of known specificities, those critical structural details may be discerned. Therefore, we are using a combination of NMR methods, mutagenesis, and thermodynamics to attempt to understand the relation between structure, stability, and RNA binding. Specifically, we are using $^{13}C/^{15}N$ NMR methods to determine the structure of the C-terminal domain to compare the structures and binding properties of the two U1A RBDs; in addition, we are using ^{13}C- and ^{15}N-labelled RNA to observe the RNA in the 102A:RNA complex.

Structure, Function and Stability of Monomeric Variants of the λ Cro Repressor

Mike Mossing
Department of Biological Sciences,
University of Notre Dame, Notre Dame, IN 46556.

A family of proteins has been constructed in which the dimer interface of the lambda Cro repressor has been replaced by a beta hairpin. The engineered proteins are both monomeric and more stable than the original Cro dimer. Specific DNA binding function is retained in the monomers, but affinity is reduced by > 1000 fold (Science 250:1712 1990). The designed structures have been confirmed by NMR and crystallography (in collaboration with Ron Albright and Brian Matthews at the University of Oregon). ^{1}H and ^{15}N resonance assignments have been completed for two of the family members which differ in the conformation of the engineered turn. Global stabilities of these proteins have been measured in thermal denaturation experiments as followed by circular dichroism. Local stabilities are being measured by amide hydrogen exchange. A second generation of proteins, engineered for greater stability and DNA binding activity based on the crystal structure of the original engineered monomer, has been constructed using combinatorial mutagenesis.

Heteronuclear Studies of the Phage R17 RNA and RNA-Coat Protein Complex

*Edward P. Nikonowicz, Department of Biochemistry, Rice University
Houston, TX. 77005

A fundamental control mechanism of gene expression involves the interaction of nucleic acids with proteins. Several proteins have now been identified that recognize and bind to a DNA or RNA molecule with remarkably high specificity. The bacteriophage R17 RNA-coat protein complex, which functions to regulate translation of the phage replicase gene, is an attractive model for probing structural elements that confer specificity to RNA-protein interactions. The coat protein binds as a homodimer to an RNA hairpin that contains a bulged purine in the stem and is capped by a four nucleotide loop. We are applying double and triple resonance ^{1}H-[^{13}C/^{15}N/^{31}P] NMR methods investigate the three-dimensional solution structure of the R17 RNA hairpin in both the free and protein bound states. ^{15}N-^{1}H NMR experiments were used to monitor the titration of protein with RNA and have confirmed the formation of a single RNA-protein complex at a stoichiometric ratio of 1:2. We have obtained resonance assignments for the RNA hairpin in the free state and partial assignments for the RNA in the protein bound state. Resonances corresponding to nucleotides in both the loop and helix regions of the hairpin show significant perturbations in chemical shift upon complex formation. We are currently working to obtain structural constraints for the free and protein bound forms of the R17 RNA hairpin.

The Effects of 5-FU Incorporation on RNA Structure, Stability, and Dynamics Parag

Sahasrabudhe, Richard T. Pon, William H. Gmeiner Eppley Cancer Institute, University of Nebraska Medical Center, Omaha, NE 68198-6805

The anti-cancer drug 5-fluorouracil (5-FU) is heavily utilized in the clinic for the treatment of solid tumors. Despite its heavy usage, its mechanism of action is incompletely understood. 5-FU is metabolically converted to 5-FdUMP, a potent inhibitor of thymidylate synthase. Cells treated with 5-FU cannot necessarily be rescued by added thymidine and alternative mechanisms of activity are operative. 5-FU is known to be converted into FUTP and incorporated into RNA. Incorporation of 5-FU into RNA changes the migration of the U4 and U6 snRNA species on non-denaturing PAGE[1]. 5-FU is also known to perturb pre-mRNA splicing and rRNA assembly. 5-FU has been incorporated into tRNA[2] with little apparent structural consequence. To investigate the RNA-mediated effects of 5-FU we have prepared a series of self-complementary RNA and DNA duplexes containing either zero, one, or two site-specifically incorporated 5-FU residues. We have evaluated the effects of 5-FU incorporation to RNA structure, stability and dynamics. All of the ^{1}H resonances for these eight duplexes have been assigned by 2D NMR. The structures are being determined by using MARDIGRAS and restrained molecular dynamics calculations. The stabilities have been determined by UV hyperchromicity and 1D NMR spectroscopy of the imino ^{1}H as a function of temperature. The dynamics of imino ^{1}H exchange have been determined by semi-selective inversion/recovery experiments at different catalyst concentrations using shaped pulses. Recent studies on a 35mer stem-loop from the U4 snRNA containing 5-FU will also be presented.

1 - Armstrong, R.D. et al. (1986) J. Biol. Chem. 261, 21-24.
2 - Chu, W.C. & Horowitz, J. (1989) Nucl. Acids Res. 17, 7241-7252.

Structural Analysis of the Intramolecular Inhibition of DNA Binding in Ets-1

Jack J. Skalicky, Logan W. Donaldson and Lawrence P. McIntosh.
The University of British Columbia, Department of Biochemisty
2146 Health Sciences Mall Vancouver, BC V6J-1Z3.

Ets proteins regulate transcription activation from a variety of cellular and viral gene promoter and enhancer elements (Review: Macleod et al., 1992, TIBS. 17, 251). Ets proteins have a conserved DNA-binding sequence, called the ETS domain, that binds to the core DNA sequence GGA. The murine Ets-1 and Ets-2 proteins also contain sequences N-terminal to the ETS domain that negatively modulate DNA binding (Waslyk et al., 1992, Genes Dev. 6, 965; Nye et al., 1992, Genes Dev. 6, 975). To provide an explanation for this modulation we are comparing the structures of two recombinant fragments of murine Ets1; the 12kDa. fragment containing the ETS domain and an 18kDa. fragment containing the ETS domain and the inhibitory domain. We have prepared uniform and selectively labeled samples of both proteins and are using triple resonance nmr spectroscopy to define their solution structures. In addition, we are studying the complexes of the proteins with GGA-containing DNA oligomers. These structural studies will provide an explanation for the attenuation of Ets binding to DNA. Elements of secondary structure in the inhibitory domain and in the ETS domain will be presented and potential modes of binding inhibition discussed.

3D MORASS: a Hybrid-Hybrid Matrix Method for 3D NOE-NOE Spectral Analysis

Qun Zhang, David G. Donne, Elliott K. Gozansky, Jiyan Chen, Frank Zhu, Patricia L. Jackson and David G. Gorenstein*, Department of Chemistry, Purdue University, W. Lafayette, IN 47907 and *Department of Human Biological Chemistry and Genetics, University of Texas Medical Branch, Galveston, TX 77555

An exact full matrix analysis has been extended to 3D homonuclear NOE-NOE spectroscopy and applied to the simulation of 3D NOE-NOE spectra of the Dickerson dodecamer d(CGCGAATTCGCGC)$_2$. The exact method has been compared with the approximation method based on Taylor expansions. It has been found that the approximations using one or two terms in the Taylor expansion series are generally inadequate for simulations of 3D NOE-NOE volumes and often fails even at very short mixing times depending on the motional properties of the biomolecule in question. Direct refinement of structure from a 3D NOE-NOE spectrum thus should utilize the complete relaxation matrix approach. A "hybrid matrix approach" to NMR structural refinement involves the direct calculation of the NOESY rate matrix (and hence distances) from a hybrid experimental and calculated 2D NOESY volume matrix. We have now extended this method to 3D NOE-NOE spectra and have developed a 3D "hybrid-hydrid" matrix methodology for refinement from 3D NOE-NOE spectra in our program *MORASS*: *M*ultiple *O*verhauser *R*elaxation *A*nalysi*S* and *S*imulation. Experimental 3D data is merged with simulated 3D data to create a hybrid 3D NOE-NOE spectrum. This is then deconvoluted into a 2D hybrid NOESY spectrum. The deconvoluted, 2D hybrid NOESY spectrum can then be merged with other 2D NOESY experimental data along with additional simulated 2D data as necessary to create a hybrid-hybrid 2D NOE volume matrix. This hybrid-hybrid volume matrix is then used in MORASS to calculate a rate matrix. The resulting distances taken from the off-diagonal cross-relaxation rates can then be utilized in a distance geometry or restrained molecular dynamics refinement of the structure. This process is repeated until a satisfactory agreement between the calculated and observed spectra is obtained. This hybrid-hybrid matrix method retains computational efficiency and utilizes the resolution of the 3D data set while allowing for the direct incorporation of any information content of the available 2D data. We have examined the accuracy and precision of structures derived from the MORASS/3D hybrid-hybrid matrix, distance restrained molecular dynamics methodology with the simulation of 3D NOE-NOE volumes for the dodecamer duplex (CGCGAATTCGCG)$_2$. Analysis shows that the 3D MORASS method is indeed much superior to the two-spin approximation analysis. Our results demonstrate that the hybrid-hybrid matrix method for analysis of 3D NOE-NOE spectra provides a viable tool in the refinement of large molecules.

A New Analysis of Proton Chemical Shifts in Nucleic Acids

Richard Bryce and David A. Case
Dept. of Molecular Biology
The Scripps Research Institute
La Jolla, California 92037

Chemical shifts of non-exchangeable protons in nucleic acids often show characteristic deviations from the corresponding resonance positions in nucleosides or sugars. We have found that many of these "structural shifts" can be understood in empirical terms similar to those used previously for proteins (Ösapay and Case, *J. Am. Chem. Soc.* **113**: 9436, 1991). These include ring current contributions from aromatic groups, electrostatic effects (primarily from the phosphate backbone) and magnetic susceptibility anisotropy contributions from sugars. We have prepared a "database" of nucleic acid structures where observed shifts are available, and will present comparisons of calculated and observed secondary structural shifts in a variety of small DNA and RNA macromolecules.

Quantum chemical calculations are now becoming accurate enough to provide quantitative insights into the origins of chemical shift dispersion for protons. We will describe a novel approach to using quantum calculations to explore and calibrate ring current and magnetic anisotropy contributions to chemical shifts.

Selective Excitation of Multiple-Quantum Coherences in Coupled Spins by Cascades of Non-commuting Selective R. F. Pulses on Connected Transitions.

Kavita Dorai and Anil Kumar, Department of Physics and Sophisticated Instruments Facility, Indian Institute of Science, Bangalore 560012, India

It is shown that symmetric cascades of selective r.f. pulses on connected single quantum transitions can be used for selective excitation of multiple-quantum coherences or for selective perturbation of the populations of multiple quantum levels (energy levels whose magnetic quantum number difference differs from ± 1). If there are n progressively connected and m regressively connected levels in the cascade then the excited coherence has an order |n - m|. Experiments in two and three coupled spin systems have been performed. It is shown that the decay rates of selected MQC can be measured by such experiments.

CALIBRATION OF ^{15}N-EDITED NOESY SPECTRA USING AN NOE RATIO.

Stéphane M. Gagné, Frank D. Sönnichsen and Brian D. Sykes.

Dept. Biochemistry, University of Alberta, Edmonton, Alberta. T6G 2H7 Canada.

The nuclear Overhauser effect (NOE) is the major tool in the determination of protein structure by NMR, nowadays often obtained in the form of 3D- and 4D-NOE experiments. The quantification of NOE peaks in such experiments must be taken with care, as a single calibration is usually not appropriate for the whole 3D- or 4D-spectrum.

We present a method based on the ratio of the intraresidue NOE [HNi - Hαi] and the sequential NOE [Hαi-1 - HNi], termed here NOE$_{(N\alpha/\alpha N)}$. By its nature, this ratio is independent of the properties of the amide in question (exchange, relaxation,...) and is perfect for the calibration of the 3D-^{15}N-edited NOESY spectrum. The NOE$_{(N\alpha/\alpha N)}$ ratio provides a powerful tool for the determination of secondary structure, as it is greater then one for α-helices and smaller than one for β-strands (see figure). Furthermore the NOE$_{(N\alpha/\alpha N)}$ ratio also has potential in the determination of psi angle. Extensive statistics made on over 200 high resolution protein structures (~ 30 000 residues) are presented to clearly show the justification of this approach. The methods are applied to the secondary and tertiary structure determinations of two examples: an helical protein (troponin-C) and a β-sheet protein (antifreeze type III).

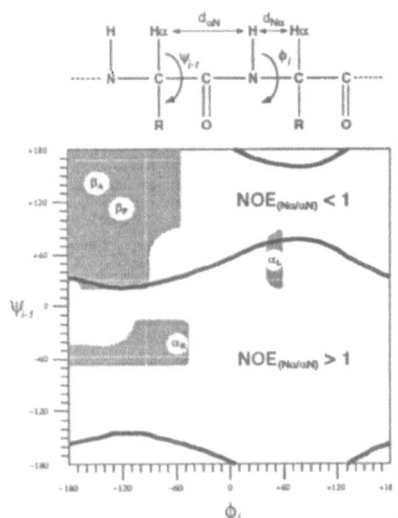

NMR Velocity Mapping of Couette Flow Using Oscillating Magnetic Field Gradients

Jeffrey A. Hopkins and John B. Grutzner

Department of Chemistry, Purdue University, West Lafayette, IN 47907-1393

The application of orthogonal, magnetic gradients oscillating in quadrature to produce a rotating, transverse gradient is studied. The calibrated gradient is applied to solid body rotations to demonstrate the equivalence of the frames of reference of a rotating sample and a rotating gradient. Modulation of the effective magnetic field leads to sidebands, most often observed as spinning sidebands in high resolution and MAS NMR. When the gradient is oscillated at a frequency different from that of the sample the sidebands appear at integral multiples of the difference between the two rotation frequencies. If the gradient is oscillated synchronously with the sample, the spectrum is a spin density projection of the sample onto the gradient. Application is made to slice selection of layers of fluid moving at a chosen angular frequency from a sample undergoing shearing Couette flow. A radial frequency map is determined to evaluate the efficiency of a proposed 2D technique of slice selection.

Isotopically Assisted Topological Editing of Cross-Relaxation Networks in Proteins

Nenad Juranić and Slobodan Macura

Department of Biochemistry and Molecular Biology, Mayo Graduate School
Mayo Clinic and Foundation, Rochester, MN 55905

Manipulation of magnetization exchange networks during the mixing time, exchange network editing[1], is very useful for removal of unwanted magnetization exchange pathways and for the accuracy increase in measurements of desired ones. The major limitation in the previously proposed experiments is limited selectivity due to spectral overlap of different classes of spins. Here we demonstrate that on isotopically labeled samples the cross-relaxation network of proteins can be edited in several new ways. For example, one can remove all carbon bound protons from cross-relaxation network, leaving only cross-relaxation among amide protons. Increased selectivity comes from the fact that ^{13}C spins are used for network decoupling. In addition, due to better spread of carbon resonances it is possible to manipulate selectively, aromatic and aliphatic side chains. Some new topological editing schemes, are demonstrated on uniformly doubly labeled *Human Ubiquitin*, a small globular protein, Mw 8565.

1. S. Macura, J. Fejzo, C. Hoogstraten, W. M. Westler and J. L. Markley, Israel J. Chem. **32**, 245-256 (1992)

Rapid Depaking *via* Fast Fourier Transform

M. Alan McCabe and Stephen R. Wassall, Department of Physics,
Indiana University-Purdue University Indianapolis, Indianapolis, Indiana 46202-3273

Lineshape analysis is a powerful tool in the interpretation of NMR spectra. In our studies we employ broadline ^2H NMR of aqueous multilamellar dispersions of phospholipids perdeuterated in an acyl chain to comprehensively probe order within membranes. As the Hamiltonian associated with the quadrupolar interaction possesses a $P_2(\cos\theta)$ dependence upon orientation, where θ is the angle between the symmetry axis for molecular motions and the applied magnetic field, the lineshape is a complicated and relatively unresolved superposition of contributions from all angles (powder pattern) and positions. Deconvolution to a spectrum characteristic of a membrane of single alignment provides the greatly enhanced resolution required to extract the ordering information from the NMR signal.

We have developed an innovative method to deconvolute. Our approach contrasts with previously devised techniques such as the depaking procedure [1], which iteratively recalculates the spectrum as if all the random distributions of molecules possess a single orientation, and a variety of methods which require the inversion of large matrices [2]. These procedures all work, but they all suffer from the time intensive nature of the calculations involved which limits their applicability. The new approach, instead, utilizes weighting functions in time and frequency domains to facilitate a solution by fast Fourier transform. Because the relative computational efficiency of the FFT is exploited, execution is rapid enough to be available online and it now supersedes our use of the other methods. We are currently generalizing the procedure to other functional dependencies of the Hamiltonian including non-zero asymmetry parameter ($\eta \neq 0$).

1. E. Sternin, M. Bloom, and A. L. Mackay, *J. Magn. Reson.* **55**, 274 (1983).
2. K. P. Whittall, E. Sternin, M. Bloom, and A. L. Mackay, *J. Magn. Reson.* **84**, 64 (1989).

Peptide and Protein Conformation from Ensemble Calculations and NMR Data

Dale F. Mierke Gustaf H. Carlson School of Chemistry, Clark University, Worcester, MA 01610

Computational methods for the development of peptide and protein conformation from ensemble calculations will be highlighted. Ensemble calculations utilize a large number of structures, in place of the single structure used in more standard refinement methods. The application of experimental restraints are then applied to the complete ensemble (as an ensemble average) and not to each individual structure. NMR observables are clearly ensemble averages and therefore it is advantageous to treat the restraints developed from these experimental measurements as average quantities. Such averaging is easily accomplished with the ensemble calculation method proposed here. The method will be illustrated for the utilization of NOEs and coupling constants.

In addition, the utilization of both NOEs and coupling constants as experimental restraints as a method to unambiguously identify conformational averaging will be discussed. This method is based on the fact that NOEs will average as a function of R^{-6} (where R is the distance between the atoms) while coupling constants will average as a function of the cosine series ($\cos^2\theta$ and $\cos\theta$). Application of the NOEs and coupling constants will therefore produce different conformations in the presence on conformational averaging fast on the NMR time scale.

COmplete Relaxation and Conformational Exchange MAtrix (CORCEMA) Analysis OF NOESY Spectra: Applications to Transferred NOESY and Protein Folding Studies

Hunter N. B. Moseley, Ernest V. Curto, and N. Rama Krishna

Department of Biochemistry and Molecular Genetics
University of Alabama at Birmingham (UAB), Birmingham, AL, 35294

Analyses of NOESY spectra to determine molecular structures commonly involve the assumption of a single conformational model. This assumption, often used in Distance Geometry or back-calculation analysis, produces inherently flawed results, because macromolecules undergo complex dynamics involving multiple conformations. Including dynamic models in spectral analysis may yield better molecular structures.

We have developed a very general program called CORCEMA, an extension of earlier work[1] from this laboratory. It incorporates the multistate conformational exchange formulation[2] and calculates resonance intensities for NOESY and/or ROESY experiments for an individual molecule or set of interacting molecules based upon models of 3D structure and arbitrary rates of conformational exchange. CORCEMA allows the study of multiple conformations (states), correlation times, and internal motions like ring flipping and methyl rotation as well. Here we present two computer simulations of CORCEMA application: (i) a three-state model of the folding pathway of penta-L-alanine (α-helix $\rightleftharpoons$ β-III-turn (3-10 helix) $\rightleftharpoons$ β-II-turn), and (ii) a three-state model of a transferred NOESY experiment depicting hinge-bending motion of a thermolysin-inhibitor complex. Analysis predicted systematic changes in NOESY intensities due to exchange kinetics which may be tested by experiment.

[1] W. Lee and N. R. Krishna, *J. Magn. Reson.* 98, 36 (1992).
[2] N. R. Krishna, G. Goldstein, and J. D. Glickson, *Biopolymers* 19, 2003 (1980).

Maximum Entropy reconstruction of non-linearly sampled data

Peter Schmieder, Jeffrey C. Hoch, Alan S. Stern, Gerhard Wagner, Department of Biological Chemistry and Molecular Pharmacology, Harvard Medical School, Boston, MA, 02115 and Rowland Institute for Science, Cambridge, MA, 02142.

MaxEnt reconstruction of non-linearly sampled data (1-3) is a versatile approach for reducing the number of data points required to obtain sufficient signal-to-noise ratio or resolution in multidimensional NMR spectra. It is especially well suited for constant time data, i.e. for most of the recently developed triple resonance experiments.

The design of the sampling schedule is crucial for a artefact-free reconstruction and several possibilities will be discussed. Applications to triple resonance experiments will be shown.

Since MaxEnt is a non-linear method the quantification of the reconstructed spectra is not straightforward. A method will be presented that allows the calibration with "injected" resonances.

1. J.C.J. Barna, E.D. Laue, M.R. Mayger, J. Skilling, S.P. Worrall, (1987) *J. Magn. Reson.* **73**, 69-77.
2. M. Robin, M.-A. Delsuc, E. Guittet, E., J.-Y. Lallemand, (1991) *J. Magn. Reson.*, **92**, 645-650.
3. P. Schmieder, A.S. Stern, G. Wagner, J.C. Hoch, (1993) *J. Biomol. NMR*, **3**, 569-576.

Selective Manipulations of Magnetization :
Quenching of Chemical Exchange and Cross-Relaxation

Sébastien J. F. Vincent, Catherine Zwahlen and Geoffrey Bodenhausen

National High Magnetic Field Laboratory (NHMFL), 1800 E. Paul Dirac Drive, Tallahassee, FL 32306, USA

Chemical exchange can be studied very effectively by 2D exchange spectroscopy ("EXSY") in which the migration of the longitudinal magnetization in the mixing interval τ_m due to chemical reactions in dynamic equilibrium leads to the appearance of cross peaks. In principle, the measurement of the intensities of these cross peaks allows one to determine the rates of all exchange processes, but the quantitative interpretation may be difficult if many sites are simultaneously exchanging with each other. We shall describe a selective approach which is designed to focus attention on two selected sites A and X that are mutually interchanging, by isolating these artificially from a larger network. This can be achieved by inverting the magnetizations of all other sites at $\tau_m/2$. This method has been dubbed QUIET-EXSY for quenching undesirable indirect external trouble in exchange spectroscopy. The time-dependence of the migration of the longitudinal magnetization is determined only by the exchange rates k_{AX} and k_{XA} between the two selected sites and by the spin-lattice relaxation rates $1/T_1^A$ and $1/T_1^X$ of these sites.

Cross-relaxation rates (nuclear Overhauser effects) can be measured selectively by a 1D method derived from NOESY, which we have called quantitative unravelling of intensities for corroborating knowledge (QUICK-NOESY). A variant called quenching undesirable indirect external trouble (QUIET-NOESY) allows one to observe cross-relaxation processes between a selected pair of sites while eliminating perturbations from surrounding spins. This can be achieved by inverting the magnetizations of both A and X at $\tau_m/2$. This makes it possible to eliminate spin diffusion via neighbouring spins K and thus obtain a more accurate measurement of internuclear distances.

Intramolecular Dynamics in Biomolecules, Possibilities and Limitations of NMR

Richard R. Ernst, Laboratorium für Physikalische Chemie, Eidgenössische Technische Hochschule, 8092 Zürich, Switzerland.

The structure determination of biomolecules by NMR has already reached a high level of accuracy and sophistication. The exploration of intramolecular dynamics, on the other hand, is still in its infancy due to the inherent difficulties. First of all, the number of structural parameters exceeds the number of structural parameters by orders of magnitude. Secondly, NMR measurements in solution are not equally sensitive to all types of motion and all ranges of motional rate constants. The possibilities and limitations of NMR studies, combined with molecular dynamics simulations, are demonstrated for peptides. The feasibility for distinguishing different motional models will be discussed for the side–chain mobility.

Structural, Dynamic and Folding Studies of SH2 and SH3 domains

Julie D. Forman-Kay, N. Farrow, L.E. Kay, S.M. Pascal, A.U. Singer, T. Yamazaki and O. Zhang; Hospital for Sick Children, Toronto, Ontario, Canada M5G 1X8 and Department of Biochemistry, University of Toronto, Toronto, Ontario, Canada M5S 1A8

NMR, in addition to enabling the determination of tertiary structures of biomolecules, can provide knowledge about their dynamic behavior and stability by a qualitative analysis of resonance multiplicities, chemical shifts and lineshapes. Resonance assignments, required for structural studies, also serve as the starting point for detailed relaxation measurements. In the course of structural studies of the C-terminal Src Homology 2 (SH2) domain of phospholipase C-γ_1 (PLC) and the N-terminal Src Homology 3 (SH3) domain of the drosophila adaptor protein drk, we have gained insights into the dynamic behavior and stability of these domains which could be important for the function of the proteins. We have also performed quantitative relaxation experiments leading to significant advances in our understanding of the behavior of these protein domains.

The structure of the C-terminal SH2 domain of PLC has been determined in complex with a 12-residue phosphopeptide from the high-affinity-binding Tyr-1021 site of the platelet-derived growth factor receptor. Broadening and unusual chemical shifts for resonances of the arginine sidechains located in the phosphotyrosine binding site, as well as ^{15}N relaxation experiments focusing on backbone NH and arginine Nε resonances, demonstrate the presence of various degrees of mobility for SH2 residues involved in binding, with certain arginines sampling multiple interactions with the phosphopeptide.

Studies of the N-terminal SH3 domain of drk in aqueous buffer at neutral pH revealed a doubling of resonances that could only be explained by slow exchange on the NMR-timescale between folded and predominantly unfolded states of the protein. Measurement of the exchange rates for this folding/unfolding transition and characterization of the relaxation behavior of the two states support the emerging description of unfolded proteins as compact states having more similarity to folded proteins than to random coil models.

Recent NMR Studies of Proteins Involved in Cell Adhesion Processes

Gerhard Wagner, Daniel Wyss, Johnathan Choi and Andrzej Krezel

Department of Biological Chemistry
and Molecular Pharmacology
Harvard Medical School
240 Longwood Avenue
Boston, MA 02115

Contacts between cell surface adhesion receptors and counter receptors on other cells or proteins of the extra cellular matrix are crucial for many biological processes. An important class of such cell adhesion receptors are the integrins. To date, no atomic structures of integrins are available yet. However, several structures of antagonists of the integrin GPIIbIIIa have been reported. GPIIbIIIa is involved in blood coagulation processes. We have solved a structure of the leech protein decorsin, a GPIIbIIIa antagonist with an RGD sequence in its active site. In contrast to several other RGD proteins analyzed previously, decorsin has a well defined RGD region. Furthermore, it was found that the overall fold is similar to hirudin, an inhibitor of thrombin. However, the active site is on the opposite end of the molecule. Sequence searches found then that decorsin has homology to antistasin, an inhibitor of factor Xa, yet a different factor of the blood coagulation cascade. The active site of antistasin is on a third distinct epitope of the protein.

Another protein studied is the T-cell adhesion receptor CD2. For this glycoprotein the carbohydrate is crucial for function. We have therefore determined the structure of the polypeptide and the glycan of the N-terminal adhesion domain of human CD2. The carbohydrate has a defined fold. Details of the carbohydrate structure determination will be reported.

Protein Relaxation Analysis via the Combination of Alternating Carbon Enrichment with Random Fractional Deuteration David M. LeMaster, Dept. of Biochemistry, Molecular Biology and Cell Biology. Northwestern University, Evanston, IL USA

Extensive mapping of the dynamical behavior of proteins in solution via ^{13}C relaxation measurements is hindered by two major factors. Sensitivity considerations necessitate isotopic enrichment in all but the smallest systems. However, uniform ^{13}C enrichment results in homonuclear scalar and dipolar interactions which preclude the straightforward use of the T_1, T_2 and NOE experiments described to date. Furthermore, cross correlation effects between the geminal ^1H-^{13}C dipoles at the methylene positions invalidate the standard dynamical analysis of the relaxation measurements.

We have approach both of these problems via the use of alternating carbon enrichment combined with random fractional deuteration. An E. coli protein expression system has been developed which utilizes selectively enriched carbon sources such as [2-^{13}C]glycerol or [1,3-^{13}C]glycerol to synthesize the constituent amino acids in a ^{13}C-^{12}C-^{13}C... pattern out along each of the various sidechains. When combined with random fractional deuterium enrichment, ^2H decoupling and selection for the IS signals, the relaxation behavior of the methylene and methyl sites can be monitored with the standard methine pulse sequences. Relaxation measurements have been carried out on the 108 residue protein E. coli thioredoxin. Results will be discussed, particularly emphasizing the utility of relaxation measurements at methylene sites as the monitoring of the carbon dynamics from two nearly perpendicular directions offers a powerful probe of motional anisotropy.

Incorporating Motional Properties into the Interpretation of Three-Dimensional Solution Structures.

W.J. Chazin
Department of Molecular Biology, The Scripps Research Institute, La Jolla CA, 92037.

A detailed understanding of biological systems at the molecular level requires characterization of structure and dynamics in atomic detail. NMR spectroscopy is uniquely suited to the determination of high resolution solution structures, and has vast potential as a means for examining molecular dynamics because a wide array of nuclear sites within a molecule can be assayed at a variety of different time scales. However, at present we are in the midst of a search to identify how the structural and dynamic information can be fused to generate a unified and comprehensible view for analyzing biomolecular function. The current status and future directions will be summarized using calbindin D_{9k}, a small protein for which solution structure and dynamics have been extensively characterized by various NMR techniques. High resolution structures have been determined for the Ca^{2+}-free, $(Cd^{2+})_1$ and $(Ca^{2+})_2$ states of the protein, and molecular motions have been probed by ^{15}N relaxation, amide proton exchange and other measurements. The combined view of the conformational and dynamical responses to ion binding have revealed insights into the molecular basis for the cooperativity in Ca^{2+}-binding. In addition, these results provide a foundation for a comparative analysis of calcium-binding proteins that is designed to elucidate how the fine-tuning of the response to calcium-binding is used to generate the extremely wide functional diversity within this family of proteins.

Phosphotyrosyl Peptide-enzyme Complexes: How Much Structure Can We Get From Transferred NOE's?

Carol Beth Post and Michael L. Schneider, Department of Medicinal Chemistry, Purdue University, West Lafayette, IN 47907-1333

The regulation of numerous cellular processes involves the transduction of extracellular signals across the membrane, and phosphorylation is often a key feature of the control mechanism. We are obtaining structural information about processes associated with phosphorylation steps by investigating a peptide fragment containing the phosphorylation site of band 3 bound to three glycolytic enzymes. Formation of the band 3-enzyme complex inhibits the catalytic activity of the enzyme and is controlled by tyrosine phosphorylation. The structure of a 15-residue peptide from band 3 complexed to aldolase and a hypothesis for the energetic basis of phosphorylation control of binding will be presented. Structure determination of peptides bound to a protein using transferred NOE data is distinguished from protein structure determination in that the peptide binding on the protein surface results in fewer structural constraints. Although the bound peptide structure is well defined in the region of the tyrosine-phosphorylation site, multiple structural solutions were found for the peptide ends. This disorder is evident in the nmr data, thus conformational heterogeneity exists in solution. The talk will address factors influencing the accuracy of 3-dimensional structure determination by transferred NOE and the results of a cluster analysis on the structures.

High Resolution Field-Cycling NMR in Liquids.
Alfred G. Redfield, Brandeis University, Waltham, MA 02254

Abstract. Field-cycling NMR consists, usually, of a preparation time at as high a field as possible, followed by a rapid decrease to a (usually) lower main field, followed by a return to high field and a readout of the nuclear magnetization. It has been achieved by moving samples, by turning electromagnets on and off, and combinations of both methods. I will briefly review prior methods and results of field cycling and then discuss possible future applications for work in macromolecules, and how they might be performed. The only small molecule studied extensively by field cycling in solution is H_2O (in interaction with a wide variety of solutes) by Brown and Koenig and others, in relatively low-resolution low-field cyclers. The only work directed toward high-field high resolution is by P. Bolton and his group. The most obvious use for such measurement would be the ability to measure spectral densities over the entire frequency range, as already demonstrated by water relaxometry.

Cross-Correlations; Obstacles or Tools for Structure Determination of Biomolecules.

Anil Kumar, Department of Physics and Sophisticated Instruments Facility, Indian Institute of Science, Bangalore 560012, India

Whenever there are multiple pathways for the relaxation of a spin (such as the chemical shift anisotropy relaxation of a spin and its dipolar relaxation with another spin or two or more dipolar relaxation partners of a spin) there can be cross-terms between these pathways - known as cross-correlations. Most analyses of biomolecular NOE driven structure determination algorithms ignore these cross-terms. An analysis is carried out to point out the effects of these cross-terms. The major effect of these cross-terms is a differential effect also known as the multiplet effect between different transitions of a spin. The multiplet effect is strongly geometry dependent and hence contains additional information on the structure of the molecule. For example the dipole-dipole cross-corrections are largest for linear geometry. The multiplet effect can be detected by the use of a small angle measuring pulse, in J-coupled spins having resolved multiplets. The effect is suppressed in weakly coupled spins if a 90° measuring pulse is used.
A second order effect in magnitude and time is a net effect. This net effect can cause errors in estimation of distances by NOE. A redeeming factor is that for short mixing times (times usually used for distance estimation) the net effect is small. Representative model calculations have been performed in three relaxation coupled spins both for the multiplet effect and for the net effect. Experimental observation of multiplet effect has been carried out in proton-proton and fluorine-proton systems. These results will be presented.

Towards the Accurate Measurement of Internuclear Distances in Biological Macromolecules by Suppression of Spin Diffusion

Catherine Zwahlen, Sébastien Vincent and Geoffrey Bodenhausen
National High Magnetic Field Laboratory
Tallahassee, Florida

It is well known that build-up plots of nuclear Overhauser effects do not simply reflect distances between pairs of spins. The interpretation is made more difficult by indirect effects, where the longitudinal magnetization migrates from a "source" spin A to a "target" spin X *via* one or several intermediate spins K, K'... . Although simulations based on the so-called "total relaxation matrix" approach correctly incorporate indirect effects, it cannot be denied that spin diffusion tends to degrade the quality of the information that is available. By using a variety of selective pulses, in particular a doubly-selective inversion pulse applied to both source and target spins A and X, it is possible to inhibit spin-diffusion effects. Experimental examples will be shown for some nucleic acids, both for conventional B DNA as well as for an unusual tetramer recently discovered by Guéron and Leroy. Similar principles can also be applied to simplify complex networks with several competing chemical exchange processes.

NMR of Symmetrical Assemblies of Self-Recognizing Oligonucleotides

Maurice Guéron, Kalle Gehring[#] and Jean-Louis Leroy; Groupe de Biophysique de l' Ecole Polytechnique et de l' URA D1254 du CNRS, 91128 Palaiseau, FRANCE.
Present address: McGill University, Montreal.

At slightly acid or even neutral pH, oligodeoxynucleotides carrying a stretch of cytidines form a tetramer in which two parallel-stranded duplexes are intimately associated, with their hemi-protonated $C \cdot C^+$ base pairs face-to-face and fully intercalated, in a so-called "i-motif". This structure, first observed in the tetramer of d- $(TCCCCC)^1$, is also formed by other sequences, including the very short d- (TCC). In all cases the four strands appear as identical in the NMR spectrum. Therefore, the spectrum does not reveal the stoichiometry, cross-peaks between non-corresponding protons are 4-times ambiguous, and cross-peaks between corresponding protons of different strands are unobservable without isotope substitution. One can nevertheless detect the i-motif and solve structures containing it. Detection relies on the determination of stoichiometry, on the recognition of $C \cdot C^+$ pairs, of 3'-endo conformation and of very slow proton exchange, and on the observation of three types of NOE cross-peaks between protons of intercalating strands. The first type is a direct H1'-H1' cross-peak. The second is a direct cross-peak between an amino proton and an H2' ribose proton. The third is a rectangular pattern of indirect NOEs created by the head-to-head stacking of bases. The amino-ribose NOEs determine the intercalation topology and provide distance constraints which are precious for structure resolution because they involve the bases: even if all other NOE constraints are ignored, they force the formation of the intercalated structure. A high-resolution structure of the tetramer of d-(TCC) will be presented, and its topology will be compared to those of analogous oligonucleotides. We shall also discuss the symmetry requirements for the formation of the NOE rectangular pattern.

1. K. Gehring, J.L. Leroy and M. Guéron (1993) Nature 363, 561-565; and J.L. Leroy, K. Gehring, A. Kettani and M. Guéron (1993) Biochem. 32, 6019-31.

NMR and Protein-Nucleic Acid Interaction

R. Kaptein
Bijvoet Center for Biomolecular Research, Utrecht University, Padualaan 8, 3584
CH Utrecht, The Netherlands

NMR has contributed considerably to our understanding of DNA sequence recognition by proteins. Several solution structures of DNA binding domains of gene regulatory proteins have been solved, an early example being the lac repressor headpiece in 1985. In particular, for several sub-families of zinc-finger proteins the structural information came first from NMR studies and was later confirmed by X-ray crystallography. We have been studying various classes of DNA-binding proteins such as prokaryotic repressors (lac and lexA repressors), and eukaryotic transcription factors including the nuclear hormone receptors (glucocorticoid and retinoic acid receptors) and POU-domains proteins.

The use of stable isotopes had been extremely fruitful, both in the structural work on the proteins themselves and in characterizing their interactions. Thus, in the case of the Arc and Mnt repressors ^{13}C and ^{15}N labeling has been used in conventional editing of NOEs, but also in the discrimination of intra-versus intersubunit NOEs in these multimeric proteins. The complex of lac repressor headpiece with an 11 bp lac operator half-site has been studied in great detail. Both structural and dynamic aspects of the protein-DNA complex will be discussed.

Using NMR to Examine the Dynamic Structure of Biopolymers

Thomas L. James, Carlos González, He Liu, Anwer Mujeeb, Uli Schmitz, and Nicolai Ulyanov, Department of Pharmaceutical Chemistry, University of California, San Francisco, CA 94143-0446.

Determination of the sequence-dependent solution structure of a nucleic acid duplex presents a challenge, since it is known *a priori* to exist as a helix — probably a right-handed, antiparallel helix in A-form if it is RNA or B-form if it is DNA. Unless the duplex contains some unusual feature such as a mismatch, the challenge is to discern the comparatively subtle structural variations which may be important factors in the binding of proteins, mutagens or drugs. In spite of the challenge, we are approaching a point where we can determine a time-averaged structure reasonably well by careful analysis of NMR spectral data related to distance and torsion angle restraints. Consequently, some genomic sequences important in recognition have been studied and their structures determined. On this basis, some tentative generalizations about duplex structures can be made. While nucleic acid duplexes do assume a "structure" in solution, the data do suggest that there is some conformational flexibility. For conformationally flexible molecules in solution, NMR-derived distance and torsion angle restraints are time-averaged. While this complicates the situation, there are ways to address this complication.

NMR Structures of Proteins Involved in Signal Transduction

S.W. Fesik, E.T. Olejniczak, A.P. Petros, P.J. Hajduk, H.S. Yoon, J.E. Harlan, T.M. Logan, M.-M. Zhou, D.G. Nettesheim, H. Liang, R.P. Meadows, and L. Yu.
Abbott Laboratories, Abbott Park, IL   USA 60064
R.L. Van Etten
Department of Chemistry, Purdue University, West Lafayette, IN 47907
X. Mao and C.B. Thompson
Howard Hughes Medical Institute and Department of Molecular Genetics and Cell Biology, The University of Chicago, Chicago, IL 60637

Signal transduction is a complicated process that may involve several steps mediated by specific intermolecular interactions.   Recently, three-dimensional structures of proteins involved in signal transduction have been obtained and have greatly aided in our understanding of these processes at the molecular level.   In this presentation, three-dimensional structures of three proteins that are involved in signal transduction will be presented.   The three-dimensional structure of a protein tyrosine phosphatase, a pleckstrin homology (PH) domain, and the DNA-binding domain of a member of the *ets* family of transcription factors in the DNA-bound form will be discussed.

Structures of Larger Proteins, Protein-Protein and Protein-DNA Complexes by Heteronuclear Multi-Dimensional NMR

G. Marius Clore and Angela M. Gronenborn

Laboratory of Chemical Physics, Building 5, National Institute of Diabetes and Digestive and Kidney Diseases, National Institutes of Health, Bethesda, MD 20892

The present lecture will discuss the application of multidimensional NMR to the study of protein-protein and protein-DNA complexes. In particular, the 3D and 4D methods that have been so successfully used to solve the three-dimensional structures of larger proteins (1) can readily be extended to protein-ligand, protein-DNA and protein-protein complexes using $^{13}C/^{15}N$ uniformly labeled protein and unlabeled ligand. By this means, it is possible to design experiments in which correlations involving only protein resonances, only ligand resonances, or only through-space interactions between the ligand and protein are observed. We will first illustrate these methods with respect to the structure determination of a complex of the transcription factor GATA-1 with its cognate DNA recognition site (2,3). We will then demonstrate their application to multimeric proteins (4), and specifically to the tetrameric oligomerization domain of the tumour suppressor p53 (5), a protein which is responsible for 50% of all human cancers.

1. Clore, G. M. & Gornenborn, A. M. (1991) *Science 252*, 1390-1399
2. Omichinski, J. G., Clore G. M., Schaad, O., Felsenfeld, G., Trainor, C., Appella, E., Stahlh, S. J. & Gronenborn, A. M. (1993) *Science 261*, 438-446
3. Clore, G. M., Bax, A., Omichinski, J. G. & Gronenborn, A. M.(1994) *Structure 2*, 89-94
4. Lodi, P. J., Garrett, D. S., Kuszweski, J., Tsang, M. L. S., Wetherbee, J. A., Leonard, W. J., Gronenborn, A. M. & Clore, G. M. (1994) *Science 263*, 1762-1767
5. Clore, G. M. Omichinski, J. G., Sakaguchi, K., Zambrano, N., Sakamoto, H., Appella, E. & Gronenborn, A. M. (1994) *Science 265*, 386-391

Structural Analysis of Flexible Molecules

Peter Wright, The Scripps Research Institute, Department of Molecular Biology - MB2, 10666 North Torrey Pines Road, La Jolla, CA 92037

Application of conventional methods of structure calculation to flexible molecules is fraught with difficulties because of averaging of the NMR parameters. However, valuable insights into the preferred structure(s) of such molecules can often be obtained by careful deconvolution of the NOE and coupling constant data or supplementation with other NMR data such as heteronuclear relaxation parameters. The problem of structural characterization of flexible, multidomain proteins will be discussed, using a protein containing three zinc fingers as an example. Relaxation measurements provide insights into both intradomain dynamics and motional coupling of the individual zinc finger domains on a time scale longer than the rotational correlation time. Deconvolution of the conformationally averaged NMR parameters and calculation of the dominant solution structure of a flexible peptide will be discussed.

NMR Approaches to the Study of Structure-Function Relationships in Iron-Sulfur Proteins

John L. Markley, Young Kee Chae, Bin Xia, Hong Cheng, Frits Abildgaard, Ed S. Mooberry, and William M. Westler, Dept. of Biochemistry, University of Wisconsin-Madison, 420 Henry Mall, Madison WI 53706, USA

Although biologically-important proteins that contain metals or stable free radicals at their active sites can be particularly amenable to EPR and ENDOR approaches, the inherent paramagnetism of their active sites makes these parts of the molecule difficult to study by conventional NMR methods. In [2Fe-2S] ferredoxins, the extreme line broadening of signals from protons near the iron atoms abrogates the common homonuclear or multinuclear approaches for signal assignment and distance analysis. We have demonstrated that uniform- and selective-labeling with stable isotopes (^{2}H, ^{13}C, and ^{15}N) afford powerful solutions to these problems. The lower magnetogyric nuclei yield sharper and better-resolved hyperfine signals than protons at similar distances to the iron atoms. Selective labeling leads to spectral simplification and assists in assignments. This approach will be illustrated by recent NMR results with wild-type and site-directed mutant [2Fe-2S] ferredoxins: *Anabaena* vegetative ferredoxin, *Anabaena* heterocyst ferredoxin, and human ferredoxin. Selective labeling has been used in conjunction with specialized NMR experiments to verify and extend previous assignments of ^{1}H signals and to provide the first assignments of hyperfine-shifted ^{13}C and ^{15}N resonances in [2Fe-2S] ferredoxins. Comparisons of the hyperfine-shifted signals in the oxidized and reduced forms of these proteins provide unique information about hydrogen-bonding and electron delocalization in their active sites. The structures of the two *Anabaena* ferredoxins have been studied in solution and have been compared with the x-ray structures of crystalline forms of the same proteins. Site-directed mutagenesis has been used in collaborative investigations of residues that are critical for electron transfer from *Anabaena* vegetative ferredoxin to *Anabaena* ferredoxin reductase (FNR).[1] Additional mutants have been designed to test ideas about the determinants of the reduction potential and the stability of the [2Fe-2S] clustser. [Supported by grants from the NSF (MCB-9215142) and USDA (92-37306-7699). NMR studies were carried out at the National Magnetic Resonance Facility at Madison which has partial support from the Biomedical Research Technology Program of the NIH (RR02301).]

1. H.M. Holden, B.L. Jacobson, J.K. Hurley, G. Tollin, B.-H. Oh, L. Skjeldal, Y.K. Chae, H. Cheng, B. Xia, & J.L. Markley (1994), Structure-Function Studies of [2Fe-2S] Ferredoxins, *J. Bioeng. Biomemb.* **26**, 67-88.

The Calcium Induced Structural Change that Triggers Muscle Contraction
Stéphane M. Gagné, Sakae Tsuda, Monica X. Li, Carolyn M. Slupsky, Larry A. Calhoun, Lawrence B. Smillie and Brian D. Sykes. Department of Biochemistry and MRC Group in Protein Structure and Function, University of Alberta, Edmonton, Alberta, Canada, T6G 2H7.

In the thin filament of skeletal muscle, the binding of calcium to the calcium-binding protein troponin-C (TnC) induces a conformational change in the protein which alters its interaction with the inhibitory protein troponin-I (TnI) and ultimately leads to muscle contraction. We have focused on several aspects of this calcium triggering including the structures of the calcium binding domains of TnC, the nature of the conformational change that occurs in TnC upon calcium binding, and the structure and location of TnI binding to TnC. To accomplish this we have studied peptide fragments and domains of TnC and TnI, along with the intact TnC, and binary and ternary complexes of these proteins. Many of these have been cloned and expressed in *E. coli* so they have been ^{13}C and ^{15}N enriched.

The structures of apo TR$_1$C and apo N-domain have been determined. The structures are similar to the N-domain in the X-ray structure and show the lack of contacts of the C-helix with the A, B and D helices of the site I•II domain. This is the helix that carries the extra negative charge postulated to be important in the stability and calcium binding properties of the III•IV heterodimer and which may be crucial in the interaction of TnC with TnI. The calcium saturated form of the N-domain has been determined by triple resonance 3D experiments of the type developed by Bax and co-workers. The comparison of the structures of the regulatory domain of TnC in the apo and calcium saturated states allow us to define the primary structural change responsible for triggering muscle contraction, and propose a mechanism for the direct coupling of calcium binding to this conformational change. The NMR results also allow us to explain the contradictory results describing this conformational change from CD spectroscopy. Ongoing work is focused on the subsequent interactions of TnC with TnI upon calcium binding. In particular, we have used isotope filtered NMR to determine the structure of the N-terminal region of TnI bound to the C-domain of TnC, and transferred NOESY to determine the interaction of the TnIp peptide with TnC.

The Structure of Lentiviral Tat Proteins - D. Willbold, P. Bayer, H. Sticht, R. Rosin-Arbesfeld[1], M. Kraft[2], R. Frank[2], P. Rösch, Universität Bayreuth, 95440 Bayreuth, Germany;[1]Zentrum für Molekulare Biologie, 69120 Heidelberg, Germany;[2]Tel Aviv University, 69978 Tel Aviv, Israel

Tat (trans activator) proteins are small, RNA binding proteins that are mandatory components in the life cycle of lentiviruses such as the equine infectious anemia virus, EIAV, and the human immunodeficiency virus, HIV. EIAV and HIV-1 Tat proteins contain regions of different homology, most prominent the HIV-1 Tat cysteine rich region, which is missing in EIAV Tat protein but is highly conserved among immunodeficiency virus Tat proteins, the highly homologous core region, and the basic domain, which is involved in RNA binding. The C-terminal glutamine rich region is conserved among HIV Tat proteins, but not in Tat proteins from other lentiviruses. We determined the structure of the EIAV and the HIV-1 Tat protein in solution by standard methods. Both proteins show similar general features: The core region forms a

```
                                          cys-rich        core
TAT$HV1Z2  MD----P--VDPNIE------PWNHPGSQ---PKTACNRCHCKKCCYHCQVCFITKGLGISY|  
TAT$EIAV   LADRRIPGTAEENLQKSSGGVPGQNTGGQEARPN----------YHCQLCFL-RSLGIDY|  
           .     *  . *.|.      *   . *.*   *.|       |****.**. ..*** *|  
               basic
TAT$HV1Z2  G----RKKRRQRRRPSQGGQTHQDPIPKQPSSQPRGDPTGPKE---
TAT$EIAV   LDASLRKKNKQRLKAIQQG--------RQP-----------QYLL
```

hydrophobic domain acting as a scaffold for the otherwise highly flexible protein [1]. In EIAV Tat, the basic sequence region has a tendency to form a helical structure, manifested by weak helix-type NOEs [2]. In HIV-1 Tat, the glutamine rich sequence region forms an additional stabilizing domain fixing the protein structure [3].

[1] D. Willbold, R. Rosin-Arbesfeld, H. Sticht, R. Frank, & P. Rösch (1994) *Science* 264, 1584-1587 [2] H. Sticht, D. Willbold, P. Bayer, A. Ejchart, F. Herrmann, R. Rosin-Arbesfeld, A. Gazit, A. Yaniv, R. Frank, & P. Rösch, (1993) *Eur. J. Biochem. 218*, 973-976; H. Sticht, D. Willbold, A. Ejchart, R. Rosin-Arbesfeld, & P. Rösch (1994) *Eur. J. Biochem.*, submitted; [3] P. Bayer, M. Kraft, R. Frank, & P. Rösch (1994): The Structure of HIV-1 Tat Protein in Solution, *Science*, submitted

Recent Developments in Protein NMR Spectroscopy

Ad Bax, Stephan Grzesiek, Hitoshi Kuboniwa, John Marquardt, Geerten W. Vuister, and Andy C. Wang
Laboratory of Chemical Physics, NIDDK, NIH, Bethesda, MD 20892.

Although uniform isotope labeling and triple resonance J connectivity experiments have largely solved the assignment problem for intermediate sized proteins, Phe ring protons frequently pose a problem with this approach due to the poor chemical shift dispersion of the aromatic ^{13}C spectrum and the presence of strong ^{13}C-^{13}C coupling. A reverse labeling approach is described that allowed all 10 Phe's in the DNA-binding domain of *Drosophila* heat shock factor to be completely assigned. Because of the favorable relaxation properties of ^{12}C-attached protons, 3D reverse filtering experiments on such a reverse labeled sample provide NOE data that are superior in quality over what can be obtained with 4D experiments, and yield an unusually large number of long range NOE distance constraints.

Accurate measurement of homo- and heteronuclear coupling constants can provide detailed information on conformation and dynamics, both in small molecules and in isotopically labeled proteins. A group of experiments that we refer to as "quantitative J correlation" has been designed in such a way that the intensity of cross peaks is related in a simple manner to the size of the J coupling. This type of experiment is shown to provide reliable ^1H-^1H, ^1H-^{13}C, ^1H-^{15}N, ^{13}C-^{13}C and ^{13}C-^{15}N coupling constant information.

Structural studies of proteins are usually conducted in H_2O solution. As the protein is typically 10^5 times more dilute than the H_2O concentration, the vast majority of nuclear spin magnetization resides on H_2O. The T_1 of the H_2O protons is typically much longer (3-5 s) than the T_1 of protein protons (~1.4 s). As there is a continuous exchange of magnetization between the solvent and the protein, maintaining the H_2O magnetization in a fully relaxed state significantly enhances the sensitivity of a large variety of experiments, including those commonly used for triple resonance backbone assignment, measurement of ^1H-^1H J couplings, and intramolecular NOEs. The gain in sensitivity is even more dramatic for the measurement of the intermolecular NOE interaction between protein and solvent.

Although the one-dimensional deuterium spectrum of a perdeuterated protein in solution shows massive overlap of the very broad ^2H resonances, it is nevertheless possible to measure the individual T_1 and T_2 ^2H relaxation times. Provided that the dynamics of the protein are known from other relaxation studies, the relaxation rates provide detailed information on the magnitude of the quadrupole interaction, i.e., on the strength of the electric field gradient at the site of the deuteron and thereby on the strength of the hydrogen bonding interaction.

A Structural Biologist's View of Precision and Accuracy of Structural Models of Proteins Based on NMR Data. A. Joshua Wand, Department of Biochemistry, University of Illinois at Urbana-Champaign, Urbana, IL 61810

The dramatic increase in the power, flexibility and efficiency of modern multimdimensional and multinuclear NMR spectroscopy has provided the structural biologist with a previously unavailable avenue to high resolution structural information in solution. In the spirit of this conference, several topics revolving around issues of precision and accuracy of NMR-based models of protein structure will be addressed. We have recently solved the structure of reduced and oxidized cytochrome c to relatively high apparent resolution. For the reduced protein, a family of structures was obtained using a total of 1940 distance constraints based on the observed magnitude of nuclear Overhauser effects and 85 torsional angle restraints based on the magnitude of determined J-coupling constants. The all-residue root mean square deviation about the average structure is 0.47 ± 0.09 Å for the backbone N, Cα and C' atoms and 0.91 ± 0.07 Å for all heavy atoms. Similar results were obtained for the oxidized protein. Together, these structures provide a template to address a fundamental consideration: How useful are they as predicative tools? To illustrate, results of calculations on the electrostatic potential, the redox potential and the solvent reorganization energy will be presented. These structures also serve to show that highly defined models can be obtained using purely homonuclear 1H methods. The second phase of the presentation will center on the potential influence of internal dynamics on a primary experimental constraint - the NOE. The character of surface and core internal motions will be illustrated by preliminary 13C and 15N NMR relaxation studies of the protein ubiquitin. Finally, a simple expression for an "NMR B-factor" will be proposed and the general challenge that it implies discussed.

Work in the author's laboratory is supported by NIH grants DK-39806 and GM-35940.

INDEX

ADP, 11
  MgADP, 11
Aldolase, 91–93, 97–98
AMBER (program), 82, 246
AMNESIA, 117–118
Anisotropic motion, 248, 318
ANSWERS (program), 115–116
Antamanide, 20–25, 27–28, 31
Arginine dynamics, 38
Aromatic rotation, 63, 77
ATP, 6–7, 11–12, 337
  MgATP, 11

B factors, 110, 112, 322
Baculovirus, 212–213
Band 3 peptide, 91–98
  -Aldolase complex, 97
Born–Oppenheimer approximation, 67–68
BPTI, 9
BSA, 11

Calbindin D_{9K}, 77–85, 87–88, 105, 110–111,
  115
Calmodulin, 105, 114–115, 118–119, 121–122,
  282
CAMELSPIN, 128
Carboxypeptidase, 107
CAT assay, 291
CD, 333–334
CHAPS (detergent), 210, 282
CHARMM (program), 24–26, 32, 108
Chemical shift index (CSI), 280–281
Chemical shifts
  structure determination, 217–218
Chemokines, 237, 239, 243
Chirality, 66, 68
CIDNP, 170, 173, 327
Cis–trans, 24, 78, 158–160, 207, 246
Cluster analysis, 94–96, 189
CMP-KDO synthetase, 212
Conformational averaging, 205
Conformer probability, 87, 201–202
CORCEMA (program), 107
CORMA (program), 107

Correlation, 253
  functions, 16
  time, 19–20, 33, 43, 71–73, 92, 101, 104–105,
    114, 119, 141, 146, 152, 194, 248, 282,
    318, 337
COSY, 117, 119, 194, 199–200, 205, 288–291, 297
Coupling constant, 67, 194–195, 198, 201–202,
  205, 207, 246, 271, 280
CPMAS, 128
Creatine kinase, 10
Cross-correlation, 6, 65, 70, 135–143
Cross-relaxation, 22, 27, 31, 92, 107, 137,
  145–147, 192–193, 253
CSA, 4, 9, 20, 27, 105, 128, 137, 143, 214
CW, 1–2, 4, 7
Cyclosporin/cyclophilin complex, 208, 213
Cytochrome c, 308–317

Database, 111
Decorsin, 57–59
Denaturation, 40
DEPT, 321
DEPT-HMQC, 56
Deuteration, 211–212
Deuterium enrichment, 65–66, 72–74
Dihedral angle, 66, 93, 115, 177, 246, 279
Dimer, 237–240
Dipolar interaction, 4, 20, 71, 105, 137, 146,
  266–267
Distance difference matrix, 80–81, 89
Disulfides, 57, 59
DNA, 150–152, 157, 168–169, 173, 175–176,
  178–186, 189, 191–192, 196–197,
  198–199, 205, 211, 214, 216, 229–230,
  231, 235–236, 244, 255, 275, 287, 302
  B-DNA, 150–153, 169, 178, 195, 231
  I-tetrad, 157
  G-tetrads, 168
DNP, 125–126
Drosophila protein Drk, 35, 39, 42
Docking, 97, 329
Double resonance, 4–6, 135, 141
Dynamic frequency shift, 68, 70, 72
Dynamics, 15, 29, 65, 77–78, 83, 103–105, 181,
  121, 123, 208, 216, 275, 280

E.COSY, 24, 194, 205
EF-hand, 78–80, 87–88
ENZYMIX (program), 316–317
Erythrocyte, 91
ET-NOESY, 91–93, 94–97
Ets protein domain, 229–231
Exchange 43–44, 78, 92–93, 148, 159–161, 165, 168, 195
   chemical, 8, 73, 148
   conformational, 35, 40, 112–114, 174, 205, 245–246
   conformational flexibility, 198, 200–201, 247
   dynamics, 22
   peaks, 42
   proton, 84, 168, 170, 195
   rates, 44–45
   slow, 40, 167
   time, 18
   two-dimensional, 8
   two-site, 18, 22
Extreme narrowing, 4
EXSY, 17, 148, 158–160

Ferredoxin
   heterocyst (HFd), 254, 264–267
   human (HuFd), 254, 261–264
   vegetative (VFd), 254–261
Field-cycling (FCNMR), 123–130
Fli-1, 229–231, 235–236
Fluorescence, 327, 332–334, 336–337
Fluorescence anisotropy, 106, 114–115, 333–334, 336–337
Folding, 35, 44,
Frequency shift, 68, 70–72
FTIR, 333–334
FT-NMR, 7

γ-globulin, 11
g-tensor, 314
Glycan, 54–57
Glyceraldehyde-3-phosphate dehydrogenase (G3PDH), 91–92
Glycoprotein receptor
   CD2, 51–56
   CD58, 51, 53–54
   GPIIbIIIa, 51
GRASP (program), 83
GROMOS (program), 26, 83, 108, 152, 179

Hartmann–Hahn effect, 148
Helix-turn-helix (HTH) proteins, 230–231
Hirudin, 58–59
HIV RNAse-H, 210
HOHAHA, 148–152, 154
Holliday junctions, 214
HSQC, 38–41, 44, 56, 66, 68, 74, 214, 282–283
Human macrophage inflammatory protein-1β (hMIP-1β), 237–239, 243
Hyperconjugation, 67–68
Hyperfine-shifted resonances, 251–254, 257–258, 261–262, 264, 266, 271

I-BURP pulses, 158
I-Motif (of DNA), 167–170, 173–174
INEPT, 74, 321
Interleukin-8 (IL-8)
   dimer, 237–239
   monomer, 243
Internal correlation time, 19, 105–106, 112, 318, 337
Iron-sulfur clusters, 251–271
Irradiation
   weak, 4
   doubly-selective, 149
Isolated Spin-Pair Approximation (ISPA), 193–194
Isotope-edited, 36, 118
Isotope-filtered, 36
Isotope shift, 66–68, 72
Isotopic labeling
   random, 319–320
   selective, 210–211, 253, 255, 258–261, 263–271, 319
   uniform, 210–213, 237, 255, 259, 263–264, 269–270

J(scalar)-Coupling, 18–19, 21–22, 24, 103, 109, 117, 119, 121, 177, 195, 280, 319
   three-bond, 119, 246

Karplus relation, 24, 194–196, 198

Lac repressor, 175–178
   headpiece, 175–177, 179, 180, 182, 184, 185
   headpiece-operator complex, 176–179, 181, 184–185
   operator, 176–178, 254
Langevin dynamics, 25
LAP motif, 59
Larger volume probes, 215–216
Larmor frequency, 24, 71, 73, 193,
Lentiviruses, 287
Leucine zippers, 175
LexA
   repressor, 185–186
   -SOS complex, 186
Light scattering, 209
Limiting fluorescence anisotropy, 333
Line-shape simulation, 17
Lipari and Szabo (model-free), 16, 19, 31–32, 38, 43–44, 104–106, 112–114, 317–318
Longitudinal relaxation (T_1), 3, 17, 24, 111, 123–125, 248, 320
   paramagnetic contributions, 266–267
Lysozyme, 107–108

Madison biomolecular NMR database, 104
Maltodextrin, 107
Marcus theory, 310, 315–317
MARDIGRAS (program), 193–194
MEDUSA (program), 22, 113, 202
Metalloproteins, 85
Mixed solvents, 210
Mixing time, 147–148, 152, 154–160, 165–166, 194

Model
 motional, 16
 three-site jump, 27
 two-site exchange, 22
 two-site jump, 27
Molecular dynamics, 78–79, 81, 128, 177, 179,
   189, 192, 195. 291, 293, 298, 299, 322
 restrained (rMD), 78–79, 111, 177, 179, 182,
   189, 200–202, 205, 245–248, 316
 simulation, 25–26, 33–34, 81–83, 93, 96–98,
   101, 151–153, 200, 253, 330–332, 337
Molecular size limits, 207–208
Molten globule, 121
MOLSCRIPT (program), 296
Monte Carlo, 87, 111, 176, 196
 docking, 182, 185
 docking program (MONTY) 176, 182–186, 189
 restrained Monte Carlo, 196–197, 199–202
 sampling, 184
 simulation, 34
Multidomain proteins, 245
Multimeric proteins, 237
Multiple conformations (conformers), 32–34, 84,
   87, 109–110, 112–113
Multiplet effect, 136, 138–141, 143
Multiwavelength anomalous diffraction
   (MAD), 336
Mutations, 240

NDEE (program), 288–292
Net effect, 141, 165
NMR spectroscopy
 Three-dimensional, 36, 38, 40, 277, 308
 Two-dimensional, 8, 308
NOE, 6, 17, 36–37, 38, 40–41, 57, 80, 93–95, 97,
   104–105, 108–109, 111, 113–116, 118,
   121, 125, 127, 135, 138, 140–141, 145,
   152, 157, 167, 177, 179, 192, 198–199,
   206. 211, 214, 219, 230, 237–240, 243,
   246, 258, 277–279, 281, 285, 291–292,
   308–312, 330
 transferred, 91, 106–107, 113, 211
 transient, 17, 152
NOESY, 8, 21, 36, 40, 107–107, 111, 118, 121, 125,
   140, 145–148, 151–157, 160, 166, 213–214,
   257, 277–278, 292–295, 297, 309–311
 cross-peaks, 168–169, 294
 spectrum, 9, 168, 246, 257
Nonlinear sampling, 216
Normal modes, 110

Oligomer duplexes, 197
Oligomerization domain, 239–240, 243
Oligonucleotides, 167, 191, 197–198
Oligosaccharide, 56
Order parameter, 19, 38, 84, 104–106, 111, 121,
   305, 318–319, 337–338
Overhauser effect, 4, 145, 147, 152, 154, 157

p53 (tumor suppressor), 239–241, 243–244
Paramagnetic centers, 218–219, 252

Paramagnetic ions, 127
Paramagnetic shift, 309, 313
PARSE (Probability Assessment via
   Relaxation rates of a Structural
   Ensemble), 200–202
PDB, 104, 110, 277
PDGFR (platelet-derived growth factor
   receptor), 35, 37
Pepsin/inhibitor complex, 208
Peptide dynamics, 21, 32
Plasmids, 254
Pleckstrin homology domain, 223, 225–229, 235
Phenylalanine, 20, 26, 33, 117–118
ϕ, ψ angles, 67, 180, 217, 277, 310–311, 318
 plot, 279
 Ramachandran, 205
Phosphatase, 92
Phosphofructokinase (PFK), 91–92
Phospholipase Cγ (PLCγ) 35–38, 39
Phosphorylation, 91–92, 98, 221, 229
Phosphotyrosine, 36, 49
Proline, 23–26, 32–33, 57, 78
Protein
 folding, 40, 44, 49
 retinol-binding, 226, 235
 tyrosine phosphatase, 221–224
 unfolded state, 39–40, 43, 45
Pseudo-contact shift, 313–315
Pure Quadrupole Resonance (PQR), 129–130
Purine nucleoside phosphorylase, 107

Q^3 Gaussian cascade, 154
Q^3 pulses, 158, 165
Quadrupolar interaction, 20, 71, 125, 129, 133
QUANTA (program), 98
Quenching, 130, 133
QUICK-NOESY, 151–153, 155–157, 160, 166
QUIET-EXSY, 148, 157–160
QUIET-NOESY, 147–148, 153–157, 166

Relaxation, 43, 124, 126, 128, 135–136, 138–139,
   156–157, 160, 193, 202, 206
 delays, 213
 experiments, 43, 74
 information, 8
 matrix, 6, 101, 116, 147, 160, 192, 199
 mechanism, 4,6
 pathway, 153–154
 processes, 4
 rates, 43, 127, 145–147, 155, 201, 248, 264,
   267
 studies, 2–3, 43, 65, 319
 time, 3, 266, 320
Redfield matrices, 4, 6
Repressor
 complex (4 -3- 4), 216
 cro, 178, 184
 λ, 178
RF field, 3, 28, 118, 149
RGD loop, 299, 301
RGD proteins, 51, 57–59, 301

Ribbon diagram, 223, 226, 230, 238
Ring pucker, 24–26
RMSD, 79–80, 84, 179, 194, 197, 201–202, 221,
    228, 240, 244, 246, 292–293, 298–299,
    311–312
RNA, 173, 191, 254–255, 275, 287, 295–296,
    300–302, 305
RNA-DNA duplex, 196, 198
ROESY, 128, 246
Rotating frame relaxation (T_{1r}), 18, 20–21, 28, 31,
    84, 181–182, 199
Rubredoxin (Rdx), 267–271

Saturation transfer, 17, 127, 133
Serine protease, 208
SH2 domain, 35–37, 39, 49, 98, 223, 235
    binding, 39
    Src Homology 2, 35
SH3 domain, 39–44, 223
    Src Homology 3, 35
Signal transduction, 35–36, 78, 221, 223, 232
SIMPLEX (program), 159
Solid state NMR, 31, 128
SOS operators, 185–186
Spectrometer
    750 MHz, 213–215
Spectroscopy
    photoacoustic, 334
    Raman, 333–334
SPHINX (program), 195
Spin diffusion, 116, 126–127, 138, 145, 147, 170
Spin labels, 127, 218
Spin-lock, 118, 128
Spin modes, 136–137
Spin-rotation interaction, 4
Spectral density, 18–19, 43–44, 70, 72, 104–105,
    127, 137–138, 141, 146, 318
Staphylococcal nuclease, 66
Stopped-flow, 17, 126
Structural refinement, 103–104, 106

Sugar pucker, 194–195, 199
Super-operators, 4

TAR (Transactivation response element), 287, 301
TAR RNA, 295–296, 300–302
Tat (transactivator) proteins, 205, 287–305
Thermolysin, 107–108
Thioredoxin, 68–69, 73–74
TOCSY, 38, 40, 56, 128, 288–291, 297, 309–311
Transverse relaxation (T_2), 3, 84, 128, 145–146,
    248, 282
Trifluoroethanol (TFE), 282–283, 285, 295, 301,
    305
TRNOESY, 10–11
Troponin-C, 276–277, 280–283
Tryptophan, 337–338
Tryptophan motion, 105–106
Type VI turn, 245–247
Tyrosine, 36–37, 91–92, 105, 221–224
    kinase, 35–37, 92
    phosphorylation, 91, 98

Ubiquitin, 308, 317–322
Ultracentrifugation, 209, 282, 285
Urokinase, 212

Van der Waals interaction, 116, 181

WALTZ-16, 118
Weak-irradiation, 4–5

X-ray, 15, 27, 77, 97, 104, 170, 175–176, 184,
    191–192, 240, 244, 253, 255, 257, 261,
    266, 277, 281, 327–329, 331, 334–337
X-PLOR (program), 93, 95, 106, 108, 290–291,
    298, 310

Zinc finger, 175, 248
    TFIIIA, 175
    zf1–3, 248